AUTOMATIC

SPRINKLER SYSTEMS

HANDBOOK

Fourth Edition

AUTOMATIC

SPRINKLER SYSTEMS

HANDBOOK

Fourth Edition

Based on
Standard for the Installation of Sprinkler Systems, NFPA 13,
1989 Edition
*Recommended Practice for the Inspection, Testing and
Maintenance of Sprinkler Systems,* NFPA 13A, 1987 Edition
*Standard for the Installation of Sprinkler Systems in
One- and Two- Family Dwellings and Mobile Homes,* NFPA 13D,
1989 Edition
*Standard for the Installation of Sprinkler Systems in
Residential Occupancies up to Four Stories in Height,* NFPA 13R,
1989 Edition

Edited by

Robert E. Solomon, P.E.
Senior Fire Protection Engineer

National Fire Protection Association, Inc.
Quincy, Massachusetts

This Handbook has not been processed in accordance with NFPA Regulations Governing Committee Projects. Therefore, the commentary in it shall not be considered the official position of NFPA or any of its committees and shall not be considered to be, nor relied upon as, a Formal Interpretation of the meaning or intent of any specific provision or provisions of NFPA 13, *Standard for the Installation of Sprinkler Systems;* NFPA 13A, *Recommended Practice for the Inspection, Testing and Maintenance of Sprinkler Systems;* NFPA 13D, *Standard for the Installation of Sprinkler Systems in One- and Two-Family Dwellings and Mobile Homes;* or NFPA 13R, *Standard for the Installation of Sprinkler Systems in Residential Occupancies up to Four Stories in Height.*

Project Manager: Jennifer Evans
Project Editor: Jennifer L. Dougherty
Composition: Muriel McArdle and Cathy Ray
Illustrations: George Nichols
Cover Design: Woodwell Graphics
Production: Donald McGonagle

NFPA No. 13HB89
ISBN: 0-87765-359-3
Library of Congress No. 89-60574
Printed in the United States of America

First Printing, May 1989

Contents

Installation of Sprinkler Systems, NFPA 13

Inspection, Testing and Maintenance
of Sprinkler Systems, NFPA 13A

Installation of Sprinkler Systems in
One- and Two-Family Dwellings
and Mobile Homes, NFPA 13D

**Installation of Sprinkler Systems in
Residential Occupancies up to
Four Stories in Height, NFPA 13R**

Supplement — Quick Response Sprinklers

Foreword

NFPA 13, *Standard for the Installation of Sprinkler Systems*, is the oldest standard published by the National Fire Protection Association and, in some ways, predates the Association itself. The first edition of the sprinkler standard was published in 1896, the year in which the National Fire Protection Association was organized.

The Technical Committee responsible for development of the standard has, over the years, been made up of a group of highly qualified individuals respresenting a large number of interests. These have included fire service personnel, insurance company and association representatives, representatives of testing laboratories, manufacturers and installers of automatic sprinkler systems, representatives of users of automatic sprinkler systems, including industrial, commercial, and governmental personnel, as well as independent consultants.

The life safety record of the automatic sprinkler system over its long history has been enviable in occupancies which have been protected. In May of 1973, the Committee, recognizing the need to reduce the annual life loss from fire in residential occupancies, established a subcommittee to deal with residential and light hazard occupancies. This subcommittee was responsible for the development of NFPA 13D, *Standard for the Installation of Sprinkler Systems in One- and Two-Family Dwellings and Mobile Homes*, which was first adopted in May 1975 and subsequently revised in 1980 as a result of extensive fire tests. Activities in the residential sprinkler field have resulted in the development of a new residential sprinkler whose prime goal is life safety. This new fire protection tool now makes it possible to protect the person in the room of origin of the fire. Residential sprinklers may, of course, be appropriately applied under NFPA 13. In 1989, NFPA 13R was developed to expand detailed residential sprinkler technology to residential occupancies up to four stories high.

During the 1960s, the National Fire Protection Association's Board of Directors charged the Sprinkler Committee with development of a standard which would encourage innovative technology and economical system costs. This was intended to make the automatic sprinkler system available on a wider basis and to further help reduce our nation's fire losses. Major changes have included the introduction of hydraulic design methodology, new joining methods, new types of pipe or tubing, and new types of sprinklers. As a result, the cost of automatic sprinkler installations has escalated at a considerably lesser rate than the various other mechanical systems in buildings. Alternate materials, and new methods and devices, have been—and continue to be—encouraged. Innovative technology continues, supported by major research projects of the National Fire Protection Research Foundation.

One of the challenges facing those involved in sprinkler technology is development of a method of providing water at an operated sprinkler fast enough to allow application of quick-response, early suppression fast response, and residential sprinkler technologies to systems which must be dry due to lack of heat in all or part of a structure.

The sprinkler standard is not written as a legal document, nor does it cover all situations which might be encountered. It must be used with engineering judgment and with the application of common sense. Assistance in the application of the standard can, of course, be found in the Formal Interpretations which have been issued by the Committee and which are collected and published in the National Fire Codes® Subscription Service. Additional information is also available through reviewing substantiation associated with the Technical Committee Reports and the Technical Committee Documentation developed under the Regulations Governing Committee Projects, which govern the standards-making system of the Association.

This Handbook is intended to assist in the application of NFPA 13, *Standard for the Installation of Sprinkler Systems*; NFPA 13A, *Recommended Practice for Inspection, Testing and Maintenance of Sprinkler Systems*; NFPA 13D, *Standard for the Installation of Sprinkler Systems in One- and Two-Family Dwellings and Mobile Homes*; and NFPA 13R, *Standard for the Installation of Sprinkler Systems in Residential Occupancies up to Four Stories in Height*. Contributors to the Handbook include members of the Technical Committee who have been responsible for the development of the individual documents who deserve our sincere thanks not only for their work on the Handbook, but also for their tireless efforts in Committee work.

Hopefully, this Handbook will help toward achieving easier and more accurate application of the standards and will contribute to the continuation of the enviable record of life safety and property protection provided by automatic sprinkler systems.

Chester W. Schirmer
Chairman
Technical Committee on
Automatic Sprinklers

Preface

The first automatic fire extinguishing system on record was patented in England in 1723 and consisted of a cask of water, a chamber of gunpowder, and a system of fuses. The first form of sprinkler system used in the United States, however, was the perforated pipe system, which was first installed about 1852.

The first automatic sprinkler was invented in 1864, but it wasn't until 1878 that Henry Parmelee invented a sprinkler that was used extensively in practice. Some 200,000 Parmelee sprinklers were installed throughout New England by the Providence Steam and Gas Pipe Company (later the Grinnell Company) between 1878 and 1882.

The first set of rules for the installation of automatic sprinklers was developed by John Wormald of the Mutual Fire Insurance Corporation of Manchester, England in 1885, based on an 1884 Factory Mutual Fire Insurance Company's study of the performance of sprinklers conducted by C. J. H. Woodbury of the Boston Manufacturers Mutual Fire Insurance Company and F. E. Cabot of the Boston Board of Fire Underwriters. In 1887, similar rules were prepared in the United States by the Factory Improvement Committee of the New England Insurance Exchange.

By 1895, the commercial growth and development of sprinkler systems was so rapid that a number of different installation rules had been adopted by various insurance organizations. Within a few hundred miles of Boston, nine radically different standards for the size of piping and sprinkler spacing were being used. To solve this problem, the following group of men met in Boston early in 1895:

> Everett U. Crosby, Underwriters Bureau of New England
> Uberto C. Crosby, New England Insurance Exchange
> W. H. Stratton, Factory Insurance Association
> John R. Freeman, Factory Mutual Fire Insurance Companies
> Frederick Grinnell, Providence Steam and Gas Pipe Company
> F. Eliott Cabot, Boston Board of Fire Underwriters

Subsequently, a small group of inspection bureau men met in New York in December of 1895 to draw up a set of uniform sprinkler rules. At a second meeting in March of 1896, these rules were completed and, at the same time, a committee was appointed to set up an association for the proper recognition of sprinklers. On November 6, 1896 this association, the National Fire Protection Association, came into being.

Thus, the first and only standard for the National Fire Protection Association in 1896 was the *Standard for the Installation of Sprinkler Systems.*

In 1900, the National Board of Fire Underwriters joined the NFPA and voted to adopt and assume the expense of publishing the NFPA standards. Thus, from 1901 to 1964, the NFPA *Standard for the Installation of Sprinkler Systems* was published by the National Board of Fire Underwriters, first as rules and requirements, later as regulations, and still later as a standard of the National Board of Fire Underwriters as recommended by the National Fire Protection Association.

When first printed in 1896, the Sprinkler Standard concerned itself principally with sprinkler pipe sizes, sprinkler spacing, and water supplies. Since 1900, the Sprinkler Standard has been subjected to considerable amplification and refinement. The 1989 Standard on which this fourth edition of the Handbook is based is the 53rd edition of NFPA 13.

In 1938, the Committee on Automatic Sprinklers began work on a *Recommended Practice for the Care and Maintenance of Sprinkler Systems* (NFPA 13A) because an improperly maintained sprinkler system may be worthless. The 1st edition was adopted in 1939. The latest revision of NFPA 13A is the 1987 edition.

Recognizing the need to reduce the annual loss of life from fire in residential occupancies, the Committee on Automatic Sprinklers in 1973 began preparing a *Standard for the Installation of Sprinkler Systems in One- and Two-Family Dwellings and Mobile Homes* (NFPA 13D), which was first adopted in 1975.

In 1980, the standard was revised to require the installation of fast response residential sprinklers which had just been developed to improve the life safety of residential sprinkler protection. The latest revision of NFPA 13D is the 1989 edition.

Throughout the 1980s, the Committee on Automatic Sprinklers debated the need for a separate standard to cover sprinkler systems in residential occupancies other than one- and two-family dwellings and mobile homes, primarily for life safety purposes. In 1988 the first edition of NFPA 13R, *Standard for the Installation of Sprinkler Systems in Residential Occupancies up to Four Stories in Height*, was adopted to cover such occupancies as hotels, motels, dormitories, apartments, lodging and rooming houses, and certain board and care facilities.

It is hoped that this Handbook will assist fire protection practitioners in obtaining a better understanding of, and appreciation for, the requirements contained in NFPA 13 (1989), NFPA 13A (1987), NFPA 13D (1989), and NFPA 13R (1989).

It should be pointed out, however, that the commentary on the text contained in this Handbook is the opinion of the Editor and Technical Consultants. The commentary does not necessarily reflect the official position of NFPA nor the Committee on Automatic Sprinklers.

Arthur E. Cote, P.E.
Assistant Vice President
and Chief Engineer

Editor's Preface

As NFPA 13 marks its 93rd year in the NFPA system, it has been a great opportunity to work with the forerunners of the first three editions of this handbook. These include current and former members of the Technical Committee on Automatic Sprinklers as well as former members of the NFPA technical staff. Many persons contributed directly to material for this 4th edition, and many others contributed indirectly to materials found herein.

The NFPA Technical Committee on Automatic Sprinklers maintains NFPA 13, 13A, 13D, and 13R. These members give of their time to enhance the useability of these documents. Time spent is not just limited to Committee meetings but also to delivering talks on specific areas of interest, conducting seminars, answering phone calls and letters, and the usual homework time necessary in preparation for a subcommittee or committee meeting. Other members act as liaisons to other technical committees or outside organizations. It is through these various processes that materials are culled together for inclusion in this handbook. In addition, the day-to-day advisory service handled by NFPA staff in this area can be centralized and subjects which make for repeat questions can be targeted for improvements during an update to the handbook — thus, there are many users of NFPA 13 and NFPA 13D who have indirectly contributed to this fourth edition.

The first edition of this handbook, published in 1983, probably required the most basic research on the historical perspectives of NFPA 13, 13A, and 13D. Contributing authors for that edition were: Wayne Ault, Robert Duke, Denison Featherstonhaugh, Russell Fleming, Richard Hughey, Rolf Jensen, Robert Retelle, Chester Schirmer, John Walsh, and Jack Wood. Most of these people are actively involved in the work of NFPA today.

I also wish to thank contributors to the fourth edition of this handbook. They are Wayne Martin, Richard Hughey, Jack Walsh, Russ Fleming, Ken Richardson, William Wilcox, and Bill Webb. Extra thanks goes to Bill Koffel, who had the task of writing the first commentary for NFPA 13R. A special thanks goes to Chet Schirmer, who reviewed the final manuscript on a most timely basis.

Special appreciation is extended to Muriel McArdle, who keyed in all the text and helped prepare typesetting codes before transmitting the book to the compositor; Jennifer Evans, project manager, who coordinated the work of the authors and kept everyone on schedule; and Jennifer Dougherty, who did a magnificent job of arranging the text and suggesting numerous editorial changes which made the commentary more precise.

It is hoped that this handbook will allow the user to establish rationale for requirements to certain aspects of sprinkler system design and installation. New materials and components appear on an almost weekly basis which make system design easier. This handbook reflects current information, available as of January 1989. I hope this handbook assists those who are just beginning to understand the installation of sprinklers as well as those who have been involved in system design, installation, or review for many years.

Robert E. Solomon, P.E.
Senior Fire Protection Engineer
National Fire Protection Association
Editor

NOTE:

The text, illustrations, and captions to photographs and drawings that make up the commentary on various sections of this Handbook are printed in color and in a column narrower than the standard text. The photographs, most of which are part of the commentary, are printed in black for clarity. The text of the standards and the recommended practice is printed in black.

The Formal Interpretations included in this Handbook were issued as a result of questions raised on specific editions of the standard or recommended practice. They apply to all previous and subsequent editions in which the text remains substantially unchanged. Formal Interpretations are not part of the standards or recommended practice and therefore are printed in color to the full column width.

NFPA 13

Standard for the Installation of

Sprinkler Systems

1989 Edition

NOTICE: An asterisk (*) following the number or letter designating a paragraph indicates explanatory material on that paragraph in Appendix A. Material from Appendices A and B is integrated with the text and is identified by the letter A or B preceding the subdivision number to which it relates.

Information on referenced publications can be found in Chapter 10 and Appendix C.

1

General Information

1-1 Scope. This standard provides the minimum requirements for the design and installation of automatic sprinkler systems and of exposure protection sprinkler systems, including the character and adequacy of water supplies and the selection of sprinklers, piping, valves, and all materials and accessories; but not including the installation of private fire service mains and their appurtenances, the installation of fire pumps, the construction and installation of gravity and pressure tanks and towers.

NOTE: Consult other NFPA standards for additional requirements relating to water supplies.

It is very important to understand that this is an installation standard for sprinkler systems having automatic and open sprinklers and using water as the extinguishing medium. It is possible to achieve different levels of performance, for example, by providing supervisory signaling systems or by providing additional protection devices and systems. This standard establishes minimum requirements for sprinkler systems that are not dependent upon these features.

The *Standard for the Installation of Sprinkler Systems* establishes the water supply needs and the types of water supply sources that are acceptable. It does not, however, contain installation or performance requirements for those water supply sources. Other NFPA standards provide requirements for pumps and storage tanks, as well as for the underground piping system.

Other NFPA standards and recommended practices utilizing fixed piping systems for application of water and water-based extinguishing media for fire extinguishment and control include:

NFPA 11, *Standard for Low Expansion Foam and Combined Agent Systems*

NFPA 11A, *Standard for Medium and High Expansion Foam Systems*

NFPA 13D, *Standard for the Installation of Sprinkler Systems in One- and Two-Family Dwellings and Mobile Homes*

NFPA 13R, *Standard for the Installation of Sprinkler Systems in Residential Occupancies up to Four Stories in Height*

NFPA 14, *Standard for the Installation of Standpipe and Hose Systems*

NFPA 15, *Standard for Water Spray Fixed Systems for Fire Protection*

NFPA 16, *Standard on Deluge Foam-Water Sprinkler and Foam-Water Spray Systems*

NFPA 16A, *Recommended Practice for the Installation of Closed-Head Foam-Water Sprinkler Systems.*

1-2 Purpose. The purpose of this standard is to provide a reasonable degree of protection for life and property from fire through standardized installation requirements for sprinkler systems based upon sound engineering principles, test data, and field experience. The standard endeavors to continue the excellent record that has been established by standard sprinkler systems and meet the needs of changing technology. Nothing in this standard is intended to restrict new technologies or alternate arrangements, providing the level of safety prescribed by the standard is not lowered. Materials or devices not specifically designated by this standard shall be utilized in complete accord with all conditions, requirements, and limitations of their listings.

NOTE 1: A sprinkler system is a specialized fire protection system and requires knowledgeable and experienced design and installation.

NOTE 2: Since its inception, this document has been developed on the basis of standardized materials, devices, and design practices. However, certain paragraphs, such as 3-1.1.5, 4-1.1.3, and this one, allow the use of materials and devices not specifically designated by this standard, provided such use is within parameters established by testing and listing agencies. In using such materials or devices, it is important that all conditions, requirements, and limitations of the listing be fully understood and accepted, and that the installation is in complete accord with such listing requirements.

The proliferation of specially listed material and products as well as the development of new technology have necessitated the above language to alert the user of the standard to the likelihood of specialized requirements or limitations associated with such listings.

It is difficult to quantify the level of protection to life and property that sprinkler systems should provide. It would be unreasonable to expect, for example, that individuals in a fire exposure area of a building with automatic sprinklers might not be injured in the event of a rapidly spreading flammable liquids fire, or that property losses can be restricted to specific dollar values or percentages of floor areas. The best that reasonably can be expected is that, by complying with this standard, protection of life and property will be greatly improved.

Historical records, captured in the following phrase, let the effectiveness of an automatic sprinkler system speak for itself:

"Automatic Sprinklers, properly installed and maintained, provide a highly effective safeguard against the loss of life and property from fire."

"The NFPA has no record of a multiple-death fire (a fire which kills three or more people) in a completely sprinklered building where the system was properly operating, except in an explosion or flash fire or where industrial fire brigade members or employees were killed during fire suppression operations." In some cases, victims of fatal fires in sprinklered properties were involved in the ignition of the fire and received their injuries prior to the operation of the sprinklers, or were unable to escape due to a physical or mental impairment.[1]

Requirements are based upon proven engineering principles, test data, and field experience that has been analyzed, evaluated, and agreed upon by knowledgeable people. The aim of this standard is to continue to strive for innovative, progressive, and economically feasible measures that promote life safety as well as property protection.

The last sentence of Section 1-2 is a policy statement that overrides any possibility that developments in sprinkler technology not currently recognized in the standard, but which provide comparable or improved levels of protection, will not be prohibited by the standard. The use of nonmetallic piping is one such example.

1-3* Definitions.

Approved.   Acceptable to the "authority having jurisdiction."

NOTE: The National Fire Protection Association does not approve, inspect or certify any installations, procedures, equipment, or materials nor does it approve or evaluate testing laboratories. In determining the acceptability of installations or procedures, equipment or materials, the authority having jurisdiction may base acceptance on compliance with NFPA or other appropriate standards. In the absence of such standards, said authority may require evidence of proper installation, procedure or use. The authority having jurisdiction may also refer to the listings or labeling practices of an organization concerned with product evaluations which is in a position to determine compliance with appropriate standards for the current production of listed items.

"Approved," which means acceptable to the authority having jurisdiction, is different from "listed." An item that is approved is not necessarily listed. Most items that are installed in a sprinkler system, such as alarm valves, dry-pipe valves, sprin-

---

[1] NFPA Fire Analysis Division

klers, hangers, etc., would be both listed and approved. Items such as drain valves that are not listed would be approved.

Authority Having Jurisdiction. The "authority having jurisdiction" is the organization, office or individual responsible for "approving" equipment, an installation or a procedure.

NOTE: The phrase "authority having jurisdiction" is used in NFPA documents in a broad manner since jurisdictions and "approval" agencies vary as do their responsibilities. Where public safety is primary, the "authority having jurisdiction" may be a federal, state, local or other regional department or individual such as a fire chief, fire marshal, chief of a fire prevention bureau, labor department, health department, building official, electrical inspector, or others having statutory authority. For insurance purposes, an insurance inspection department, rating bureau, or other insurance company representative may be the "authority having jurisdiction." In many circumstances the property owner or his designated agent assumes the role of the "authority having jurisdiction"; at government installations, the commanding officer or departmental official may be the "authority having jurisdiction."

Dwelling Unit. One or more rooms arranged for the use of one or more individuals living together as in a single housekeeping unit normally having cooking, living, sanitary, and sleeping facilities.

For purposes of this standard, dwelling unit includes hotel rooms, dormitory rooms, apartments, condominiums, sleeping rooms in nursing homes, and similar living units.

Paragraph A-1-7.2.1 lists residential occupancies as being Light Hazard. The residential classification is meant to cover all dwelling units as defined here.

Listed. Equipment or materials included in a list published by an organization acceptable to the "authority having jurisdiction" and concerned with product evaluation, that maintains periodic inspection of production of listed equipment or materials and whose listing states either that the equipment or material meets appropriate standards or has been tested and found suitable for use in a specified manner.

NOTE: The means for identifying listed equipment may vary for each organization concerned with product evaluation, some of which do not recognize equipment as listed unless it is also labeled. The "authority having jurisdiction" should utilize the system employed by the listing organization to identify a listed product.

One prominent listing agency uses the designation "classified" for pipe. Material with this designation meets the intent of "listed." Most components that are required to be listed are critical to assure proper system effectiveness. This would include alarm valves, hangers, and sprinklers. Items that are required to be approved are necessary for maintaining a high level of reliability, but their impairment would not render the

system out of service. This would include pressure gages, drain valves, and water meters.

Shall. Indicates a mandatory requirement.

Should. Indicates a recommendation or that which is advised but not required.

Sprinkler System. A sprinkler system, for fire protection purposes, is an integrated system of underground and overhead piping designed in accordance with fire protection engineering standards. The installation includes one or more automatic water supplies. The portion of the sprinkler system aboveground is a network of specially sized or hydraulically designed piping installed in a building, structure, or area, generally overhead, and to which sprinklers are attached in a systematic pattern. The valve controlling each system riser is located in the system riser or its supply piping. Each sprinkler system riser includes a device for actuating an alarm when the system is in operation. The system is usually activated by heat from a fire and discharges water over the fire area.

NOTE: The design and installation of water supply facilities such as gravity tanks, fire pumps, reservoirs or pressure tanks, and underground piping are covered by the following NFPA standards: NFPA 22, *Water Tanks for Private Fire Protection*; NFPA 20, *Installation of Centrifugal Fire Pumps*; and NFPA 24, *Installation of Private Fire Service Mains and Their Appurtenances.*

The sprinkler system is denoted by the presence of an alarm check valve or waterflow device and a control valve.

It is important to note that the sprinkler system definition includes water supplies and underground piping. While tanks, pumps, and underground piping design and installation requirements are covered by other standards, when they are used in connection with overhead sprinkler piping, they become an integral part of the sprinkler system. They are so important to sprinkler system performance that they must be treated as parts of the system.

Systems can be designed according to hydraulic calculation techniques, as discussed in Chapter 7, or predetermined pipe sizes conforming to pipe schedules, as in Chapter 8. The standard provides the designer the option of designing the system hydraulically or according to the pipe schedules, as long as system experience continues to be satisfactory.

A-1-3 A sprinkler system is considered to have a single system riser control valve.

Standard. A document containing only mandatory provisions, using the word shall to indicate requirements. Explanatory material may be included only in the form of fine print notes, in footnotes, or in an appendix.

1-4 Other Publications.   A selected list of other publications related to the installation of sprinkler systems is published at the end of this standard.

1-5 Maintenance.

1-5.1*   A sprinkler system installed under this standard shall be properly maintained for efficient service. The owner is responsible for the condition of the sprinkler system and shall use due diligence in keeping the system in good operating condition.

A-1-5.1 Impairments.   Before shutting off a section of the fire service system to make sprinkler system connections, notify the authority having jurisdiction, plan the work carefully, and assemble all materials to enable completion in the shortest possible time. Work started on connections should be rushed to completion without interruption, and protection restored as promptly as possible. During the impairment, provide emergency hose lines, additional fire pails and extinguishers, and maintain extra watch service in the areas affected.

When changes involve shutting off water from any considerable number of sprinklers for more than a few hours, temporary water supply connections should be made to sprinkler systems so that reasonable protection can be maintained. In adding to old systems or revamping them, protection should be restored each night so far as possible. The members of the private fire brigade as well as public fire department should be notified as to conditions.

> **While maintenance may appear to be beyond the scope of this standard, it is important for effective sprinkler system performance at the time fire occurs. This is especially true because of the passive nature of a sprinkler system. It is mandated, therefore, that the owner maintain and service the system properly. A system not maintained in accordance with instructions is not in compliance with this standard. An improperly maintained system with severe impairments is subject to failure, and the net result may be the same as having no sprinkler protection.**

1-5.2   The installing contractor shall provide the owner with:

(a) Instruction charts describing operation and proper maintenance of sprinkler devices.

(b) Publication titled NFPA 13A, *Recommended Practice for the Inspection, Testing and Maintenance of Sprinkler Systems.*

> **It is quite important that the requirements of this section are followed to ensure that the system will be operable as long as is required. There are many properly maintained sprinkler sys-**

tems that are still protecting lives and property even though they were installed 50 or more years ago.

NFPA 13A contains extensive material on the care and maintenance of automatic sprinklers and sprinkler system components. It also has sections on flushing, impairments, and fire records. *Automatic Sprinkler Systems Handbook* users are referred to page 403 for the text of NFPA 13A and invaluable commentary on its application.

1-6 Classification of Sprinkler Systems.

1-6.1 This standard covers automatic sprinkler systems of the types listed below, and systems of outside sprinklers for protection against exposure fires covered specifically in Chapter 6. Manually operated deluge systems, used for certain special hazard conditions, are not specifically covered in this standard but certain provisions of this standard will be found applicable.

Wet-Pipe Systems (*See Section 5-1.*)

Dry-Pipe Systems (*See Section 5-2.*)

Preaction Systems (*See Section 5-3.*)

Deluge Systems (*See Section 5-3.*)

Combined Dry-Pipe and Preaction Systems (*See Section 5-4.*)

Sprinkler Systems—Special Types. Special purpose systems employing departures from the requirements of this standard, such as special water supplies and reduced pipe sizing, shall be installed in accordance with their listing.

1-7 Classification of Occupancies.

1-7.1* Occupancy classifications for this standard relate to sprinkler installations and their water supplies only. They are not intended to be a general classification of occupancy hazards.

A-1-7.1 Occupancy examples in the listings as shown in the various hazard classifications are intended to represent the norm for those occupancy types. Unusual or abnormal fuel loadings or combustible characteristics and susceptibility for changes in these characteristics, for a particular occupancy, are considerations that should be weighed in the selection and classification.

The Light Hazard classification is intended to encompass residential occupancies; however, this is not intended to preclude the use of listed residential sprinklers in residential occupancies or residential portions of other occupancies.

The occupancies listed under each category are meant to be general. Some of the occupancies listed could be of a more or less hazardous classification, depending on the combustibility of their contents.

It is recognized that interior arrangements of a given occupancy can vary markedly, changing the fire challenge to sprinkler systems. Therefore, consideration must be taken for each occupancy example to determine whether it is within the norm for that particular nature of business or operation. If it is not, then the proper occupancy classification must be selected by comparing with other occupancy classifications the quantity, combustibility, and heat release potential of the contents of the building or portion of building being studied.

It is also recognized that experience and judgment are factors critical to the proper determination of the hazard classification, which is very important to the adequacy of the design.

It should be noted that occupancy classifications in this standard consider the fuel load and expected severity of a fire. Unlike building codes, this standard does not consider the likelihood of an ignition; therefore, caution must be exercised when comparing occupancy classifications within the building codes and NFPA 13.

1-7.2 Light Hazard Occupancies.

1-7.2.1* Light Hazard.   Occupancies or portions of other occupancies where the quantity and/or combustibility of contents is low and fires with relatively low rates of heat release are expected.

A-1-7.2.1   Light Hazard Occupancies include occupancies having conditions similar to:

Churches
Clubs
Eaves and overhangs, if combustible construction with
    no combustibles beneath
Educational
Hospitals
Institutional
Libraries, except large stack rooms
Museums
Nursing or Convalescent Homes
Office, including Data Processing
Residential
Restaurant seating areas
Theaters and Auditoriums excluding stages and prosceniums
Unused attics.

Light Hazard Occupancies, as defined in this section, offer the least challenge to sprinkler systems. The nature of occupancies in this classification results in low fuel contribution. Generally there is no processing, manufacturing, or large storage accumulation. Fixtures and furniture remain in fairly fixed arrangements in this group of occupancies. Mainly, these are institutional, educational, religious, residential, and office-type properties.

1-7.3 Ordinary Hazard Occupancies.

1-7.3.1* Ordinary Hazard (Group 1). Occupancies or portions of other occupancies where combustibility is low, quantity of combustibles is moderate, stockpiles of combustibles do not exceed 8 ft (2.4 m), and fires with moderate rates of heat release are expected.

A-1-7.3.1 Ordinary Hazard Occupancies (Group 1) include occupancies having conditions similar to:

Automobile parking garages
Bakeries
Beverage manufacturing
Canneries
Dairy products manufacturing and processing
Electronic plants
Glass and glass products manufacturing
Laundries
Restaurant service areas.

Ordinary Hazard Occupancies include a wide range of occupancy types. Because of the various water supply demands put upon sprinkler systems in this classification, it is divided into three groups.

Group 1 occupancies offer the lowest sprinkler system challenge of the Ordinary Hazard Occupancies, and are comprised mostly of light manufacturing and servicing industries. Use of flammable and combustible liquids or gases is either very limited or so arranged that it presents low challenge to sprinklers. Stockpiles of combustible commodities having Group 1 characteristics that are not within the scope of NFPA 231, *Standard for General Storage*, and are not over 8 ft (2.4 m) high are Ordinary Hazard (Group 1) Occupancies.

1-7.3.2* Ordinary Hazard (Group 2). Occupancies or portions of other occupancies where quantity and combustibility of contents is moderate, stockpiles do not exceed 12 ft (3.7 m), and fires with moderate rate of heat release are expected.

A-1-7.3.2   Ordinary Hazard Occupancies (Group 2) include occupancies having conditions similar to:

Cereal Mills
Chemical Plants — Ordinary
Cold Storage warehouses
Confectionery products
Distilleries
Horse Stables
Leather goods manufacturing
Libraries — large stack room areas
Machine shops
Metal working
Mercantiles
Printing and publishing
Textile manufacturing
Tobacco products manufacturing
Wood product assembly.

Ordinary Hazard (Group 2) Occupancies cover the intermediate range of the Ordinary Hazard classes. Most of the Ordinary Hazard Occupancies fall into this group. They represent the norm or the average occupancy for manufacturing and processing industries.

When attempting to classify occupancies that decidedly are not Light Hazard or Extra Hazard, the Group 2 classification should be considered first. Those that do not compare well with the examples of Group 1 and Group 3 occupancies are most probably of the Group 2 classification.

Stockpiles of combustibles for Group 2 are the same as for Group 1, except the height limitation is 12 ft (3.6 m) rather than 8 ft (2.4 m).

1-7.3.3* Ordinary Hazard (Group 3).   Occupancies or portions of other occupancies where quantity and/or combustibility of contents is high, and fires of high rate of heat release are expected.

A-1-7.3.3   Ordinary Hazard Occupancies (Group 3) include occupancies having conditions similar to:

Feed Mills
Paper and pulp mills
Paper process plants
Piers and wharves
Repair garages
Tire manufacturing

Warehouses (having moderate to higher combustibility
  of content, such as paper, household furniture,
  paint, general storage, whiskey, etc.)[1]
Wood machining.

When hazards in those buildings or portions of buildings of this occupancy group are severe, the authority having jurisdiction should be consulted for special rulings regarding water supplies, types of equipment, pipe sizes, types of sprinklers, and sprinkler spacing.

The occupancies in Group 3 border on being classified as Extra Hazard. These occupancies can typically include dust or residue problems or include materials with higher rates of heat release than contemplated in the other two groups. They are usually the troublesome occupancies of the Ordinary classes, with which fire experience has shown a need for strengthened water supplies.

1-7.4* Extra Hazard Occupancies.

A-1-7.4 New installations protecting Extra Hazard Occupancies should be hydraulically designed where standards giving design criteria are available.

1-7.4.1* Extra Hazard Occupancies or portions of other occupancies where quantity and combustibility of contents is very high, and flammable and combustible liquids, dust, lint, or other materials are present introducing the probability of rapidly developing fires with high rates of heat release.

A-1-7.4.1 Extra Hazard Occupancies (Group 1) include occupancies having conditions similar to:

Combustible Hydraulic Fluid use areas
Die Casting
Metal Extruding
Plywood and particle board manufacturing
Printing [using inks with below 100°F (37.8°C)
  flash points]
Rubber reclaiming, compounding, drying, milling,
  vulcanizing
Saw Mills
Textile picking, opening, blending, garnetting, carding,
  combining of cotton, synthetics, wool shoddy, or burlap
Upholstering with plastic foams.

---

[1] For high-piled storage as defined in 4-1.3.10, see Appendix C for separately published NFPA standards relating to water supply requirements, particularly NFPA 231, *Standard for General Storage*, and NFPA 231C, *Standard for Rack Storage of Materials*.

Extra Hazard Occupancies (Group 2) include occupancies having conditions similar to:

Asphalt saturating
Flammable liquids spraying
Flow coating
Mobile Home or Modular Building assemblies (where finished
 enclosure is present and has combustible interiors)
Open oil quenching
Solvent cleaning
Varnish and paint dipping.

1-7.4.2 Extra Hazard Occupancies involve a wide range of variables that may produce severe fires. The following shall be used to evaluate the severity of Extra Hazard Occupancies:

Extra Hazard (Group 1) includes occupancies described in 1-7.4.1 with little or no flammable or combustible liquids.

Extra Hazard (Group 2) includes occupancies described in 1-7.4.1 with moderate to substantial amounts of flammable or combustible liquids or where shielding of combustibles is extensive.

Extra Hazard Occupancies represent the occupancy fire conditions covered by this standard which provide the most severe challenges to sprinkler protection. The Extra Hazard (Group 1) Occupancies include those in which hydraulic machinery or systems with flammable or combustible hydraulic fluids under pressure are present. Ruptures and leaks in piping or fittings have resulted in fine spray discharge of such liquids, and intense fires result. Those occupancies with process machinery that use flammable or combustible fluids in closed systems are Extra Hazard (Group 1). Also in this group are occupancies that have dust and lint in suspension or that contain moderate amounts of combustible cellular foam materials.

The Extra Hazard (Group 2) Occupancies contain more than small amounts of flammable or combustible liquids, usually in open systems where rapid evaporation can occur when subjected to high temperatures.

This occupancy classification also applies when ceiling sprinklers are obstructed by occupancy conditions, not structural conditions, and water discharged by sprinklers may not reach the burning material because of the shielding.

The Extra Hazard Occupancy examples were classified on the basis of actual field experience with sprinkler system operations in those kinds of occupancies shown in the examples.

1-8* Design and Installation.

A-1-8 Sprinkler Systems in Buildings Subject to Flood. When sprinkler systems are installed in buildings subject to recurring floods, the location of control valves, alarm devices, dry-pipe valves, pumps, compressors, power and fuel supplies should be such that system operation will be uninterrupted by high water.

1-8.1 Devices and Materials.

1-8.1.1 Only new sprinklers shall be employed in the installation of sprinkler systems.

The sprinkler is one component of the system that must be depended upon to operate efficiently and effectively when fire occurs. Consequently, it should not leak or rupture, or operate for any reason other than to extinguish fire.

1-8.1.2* When a sprinkler system is installed, only approved materials and devices shall be used.

A-1-8.1.2 Under special conditions used equipment may be reused by the original owner, subject to the approval of the authority having jurisdiction. Second-hand alarm valves, retarding chambers, circuit closers, water-motor alarms, dry-pipe valves, quick-opening devices, and other devices may be used as replacement equipment in existing systems subject to the approval of the authority having jurisdiction.

The key word here is "approved" (acceptable to the authority having jurisdiction). Some components of sprinkler systems are required to be "listed." (*See Section 1-3, Definitions.*) This is a quality control that ensures that materials such as pipe, fittings, sprinklers, hangers, and other accessory equipment meet a certain level of performance and reliability. When listed materials are not specifically required by the standard, it means that they are not necessarily evaluated by Underwriters Laboratories Inc., Factory Mutual, or other agencies with similar testing and follow-up programs. In that event, the materials must be approved (acceptable to the authority having jurisdiction), meaning the authority having jurisdiction would require materials and devices equivalent to materials specified in standards such as the ANSI/ASTM standard for piping (as shown in Table 3-1.1.1) or for fittings (as shown in Table 3-8.1.1). If listed material is not required by specific reference and approved material is required, it will be left to the authority having jurisdiction to determine the suitability of the product or component. An example of the latter is the nonindicating valve

referred to in 3-9.1.1. In that case the nonindicating valve is not listed or required by this standard to meet the equivalency of a specified ANSI/ASTM standard. The authority having jurisdiction must then decide if it is to be considered approved.

Formal Interpretation

Question: In the second sentence, are the words "and other devices" intended to include equipments comprising the water supply, such as a fire pump assembly complete with driver and controls, a water tank, and the heating systems for the valve room, pump room, and water tank?

Answer: No. While the scope of NFPA 13 includes the character and adequacy of water supplies, "equipments" of the nature described are within the scopes of the NFPA Standards 20 and 22.

1-8.1.3 Sprinkler systems shall be designed for a maximum working pressure of 175 psi (12.1 bars).

Exception: Higher design pressures may be used when all system components are rated for pressures higher than 175 psi (12.1 bars).

Sprinklers are currently rated for a 175-psi (12.1-bar) working pressure. Until and unless all components that make up sprinkler systems are approved for working pressures higher than 175 psi, systems shall be designed for not more than 175 psi working pressure.
The Exception merely removes the 175-psi restriction in the event that higher pressure rated sprinklers are made available after the publication of this edition of the standard.

1-8.1.3.1 Interior system components subject to pressure shall be designed for a working pressure not less than 175 psi (12.1 bars).

1-9* Working Plans.

A-1-9 Preliminary layouts should be submitted for review to the authority having jurisdiction before any equipment is installed or remodeled in order to avoid error or subsequent misunderstanding. Any material deviation from approved plans will require permission of the authority having jurisdiction.

Preliminary layouts should show:

(a) Name of owner and occupant

(b) Location, including street address

(c) Point of compass

(d) Construction and occupancy of each building

NOTE: Date on special hazards should be submitted as they may require special rulings.

(e) Building height in feet

(f) If it is proposed to use a city main as a supply, whether the main is dead-end or circulating, size of the main and pressure in psi; and if dead-end, direction and distance to nearest circulating main

(g) Distance from nearest pumping station or reservoir

(h) In cases where reliable up-to-date information is not available, a waterflow test of the city main should be conducted in accordance with B-2-1.1. (The preliminary plan should specify who conducted the test, date and time, the location of the hydrants where flow was taken, and where static and residual pressure readings were recorded; the size of main supplying these hydrants, and the results of the test, giving size and number of open hydrant butts flowed; also data covering minimum pressure in the connection with the city main should be included.)

(i) Data covering waterworks systems in small towns in order to expedite the review of plans

(j) Fire walls, fire doors, unprotected window openings, large unprotected floor openings, and blind spaces

(k) Distance to and construction and occupancy of exposing buildings — e.g., lumber yards, brick mercantiles, fire-resistive office buildings, etc.

(l) Spacing of sprinklers, number of sprinklers in each story or fire area and total number of sprinklers, number of sprinklers on each riser and on each system by floors, total area protected by each system on each floor, total number of sprinklers on each dry-pipe system or preaction or deluge system and if extension to present equipment, number of sprinklers on riser per floor, sprinklers already installed

(m) Capacities of dry-pipe systems with the bulk pipe included (see Table A-5-2.3), and if an extension is made to an existing dry-pipe system. The total capacity of the existing and also extended portion of the system

(n) Weight or class, size, and material of any proposed underground pipe

(o) Whether property is located in a flood area requiring consideration in the design of sprinkler system

(p) Name and address of party submitting the layout.

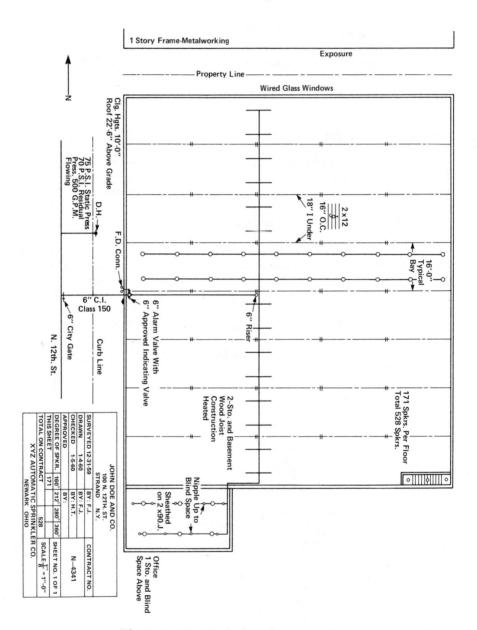

Figure A-1-9   Typical Preliminary Plan.

1-9.1 Working plans shall be submitted for approval to the authority having jurisdiction before any equipment is installed or remodeled. Deviation from approved plans will require permission of the authority having jurisdiction.

Working plans are prepared primarily for the mechanics who do the on-the-job installation. They also serve to protect the interest of the owner, who usually is not knowledgeable about sprinkler system installations and consequently will rely upon others to check the plans for conformance to this standard. The owner determines to which "authorities having jurisdiction" — be they code enforcers, consultants, insurers, or even the owner himself — the plans should be submitted for examination and approval.

Development and approval of plans may be equally difficult for the contractor and the authority having jurisdiction in the case of a speculative-type building. Oftentimes, building owners may not have a particular tenant in mind and everything from the occupancy classification to the sprinkler spacing criteria may be questioned unless specific information on the future tenants can be established. This requires good communication between the owner, the contractor, and the authority having jurisdiction.

When the installation is completed, it is checked against the plans in order to determine compliance with approved plans and the sprinkler standard.

The property owner should retain working plans and specifications for reference. In the event that alterations are undertaken sometime in the future, a good deal of expense reduction and time saving, especially for systems of hydraulic design, can result if the plans are available at that time.

The symbols commonly used in plan drawings can be found in NFPA 172, *Fire Protection Symbols for Architectural and Engineering Drawings.*

1-9.2* Working plans shall be drawn to an indicated scale, on sheets of uniform size, with a plan of each floor, made so that they can be easily duplicated, and shall show the following data:

(a) Name of owner and occupant

(b) Location, including street address

(c) Point of compass

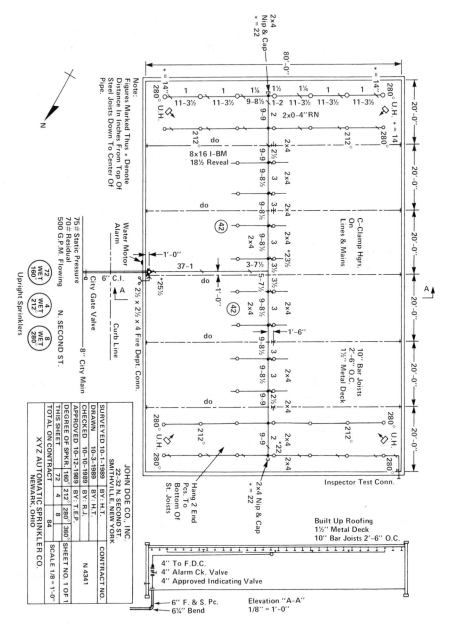

Figure A-1-9.2   Typical Working Plans.

(d) Ceiling construction

(e) Full height cross section

(f) Location of fire walls

(g) Location of partitions

(h) Occupancy of each area or room

(i) Location and size of concealed spaces, closets, attics, and bathrooms [*see 4-4.3 to 4-4.16 inclusive (except 4-4.5 and 4-4.6), and 4-4.19*]

(j) Any questionable small enclosures in which no sprinklers are to be installed

(k) Size of city main in street, pressure, and whether dead-end or circulating and, if dead-end, direction and distance to nearest circulating main, city main test results, and elevation relative to the test hydrant (*see B-2-1*)

(l) Other sources of water supply, with pressure or elevation

(m) Make, type, and nominal orifice size of sprinkler

(n) Temperature rating and location of high-temperature sprinklers

(o) Total area protected by each system on each floor

(p) Number of sprinklers on each riser per floor

(q) Make, type, model, and size of alarm or dry-pipe valve

(r) Make, type, model, and size of preaction or deluge valve

(s) Kind and location of alarm bells

(t) Total number of sprinklers on each dry-pipe system, preaction system, combined dry-pipe/preaction system, or deluge system

(u) Approximate capacity in gallons of each dry-pipe system

(v) Pipe type and schedule of wall thickness

(w) Nominal pipe size and cutting lengths of pipe (or center to center dimensions)

NOTE: Where typical branch lines prevail, it will be necessary to size only one line.

(x) Location and size of riser nipples

(y) Type of fittings and joints and location of all welds and bends

(z) Type and locations of hangers, sleeves, braces, and methods of securing sprinklers when applicable

(aa) All control valves, check valves, drain pipes, and test connections

(bb) Size and location of hand hose, hose outlets, and related equipment

(cc) Underground pipe size, length, location, weight, material, point of connection to city main; the type of valves, meters, and valve pits; and the depth that top of the pipe is laid below grade

(dd) Provision for flushing (*see 3-3.2*)

(ee) When the equipment is to be installed as an addition to an existing system enough of the existing system shall be indicated on the plans to make all conditions clear

(ff) For hydraulically designed systems, the material to be included on the hydraulic data nameplate

(gg) Name and address of contractor

(hh) Method of protection for nonmetallic piping

(ii) A graphical representation of the scale shall be provided on all plans.

Formal Interpretation

Question: Is it the intent of 1-9.2(z) that hanger locations be dimensioned?

Answer: No, but hanger locations are required to be in accordance with 3-10.5 and 3-10.6.

1-10 Approval of Sprinkler Systems.

1-10.1 The installer shall perform all required acceptance tests (*see Section 1-11*), complete the Contractor's Material and Test Certificate(s) [*see Figures 1-10.1(a) and 1-10.1(b)*], and forward the certificate(s) to the authority having jurisdiction prior to asking for approval of the installation.

Material and test certificates, one for aboveground installations and one for underground installations, are shown in Figures 1-10.1(a) and 1-10.1(b).

The certificates are acknowledgments that materials used and tests performed are in accordance with the requirements of this standard. They also provide a record of the test results that can be used for comparison with tests that will or may take place in the future.

CONTRACTOR'S MATERIAL & TEST CERTIFICATE FOR A BOVEGROUND PIPING

PROCEDURE

Upon completion of work, inspection and tests shall be made by the contractor's representative and witnessed by an owner's representative. All defects shall be corrected and system left in service before contractor's personnel finally leave the job.

A certificate shall be filled out and signed by both representatives. Copies shall be prepared for approving authorities, owners, and contractor. It is understood the owner's representative's signature in no way prejudices any claim against contractor for faulty material, poor workmanship, or failure to comply with approving authority's requirements or local ordinances.

PROPERTY NAME		DATE

PROPERTY ADDRESS

PLANS	ACCEPTED BY APPROVING AUTHORITIES (NAMES)	
	ADDRESS	
	INSTALLATION CONFORMS TO ACCEPTED PLANS	☐ YES ☐ NO
	EQUIPMENT USED IS APPROVED	☐ YES ☐ NO
	IF NO, EXPLAIN DEVIATIONS	
INSTRUCTIONS	HAS PERSON IN CHARGE OF FIRE EQUIPMENT BEEN INSTRUCTED AS TO LOCATION OF CONTROL VALVES AND CARE AND MAINTENANCE OF THIS NEW EQUIPMENT?	☐ YES ☐ NO
	IF NO, EXPLAIN	
	HAVE COPIES OF THE FOLLOWING BEEN LEFT ON THE PREMISES:	☐ YES ☐ NO
	1. SYSTEM COMPONENTS INSTRUCTIONS	☐ YES ☐ NO
	2. CARE AND MAINTENANCE INSTRUCTIONS	☐ YES ☐ NO
	3. NFPA 13A	☐ YES ☐ NO
LOCATION OF SYSTEM	SUPPLIES BUILDINGS	

SPRINKLERS	MAKE	MODEL	YEAR OF MANUFACTURE	ORIFICE SIZE	QUANTITY	TEMPERATURE RATING

PIPE AND FITTINGS	Type of Pipe _____
	Type of Fittings _____

ALARM VALVE OR FLOW INDICATOR	ALARM DEVICE			MAXIMUM TIME TO OPERATE THROUGH TEST CONNECTION	
	TYPE	MAKE	MODEL	MIN.	SEC.

	DRY VALVE			Q.O.D.		
	MAKE	MODEL	SERIAL NO.	MAKE	MODEL	SERIAL NO.

DRY PIPE OPERATING TEST		TIME TO TRIP THRU TEST CONNECTION*		WATER PRESSURE	AIR PRESSURE	TRIP POINT AIR PRESSURE	TIME WATER REACHED TEST OUTLET*		ALARM OPERATED PROPERLY	
		MIN.	SEC.	PSI	PSI	PSI	MIN.	SEC.	YES	NO
	Without Q.O.D.									
	With Q.O.D.									
	IF NO, EXPLAIN									

---

* MEASURED FROM TIME INSPECTOR'S TEST CONNECTION IS OPENED.

85A (10-88)   PRINTED IN U.S.A.   (OVER)

Figure 1-10.1(a)   Contractor's Material and Test Certificate for Aboveground Piping.

	OPERATION								

	OPERATION ☐ PNEUMATIC ☐ ELECTRIC ☐ HYDRAULIC
DELUGE & PREACTION VALVES	PIPING SUPERVISED ☐ YES ☐ NO DETECTING MEDIA SUPERVISED ☐ YES ☐ NO
	DOES VALVE OPERATE FROM THE MANUAL TRIP AND/OR REMOTE CONTROL STATIONS ☐ YES ☐ NO
	IS THERE AN ACCESSIBLE FACILITY IN EACH CIRCUIT FOR TESTING IF NO, EXPLAIN ☐ YES ☐ NO

	MAKE	MODEL	DOES EACH CIRCUIT OPERATE SUPERVISION LOSS ALARM		DOES EACH CIRCUIT OPERATE VALVE RELEASE		MAXIMUM TIME TO OPERATE RELEASE	
			YES	NO	YES	NO	MIN.	SEC.

TEST DESCRIPTION	HYDROSTATIC: Hydrostatic tests shall be made at not less than 200 psi (13.6 bars) for two hours or 50 psi (3.4 bars) above static pressure in excess of 150 psi (10.2 bars) for two hours. Differential dry-pipe valve clappers shall be left open during test to prevent damage. All aboveground piping leakage shall be stopped. PNEUMATIC: Establish 40 psi (2.7 bars) air pressure and measure drop which shall not exceed 1-1/2 psi (0.1 bars) in 24 hours. Test pressure tanks at normal water level and air pressure and measure air pressure drop which shall not exceed 1-1/2 psi (0.1 bars) in 24 hours.

TESTS	ALL PIPING HYDROSTATICALLY TESTED AT_____ PSI FOR_____ HRS. IF NO, STATE REASON
	DRY PIPING PNEUMATICALLY TESTED ☐ YES ☐ NO
	EQUIPMENT OPERATES PROPERLY ☐ YES ☐ NO
	DO YOU CERTIFY AS THE SPRINKLER CONTRACTOR THAT ADDITIVES AND CORROSIVE CHEMICALS, SODIUM SILICATE OR DERIVATIVES OF SODIUM SILICATE, BRINE, OR OTHER CORROSIVE CHEMICALS WERE NOT USED FOR TESTING SYSTEMS OR STOPPING LEAKS? ☐ YES ☐ NO
	DRAIN TEST: READING OF GAGE LOCATED NEAR WATER SUPPLY TEST CONNECTION: _____ PSI RESIDUAL PRESSURE WITH VALVE IN TEST CONNECTION OPEN WIDE _____ PSI
	UNDERGROUND MAINS AND LEAD IN CONNECTIONS TO SYSTEM RISERS FLUSHED BEFORE CONNECTION MADE TO SPRINKLER PIPING. VERIFIED BY COPY OF THE U FORM NO. 85B ☐ YES ☐ NO OTHER EXPLAIN FLUSHED BY INSTALLER OF UNDER- GROUND SPRINKLER PIPING ☐ YES ☐ NO

BLANK TESTING GASKETS	NUMBER USED	LOCATIONS	NUMBER REMOVED

WELDING	WELDED PIPING ☐ YES ☐ NO
	IF YES . . .
	DO YOU CERTIFY AS THE SPRINKLER CONTRACTOR THAT WELDING PROCEDURES COMPLY WITH THE REQUIREMENTS OF AT LEAST AWS D10.9, LEVEL AR-3 ☐ YES ☐ NO
	DO YOU CERTIFY THAT THE WELDING WAS PERFORMED BY WELDERS QUALIFIED IN COMPLIANCE WITH THE REQUIREMENTS OF AT LEAST AWS D10.9, LEVEL AR-3 ☐ YES ☐ NO
	DO YOU CERTIFY THAT WELDING WAS CARRIED OUT IN COMPLIANCE WITH A DOCUMENTED QUALITY CONTROL PROCEDURE TO INSURE THAT ALL DISCS ARE RETRIEVED, THAT OPENINGS IN PIPING ARE SMOOTH, THAT SLAG AND OTHER WELDING RESIDUE ARE REMOVED, AND THAT THE INTERNAL DIAMETERS OF PIPING ARE NOT PENETRATED ☐ YES ☐ NO

CUTOUTS (DISCS)	DO YOU CERTIFY THAT YOU HAVE A CONTROL FEATURE TO ENSURE THAT ALL CUTOUTS (DISCS) ARE RETRIEVED? ☐ YES ☐ NO

HYDRAULIC DATA NAMEPLATE	NAME PLATE PROVIDED ☐ YES ☐ NO	IF NO, EXPLAIN

REMARKS	DATE LEFT IN SERVICE WITH ALL CONTROL VALVES OPEN:

SIGNATURES	NAME OF SPRINKLER CONTRACTOR		
	TESTS WITNESSED BY		
	FOR PROPERTY OWNER (SIGNED)	TITLE	DATE
	FOR SPRINKLER CONTRACTOR (SIGNED)	TITLE	DATE

ADDITIONAL EXPLANATION AND NOTES

85A BACK

Figure 1-10.1(a) (Continued)   Contractor's Material and Test Certificate for Aboveground Piping.

CONTRACTOR'S MATERIAL & TEST CERTIFICATE FOR **U**NDERGROUND PIPING

PROCEDURE

Upon completion of work, inspection and tests shall be made by the contractor's representative and witnessed by an owner's representative. All defects shall be corrected and system left in service before contractor's personnel finally leave the job.

A certificate shall be filled out and signed by both representatives. Copies shall be prepared for approving authorities, owners, and contractor. It is understood the owner's representative's signature in no way prejudices any claim against contractor for faulty material, poor workmanship, or failure to comply with approving authority's requirements or local ordinances.

PROPERTY NAME	DATE

PROPERTY ADDRESS

PLANS	ACCEPTED BY APPROVING AUTHORITIES (NAMES)		
	ADDRESS		
	INSTALLATION CONFORMS TO ACCEPTED PLANS	☐ YES	☐ NO
	EQUIPMENT USED IS APPROVED	☐ YES	☐ NO
	IF NO, STATE DEVIATIONS		
INSTRUCTIONS	HAS PERSON IN CHARGE OF FIRE EQUIPMENT BEEN INSTRUCTED AS TO LOCATION OF CONTROL VALVES AND CARE AND MAINTENANCE OF THIS NEW EQUIPMENT? IF NO, EXPLAIN	☐ YES	☐ NO
	HAVE COPIES OF APPROPRIATE INSTRUCTIONS AND CARE AND MAINTENANCE CHARTS BEEN LEFT ON PREMISES? IF NO, EXPLAIN	☐ YES	☐ NO
LOCATION	SUPPLIES BUILDINGS		

UNDERGROUND PIPES AND JOINTS	PIPE TYPES AND CLASS	TYPE JOINT		
	PIPE CONFORMS TO _____ STANDARD		☐ YES	☐ NO
	FITTINGS CONFORM TO _____ STANDARD		☐ YES	☐ NO
	IF NO, EXPLAIN			
	JOINTS NEEDING ANCHORAGE CLAMPED, STRAPPED, OR BLOCKED IN		☐ YES	☐ NO
	ACCORDANCE WITH _____ STANDARD			
	IF NO, EXPLAIN			

TEST DESCRIPTION	FLUSHING. Flow the required rate until water is clear as indicated by no collection of foreign material in burlap bags at outlets such as hydrants and blow-offs. Flush at flows not less than 390 GPM (1476 L/min) for 4-inch pipe, 880 GPM (3331 L/min) for 6-inch pipe, 1560 GPM (5905 L/min) for 8-inch pipe, 2440 GPM (9235 L/min) for 10-inch pipe, and 3520 GPM (13323 L/min) for 12-inch pipe. When supply cannot produce stipulated flow rates, obtain maximum available. HYDROSTATIC. Hydrostatic tests shall be made at not less than 200 psi (13.8 bars) for two hours or 50 psi (3.4 bars) above static pressure in excess of 150 psi (10.3 bars) for two hours. LEAKAGE. New pipe laid with rubber gasketed joints shall, if the workmanship is satisfactory, have little or no leakage at the joints. The amount of leakage at the joints shall not exceed 2 qts. per hr. (1.89 L/h) per 100 joints irrespective of pipe diameter. The leakage shall be distributed over all joints. If such leakage occurs at a few joints the installation shall be considered unsatisfactory and necessary repairs made. The amount of allowable leakage specified above may be increased by 1 fl oz per in. valve diameter per hr. (30 mL/25 mm/h) for each metal seated valve isolating the test section. If dry barrel hydrants are tested with the main valve open, so the hydrants are under pressure, an additional 5 oz per minute (150 mL/min) leakage is permitted for each hydrant.

FLUSHING TESTS	NEW UNDERGROUND PIPING FLUSHED ACCORDING TO _____ STANDARD		☐ YES	☐ NO
	BY (COMPANY)			
	IF NO, EXPLAIN			
	HOW FLUSHING FLOW WAS OBTAINED ☐ PUBLIC WATER ☐ TANK OR RESERVOIR ☐ FIRE PUMP	THROUGH WHAT TYPE OPENING ☐ HYDRANT BUTT.	☐ OPEN PIPE	
	LEAD-INS FLUSHED ACCORDING TO _____ STANDARD		☐ YES	☐ NO
	BY (COMPANY)			
	IF NO, EXPLAIN			
	HOW FLUSHING FLOW WAS OBTAINED ☐ PUBLIC WATER ☐ TANK OR RESERVOIR ☐ FIRE PUMP	THROUGH WHAT TYPE OPENING ☐ Y CONN. TO FLANGE & SPIGOT	☐ OPEN PIPE	

85B(10-88)   PRINTED IN USA   (OVER)

Figure 1-10.1(b)   Contractor's Material and Test Certificate for Underground Piping.

HYDROSTATIC TEST	ALL NEW UNDERGROUND PIPING HYDROSTATICALLY TESTED AT	JOINTS COVERED
	_____ PSI FOR_____ HOURS	☐ YES ☐ NO

LEAKAGE TEST	TOTAL AMOUNT OF LEAKAGE MEASURED	
	_____ GALS. _____ HOURS	
	ALLOWABLE LEAKAGE	
	_____ GALS. _____ HOURS	

HYDRANTS	NUMBER INSTALLED	TYPE AND MAKE	ALL OPERATE SATISFACTORILY
			☐ YES ☐ NO

CONTROL VALVES	WATER CONTROL VALVES LEFT WIDE OPEN IF NO, STATE REASON	☐ YES ☐ NO
	HOSE THREADS OF FIRE DEPARTMENT CONNECTIONS AND HYDRANTS INTERCHANGEABLE WITH THOSE OF FIRE DEPARTMENT ANSWERING ALARM	☐ YES ☐ NO

REMARKS	DATE LEFT IN SERVICE

SIGNATURES	NAME OF INSTALLING CONTRACTOR		
	TESTS WITNESSED BY		
	FOR PROPERTY OWNER (SIGNED)	TITLE	DATE
	FOR INSTALLING CONTRACTOR (SIGNED)	TITLE	DATE

ADDITIONAL EXPLANATION AND NOTES

Figure 1-10.1(b) (Continued)   Contractor's Material and Test Certificate for Underground Piping.

1-10.2   When the authority having jurisdiction desires to be present during the conduct of acceptance tests, the installer shall give advance notification of the time and date the testing will be performed.

Occasionally the authority having jurisdiction wishes to be present at the time performance tests are conducted. It is the responsibility of the authority having jurisdiction to inform the installer that he desires to be present. This requirement then obligates the installer to comply with such a request and to give the authority having jurisdiction advance notice. Failure to do so may make it necessary to rerun tests.

1-11 Acceptance Tests.

1-11.1* Flushing of Piping.   Underground mains and lead-in connections to system risers shall be flushed thoroughly before connection is made to sprinkler piping, in order to remove foreign materials which may have entered the underground during the course of the installation or which may have been present in existing piping. The minimum rate of flow shall be not less than the water demand rate of the system which is determined by the system design, or not less than that necessary to provide a velocity of 10 ft per second (3 m/s), whichever is greater. For all systems the flushing operations shall be continued for a sufficient time to ensure thorough cleaning. When planning the flushing operations, consideration shall be given to disposal of the water issuing from the test outlets.

Exception: When the flow rate, as listed in Table 1-11.1, cannot be verified or met, supply piping shall be flushed at the maximum flow rate available to the system under fire conditions.

This Exception recognizes that a quantity of water producing velocities of 10 ft per second (3 m/s) may simply not be available under all circumstances. If this is the case, then the anticipated actual demand of the system should be simulated in order to determine if any foreign matter is obstructing the underground pipe network.

Table 1-11.1   Flow Required to Produce a Velocity
of 10 Ft per Second (3 m/s) in Pipes

Pipe Size (in.)	Flow Rate (gpm)	Flow Rate (L/min)
4	390	1476
6	880	3331
8	1560	5905
10	2440	9235
12	3520	13323

For SI Units: 1 in. = 25.4 mm; 1 gpm = 3.785 L/min.

A-1-11.1 Underground mains and lead-in connections to system risers should be flushed through hydrants at dead ends of the system or through accessible aboveground flushing outlets allowing the water to run until clear. If water is supplied from more than one source or from a looped system, divisional valves should be closed to produce a high-velocity flow through each single line. The flows specified in Table 1-11.1 will produce a velocity of at least 10 ft/sec (3 m/s), which is necessary for cleaning the pipe and for lifting foreign material to an aboveground flushing outlet.

Experience has shown that stones, gravel, blocks of wood, bottles, work tools, work clothes, and other objects have been found in piping when flushing was performed. Also, objects in underground piping quite remote from the sprinkler installation, that otherwise would remain stationary, will sometimes be transported into sprinkler system piping when sprinkler systems operate. Sprinkler systems can draw greater flows than normal for domestic or process uses. Fire department pumpers,

when taking suction from hydrants for pumping into sprinkler systems, may compound the problem by increasing the velocity of water flow in underground piping.

Because of the inherent nature of sprinkler system design in which pipe sizes diminish from the point of connection of underground piping, objects that move from underground piping and enter sprinkler system risers will lodge at the point in the system where they may totally obstruct passage of water.

Previous studies by Factory Mutual Research determined that the size of particles that will move upward in piped water streams can be determined if the specific gravity of the particle and the velocity of the water stream are known. For example, granite that is 2 in. (51 mm) in diameter will move upward in piping if the water stream velocity is 5.64 ft per second (1.72 m/s), which is equivalent to the minimum flows this section used to require. The flow rates were increased to reflect velocities of 10 ft per second (3 m/s) in this edition since this number is more in line with requirements of other accepted standards. 10 ft per second is also in agreement with flushing requirements of NFPA 15, *Water Spray Fixed Systems for Fire Protection*, and NFPA 24, *Private Fire Service Mains and Their Appurtenances*.

It should be recognized that overall field experience is such that actual flushing is normally done without a measurement of the flow rate. Flushing is normally accomplished at the maximum flow rate available from the water supply.

1-11.1.1 Provision shall be made for the disposal of water issuing from test outlets to avoid property damage.

1-11.2 Hydrostatic Tests.

1-11.2.1* All new systems including yard piping shall be hydrostatically tested at not less than 200 psi (13.8 bars) pressure for 2 hours, or at 50 psi (3.4 bars) in excess of the maximum pressure, when the maximum pressure to be maintained in the system is in excess of 150 psi (10.3 bars).

The test pressure shall be read from a gage located at the low elevation point of the individual system or portion of the system being tested.

Exception: At seasons of the year that will not permit testing with water an interim test may be conducted with air pressure of at least 40 psi (2.8 bars) allowed to stand for 24 hours. The standard hydrostatic test shall be conducted when weather permits.

Part of the purpose of this test requirement is to check the quality of the original workmanship. When an addition is made

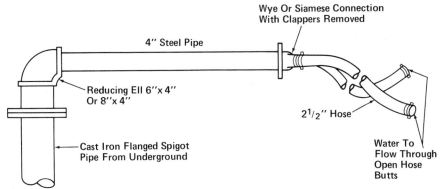

Wye Or Siamese Connection
With Clappers Removed

4" Steel Pipe

Reducing Ell 6"x 4"
Or 8"x 4"

2¹/₂" Hose

Cast Iron Flanged Spigot
Pipe From Underground

Water To
Flow Through
Open Hose
Butts

Employing Horizontal Run Of 4-Inch Pipe And Reducing Fitting Near Base Of Riser.

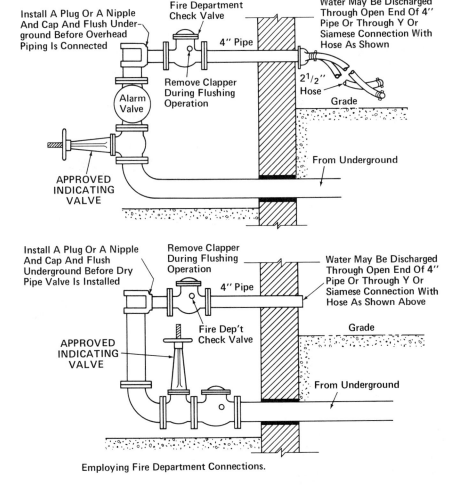

Install A Plug Or A Nipple
And Cap And Flush Under-
ground Before Overhead
Piping Is Connected

Fire Department
Check Valve

Water May Be Discharged
Through Open End Of 4"
Pipe Or Through Y Or
Siamese Connection With
Hose As Shown

4" Pipe

Remove Clapper
During Flushing
Operation

Alarm
Valve

2¹/₂"
Hose

Grade

APPROVED
INDICATING
VALVE

From Underground

Install A Plug Or A Nipple
And Cap And Flush
Underground Before Dry
Pipe Valve Is Installed

Remove Clapper
During Flushing
Operation

Water May Be Discharged
Through Open End Of 4"
Pipe Or Through Y Or
Siamese Connection With
Hose As Shown Above

4" Pipe

Fire Dep't
Check Valve

Grade

APPROVED
INDICATING
VALVE

From Underground

Employing Fire Department Connections.

Figure A-1-11.1   Methods of Flushing Water Supply Connections.

or an existing system is expanded, the new installation should be isolated and tested per 1-11.2.1. The existing system need not be subjected to a new hydrostatic test.

A-1-11.2.1 Example. A sprinkler system has for its water supply a connection to a public water service main. A 100 psi (6.9 bar) rated pump is installed in the connection. With a maximum normal public water supply of 70 psi (4.8 bars) at the low elevation point of the individual system or portion of the system being tested and a 120 psi (8.3 bars) pump (churn) pressure, the hydrostatic test pressure is 70 + 120 + 50 or 240 psi (16.5 bars).

Systems that have been modified or repaired to any appreciable extent should be hydrostatically tested at not less than 50 psi (3.4 bars) in excess of normal static pressure for 2 hours.

To reduce the possibility of serious water damage in case of a break, pressure may be maintained by a small pump, the main controlling gate meanwhile being kept shut during the test.

Polybutylene pipe will undergo expansion during initial pressurization. In this case a reduction in gage pressure may not necessarily indicate a leak. The pressure reduction should not exceed the manufacturer's specifications and listing criteria.

Formal Interpretation

Question: Is hydropneumatic the substitute or equal of hydrostatic procedure according to Testing Laboratory or NFPA Standards?

Answer: **No. The NFPA** *Standard for Installation of Automatic Sprinkler Systems* **requires hydrostatic pressure testing of new systems under 1-11.2. During seasons in which freezing may occur, interim testing of dry-pipe systems with air is required with the customary hydrostatic testing required when weather permits in accordance with 1-11.2.1.**

Hydropneumatic procedures are associated with flushing of foreign materials that may obstruct waterways in piping and not with the pressure tests required for leakage testing in new systems.

Formal Interpretation

Question: Is it the intent of 1-11.2.1 and 1-11.2.2 that the sprinkler pipings at the top of a system fed by a riser 105 ft (32 m) in height be pressurized to 200 psi (14 bar) for 2 hours when the sprinkler piping and riser are considered as "the individual system or zone"?

Answer: No.

Formal Interpretation

Question: In determining hydrostatic test pressure, when a booster pump is installed, should the shut off pressure or normal operating pressure be considered in determining hydrostatic test pressure? If a relief valve is installed, should the relief valve setting be considered in determining the hydrostatic pressure test?

Answer: The hydrostatic pressure should, in the case of new installations, be based upon the maximum pressure exerted upon the new system piping and components. When a booster pump is provided, the maximum pressure is shut off pressure or "churn" pressure without consideration of relief valve pressure settings.

Formal Interpretation

Question: Is there any NFPA requirement to exhaust the trapped air from the spring-ups, armover, drops, or branch lines in a sprinkler system in the performance of a hydrostatic test?

Answer: No.

All new systems are tested hydrostatically to at least 200 psi (13.8 bars). This standard is set to ensure that pipe joints are made to withstand that pressure without coming apart or leaking. It is primarily a workmanship test and not a materials performance test. However, damaged materials (cracked fittings, leaky sprinklers, etc.) are routinely discovered during the hydrostatic test. As noted previously, all materials used must be rated for at least 175 psi (12.1 bars) working pressure, the maximum pressure at which the system is normally maintained.

The measurement of the hydrostatic test pressure is taken at the lowest elevation within the system or portion of the system being tested. It is not considered necessary to test at the high point of the system (which, due to static head, would increase the test pressure significantly) due to the fact that application of pressure typically occurs at the lower elevation, and these high pressures would not be anticipated at the higher elevations within the system.

The air test for dry-pipe systems in 1-11.3.2 is permitted, by the Exception to 1-11.2.1, as an interim substitution for the hydrostatic test when there is a danger of water freezing in the system. The standard hydrostatic test must be made, however, when it is possible to do so without danger of freezing.

The measure of success for inside sprinkler piping under hydrostatic test is that there is no visible leakage. Very often a very small bead of water may form on a fitting during the test. Unless the bead continues to grow and drip, this is not considered visible leakage.

1-11.2.2* **Permissible Leakage.**   The inside sprinkler piping shall be installed in such a manner that there will be no visible leakage when the system is subjected to the hydrostatic pressure test. Refer to NFPA 24, *Standard for the Installation of Private Fire Service Mains and Their Appurtenances*, for permissible leakage in underground piping. The amount of leakage shall be measured by pumping from a calibrated container.

A-1-11.2.2   Valves isolating the section to be tested may not be "drop tight." When such leakage is suspected, test blanks of the type recommended in 1-11.2.5 should be used in a manner that includes the valve in the section being tested.

1-11.2.3 Fire Department Connection.   Piping between the check valve in the fire department inlet pipe and the outside connection shall be tested in the same manner as the balance of the system.

This section is intended as a reminder that the fire department pumping connection is required to be tested. The fact that the piping from the check valve to the hose connections is not normally subjected to water pressure may lead the contractor to believe testing of that portion is not required. Other portions of pipe, such as the inspectors' test connection drain and auxiliary drains, are not typically subject to high pressures and do not have to be tested in accordance with this paragraph.

1-11.2.4 Corrosive Chemicals.   Additives and corrosive chemicals, sodium silicate or derivatives of sodium silicate, brine or other corrosive chemicals shall not be used for testing systems or stopping leaks.

Water additives, such as sodium silicate, that are intended to plug small system leaks during hydrostatic testing may clog small orifices, bind sprinkler parts, and prevent or delay operation. These chemicals tend to harden when exposed to air and may create significant obstructions later, when they detach themselves from the internal portions of the pipe.

1-11.2.5 Test Blanks. Whenever a test blank is used it shall be of the self-indicating type. Test blanks shall have red painted lugs protruding beyond the flange in such a way as to clearly indicate their presence. The installer shall have all test blanks numbered so as to keep track of their use and assure their removal after the work is completed.

It is extremely important that all test blanks are removed after the work is completed. A test blank without a protruding lug is virtually impossible to detect by visual inspection. One such test blank is commonly referred to as a "frying pan" because of its characteristic shape.

1-11.3 Test of Dry-Pipe Systems.

1-11.3.1 Differential Dry-Pipe Valves. The clapper of a differential type dry-pipe valve shall be held off its seat during any test in excess of 50 psi (3.4 bars) to prevent damaging the valve.

The design of differential dry-pipe valves incorporates flexible gaskets, which prevent water from entering the system. Pressures in excess of 50 psi (3.4 bars) may cause damage to the mechanism of the valve.

1-11.3.2 Air Test. In dry-pipe systems an air pressure of 40 psi (2.8 bars) shall be pumped up, allowed to stand 24 hours, and all leaks that allow a loss of pressure over 1½ psi (0.1 bar) for the 24 hours shall be stopped.

Formal Interpretation

Question: If a new dry-pipe system is hydrostatically tested in accordance with 1-11.2.1 is it also required to conduct an air test of the system in accordance with 1-11.3.2?

Answer: Yes, both tests are required. The hydrostatic test is for visible signs of water leakage providing watertight integrity, while the air test, limiting air pressure loss to 1½ psi (0.1 bar) minimum over a 24-hour period, is for proof of airtight integrity. One test does not guarantee that the system will perform satisfactorily when subjected to the other test.

The air test is a test for airtight integrity of the system. Under normal conditions, dry-pipe systems are subjected to lower pressures than those existing when the system has tripped and filled with water. Since dry systems are subjected to pressures from both air and water, they must be pressure tested by both air and water. Characteristic differences between air and water make it impossible to judge system integrity based upon only one of these tests. The 200 psi (13.8 bars) for water and 40 psi (2.8 bars) for air are considered the extreme pressures for normal use in systems.

1-11.3.3 Operating Test of Dry-Pipe Valve. A working test of the dry-pipe valve alone and with a quick-opening device, if installed, shall be made before acceptance by opening the system test connection. Trip and water delivery times shall be measured from the time the inspector's test connection is opened and shall be recorded using the Contractor's Material and Test Certificate for Aboveground Piping.

Full operational tests of dry-pipe valves are required. The Contractor's Material and Test Certificate has provisions for recording the time it takes for the valve to trip after opening the system test valve, the air and water pressures before the test, the air pressure at trip point, the delivery time for water to reach the test outlet, and whether or not alarms operated properly. The valve should operate in accordance with the tolerance specified by the manufacturer for the trip point.

1-11.4 Tests of Drainage Facilities. Tests of drainage facilities shall be made while the control valve is wide open. The main drain valve shall be opened and remain open until the system pressure stabilizes. (*See 2-9.1.*)

1-11.5 Each pressure reducing valve shall be tested upon completion of the initial installation to ensure proper pressure reduction at both maximum and normal inlet pressures.

1-12 Operation of Sprinkler System Control Valves by Contractors. When work on a sprinkler system requires that a contractor operate a valve controlling water supplies to a sprinkler system, the contractor shall inform the owner so that the owner may follow the normal valve supervision procedure.

1-13 Units. Metric units of measurement in this standard are in accordance with the modernized metric system known as the International System of Units (SI). Two units (liter and bar), outside of but recognized by SI, are commonly used in international fire protection. These units are listed in Table 1-13 with conversion factors.

1-13.1 If a value for measurement as given in this standard is followed by an equivalent value in other units, the first stated is to be regarded as the requirement. A given equivalent value may be approximate.

Table 1-13   Metric Units of Measurement

Name of Unit	Unit Symbol	Conversion Factor
liter	L	1 gal = 3.785 L
liter per minute per square meter	(L/min)/m²	1 gpm/ft² = 40.746 (L/min)/m²
millimeter per minute	1 mm/min	1 gpm/ft² = 40.746 mm/min
cubic decimeter	dm³	1 gal = 3.785 dm³
pascal	Pa	1 psi = 6894.757 Pa
bar	bar	1 psi = 0.0689 bar
bar	bar	1 bar = 10⁵ Pa

For additional conversions and information see ASTM E380, *Standard for Metric Practice.*

When and if the United States converts to SI, some of the values will probably be rounded off to the nearest whole number. A value such as 0.15 gpm/sq ft would become (6.0 L/min)/m² or 6.0 mm/min instead of (6.1119 L/min)/m² or 6.1119 mm/min.

1-13.2   The conversion procedure for the SI units has been to multiply the quantity by the conversion factor and then round the result to the appropriate number of significant digits.

REFERENCES CITED IN COMMENTARY

The following publications are available from the National Fire Protection Association, Batterymarch Park, Quincy, MA 02269.

NFPA 11-1988, *Standard for Low Expansion Foam and Combined Agent Systems*

NFPA 11A-1988, *Standard for Medium and High Expansion Foam Systems*

NFPA 13A-1987, *Recommended Practice for the Inspection, Testing and Maintenance of Sprinkler Systems*

NFPA 13D-1989, *Standard for the Installation of Sprinkler Systems in One- and Two-Family Dwellings and Mobile Homes*

NFPA 13R-1989, *Standard for the Installation of Sprinkler Systems in Residential Occupancies up to Four Stories in Height*

NFPA 14-1986, *Standard for the Installation of Standpipe and Hose Systems*

NFPA 15-1985, *Standard for Water Spray Fixed Systems for Fire Protection*

NFPA 16-1986, *Standard on Deluge Foam-Water Sprinkler and Foam-Water Spray Systems*

NFPA 16A-1988, *Recommended Practice for the Installation of Closed-Head Foam-Water Sprinkler Systems*

NFPA 20-1987, *Standard for the Installation of Centrifugal Fire Pumps*

NFPA 22-1987, *Standard for Water Tanks for Private Fire Protection*

NFPA 24-1987, *Standard for the Installation of Private Fire Service Mains and Their Appurtenances*

NFPA 172-1986, *Standard Fire Protection Symbols for Architectural and Engineering Drawings*

NFPA 231-1987, *Standard for General Storage.*

2

Water Supplies

2-1* General Provisions. Every automatic sprinkler system shall have at least one automatic water supply.

An "automatic" supply is one that is not dependent on any manual operation, such as making connections, operating valves, or starting pumps, to supply water at the time of a fire.

A-2-1 Water supplies should have adequate pressure, capacity, and reliability.

The water supply needed for various occupancies, including Extra Hazard Occupancies, is determined by evaluating the number of sprinklers that may be expected to operate from any one fire plus quantities needed simultaneously for hose streams.

Determination of the water supply needed for Extra Hazard Occupancies will require special consideration of four factors: (1) area of sprinkler operation, (2) density of discharge, (3) required time of discharge, and (4) amount of water needed simultaneously for hose streams.

When the occupancy presents a possibility of intense fires requiring extra heavy discharge, this may be obtained by an increase in the pressure and volume of the water supply, by a closer spacing of sprinklers, by the use of larger pipe sizing, or by a combination of these methods. In such cases, consideration should be given to hydraulically designed systems. (*See Chapter 7.*)

When separately published standards on various subjects contain specific provisions for water supplies, these should be consulted. (*See Chapter 10 for availability of standards.*)

B-2 Water Supplies.

B-2-1 Testing of Water Supply.

B-2-1.1 To determine the value of public water as a supply for automatic sprinkler systems, it is generally necessary to make a flow test to determine how much water can be discharged at a residual pressure at a rate sufficient to give the required residual pressure under the roof (with the volume flow hydraulically translated to the base of the riser) — i.e., a pressure head represented by the height of the building plus the required residual pressure.

Fire flow tests are the subject of NFPA 291, *Recommended Practice for Fire Flow Testing and Marking of Hydrants. (Also see A-2-3.1.1.)*

B-2-1.2 The proper method of conducting this test is to use two hydrants in the vicinity of the property. The static pressure should be measured on the hydrant in front of or nearest to the property and the water allowed to flow from the hydrant next nearest the property, preferably the one farthest from the source of supply if the main is fed only one way. The residual pressure will be that indicated at the hydrant where water is not flowing.

B-2-1.3 Referring to Figure B-2, the method of conducting the flow tests is as follows:

1. Attach gage to hydrant (A) and obtain static pressure.

2. Either attach second gage to hydrant (B) or use pitot tube at outlet. Have hydrant (B) opened wide and read pressure at both hydrants.

3. Use the pressure at (B) to compute the gallons flowing and read the gage on (A) to determine the residual pressure or that which will be available on the top line of sprinklers in the property.

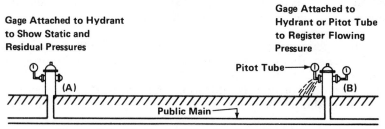

Figure B-2   Method of Conducting Flow Tests.

B-2-1.4 Water pressure in pounds for a given height in feet equals height multiplied by 0.434.

B-2-1.5 In making flow tests, whether from hydrants or from nozzles attached to hose, always measure the size of the orifice. While hydrant outlets are usually 2½ in. (64 mm) they are sometimes smaller and occasionally larger. The UL play pipe is 1⅛ in. (29 mm) and 1¾ in. (44 mm) with tip removed, but occasionally nozzles will be 1 in. (25 mm) or 1¼ in. (33 mm), and with the tip removed the opening may be only 1½ in. (38 mm).

B-2-1.6 The pitot tube should be held approximately one-half the diameter of the hydrant or nozzle opening away from the opening. It should be held in the center of the stream, except that in using hydrant outlets the stream should be explored to get the average pressure.

2-2* Water Supply Requirements for Sprinkler Systems.

A-2-2 The water supply requirement for sprinkler protection is determined by the number of sprinklers expected to operate in the event of fire. The primary factors affecting the number of sprinklers that might open are:

1. Occupancy

2. Combustibility of contents

3. Area shielded from proper distribution of water

4. Height of stock piles

5. Combustibility of construction (ceilings and blind spaces)

6. Ceiling heights and draft conditions

7. Horizontal and vertical cutoffs

8. Wet or dry sprinkler system

9. High water pressure

10. Housekeeping

11. Temperature rating of sprinklers

12. Waterflow alarm and response thereto.

Many of these items may be interdependent, but all are individual factors which, if adverse, affect the ability of sprinklers to control fire in its incipient stage. The proper design of systems, including density and operating areas, normally considers Items 1, 2, 4, 8, 9, 11, and 12. However, the possibility of all sprinklers operating in a room or compartment is very real when some of the other serious factors, such as Items 3, 5, 6, 7, and 10, are considered.

B-2-2 Interconnection of Water Supplies.

B-2-2.1 All main water supplies should be connected with the sprinkler system at the base of the riser, except that where a gravity or pressure tank or both constitute the only automatic source of water supply, permission may be given to connect the tank or tanks with the sprinkler system at the top of the riser.

B-2-2.2 Where a gravity tank and a pressure tank are connected to a common riser, approved means should be provided to prevent residual air pressure in the pressure tank (after water has been drained from it) from holding the gravity tank check valve closed, a condition known as air lock. Under normal conditions, air lock may be conveniently prevented in new equipment by connecting the gravity tank and pressure tank discharge pipes 45 ft (13.7 m) or more below the bottom of the gravity tank and placing the gravity tank check valve at the level of this connection.

2-2.1 General.

2-2.1.1 Water supply requirement tables shall be used in determining the minimum water supply requirements for Light, Ordinary, and Extra Hazard Occupancies. Occupancy classification shall be determined from Section 1-7.

Exception No. 1:   Water supply requirements for dwelling units protected by residential sprinklers shall be in accordance with 7-4.4 in wet systems only.

Exception No. 2:   The water supply requirements for ESFR and large-drop sprinkler systems shall be in accordance with Chapter 9.

Exception No. 3:   The water supply requirements for exposure protection systems shall be in accordance with Chapter 6.

Exception No. 4:   For hazard classifications other than those indicated, see appropriate NFPA standards for design criteria. When other NFPA standards have developed sprinkler system design criteria, they shall take precedence.

Formal Interpretation

Question: Where paragraph 2-2.1.1 Exception No. 4 indicates that other NFPA standards have developed specific sprinkler system design criteria, they shall take precedence over NFPA 13 criteria, does this include commodity classification, spacing, density, and area of application?

Answer: Yes.

This section covers the minimum volume and pressure requirements for average occupancies within the types classified. Many occupancies may have a need for increased requirements, especially when occupancy hazards are above normal or a very high degree of reliability is desired. Through the use of the classifications defined in Section 1-7, the appropriate water supply requirement can be readily determined. The Exceptions note that design criteria for residential sprinklers, large-drop sprinklers, and Early Suppression Fast Response (ESFR) sprinklers are considerably different than for standard spray sprinklers. Other NFPA documents such as NFPA 231, *Standard for General Storage*, NFPA 231C, *Standard for Rack Storage of Materials*, or NFPA 409, *Standard on Aircraft Hangars*, provide density requirements for certain specialized structures. While still deferring spacing requirements, hydraulic calculation methods, and in many cases, materials permitted by NFPA 13, the other documents will take precedence, for particular requirements, as far as densities and areas of application are concerned.

(a) Table 2-2.1.1(a) shall be used to determine the minimum volume of water and pressure normally required for a pipe schedule sprinkler system. The table is to be used only with experienced judgment.

(b) Table 2-2.1.1(b) shall be used to determine the minimum volume of water and pressure normally required for a hydraulically designed sprinkler system.

2-2.1.2 The following shall be used in applying Table 2-2.1.1(b).

2-2.1.3 The water supply for sprinklers only shall be calculated either from the area/density curves in Figure 2-2.1.1(b) in accordance with 2-2.2 or be based upon the room design method in accordance with 2-2.3.1 at the discretion of the designer.

Two generally accepted methods of determining water supply requirements for sprinklers in hydraulically designed systems are available to the designer. The area/density method is the most widely used, but the room design method is also acceptable.

2-2.2* Area/Density Method.

A-2-2.2 For occupancies with the potential for fast-spreading fire due to the presence of lint, combustible residue, combustible hydraulic fluids under high pressure with ignition sources nearby, etc., the minimum area of operation should encompass the entire area likely to be involved in such a fire.

2-2.2.1 The water supply requirement for sprinklers only shall be calculated from the density curves in Figure 2-2.1.1(b). The calculations shall

Table 2-2.1.1(a)   Guide to Water Supply Requirements for Pipe Schedule Sprinkler Systems

Occupancy Classification	Residual Pressure Required (see Note 1)	Acceptable Flow at Base of Riser (see Note 2)	Duration in Minutes
Light Hazard	15 psi	500-750 gpm (see Note 3)	30-60
Ordinary Hazard (Group 1)	15 psi or higher	700-1000 gpm	60-90
Ordinary Hazard (Group 2)	15 psi or higher	850-1500 gpm	60-90
Ordinary Hazard (Group 3)	Pressure and flow requirements for sprinklers and hose streams to be determined by authority having jurisdiction.		60-120
High-Piled Storage (see 4.1.3.10)	Pressure and flow requirements for sprinklers and hose streams to be determined by authority having jurisdiction. (See Chapter 7 and NFPA 231 and NFPA 231C.)		
High-Rise Buildings	Pressure and flow requirements for sprinklers and hose streams to be determined by authority having jurisdiction.		
Extra Hazard	Pressure and flow requirements for sprinklers and hose streams to be determined by authority having jurisdiction.		

For SI Units: 1 psi = 0.0689 bar; 1 gpm = 3.785 L/min.
Notes:
   1. The pressure required at the base of the sprinkler riser(s) is defined as the residual pressure required at the elevation of the highest sprinkler plus the pressure required to reach this elevation.
   2. The lower figure is the minimum flow including hose streams ordinarily acceptable for pipe schedule sprinkler systems. The higher flow should normally suffice for all cases under each group.
   3. The requirement may be reduced to 250 gpm if building area is limited by size or compartmentation or if building (including roof) is noncombustible construction.

Table 2-2.1.1(b)   Table and Design Curves for Determining Density, Area of Sprinkler Operation, and Water Supply Requirements for Hydraulically Designed Sprinkler Systems

Minimum Water Supply Requirements

Hazard Classification	Sprinklers Only—gpm	Inside Hose—gpm	Total Combined Inside and Outside Hose—gpm	Duration in Minutes
Light	See 2-2.1.3	0, 50, or 100	100	30
Ord.—Gp. 1	See 2-2.1.3	0, 50, or 100	250	60-90
Ord.—Gp. 2	See 2-2.1.3	0, 50, or 100	250	60-90
Ord.—Gp. 3	See 2-2.1.3	0, 50, or 100	500	60-120
Ex. Haz.—Gp. 1	See 2-2.1.3	0, 50, or 100	500	90-120
Ex. Haz.—Gp. 2	See 2-2.1.3	0, 50, or 100	1000	120

For SI Units: 1 gpm = 3.785 L/min.

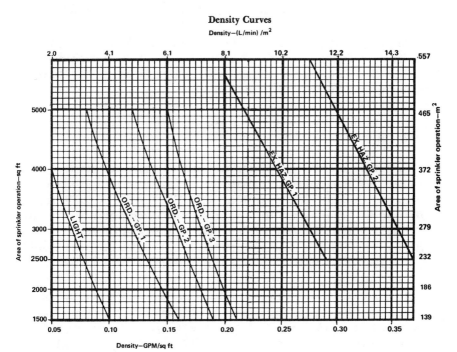

Figure 2-2.1.1(b).

satisfy a single point on the appropriate design curve. It is not necessary to meet all points on the selected curve.

Design of a sprinkler system in accordance with this standard only requires that a single point on the curve be met. Selecting a point on the low end (smaller operating area) of the curve will result in a somewhat higher density but a lower overall water supply. The higher density also requires a higher pressure but is superior in terms of fire control and economics. Selection of a point on the high end (larger operating area) of the curve allows for a lower density but a higher overall water supply.

2-2.2.2 The densities and areas provided in Figure 2-2.1.1(b) are for use only with standard response sprinklers. For use of other types of sprinklers see 4-1.1.3.

The areas and densities shown in Figure 2-2.1.1(b) are based upon the use of standard response sprinklers. Residential units protected by residential sprinklers are designed in accordance with the requirements of 7-4.4. Storage facilities protected by large-drop or ESFR sprinklers are designed in accordance with Chapter 9.

For special types of sprinklers, such as quick-response, their respective design parameters from the manufacturer must be relied on. These parameters will also appear in the listing information for the particular product.

2-2.2.3 For dry-pipe systems, increase the area of sprinkler operation by 30 percent without revising density.

Dry-pipe sprinkler systems will usually take longer to apply water to a fire condition than wet-pipe or preaction systems. Because of this delay, system size (in terms of gallons) is limited by Chapter 5. Other adjustments to compensate for the delay are incorporated, such as the 30 percent increase in the design area. In this case, more sprinklers in the fire area would be expected to operate while the pipe is being charged with water. Adjusting the area but not the density will result in better fire control. As an example, if the density and area for an Ordinary Hazard (Group 3) Occupancy are selected at 0.21 gpm per sq ft and 1500 sq ft [*see Figure 2-2.1.1(b)*], this would be adjusted to: 30 percent × 1500 + 1500 = 1950 sq ft if you were designing a dry-pipe system. The density remains unchanged (0.21 gpm per sq ft) but must now be distributed over a larger area, 1950 sq ft.

It should be noted that this rule does not apply to preaction systems. Even though the pipe has no water in it, the delay in operation of these systems will be minimal since the detection device will typically respond to the fire sooner than the sprinkler.

Formal Interpretation

Question: For a dry-pipe sprinkler system in an Ordinary Hazard Occupancy having unsprinklered combustible concealed spaces, would 2-2.2.3 and 2-2.4(b) permit a density based on less than 3000 sq ft area of sprinkler operation provided the 30 percent increase in the area of sprinkler operation required by 2-2.2.3 brings the area of sprinkler operation to over 3000 sq ft as required by 2-2.4(b)?

Answer: Yes.

2-2.2.4 When high-temperature sprinklers are used for Extra Hazard Occupancies, the area of sprinkler operation may be reduced by 25 percent without revising the density, but to not less than 2000 sq ft (186 m²).

Use of high-temperature sprinklers in full-scale fire tests has shown that, although fewer sprinklers tend to operate, the result is a reduced level of fire damage. Use of high-temperature sprinklers is still restricted to those conditions defined in Section 3-11.

2-2.3* Room Design Method.

A-2-2.3 Corridors are rooms and should be considered as such. This section allows for calculation of the sprinklers in the largest room, so long as the calculation produces the greatest hydraulic demand among selection of rooms and communicating spaces. For example, in a case where the largest room has four sprinklers and a smaller room has two sprinklers but communicates through unprotected openings with three other rooms, each having two sprinklers, the smaller room and group of communicating spaces should also be calculated. Another example in which the room which creates the greatest demand is not the largest room would be where a large room is located near the system riser but a small room with the same hazard classification is located at the opposite end of the building or on an upper floor so as to produce a significantly higher demand.

When a corridor protected by a single row of sprinklers is considered the largest room, a maximum of 5 sprinklers (*see 7-4.3.1, Exception No. 1*), plus those sprinklers in adjacent rooms with unprotected openings, are calculated.

2-2.3.1 The water supply requirements for sprinklers only shall be based upon the room which creates the greatest demand. The density selected shall be that from Figure 2-2.1.1(b) corresponding to the room size. If the room is smaller than the smallest area shown in the applicable curve in Figure 2-2.1.1(b) see 2-2.4.1(a). If the room is larger than the largest area shown in the applicable curve, use the density for the largest area shown for all sprinklers in the room. All rooms shall be enclosed with construction having a fire-resistance rating equal to the water supply duration indicated in Table 2-2.1.1(b) with minimum protection of openings as follows:

(a) Light Hazard — automatic or self-closing doors.

Exception: When openings are not protected, calculations shall include the sprinklers in the room plus two sprinklers in the communicating space nearest each such unprotected opening unless the communicating space has only one sprinkler, in which case calculations shall be extended to the operation of that sprinkler. The selection of the room and communicating space sprinklers to be calculated shall be that which produces the greatest hydraulic demand.

(b) Ordinary and Extra Hazard — automatic or self-closing doors with appropriate fire-resistance ratings for the enclosure.

This method has also been referred to as the "largest room design" method. When opting for this approach, the designer must use the room that is also the most hydraulically demanding in terms of water supply and pressure. Enclosure of such a room must comply with minimum fire-resistance ratings of walls as follows:

Hazard Classification	Fire-Resistance Rating
Light	½ Hour
Ordinary (1, 2, 3)	1 Hour
Extra (Group 1)	1½ Hour
Extra (Group 2)	2 Hour

Any communicating openings into such a room are to be protected in the same manner as any opening in a fire-resistive partition.

Formal Interpretation

Question: When using Table 2-2.1.1(b) for calculating the minimum water supply for sprinklers only for buildings having individual room areas less than 1500 sq ft and all rooms constructed in accordance with the requirements set forth in 2-2.3.1, can areas of sprinkler operation less than 1500 sq ft be used?

Answer: Yes. The minimum water supply requirement for the above described conditions is then calculated on the basis of the area of the largest room; however, the sprinkler discharge density for 1500 sq ft must be used [*see 2-2.4(a)*] provided it is not a dry-pipe system (*see 2-2.2.3*) and there are no unsprinklered combustible spaces [*see 2-2.4(b)*].

Formal Interpretation

Question: In a building classified as Light Hazard (such as an apartment building), where all walls are at least 30-minute fire rated, but without self-closing doors, would it be correct to base the calculations on the largest room and adjacent communicating rooms even though the combined area of these rooms total less than 1500 sq ft?

Answer: Yes. The Exception to 2-2.3.1(a) would apply when doors are not self-closing. The density for 1500 sq ft should be used although the calculated area may be smaller.

Formal Interpretation

Question: Is it the intent of the Committee to permit the omission from the hydraulic calculations, sprinkler discharge in closets and washrooms (not protected by self-closing or automatic doors), when the Exception to 2-2.3.1(a) is used to determine the area of operation to be used?

Answer: No. The Exception to 2-2.3.1(a) says "communicating space" and does not differentiate between closets, washrooms, or other rooms. Paragraph 7-4.3.1.6 allows the omission of sprinklers in closets, washrooms, and similar small compartments when the areas in Table 2-2.1.1(b) are being calculated.

2-2.4* Additional Requirements Regardless of which of the above methods is used, the following restrictions apply:

(a) For areas of sprinkler operation less than 1500 sq ft (139 m^2) used for Light and Ordinary Hazard Occupancies, the density for 1500 sq ft (139 m^2) shall be used. For areas of sprinkler operation less than 2500 sq ft (232 m^2) for Extra Hazard Occupancies, the density for 2500 sq ft (232 m^2) shall be used.

A minimum density is necessary in order to achieve timely fire control or extinguishment. When the area of sprinkler operation selected conforms to 2-2.3.1, the densities for the low end of the curves in Figure 2-2.1.1(b) are to be used. These densities range from 0.10 gpm per sq ft [(4.1 L/min)/m²] for a Light Hazard Occupancy and extend to 0.37 gpm per sq ft [(15.1 L/min)/m²] for an Extra Hazard (Group 2) Occupancy. The resultant quantity of water for sprinklers should be approximately equivalent to the density multiplied by the area of the room or of the hydraulically selected area.

Formal Interpretation

Question: When applying Sections 2-2.2.1 and 2-2.3.1 and Table 2-2.1.1(b) to the hydraulic design of a room covering less than 1500 sq ft in area, is it the intent of the standard to impose a minimum water flow demand requirement for sprinklers only of 150 U.S. gpm (i.e., 0.1 × 1500 = 150) and, if so, should this minimum water flow be carried through the sprinkler piping from the area of application back to the alarm valve header?

Answer: No.

(b) For construction having unsprinklered combustible concealed spaces (as described in 4-4.4) the minimum area of sprinkler operation shall be 3000 sq ft (279 m²).

Exception No. 1: Combustible concealed spaces filled entirely with non-combustible insulation.

Exception No. 2: Light or Ordinary Hazard Occupancies where noncombustible ceilings are directly attached to the bottom of solid wood joists so as to create enclosed joist spaces, each less than 160 cu ft (4.8 m³) in volume.

A-2-2.4 This section is included to compensate for possible delay in operation of sprinklers from fires in combustible concealed spaces found in

The room design method could still be considered for use provided (1) any combustible concealed spaces are sprinklered, or (2) one of the two Exceptions to 2-2.4(b) is satisfied. When the area/density method is selected, the design area may have to be increased to 3000 sq ft (279 m²).

It is not the intent of this section to allow permission to omit sprinklers from concealed combustible spaces, but rather to compensate for the anticipated delay of sprinkler operation for fires originating in an unprotected concealed combustible space.

If residential sprinklers are utilized in a dwelling unit and an unprotected combustible concealed space is present, sprinkler design must still comply with 7-4.4.

Formal Interpretation

Question: Is it the intent of 2-2.4(b) to impose the 3000 sq ft design area in a fully sprinklered wood frame construction building even though there exists no unsprinklered combustible concealed spaces?

Answer: No. 2-2.4(b) applies only when you have unsprinklered combustible concealed spaces.

Formal Interpretation

Question: Section 2-2.4. Can unsprinklered combustible concealed spaces as described in 4-4.4.1 be permitted, provided the below ceiling design area is increased to a minimum of 3000 sq ft?

Answer: No. Permitted unsprinklered combustible concealed spaces as provided in 4-4.4.1 are not contingent upon use of hydraulically designed systems. However, when hydraulically designed systems are installed and such conditions exist, the minimum design area of sprinkler operation is 3000 sq ft.

2-2.4.1* When inside hose stations are planned or are required by other standards, a water allowance of 50 gpm (189 L/min) for a one hose station installation and 100 gpm (378 L/min) for a two or more hose station installation shall be added to the sprinkler requirements at the point of connection to the system at the pressure required by the sprinkler system design.

Formal Interpretation

Question: When considering the 100 gpm allowance for several hose streams, is it the intent of 2-2.4.1 to apply a 50 gpm allowance to each of the two most remote area connections to the sprinkler system, for an aggregate of 100 gpm?

Answer: Yes.

The size and type of hose connections contemplated by this paragraph are basically limited to those necessary for first-aid fire fighting or final extinguishment. Even though more than two such hose stations may be connected to the system, operation of more than two would be considered unlikely and thus a maximum of two hose stations, simultaneously operating, are to be included in the total water supply. Table 2-2.1.1(b) highlights this point. Accounting for this potential use of water will permit the sprinkler system to operate as intended while at the same time allowing some form of first-aid fire fighting.

NFPA 14, *Standard for the Installation of Standpipe and Hose Systems*, provides specific requirements for Class II standpipe systems, intended for building occupant use. These requirements are considerably more conservative in terms of flow and pressure for the standpipe system than the values given in this standard.

For all practical purposes, these hose connections should be viewed as "large sprinklers" in the hydraulically most remote area and should be carried through the calculations back to the system riser.

A-2-2.4.1 When considering the 100 gpm allowance for several hose stations, it is the intent of 2-2.4.1 to apply a 50 gpm allowance to each of the two most remote area connections to the sprinkler system, for an aggregate of 100 gpm.

2-2.4.2 When hose valves for fire department use are attached to wet-pipe sprinkler system risers in accordance with 3-3.8, the water supply requirements shall be as follows:

(a) For buildings protected in accordance with this standard, the water supply for sprinklers need not be added to standpipe demand as determined from NFPA 14, *Standard for the Installation of Standpipe and Hose Systems*.

Exception: When the sprinkler system demand, including hose stream allowance indicated in Table 2-2.1.1(b), exceeds the requirements of NFPA 14, Standard for the Installation of Standpipe and Hose Systems, the values in Table 2-2.1.1(b) shall be used.

(b) For partially sprinklered buildings, the sprinkler demand, not including hose stream allowance, as indicated in Table 2-2.1.1(b) shall be added to the requirements given in NFPA 14, *Standard for the Installation of Standpipe and Hose Systems*.

For a completely sprinklered building, the demand (pressure and flow) from a standpipe system will, in most cases, exceed the demand of the sprinkler system, resulting in exclusive adherence to NFPA 14, *Standard for the Installation of Standpipe and Hose Systems.*

If a building is only partially sprinklered (a practice strongly discouraged by 4-1.1.1), then the combined effect of sprinkler demand and hose demand from NFPA 13 and NFPA 14 must be summed. This is necessary since there is no way to prevent fires from originating in the nonsprinklered area.

It should be recognized that, in many cases, 2½-in. (64-mm) hose valves will in fact be used by the fire department, in which case either on-site water supplies or pumping capacity for such hose valves should not be considered necessary. This additional hose stream usage within the building would normally be provided by the fire department pumper pumping into the fire department connection.

2-2.4.3 Water demand of sprinklers installed in racks shall be added to the ceiling sprinkler water demand at the point of connection. Demands shall be balanced to the higher pressure.

As noted in 2-2.1.1 Exception No. 4, the designer may defer to some other standard for special protection requirements. NFPA 231C, *Standard for Rack Storage of Materials*, is one such document and, in many cases, will mandate the installation of rack sprinklers. NFPA 231C will stipulate the number of rack sprinklers to calculate and the minimum operating pressure of such sprinklers. NFPA 231C presumes that simultaneous operation of in-rack sprinklers as well as ceiling sprinklers will be the rule rather than the exception, and this should be considered in the hydraulic calculation procedure.

2-2.4.4 Water allowance for outside hose shall be added to the sprinkler and inside hose requirement at the connection to the city water main, or at a yard hydrant, whichever is closer to the system riser.

If no inside hose connections are provided, the outside hose requirement will be that given in the fourth column (from the left) of Table 2-2.1.1(b).

2-2.4.5 The lower duration figures in Tables 2-2.1.1(a) and 2-2.1.1(b) are ordinarily acceptable where remote station waterflow alarm service or equivalent is provided.

2-2.4.6 When pumps, gravity tanks, or pressure tanks supply sprinklers only, requirements for inside and outside hose need not be considered in determining the size of such pumps or tanks.

2-3 Connections to Water Works Systems.

2-3.1 Acceptability.

2-3.1.1* General. A connection to a reliable water works system shall be an acceptable water supply source. The volume and pressure of a public water supply shall be determined from waterflow test data.

Some health officials and water departments are now prescribing the use of reduced pressure backflow prevention devices on connections between fire protection systems and water supply systems. Since these devices are not installed for fire protection purposes but rather out of a concern for public well-being, it is not appropriate for this standard to determine requirements for such devices. While nothing in this standard prohibits the use of such devices, their presence will have an adverse impact on available operating pressures to the sprinkler system and will also introduce maintenance factors which must be considered. NFPA 24, *Standard for the Installation of Private Fire Service Mains and Their Appurtenances,* provides some additional discussion on backflow prevention devices.

A-2-3.1.1 Care should be taken in making water tests to be used in designing or evaluating the capability of sprinkler systems. The water supply tested should be representative of the supply that may be available at the time of a fire. For example, testing of public water supplies should be done at times of normal demand on the system. Public water supplies are likely to fluctuate widely from season to season and even within a 24-hour period. Allowance should be made for seasonal or daily fluctuations, for drought conditions, for possibility of interruption by flood, or for ice conditions in winter. Testing of water supplies also normally used for industrial use should be done while water is being drawn for industrial use. The range of industrial-use demand should be taken into account.

Future changes in water supplies should be considered. For example a large, established, urban supply is not likely to change greatly within a few years. However, the supply in a growing suburban industrial park may deteriorate quite rapidly as greater numbers of plants draw more water.

For further information about performing flow tests see B-2-1 and NFPA 291, *Recommended Practice for Fire Flow Testing and Marking of Hydrants.*

2-3.1.2 Meters. Meters are not recommended for use on sprinkler systems; however, where required by other authorities, they shall be of an approved type.

Meters are not considered necessary on fire protection connections. Because of the pressure loss they introduce, as well as the additional cost associated with them, their use with automatic sprinkler systems should be discouraged. The arrangements indicated in Figure B-2-3.1 provide for proper metering of water that is used for other than fire protection purposes. Where there is a concern about possible leakage or loss from the fire protection lines, a detector check valve can be substituted for the check valve indicated.

2-3.2* **Capacity.** The connection and arrangement of underground supply piping shall be capable of supplying the volume as required in Table 2-2.1.1(a) or 2-2.1.1(b). Pipe size shall be at least as large as the system riser. (*See NFPA 24, Standard for the Installation of Private Fire Service Mains and Their Appurtenances.*)

Exception: Unlined cast or ductile iron pipe shall not be less than 4 in. (102 mm) in size.

A-2-3.2   In private underground piping systems for buildings of other than Light Hazard Occupancy, any dead-end pipe supplying both sprinklers and hydrants should not be less than 8 in. (203 mm) in size. Also see NFPA 24, *Standard for the Installation of Private Fire Service Mains and Their Appurtenances.*

In order to provide adequate water capacity, yard systems that supply at least one sprinkler system, in addition to at least one hydrant, may be sized hydraulically to comply with 2-2.3.6. The capacity should be calculated to supply the most hydraulically demanding system and the closest hydrant to that system. When underground systems are not hydraulically calculated, larger pipe sizing, such as a minimum of 6 in. (152 mm) for looped and a minimum of 8 in. (203 mm) for dead-end underground mains, may be necessary. Note that the capacity required may include inside hose, outside hose, and other water demands.

B-2-3 Special Provisions.

B-2-3.1 Domestic Connections.   Connections for domestic water service should be made on the water supply side of the check valve in the water supply main so that the use of the fire department connection will not subject the domestic water system to high pressure. If the domestic consumption will significantly reduce the sprinkler water supply, an increase in the size of the pipe supplying both the domestic and sprinkler water may be justified. Circulation of water in sprinkler pipes is objectionable, owing to increased corrosion, deposits of sediment, and condensation drip from pipes.

Aboveground domestic connections to sprinkler system piping should be metallic pipe or suitably protected. Nonmetallic aboveground domestic pipe connections may fail during a fire, diverting water from the sprinklers.

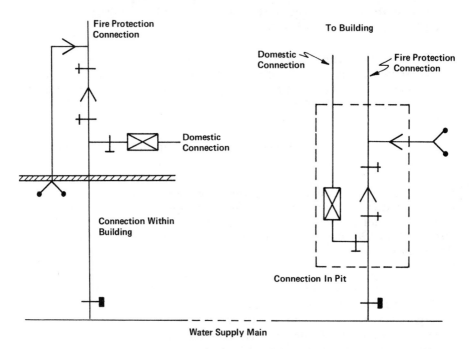

Figure B-2-3.1   Connection for Domestic Water.

B-2-3.2 Water Hammer. When connections are made from water mains subject to severe water hammer [especially when pressure is in excess of 100 psi (6.9 bars)], it may be desirable to provide either a relief valve, properly connected to a drain, or an air chamber in the connection. If an air chamber is used, it should be located close to where the pipe comes through the wall and on the supply side of all other valves and so located as to take the full force of water hammer. Air chambers should have a capacity of not less than 4 cu ft (0.12 m³), should be controlled by an approved indicating valve, and should be provided with a drain at the bottom and an air vent with control valve and plug to permit inspection.

B-2-3.3 Penstocks, Flumes, etc. Water supply connections from penstocks, flumes, rivers, or lakes should be arranged to avoid mud and sediment, and should be provided with approved double removable screens or approved strainers installed in an approved manner.

2-4 Gravity Tanks.

2-4.1 Acceptability. An elevated tank sized in accordance with Table 2-2.1.1(a) or 2-2.1.1(b) shall be an acceptable water supply source. (*See NFPA 22, Standard for Water Tanks for Private Fire Protection.*)

2-4.2 Capacity and Elevation. The capacity and elevation of the tank and the arrangement of the underground supply piping shall provide the volume and pressure required by Table 2-2.1.1(a) or 2-2.1.1(b) designs.

2-5 Pumps.

2-5.1* Acceptability. A single automatically controlled fire pump sized in accordance with Table 2-2.1.1(a) or 2-2.1.1(b) supplied under positive head shall be an acceptable water supply source. (*See NFPA 20, Standard for the Installation of Centrifugal Fire Pumps.*)

A-2-5.1 An automatically controlled vertical turbine pump taking suction from a reservoir, pond, lake, river, or well complies with 2-5.1.

A vertical turbine pump is a type of centrifugal fire pump that operates in a vertical position. In that position the first impeller, or bowl, of the pump is always submerged; therefore, the first pump impeller has positive suction and the pump complies with the requirements of 2-5.1. This type of pump, with its first impeller submerged, does not actually "take suction from," but rather "pumps from," an impounded water source.

2-5.2* Supervision. When a single fire pump constitutes the sole sprinkler supply, it shall be provided with supervisory service from an approved central station, proprietary, remote station system or equivalent.

A-2-5.2 See sections dealing with sprinkler equipment supervisory and waterflow alarm services in NFPA 71, *Standard for the Installation, Maintenance, and Use of Signaling Systems for Central Station Service*; NFPA 72A, *Standard for the Installation, Maintenance, and Use of Local Protective Signaling Systems for Guard's Tour Fire Alarm, and Supervisory Service*; NFPA 72B, *Standard for the Installation, Maintenance, and Use of Auxiliary Protective Signaling Systems for Fire Alarm Service*; NFPA 72C, *Standard for the Installation, Maintenance, and Use of Remote Station Protective Signaling Systems*; or NFPA 72D, *Standard for the Installation, Maintenance, and Use of Proprietary Protective Signaling Systems*.

In order to improve the reliability of a single fire pump so it is acceptable as the sole supply for sprinkler systems, supervisory service to a constantly attended location is required. Conditions to be supervised should include pump-running indication, pump power supply, and other conditions that would render the pump inoperative.

2-6 Pressure Tanks.

2-6.1 Acceptability.

2-6.1.1 A pressure tank sized in accordance with Table 2-2.1.1(a) or 2-2.1.1(b) is an acceptable water supply source. (*See NFPA 22, Standard for Water Tanks for Private Fire Protection.*)

2-6.1.2 Pressure tanks shall be provided with an approved means for automatically maintaining the required air pressure. When a pressure tank is the sole water supply there shall also be provided an approved trouble alarm to indicate low air pressure and low water level with the alarm supplied from an electrical branch circuit independent of the air compressor.

As in the case with a single pump, a pressure tank that serves as the sole water supply should be supervised for reliability in line with 2-5.2. Trouble alarms that indicate low air pressure and low water levels should be received at a constantly attended location.

2-6.1.3 Pressure tanks shall not be used to supply other than sprinklers and hand hose attached to sprinkler piping.

2-6.2 Capacity. The required water capacity of a pressure tank shall be in accordance with 2-2.1.1 and shall include the extra capacity needed to fill dry-pipe or preaction systems when installed. The total volume shall be based on the water capacity, plus the air capacity required by 2-6.3.

2-6.3* Water Level and Air Pressure. Unless otherwise approved by the authority having jurisdiction, the pressure tank shall be kept two-thirds full of water, and an air pressure of at least 75 psi (5.2 bars) by the gage shall be maintained. When the bottom of the tank is located below the highest sprinklers served, the air pressure by the gage shall be at least 75 psi (5.2 bars) plus three times the pressure caused by the column of water in the sprinkler system above the tank bottom.

A-2-6.3 The air pressure to be carried and the proper proportion of air in the tank may be determined from the following formulas, in which,

P = Air pressure carried in pressure tank
A = Proportion of air in tank
H = Height of highest sprinkler above tank bottom

The formulae for determining the pressures in tanks in pipe schedule systems are based on 15 psi (1.03 bars) atmospheric pressure as well as 15 psi remaining pressure at the height of the highest sprinkler.

When tank is placed above highest sprinkler

$$P = \frac{30}{A} - 15.$$

$A = \frac{1}{3}$ then $P = 90 - 15 = 75$ lb per sq in.
$A = \frac{1}{2}$ then $P = 60 - 15 = 45$ lb per sq in.
$A = \frac{2}{3}$ then $P = 45 - 15 = 30$ lb per sq in.

When tank is below level of the highest sprinkler

$$P = \frac{30}{A} - 15 + \frac{0.434H}{A}$$

$A = \frac{1}{3}$ then $P = 75 + 1.30H.$
$A = \frac{1}{2}$ then $P = 45 + 0.87H.$
$A = \frac{2}{3}$ then $P = 30 + 0.65H.$

The respective air pressures above are calculated to ensure that the last water will leave the tank at a pressure of 15 psi (1.03 bars) when the base of the tank is on a level with the highest sprinkler, or at such additional pressure as is equivalent to a head corresponding to the distance between the base of the tank and the highest sprinkler when the latter is above the tank.

The final pressure required at the pressure tank for systems designed from Table 2-2.1.1(b) will normally be higher than the 15 psi (1.03 bars) anticipated in the previous paragraph. The following formula should be used to determine the tank pressure and ratio of air to water in hydraulically designed systems.

$$P_i = \frac{P_f + 15}{A} - 15$$

where
P_i = Tank pressure
P_f = Pressure required from hydraulic calculations
A = Proportion of air

Example: Hydraulic calculations indicate 75 psi is required to supply the system. What tank pressure will be required?

$$P_i = \frac{75 + 15}{.5} - 15$$

$P_i = 180 - 15 = 165$ psi

For SI Units: 1 ft = 0.3048 m; 1 psi = 0.0689 bar.

In this case the tank would be filled with 50 percent air and 50 percent water and the tank pressure would be 165 psi (11.4 bars). If the pressure is too high, the amount of air carried in the tank will have to be increased.

Location of Pressure Tanks. Pressure tanks should be located above the top level of sprinklers but may be located in the basement or elsewhere.

See B-2-2.2 for the proper location of a pressure tank used in conjunction with a gravity tank in order to prevent air lock.

When water supplies other than circulating public systems are used, care should be taken that components (piping, pumps, or tanks) are reliably installed and maintained. Even though the capacities of private systems may be adequate, the reliability of these emergency supply systems may not prove equal to public systems without regular periodic testing, maintenance, and supervision. Public systems are "tested" and "supervised" continually by domestic usage of water, while private systems sometimes remain idle for long periods of time.

2-7 Fire Department Connections.

A fire department connection to any sprinkler system provides a desirable, auxiliary water supply to that system. By supplementing the "automatic" supply required by Section 2-1, some increased reliability in system performance should be recognized. Fire department pump operators can also determine if any sprinklers have actually operated by using the fire department connection.

2-7.1* A fire department connection shall be provided as described in this section.

Exception No. 1: The fire department connection may be omitted for systems having 20 sprinklers or less.

Exception No. 2: When permission of the authority having jurisdiction has been obtained for its omission.

Seldom, if ever, should fire department connections be omitted from sprinkler systems, except as noted in the two Exceptions. In most cases they allow the fire department to bypass a closed control valve. When omissions are considered, they are normally for very small buildings that are completely without interior partitions and have windows and doors that provide easy access for manual fire fighting as a supplement to normal sprinkler operation. Even when gravity tanks or pressure tanks are the sole source of supply, fire department connections should be included. This allows fire departments to pump from tanker trucks directly into the system, thereby either increasing the supply or adding to the tank's supply prior to its depletion, or both. (*See commentary to 2-7.3.4.*)

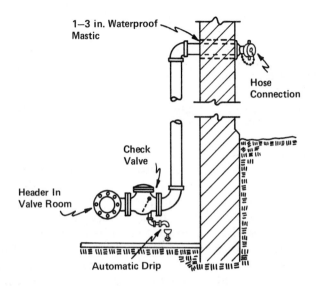

1–3 in. Waterproof Mastic

Hose Connection

Check Valve

Header In Valve Room

Automatic Drip

For SI Units: 1 in. = 25.4 mm.

Figure 2-7.1   Fire Department Connection.

A-2-7.1   The fire department connection should be located not less than 18 in. (457 mm) and not more than 5 ft (1.5 m) above the level of the adjacent grade or access level.

2-7.2* Size.   Pipe size shall be 4 in. (102 mm) for fire engine connections and 6 in. (152 mm) for fire boat connections.

Exception No. 1:   For hydraulically calculated systems, fire department connection pipe serving one system riser may be as small as the system riser.

Exception No. 2:   A single fire department hose connection may be piped to a 3-in. (76-mm) or smaller connection.

Use of a typical siamese fire department connection (with two 2½-in. connections) is not necessary on system risers that are 3 in. (76 mm) or less in diameter.

A-2-7.2   For hydraulically designed sprinkler systems, the size of the fire department connection should be sufficient to supply the sprinkler water demand developed from Table 2-2.1.1(b).

2-7.3* Arrangement.   *(See 3-9.2.5 and 3-9.2.6.)*

A-2-7.3 Fire department connections should be located and arranged so that hose lines can be readily and conveniently attached without interference from nearby objects including buildings, fences, posts, or other fire department connections. When a hydrant is not available, other water supply sources such as a natural body of water, a tank, or reservoir should be utilized. The water authority should be consulted when a nonpotable water supply is proposed as a suction source for the fire department.

2-7.3.1 The fire department connection shall be made on the system side of a check valve in the water supply piping.

Each separate water supply connection for sprinkler systems, whether automatic or manual, should be made on the system side of a check valve. Only in this way can pressure differentials between municipal, fire pump, pressure tank, or fire department connections effectively supply the sprinklers. The highest pressure supply will come in first until its capacity is reached. As capacity is exceeded the pressure will drop until the next highest pressure supply comes in. The check valve prevents the higher pressure supply from backing up or circulating into lower pressure supplies. *(See text and commentary on 3-9.2.4 and A-3-9.2.5.)* For example:

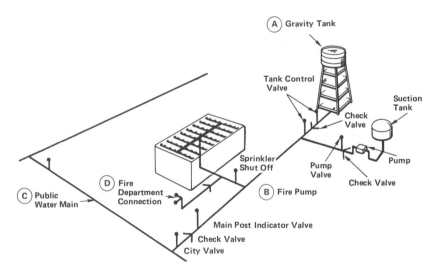

Figure 2.1.   Alternative water supplies to sprinkler system.

Sprinklers in the example are normally supplied by a public main static pressure of 50 psi (3.4 bars). Upon arrival at a fire

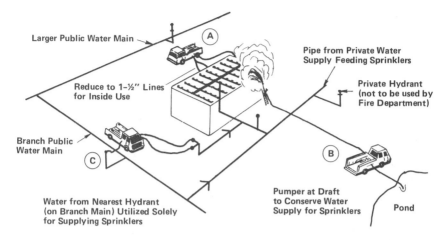

Figure 2.2.   Fire department water supply connections to sprinkler
system.

in this building, the fire officer would order a pumper to connect to the hydrant across the street and to pump at 100 psi
(6.9 bars) into the yard fire department connection, thus
adding supplementary pressure to the sprinkler system. If the
connection were not on the system side of a check valve, the
pumped water would circulate back into the same main to the
suction side of the pumper, and no supplementary pressure
could be added.

2-7.3.2   On wet-pipe systems with a single riser the connection shall be
made on the system side of approved indicating, check, and alarm valves to
the riser, unless the system is supplied by a fire department pumper
connection in the yard. (*See 3-9.2.6.*)

Formal Interpretation

Section 2-7.3.2 indicates that the fire department connection shall
be made on the system side of an approved indicating valve, unless
the sprinklers are supplied by the fire department pumper connection in the yard.

Question: Is it the intent of paragraph 2-7.3.2 to permit a post
indicator valve to be present, in line, between the fire department
connection and the main riser? (Please refer to the diagram on the
following page for clarity on this matter):

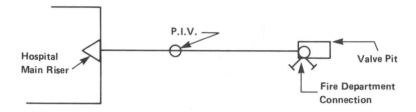

Answer: Yes.

Formal Interpretation

Question: In a sprinkler system having the fire department connection in the yard, may the system indicating valve be of the
O.S. and Y. type located on the system riser inside the building?

Answer: Yes.

2-7.3.3 On dry-pipe systems with a single riser the connection shall be
made between the approved indicating valve and the dry-pipe valve, unless
the system is supplied by a fire department pumper connection in the yard.

2-7.3.4 On systems with two or more risers, the connection shall be made
on the system side of all shutoff valves controlling other water supplies, but
on the supply side of the riser shutoff valves so that, with any one riser off,
the connection will feed the remaining sprinklers, unless the sprinklers are
supplied by a fire department pumper connection in the yard.

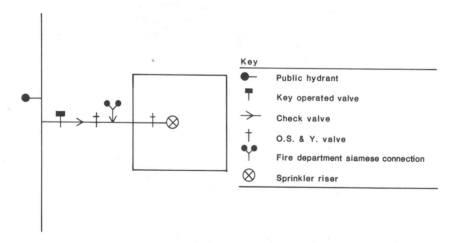

**Figure 2.3. Fire department yard connection to a system with one or
more multiple risers.**

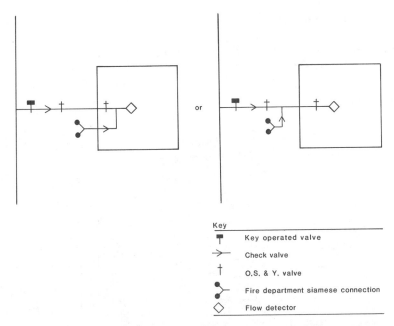

Key

   Key operated valve

   Check valve

   O.S. & Y. valve

   Fire department siamese connection

   Flow detector

Figure 2.4.   Fire department connections to a dry-pipe sprinkler system having a single riser. Either method is satisfactory according to the standard.

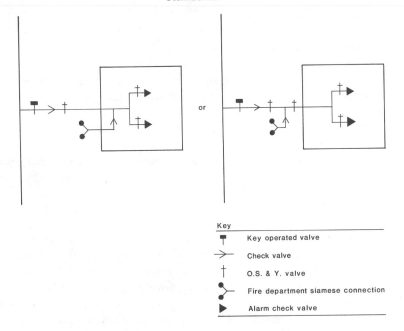

Key

   Key operated valve

   Check valve

   O.S. & Y. valve

   Fire department siamese connection

   Alarm check valve

Figure 2.5.   Alternative fire department connections to systems having two or more risers. Either method is satisfactory according to the standard.

Formal Interpretation

Question: Are control valves required on the individual risers shown in Figure 1?

Answer: Yes.

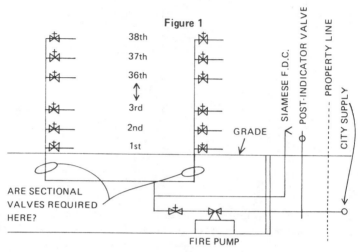

Figure 1.

Formal Interpretation

Question 1: Does 2-7.3.4 require that a fire department pumper connection be provided on each riser in a multiple riser system?

Answer: No.

Question 2: Is a fire department pumper connection installed at the city connection valve pit satisfactory to comply with 2-7.3.4?

Answer: Yes.

Fire department connections have long presented difficulty between the concept of protection and the feasibility of installation. On the one hand, fire departments should be able to pump into any system whether the final control valve is open or closed, thereby assisting low-pressure water supplies, or supplementing water supplies from other sources such as tanks or captured storage. However, when multiple risers are in a single building system or when single risers are fed directly from a private yard-main supply, it is usually not feasible to run individual fire department connections to the system side of each riser control valve. It would also be difficult to identify these multiple fire department connections for effective use. Therefore, the provisions of Section 2-7, including Exceptions, drawings, and Formal Interpretations, are intended to clarify the requirements of the standard in this area.

Formal Interpretation

Question: Does paragraph 2-7.3.4 allow a single fire department connection to supply a multi-building system where individual sprinkler risers are supplied by a common underground supply system?

Answer: Yes.

This Formal Interpretation merely reiterates the Committee's position that a single yard type fire department connection can supply multiple risers, in multiple buildings. The specific arrangement of the fire department connection with respect to the multiple riser configuration must still comply with 2-7.3.4.

2-7.3.5 Fire department connections shall not be connected on the suction side of booster pumps.

According to 2-7.3.1 through 2-7.3.4, the fire department connection is made on the system side of a check valve. NFPA 20, *Standard for the Installation of Centrifugal Fire Pumps,* does not permit a check valve on the suction side of a pump. It should be clear then that in order to be on the system side of the check valve, a fire department connection could not be made on the suction side of the pump. In addition, full efficiency and reliability cannot be obtained for a suction side connection because of losses through pump and valves and because of the inability to supplement the system if the discharge side control valves are closed. It is important to note that pumping into the suction side of a fire pump or booster pump will increase the discharge pressure of the pump, frequently beyond the pressure limits of the system. The fire department connection should be made on the system side of a pump, and on the system side of the pump discharge check valve and control valves.

2-7.3.6 Fire department connections to sprinkler systems shall be designated by a sign having raised letters at least 1 in. (25 mm) in size cast on plate or fitting reading for service designated: e.g. — "AUTOSPKR.," "OPEN SPKR.," or "AUTOSPKR. and STANDPIPE."

2-7.4 Valves.

2-7.4.1   An approved check valve shall be installed in each fire department connection, located as near as practicable to the point where it joins the system.

2-7.4.2   There shall be no shutoff valve in the fire department connection.

Formal Interpretation

Question: What is the Official Interpretation of NFPA 13, 2-7.4.2 "There shall be no shutoff valve in the fire department connection"?

Answer: A shutoff valve shall not be installed in the connection between the fire department siamese and the underground or yard system (Standards 13 and 24). This is an emergency checked connection for water supply to hydrants and sprinkler systems connected to the underground or yard system. Sectional valves are recommended in large yard systems (Standard 24, 3-5.1) and post indicator valves, wall post indicator valves, or inside control valves are recommended for sprinkler systems with two or more risers (Standard 13, 2-7.3.4). This also means that single riser building connections to a large yard system will be valved.

Paragraph 2-7.4.2 is needed only because it is usually considered good practice to install a shutoff valve on both sides of a check valve to facilitate repair. The check valve in the fire department connection is considered to be an exception to that general good practice, and no shutoff valve is permitted at that point in the emergency connection.

2-7.5 Drainage.   The piping between the check valve and the outside hose coupling shall be equipped with an approved automatic drip.

In the event that the check valve in the siamese connection leaks, it is the purpose of the automatic drip to drain this water to a safe location and maintain the piping between the check valve and the hose couplings without water collection. Without the automatic drip, such collected water might freeze and prevent the fire department from being able to pump into the system under fire conditions. This automatic drip also facilitates any maintenance of the fire department connection piping, since this portion of the pipe will already be free of water.

The drain should be located at the lowest point of the fire department connection piping to allow complete drainage.

2-7.6 Hose Connections.

2-7.6.1 The fire department connection(s) shall be internal threaded swivel fitting(s) having the NH standard thread, at least one of which shall be 2.5 — 7.5 NH standard thread, as specified in NFPA 1963, *Standard for Screw Threads and Gaskets for Fire Hose Connections.*

Exception: When local fire department connections do not conform to NFPA 1963, the authority having jurisdiction shall designate the connection to be used.

All hose coupling threads in sprinkler systems and threads for hydrants on yard mains supplying sprinkler systems should match those of the first responding fire department. If fire department hose couplings are of the unthreaded type, such as a quick connect, the sprinkler system connections should be compatible. Recognizing that not all fire departments have switched to the NH standard for couplings, the Exception permits other types of compatible connections.

2-7.6.2 Hose connections shall be equipped with listed plugs or caps.

It should be noted that the fire department connection itself is not required to be listed, only the plugs or caps.

2-8 Arrangement of Water Supply Connections.

2-8.1 Connection Between Underground and Aboveground Piping. The connection between the system piping and underground piping shall be made with a suitable transition piece and shall be properly strapped or fastened by approved devices. The transition piece shall be protected against possible damage from corrosive agents, solvent attack, or mechanical damage.

This section indicates that piping extending above the floor should be metallic.

2-8.2* Connection Passing Through or Under Foundation Walls. When system piping pierces a foundation wall below grade or is located under the foundation wall, clearance shall be provided to prevent breakage of the piping due to building settlement.

A-2-8.2 When the system riser is close to an outside wall, underground fittings of proper length should be used in order to avoid pipe joints located in or under the wall. When the connection passes through the foundation wall below grade, a 1- to 3-in. (25- to 76-mm) clearance should be provided around the pipe and the clear space filled with asphalt mastic or similar flexible waterproofing material. (*Also see Appendix B-3-1.*)

Appendix B-3-1 appears on page 166 of this handbook, following 3-10.7.4.

2-9 Water Supply Test Connections and Gages.

2-9.1* Test Connections. Test connections, which may also be used as drain pipes, shall be provided at locations that will permit flow tests to be made to determine whether water supplies and connections are in order. Such test connections shall be not less than the sizes specified in 3-6.2 and equipped with a shutoff valve. They shall be so installed that the valve may be opened wide for a sufficient time to assure a proper test without causing water damage. (*See 3-6.2 and 3-6.4.*)

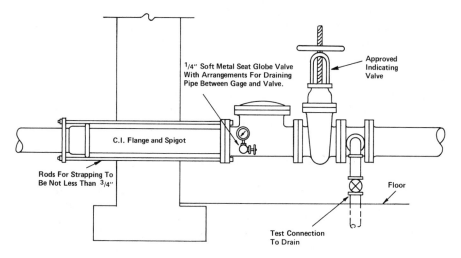

1/4" Soft Metal Seat Globe Valve With Arrangements For Draining Pipe Between Gage and Valve.

Approved Indicating Valve

C.I. Flange and Spigot

Rods For Strapping To Be Not Less Than 3/4"

Floor

Test Connection To Drain

For SI Units: 1 in. = 25.4 mm.

Figure 2-9.1   Water Supply Connection with Test Connection.

Test connections or drain pipes are used for both stated purposes. They are used to drain systems when repairs are necessary or when dry-pipe valves have tripped. They are also used to indicate that an adequate water supply is available at the system riser and on the system side of all check valves, control valves, and underground piping.

Test connections or drain pipes are normally smaller than all other water supply piping. Therefore, they are not capable of the same large-scale flow tests discussed in B-2-1. However, when the system demand is small, a drain test can be used; the test pressures and flow at the test pipe represent the exact water supply available for the amount of water flowed. These tests should not be extrapolated to larger flows or system demands because of the unknown friction values of check valves under low flow conditions that are included in the drain tests. Two-inch drain tests, as they are known, normally consist of opening the main 2-in. (51-mm) drain wide until the pressure stabilizes. Records are kept of these tests to detect possible deterioration of water supplies or to detect valves in the overall water supply system that may have been closed. The 2-in. drain test on new systems necessitates the establishment of a benchmark when the system is first approved during the acceptance tests.

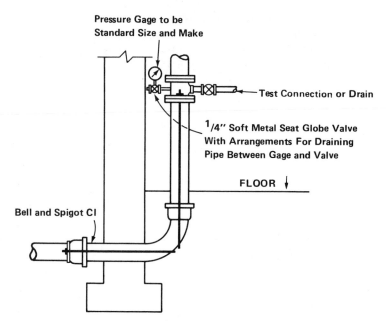

Pressure Gage to be
Standard Size and Make

Test Connection or Drain

1/4" Soft Metal Seat Globe Valve
With Arrangements For Draining
Pipe Between Gage and Valve

FLOOR

Bell and Spigot CI

Figure A-2-9.1 Test Connection on Water Supply with Outside Control. (Also applicable to an interior riser.)

2-9.2 Gages.

2-9.2.1 A pressure gage with a connection not smaller than ¼ in. (6.4 mm) shall be installed on the riser or feed main at or near each test connection. This gage connection shall be equipped with a shutoff valve and with provision for draining.

2-9.2.2   The required pressure gages shall be of an approved type and shall have a maximum limit not less than twice the normal working pressure at the point where installed. They shall be installed to permit removal, and shall be located where they will not be subject to freezing.

REFERENCES CITED IN COMMENTARY

The following publications are available from the National Fire Protection Association, Batterymarch Park, Quincy, MA 02269.

NFPA 14-1986, Standard for the Installation of Standpipe and Hose Systems
NFPA 20-1987, Standard for the Installation of Centrifugal Fire Pumps
NFPA 24-1987, Standard for the Installation of Private Fire Service Mains and Their Appurtenances
NFPA 231-1987, Standard for General Storage
NFPA 231C-1986, Standrad for Rack Storage of Materials
NFPA 291-1988, Recommended Practice for Fire Flow Testing and Marking of Hydrants
NFPA 409-1985, Standard on Aircraft Hangars.

3

System Components

3-1 Piping.

3-1.1 Piping Specifications.

3-1.1.1 Pipe or tube used in sprinkler systems shall meet or exceed the standards in Table 3-1.1.1 or be in accordance with 3-1.1.2 through 3-1.1.6. Pipe and tube used in sprinkler systems shall be designed to withstand a working pressure of not less than 175 psi (12.1 bars).

In addition to specifying piping materials for use in sprinkler systems (*see Table 3-1.1.1*), this section also permits the use of

Table 3-1.1.1   Pipe or Tube Materials and Dimensions

Materials and Dimensions	Standard
Ferrous Piping (Welded and Seamless)	
†Spec. for Black and Hot-Dipped Zinc Coated (Galvanized) Welded and Seamless Steel Pipe for Fire Protection Use	ASTM A795
†Spec. for Welded and Seamless Steel Pipe	ANSI/ASTM A53
Wrought Steel Pipe	ANSI B36.10M
Spec. for Elec.-Resistance Welded Steel Pipe	ASTM A135
Copper Tube (Drawn, Seamless)	
†Spec. for Seamless Copper Tube	ASTM B75
†Spec. for Seamless Copper Water Tube	ASTM B88
Spec. for General Requirements for Wrought Seamless Copper and Copper-Alloy Tube	ASTM B251
Brazing Filler Metal (Classification BCuP-3 or BCuP-4)	AWS A5.8
Solder Metal, 95-5 (Tin-Antimony-Grade 95TA)	ASTM B32

†Denotes pipe or tubing suitable for bending (*see 3-1.1.7)* according to ASTM standards.

pipe or tube of materials made to standards other than those shown in the table, provided the requirements of the other standards meet or exceed the cited standards.

Selection of an appropriate pipe or tube may be subject to the environment in which the pipe will be used (corrosive vs. noncorrosive) as well as the ability to achieve a certain aesthetic result. For example, copper tubing is used on some projects to blend into the architectural features when the piping must be run in an exposed manner.

3-1.1.2* When welded and seamless steel pipe listed in Table 3-1.1.1 is used and joined by welding as referenced in 3-7.2 or by roll grooved pipe and couplings as referenced in 3-7.3, the minimum nominal wall thickness for pressures up to 300 psi (20.7 bars) shall be in accordance with Schedule 10 for sizes up to 5 in. (127 mm); 0.134 in. (3.40 mm) for 6 in. (152 mm); and 0.188 in. (4.78 mm) for 8- and 10-in. (203- and 254-mm) pipe; or as modified in 3-1.1.5, or as defined in 3-1.1.6.

When the thin-wall pipe is joined by welding or roll-grooved pipe and couplings, the groove forming or welding process should not result in a thinner wall thickness, other than that due to the normal tolerances associated with the groove forming or welding process. The expression "minimum nominal wall thickness" recognizes that pipe standards permit plus or minus tolerance variations from the stated wall thickness.

3-1.1.3 When steel pipe listed in Table 3-1.1.1 is used and joined by threaded fittings referenced in 3-7.1 or by couplings used with pipe having cut grooves, the minimum wall thickness shall be in accordance with Schedule 30 [in sizes 8 in. (203 mm) and larger] or Schedule 40 [in sizes less than 8 in. (203 mm)] pipe for pressures up to 300 psi (20.7 bars).

When pipe is threaded or grooves are cut, material is lost and the use of thin-wall pipe could result in too little material remaining between the inside diameter and the root diameter of the thread or groove, which may result in failed pipe fittings.

A threaded assembly that has been investigated for suitability in automatic sprinkler installations and listed for this service is acceptable. (*See 3-7.1.2 and A-3-7.1.2.*)

3-1.1.4* Copper tube as specified in the standards listed in Table 3-1.1.1, used in sprinkler systems, shall have a wall thickness of Type K, L, or M.

Table A-3-1.1.2 Steel Pipe Dimensions

Nominal Pipe Size	Outside Diameter		Schedule 10[1]				Schedule 30				Schedule 40			
			Inside Diameter		Wall Thickness		Inside Diameter		Wall Thickness		Inside Diameter		Wall Thickness	
in.	in.	(mm)	in.	(mm)	in.	(mm)	in.	(mm)	in.	(mm)	in.	(mm)	in.	(mm)
1	1.315	(33.4)	1.097	(27.9)	0.109	(2.8)	—	—	—	—	1.049	(26.6)	0.133	(3.4)
1¼	1.660	(42.2)	1.442	(36.6)	0.109	(2.8)	—	—	—	—	1.380	(35.1)	0.140	(3.6)
1½	1.900	(48.3)	1.682	(42.7)	0.109	(2.8)	—	—	—	—	1.610	(40.9)	0.145	(3.7)
2	2.375	(60.3)	2.157	(54.8)	0.109	(2.8)	—	—	—	—	2.067	(52.5)	0.154	(3.9)
2½	2.875	(73.0)	2.635	(66.9)	0.120	(3.0)	—	—	—	—	2.469	(62.7)	0.203	(5.2)
3	3.500	(88.9)	3.260	(82.8)	0.120	(3.0)	—	—	—	—	3.068	(77.9)	0.216	(5.5)
3½	4.000	(101.6)	3.760	(95.5)	0.120	(3.0)	—	—	—	—	3.548	(90.1)	0.226	(5.7)
4	4.500	(114.3)	4.260	(108.2)	0.120	(3.0)	—	—	—	—	4.026	(102.3)	0.237	(6.0)
5	5.563	(141.3)	5.295	(134.5)	0.134	(3.4)	—	—	—	—	5.047	(128.2)	0.258	(6.6)
6	6.625	(168.3)	6.357	(161.5)	0.134[2]	(3.4)	—	—	—	—	6.065	(154.1)	0.280	(7.1)
8	8.625	(219.1)	8.249	(209.5)	0.188[2]	(4.8)	8.071	(205.0)	0.277	(7.0)	—	—	—	—
10	10.75	(273.1)	10.37	(263.4)	0.188[2]	(4.8)	10.14	(257.6)	0.307	(7.8)	—	—	—	—

NOTE 1: Schedule 10 defined to 5 in. (127 mm) nominal pipe size by ASTM A135.
NOTE 2: Wall thickness specified in 3-1.1.2.

Table A-3-1.1.4 Copper Tube Dimensions

Nominal Tube Size in.	Outside Diameter in.	(mm)	Type K Inside Diameter in.	(mm)	Type K Wall Thickness in.	(mm)	Type L Inside Diameter in.	(mm)	Type L Wall Thickness in.	(mm)	Type M Inside Diameter in.	(mm)	Type M Wall Thickness in.	(mm)
¾	0.875	(22.2)	0.745	(18.9)	0.065	(1.7)	0.785	(19.9)	0.045	(1.1)	0.811	(20.6)	0.032	(0.8)
1	1.125	(28.6)	0.995	(25.3)	0.065	(1.7)	1.025	(26.0)	0.050	(1.3)	1.055	(26.8)	0.035	(0.9)
1¼	1.375	(34.9)	1.245	(31.6)	0.065	(1.7)	1.265	(32.1)	0.055	(1.4)	1.291	(32.8)	0.042	(1.1)
1½	1.625	(41.3)	1.481	(37.6)	0.072	(1.8)	1.505	(38.2)	0.060	(1.5)	1.527	(38.8)	0.049	(1.2)
2	2.125	(54.0)	1.959	(49.8)	0.083	(2.1)	1.985	(50.4)	0.070	(1.8)	2.009	(51.0)	0.058	(1.5)
2½	2.625	(66.7)	2.435	(61.8)	0.095	(2.4)	2.465	(62.6)	0.080	(2.0)	2.495	(63.4)	0.065	(1.7)
3	3.125	(79.4)	2.907	(73.8)	0.109	(2.8)	2.945	(74.8)	0.090	(2.3)	2.981	(75.7)	0.072	(1.8)
3½	3.625	(92.1)	3.385	(86.0)	0.120	(3.0)	3.425	(87.0)	0.100	(2.5)	3.459	(87.9)	0.083	(2.1)
4	4.125	(104.8)	3.857	(98.0)	0.134	(3.4)	3.905	(99.2)	0.110	(2.8)	3.935	(99.9)	0.095	(2.4)
5	5.125	(130.2)	4.805	(122.0)	0.160	(4.1)	4.875	(123.8)	0.125	(3.2)	4.907	(124.6)	0.109	(2.8)
6	6.125	(155.6)	5.741	(145.8)	0.192	(4.9)	5.845	(148.5)	0.140	(3.6)	5.881	(149.4)	0.122	(3.1)
8	8.125	(206.4)	7.583	(192.6)	0.271	(6.9)	7.725	(196.2)	0.200	(5.1)	7.785	(197.7)	0.170	(4.3)
10	10.13	(257.3)	9.449	(240.0)	0.338	(8.6)	9.625	(244.5)	0.250	(6.4)	9.701	(246.4)	0.212	(5.4)

Formal Interpretation

Question: Is use of copper tubing (Type K) with brazed joints acceptable for underground service from a city water main to a sprinkler system?

Answer: Yes—in accordance with 3-1.1.1 and 3-7.4. Copper tubing conforming with Table 3-1.1.1 and brazed joints conforming with 3-7.4 are acceptable.

Since Type M tube is the thinnest walled, it has the largest inside diameter, which is an advantage in hydraulically designed systems. It is also the most inexpensive. Therefore, Type M tube is most commonly used when bending is not required. (*See 3-1.1.7.*)

3-1.1.5* Other types of pipe or tube may be used if investigated and listed for this service and installed in accordance with their listing limitations, including installation instructions. Pipe or tube shall not be listed for portions of an occupancy classification.

Formal Interpretation

Question 1: Is it the intent of the Committee that paragraphs 3-1.1.5 and 3-8.1.2 permit the use of piping products that differ from those specifically described in NFPA 13, if they have been investigated and listed for this service by the nationally recognized testing and inspection agency laboratory?

Answer: Yes. It permits use of different materials and different dimensions than those specifically described elsewhere.

Question 2: Is it the intent of the Committee that paragraphs 3-1.1.5 and 3-8.1.2 permit a manufacturer to pursue a testing program and obtain a listing of this product by a nationally recognized testing and inspection agency laboratory, even though that product may differ from those detailed by specifications or piping products described in NFPA 13?

Answer: Yes.

There is no prohibition of the use of any material, as long as it is listed for use in sprinkler systems. This section is intended to encourage development of either more efficient or cost-effective materials. Historically, this has been the case with copper tubing, and more recently (since 1984), nonmetallic pipe.

As indicated in Section 1-2, it is required that special listed pipe be installed in complete accord with all conditions, requirements, and limitations of its listing.

Not all pipe made to a particular standard is listed. Listed piping is identified by the logo of the listing agency. Similar (unlisted) piping manufactured with less exacting quality control must not be used in sprinkler systems.

At this time, two synthetic piping materials are listed for sprinkler system applications. Chlorinated Polyvinyl Chloride (CPVC) pipe produced from a resin manufactured by the B.F. Goodrich Company and polybutylene pipe produced from a resin manufactured by the Shell Oil Company are presently listed for residential and Light Hazard Occupancies. When a CPVC or a polybutylene system enters Ordinary or Extra Hazard areas, other types of piping must be installed.

The listings of these materials include restrictions. The user must refer to the extensive listing information, including installation instructions, to accomplish correct installation. (*Also see commentary to 3-1.1.7.*)

When the product requires a protective membrane, readily available materials such as gypsum wallboard, ½-in. (13-mm) plywood, or acoustical panels would constitute acceptable protection. Throughout the life of the building, care should be taken to assure that the protective membrane materials are not discarded or violated with poke-through penetrations.

A-3-1.1.5   The investigation of pipe and tube other than that described in Table 3-1.1.1 should involve consideration of many factors:

(a) Pressure rating.

(b) Beam strength (hangers).

(c) Unsupported vertical stability.

(d) Movement during sprinkler operation (affecting water distribution).

(e) Corrosion (internal and external; chemical and electrolytic).

(f) Resistance to failure when exposed to elevated temperatures.

(g) Methods of joining (strength, permanence, fire hazard).

(h) Physical characteristics related to integrity during earthquakes.

3-1.1.6 Whenever the word pipe is used in this standard it shall be understood to also mean tube.

For the purposes of this standard the word *pipe* refers to any conduit for transporting water.

3-1.1.7 Pipe Bending. Bending of steel pipe (Schedule 40) and copper tube (Types K and L) may be accomplished when bends are made in conformance with good installation practices and show no kinks, ripples, distortions, reduction in diameter, or any noticeable deviations from round. The minimum radius of a bend shall be 6 pipe diameters for pipe sizes 2 in. (51 mm) and smaller, and 5 pipe diameters for pipe sizes 2½ in. (64 mm) and larger.

ASTM A135 pipe cannot be used for bending because the bending operation might cause the pipe to split along the seam. This does not occur with seamless pipe, such as ASTM A53 and some types of ASTM A795 pipe. Type M copper tube and Schedule 10 steel pipe cannot be used for bending because the bending results in a wall thickness that is too thin, whereas with the heavier wall Types L and K copper tubing and Schedule 40 steel pipe there is adequate material remaining after bending. The listing of polybutylene pipe permits it to be bent to a minimum radius of 10 actual pipe diameters when bent in accordance with listing limitations and manufacturer's installation instructions. In addition, without any control over the bending of pipe, excessive friction loss values for a given elbow (or similar segment of pipe that would change the direction of flow) can be expected.

3-2* Definitions. *(See Figure A-3-2.)*

Risers. The vertical pipes in a sprinkler system.

The section of aboveground pipe directly connected to the water supply is referred to as the system riser and is equipped with an alarm actuating device such as an alarm check valve or waterflow switch. The other vertical pipes are also referred to as risers.

System Riser. The aboveground supply pipe directly connected to the water supply.

Feed Mains. Mains supplying risers or cross mains.

Cross Mains. Pipes directly supplying the lines in which the sprinklers are placed.

Branch Lines. Lines of pipe, from the point of attachment to the cross main (or similar connection) to the end sprinkler, in which the sprinklers are directly placed.

Formal Interpretation

Question: On a hydraulically calculated system, does a 1½-in. and 2-in. (pipe size) loop with sprinklers fed directly off the loop constitute a "branch line" as specified in Section 3-2 and 8-2.1?

Answer: Yes. A calculated system, as indicated in Figure 1, complies with the definition of a branch line in Section 3-2. The reference to branch lines in Section 8-2.1 is for a limitation to the number of sprinklers on a branch line for pipe schedule systems. Figure 1 is for a hydraulically calculated system and the requirements for number of sprinklers per branch line are superseded. (*See Section 7-1.1.2.*)

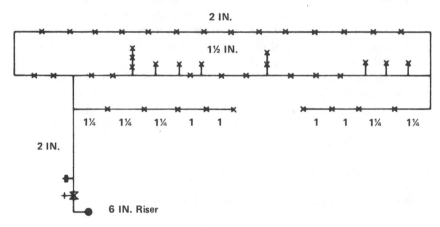

Figure 1.

For purposes of this standard, particularly sections such as 3-5.3, the horizontal 2-in. (51-mm) and 1½-in. (38-mm) lines in the figure are also cross mains, since they serve other lines in which sprinklers are placed.

3-3 Special Provisions Applicable to Piping.

3-3.1 Rack Storage. For sprinklers in storage racks see NFPA 231C, *Standard for Rack Storage of Materials.*

Rack storage fires present unique, rapidly developing, high-heat-release shielded fires that are addressed in Chapter 9 and NFPA 231C, *Standard for Rack Storage of Materials.* NFPA 231C addresses the necessity for sprinklers in the racks, sprinkler spacing, and water densities, based on a large-scale testing program dealing with rack storage.

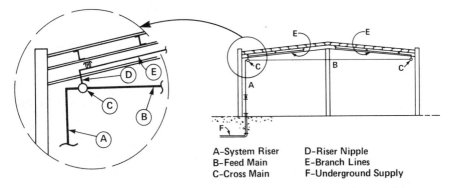

A–System Riser	D–Riser Nipple
B–Feed Main	E–Branch Lines
C–Cross Main	F–Underground Supply

Figure A-3-2   Building Elevation Showing Parts of Sprinkler Piping System.

3-3.2* Provision for Flushing Systems. All sprinkler systems shall be arranged for flushing. Readily removable fittings shall be provided at the end of all cross mains. All cross mains shall terminate in 1¼-in. (33-mm) or larger pipe. All branch lines on gridded systems shall be arranged to facilitate flushing.

The readily removable fitting at the end of each cross main is not limited with regard to maximum size. Branch lines in gridded systems must be installed in such a manner that the piping can be readily disconnected. Further guidance is provided in NFPA 13A, *Recommended Practice for the Inspection, Testing and Maintenance of Sprinkler Systems.*

It is important to provide for flushing, particularly where the system is supplied from a nonpotable water supply such as a pond, stream, or lake, or where the system has alternated or might alternate between wet- and dry-pipe operation (a practice that is discouraged). In either of these cases, foreign material, which can block the system or block individual sprinklers on the system, may be introduced, and it is necessary that provisions be made for flushing and removal of this material.

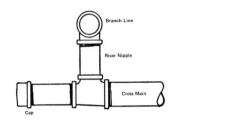

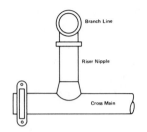

Figure A-3-3.2(a)
Screw-type Cap.

Figure A-3-3.2(b)
Groove-type Cap.

A-3-3.2 Also see NFPA 13A, *Recommended Practice for the Inspection, Testing and Maintenance of Sprinkler Systems.*

3-3.3 Stair Towers. Stairs, towers, or other construction with incomplete floors, if piped on independent risers, shall be treated as one area with reference to pipe sizes.

A fire in a stair or tower may tend to open a large percentage of sprinklers in the area. An independent riser supplying such an area must be sized to supply all the sprinklers in the area.

3-3.4 Return Bends. Return bends shall be used when pendent sprinklers are supplied from a raw water source, mill pond, or from open-top reservoirs. Return bends shall be connected to the top of branch lines in order to avoid accumulation of sediment in the drop nipples.

Exception No. 1:   Return bends are not required for deluge systems.

Exception No. 2:   Return bends are not required when dry-pendent sprinklers are used.

Formal Interpretation

Question: Section 3-3.4. Would a filtered and chlorine-treated swimming pool be classified as a raw source of water thus requiring return bends to avoid accumulation of sediment in drop nipples?

Answer: No.

Formal Interpretation

Question: In revamping an existing wet-pipe fire sprinkler system that has a potable water supply, and where it is not necessary to retain sprinklers in the concealed space above a noncombustible ceiling, would it be correct to use close nipples of the sprinkler thread size inserted in the existing sprinkler fittings with 1-in. pipe and fittings for the other portions of the drop?

Answer: Yes.

The purpose of return bends is to prevent collection of sediment in drop nipples which might be taken directly off the bottom of branch lines in wet-pipe sprinkler systems, particularly where the systems are supplied from water sources that might contain excessive sediment. Return bends are not necessary where a potable water supply is used.

Return bends are also useful where centering of sprinklers in ceiling tile or exact positioning of sprinklers is desirable from an aesthetic or architectural viewpoint. When standard sprinklers, rather than dry-pendent sprinklers, are installed on dry systems, return bends are required in all cases due to the possibility of scale build-up. In such instances the sprinklers should be installed in accordance with 5-2.2.

3-3.4.1 In revamping existing systems, when it is not necessary to retain sprinklers in the concealed space, a nipple not exceeding 4 in. (102 mm) in length and of the same pipe thread size as the sprinkler being removed, may be used with 1-in. (25-mm) pipe and fittings for the other portions of the return bend to a single sprinkler in an area.

3-3.4.2 In revamping existing systems when it is necessary to retain sprinklers in the concealed space, the return bend shall be not less than 1 in. (25 mm) throughout to a single sprinkler in each area.

Figure 3-3.4   Return Bend Arrangement.

3-3.5 Piping to Sprinklers Below Ceilings.

3-3.5.1   In new installations expected to supply sprinklers below a ceiling, minimum 1 in. (25 mm) outlets shall be provided.

Exception: Hexagonal bushings may be utilized to accommodate the temporary sprinklers and are to be removed with the sprinklers.

3-3.5.2   In revamping existing systems, a nipple not exceeding 4 in. (102 mm) in length and of the same pipe thread size as the sprinkler being removed may be used. All other piping shall be 1 in. (25 mm) which supplies a single sprinkler in an area.

When a system is to be installed exposed but it is anticipated that a false ceiling will be installed, the openings for sprinklers are to be 1 in. (25 mm) and reduced to the thread size of the sprinklers. In revamping existing systems, changing the existing line fitting is not required. Since a sprinkler retained in the concealed space and a sprinkler below the ceiling are in separate areas, both may be supplied by a single connection. [*See Figure 3-3.5.2(b)*.]

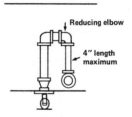

Figure 3-3.5.2(a)   Nipple and Reducing Elbow Supplying Sprinkler Below Ceiling.

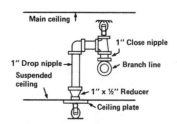

Figure 3-3.5.2(b)   Sprinklers in Concealed Space and Below Ceiling.

3-3.6 Dry-Pipe Underground. When necessary to place pipe that will be under air pressure underground, the pipe shall be protected against corrosion (*see 3-5.2*), or unprotected cast or ductile iron pipe may be used when joined with a gasketed joint listed for air service underground.

Corrosion that forms on the exterior of cast iron or ductile iron pipe insulates that pipe from further corrosion. Such pipe is normally used to transport water, which is less likely to leak than air. When the pipe is subject to air pressure it must have gasketed joints, and these joints must be specifically listed for use under air pressure.

3-3.7* One and One-Half-Inch Hose Connections. One and one-half-inch [1½-in. (38-mm)] hose used for fire purposes only may be connected to wet sprinkler systems only, subject to the following restrictions:

(a) Hose station's supply pipes shall not be connected to any pipe smaller than 2½ in. (64 mm).

Exception: For hydraulically designed loops and grids the minimum size pipe between the hose station's supply pipe and the source may be 2 in. (51 mm).

(b) For piping serving a single hose station, pipe shall be minimum 1 in. (25 mm) for horizontal runs up to 20 ft (6.1 m), minimum 1¼ in. (33 mm) for the entire run for runs between 20 and 80 ft (6.1 and 24.4 m), and minimum 1½ in. (38 mm) for the entire run for runs greater than 80 ft (24.4 m). For piping serving multiple hose stations, runs shall be a minimum of 1½ in. (38 mm) throughout.

(c) Piping shall be at least 1 in. (25 mm) for vertical runs.

(d) When the pressure at any hose station outlet exceeds 100 psi (6.9 bars), an approved device shall be installed at the outlet to reduce the pressure at the outlet to 100 psi (6.9 bars).

Formal Interpretation

Question: When a horizontal run of pipe 50 ft long supplies a 1½-inch hose station, is the Committee's intent that:

(a) The first 20 ft of the run be 1-inch and the remainder 1¼-inch pipe?

(b) The entire 50 ft be 1¼-inch pipe?

Answer: The entire 50 ft should be 1¼-inch pipe.

One and one-half-inch hose connections are restricted to wet systems. The maintenance problems due to loss of air pressure through the hose valves in other types of systems would outweigh their value.

The pressure at the outlet is restricted to 100 psi (6.9 bars) due to the danger to the operator in using a hose subject to high pressure, either static or under flow conditions. Therefore, 100 psi at the hose outlet is the maximum pressure under both conditions.

A-3-3.7   One and one-half (1½) in. hose connections for use in storage occupancies and other locations where standpipe systems are not required are covered by this standard. When Class II standpipe systems are required see the appropriate provisions of NFPA 14, *Standard for the Installation of Standpipe and Hose Systems*, with respect to hose stations and water supply for hose connections from sprinkler systems.

One and one-half-inch hose connections supplied from sprinkler systems have been successful in extinguishment and fire control for many years. It is felt that the requirements for Class II standpipe systems in NFPA 14, *Standard for the Installation of Standpipe and Hose Systems*, should not apply to hose connections attached to sprinkler systems inasmuch as this would impose severe water supply and pressure requirements on the sprinkler system.

3-3.8* Hose Connections for Fire Department Use.   In buildings of Light or Ordinary Hazard Occupancy, 2½-in. (64-mm) hose valves for fire department use may be attached to wet-pipe sprinkler system risers. (*See 2-2.4.2*) The following restrictions apply:

(a) Sprinklers shall be under separate floor control valves.

(b) The minimum size of the riser shall be 4 in. (102 mm) unless hydraulic calculations indicate a smaller size riser will satisfy sprinkler and hose stream demands.

(c) Each combined sprinkler and standpipe riser shall be equipped with a riser control valve to permit isolating a riser without interrupting the supply to other risers from the same source of supply.

(d) For fire department connections serving standpipe and sprinkler systems, refer to Section 2-7.

This section allows 2½-in. (64-mm) hose connections to be attached to wet-pipe sprinkler systems. These connections may be used for final extinguishment of the fire or to cool residual heat. Paragraph 2-2.4.2 provides requirements for calculating the water supply under these circumstances.

A-3-3.8   Combined automatic sprinkler and standpipe risers should not be interconnected by sprinkler system piping.

When risers are used to supply combination sprinkler and standpipe systems, guidance is needed to avoid confusion where systems in remotely located areas are controlled by more than one valve. No appreciable improvement in reliability is realized by cross connection.

3-4 System Test Connections.

3-4.1 Wet Systems.

3-4.1.1* A test connection not less than 1 in. (25 mm) in diameter, terminating in a smooth bore corrosion-resistant orifice, giving a flow equivalent to one sprinkler of a type having the smallest orifice installed on the particular system, shall be provided to test each waterflow alarm device for each system. The test connection valve shall be readily accessible. The discharge shall be to the outside, to a drain connection capable of accepting full flow under system pressure or to another location where water damage will not result.

The primary function of the wet system inspector's test is to verify the operation of the waterflow alarm device(s) (water-motor gong, pressure switch, flow switch) at a flow equivalent to that of one operating sprinkler.

A-3-4.1.1   This test connection should be in the upper story, and the connection should preferably be piped from the end of the most remote branch line. The discharge should be at a point where it can be readily observed. In locations where it is not practical to terminate the test connection outside the building, the test connection may terminate into a drain capable of accepting full flow under system pressure. (*See A-3-6.4.1.*) In this event, the test connection should be made using an approved sight test connection containing a smooth bore corrosion-resistant orifice giving a flow equivalent to one sprinkler simulating the least flow from an individual sprinkler in the system. [*See Figures A-3-4.1.1(a) and A-3-4.1.1(b)*.] The test valve should be located at an accessible point, and preferably not over 7 ft (2.1 m) above the floor. The control valve on the test connection should be located at a point not exposed to freezing.

The intent of this section is to require inspector's test connections in a manner that will provide for the testing of all waterflow alarm devices. This section permits both an electrically operated alarm and a water motor alarm supplied from an alarm valve to be tested through a single test connection.

Ideally the inspector's test connection is located at the end of the most remote branch line in the upper story. Such a location is not, however, required. (*See 3-12.1.*)

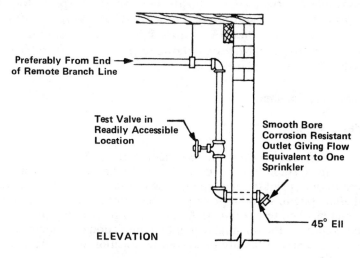

Preferably From End →
of Remote Branch Line

Test Valve in
Readily Accessible
Location

Smooth Bore
Corrosion Resistant
Outlet Giving Flow
Equivalent to One
Sprinkler

45° Ell

ELEVATION

For SI Units: 1 ft = 0.3048 m.
NOTE: Not less than 4 ft (1.2 m) of Exposed Test Pipe in Warm Room Beyond Valve
When Pipe Extends Through Wall to Outside.

Figure A-3-4.1.1(a)   System Test Connection on Wet-Pipe System.

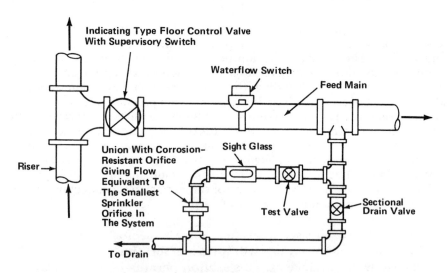

Indicating Type Floor Control Valve
With Supervisory Switch

Waterflow Switch

Feed Main

Union With Corrosion-
Resistant Orifice
Giving Flow
Equivalent To
The Smallest
Sprinkler
Orifice In
The System

Sight Glass

Riser →

Test Valve

Sectional
Drain Valve

To Drain

Figure A-3-4.1.1(b)   Floor Control Valve.

3-4.2* Dry-Pipe Systems. A test connection not less than 1 in. (25 mm) in diameter, terminating in a smooth bore corrosion-resistant orifice to provide a flow equivalent to one sprinkler of a type installed on the particular system, shall be installed on the end of the most distant sprinkler pipe in the upper story and be equipped with a readily accessible 1-in. (25-mm) shutoff valve and plug, at least one of which shall be brass. In lieu of a plug, a nipple and cap may be used.

The dry-pipe system inspector's test connection is used to measure the approximate time from the opening of the most distant sprinkler in the system until water flows from that sprinkler. For that reason it must be located at the end of the farthest line in the top story of the protected occupancy. The valve must be sealed with a plug or nipple and cap when not in use to avoid leakage of air and to avoid accidental tripping of the dry-pipe valve.

It is permissible to install the test valve on the drop pipe for the purpose of accessibility. In this case, the connection should

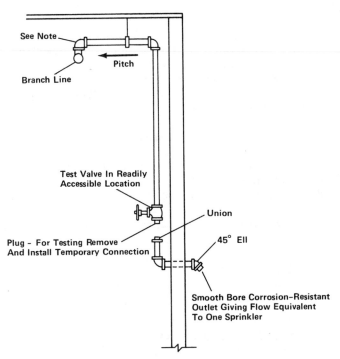

NOTE: To minimize condensation of water in the drop to the test connection, provide a nipple-up off of the branch line.

Figure A-3-4.2   System Test Connection on Dry-Pipe System.

be made to the top of the branch line to minimize condensation build-up. Moisture should be drained periodically as is done with low point drains in areas subject to freezing. Manufacturer's instructions should be followed to avoid accidental operation of dry-pipe valves equipped with quick-opening devices.

3-4.3 Preaction Systems. A test connection shall be used on a preaction system using supervisory air.

3-4.4 Deluge Systems. A test connection is not required on a deluge system.

3-5* Protection of Piping.

The current special listings for CPVC piping in most instances and for polybutylene piping specify requirements to protect the piping from fire exposure. Each type of piping must be protected in accordance with its specific listing limitation. (*See commentary to 3-1.1.5.*)

A-3-5 Protection of Piping Against Damage Due to Impact. Sprinkler piping should be located so as to minimize the possibility of damage due to impact by mobile material handling equipment and other vehicles. For example, risers adjacent to structural columns and out of vehicle travel routes are generally safe, as are feed mains and cross mains shielded by heavy structural members such as girders.

3-5.1 Protection of Piping Against Freezing.

3-5.1.1 When portions of systems are subject to freezing and temperatures cannot be reliably maintained at or above 40°F (4°C) sprinklers shall be installed as a dry-pipe or preaction system.

Exception: Small unheated areas may be protected by antifreeze systems. (See Section 5-5.)

A wet-pipe system should always be the first choice when selecting a system design. Dry-pipe systems may be considered when the 40°F (4°C) minimum temperature cannot be maintained. The use of a dry-pipe system for any unheated area should be limited to that area, while a separate wet-pipe system should be used for the heated portions of a building. Dry-pipe systems tend to have slower operating times as well as increased corrosion rates on the internal portions of the pipe in comparison to wet-pipe systems.

3-5.1.2* When water-filled supply pipes, risers, system risers, or feed mains pass through open areas, cold rooms, passageways, or other areas exposed to freezing, the pipe shall be protected against freezing by insulating coverings, frostproof casings, or other reliable means capable of maintaining a minimum temperature of 40°F (4°C).

A-3-5.1.2 In areas subject to freezing climates, when piping extends through an exterior wall, as for fire department connections, system test connection, or drains, a minimum of 4 ft (1.2 m) of pipe should be maintained between the wall and the section of piping containing water.

The recommended 4 ft (1.2 m) of pipe to be maintained between the exterior wall and the piping containing water is for the purpose of providing a frost break.

3-5.2 Protection of Piping Against Corrosion.

3-5.2.1* When corrosive conditions are known to exist due to moisture or fumes from corrosive chemicals, or both, types of piping, fittings, and hangers that resist corrosion shall be used or a protective coating shall be applied to all unprotected exposed surfaces of the sprinkler system to resist corrosion. (*See 3-11.4.*)

A-3-5.2.1 Types of locations where corrosive conditions may exist include bleacheries, dye houses, metalplating processes, animal pens, and certain chemical plants.

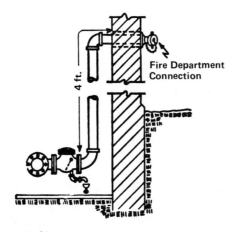

For SI Units: 1 ft = 0.3048 m.

Figure A-3-5.1.2   Minimum Clearance to Avoid Freezing.

If corrosive conditions are not of great intensity and humidity is not abnormally high, good results can be obtained by a protective coating of red lead and varnish or by a good grade of commercial acid-resisting paint. The paint manufacturer's instructions should be followed in the preparation of the surface and in the method of application.

Where moisture conditions are severe but corrosive conditions are not of great intensity, copper tube or galvanized steel pipe, fittings, and hangers may be suitable. The threaded ends of steel pipe should be painted.

In instances where the piping is not readily accessible and where the exposure to corrosive fumes is severe, either a protective coating of high quality may be employed or some form of corrosion-resistant material used.

In addition to piping in "manmade" corrosive environments, piping installed along ocean front facilities will be subject to higher than normal corrosive air.

The referenced protective paints and coatings are not applied to the automatic sprinklers. Automatic sprinklers must be provided with corrosion-resistant coatings applied by the manufacturer in accordance with 3-11.4.

3-5.2.2 When water supplies are known to have unusual corrosive properties and threaded or cut grooved steel pipe is to be used, wall thickness shall be in accordance with Schedule 30 [in sizes 8 in. (203 mm) or larger] or Schedule 40 [in sizes less than 8 in. (203 mm)].

Caution must be exercised when considering corrosion control additives to sprinkler systems, and the authority having jurisdiction should be consulted. There is the possibility of contaminating the domestic supply (backflow), and some additives may solidify when exposed to air, causing obstruction to water flow. Such chemicals may be added to poor quality water or as a corrosion inhibitor for the circulating closed-loop systems discussed in Chapter 5.

While corrosion protection for exposed surfaces is independent of wall thickness and acceptable pipe joining methods, the use of threaded thin-wall pipe must be avoided where subject internally to water having unusual corrosive properties because of the lack of material between the inside diameter and the root diameter of the threads. If threaded pipe is used, it should be Schedule 30 or heavier for pipe sizes 8 in. (203 mm) or larger and Schedule 40 for pipe sizes less than 8 in. The use of welded or roll-grooved Schedule 10 pipe would be satisfactory.

3-5.2.3*   Steel pipe, when exposed to weather, shall be externally galvanized or otherwise protected against corrosion.

A-3-5.2.3   It is important when protected steel pipe (galvanized, dipped and wrapped, coated, etc.) is used that particular care is taken to see that all exposed threads, wrench marks, or abrasions that have penetrated through the protection be repaired, sealed, and/or properly coated.

3-5.2.4   When steel pipe is used underground as a connection from a system to sprinklers in a detached building, the pipe shall be protected against corrosion before being buried.

3-5.3 Protection of Piping Against Damage Where Subject to Earthquakes.

Earthquake design criteria have been included in this standard for more than 50 years. The performance of automatic sprinkler systems in earthquakes has been good. Following the San Fernando Earthquake in 1971 (6.6 Richter magnitude), the Pacific Fire Rating Bureau surveyed 973 sprinklered buildings and reported that "if a sprinklered building fared well, so did the sprinkler system." Earthquakes of the past several years have proven the advantages of the additional flexibility which earthquake bracing provides to the piping.

It is not the intent of this standard to specify where earthquake design provisions must be used, but to provide design criteria for those systems which may be subject to earthquakes.

3-5.3.1* General.   Sprinkler systems shall be protected to minimize or prevent pipe breakage where subject to earthquakes in accordance with the requirements of 3-5.3.

Exception: Alternative methods of providing earthquake protection of sprinkler systems based on a dynamic seismic analysis certified by a registered professional engineer such that system performance will be at least equal to that of the building structure under expected seismic forces.

A-3-5.3.1   Sprinkler systems are protected against earthquake damage by means of the following:

(a) Stresses that would develop in the piping due to differential building movement are minimized through the use of flexible joints or clearances.

(b) Bracing is used to keep the piping fairly rigid when supported from a building component expected to move as a unit, such as a ceiling.

Areas known to have a potential for earthquakes have been identified in building code and insurance maps. An example of such a map is shown in Figure A-3-5.3.1.

Flexibility is provided by requirements for flexible couplings and swing joints in 3-5.3.2 and 3-5.3.3, and clearances in 3-5.3.4. Dampening is provided by the requirements for limited sway bracing in 3-5.3.5.

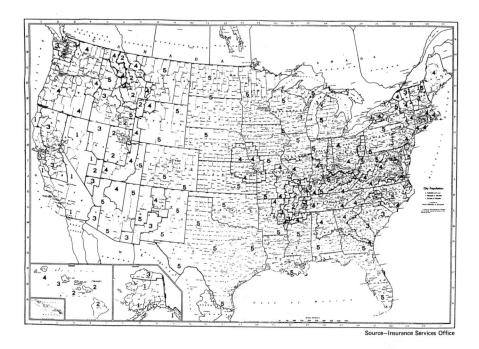

EARTHQUAKE ZONES

1 — Maximum potential for earthquake damage
2 — Reasonable potential
3 — Slight potential
4 and 5 — Earthquake protection not required

Figure A-3-5.3.1   Seismic Map.

3-5.3.2* Couplings.   Listed flexible pipe couplings joining grooved end pipe shall be provided as flexure joints to allow individual sections of piping 3½ in. (89 mm) or larger to move differentially with the individual sections of the building to which it is attached. Couplings shall be arranged to coincide with structural separations within a building. They shall be installed:

(a) Within 24 in. (610 mm) of the top and bottom of all risers.

Exception No. 1: In risers less than 3 ft (0.9 m) in length flexible couplings may be omitted.

Exception No. 2: In risers 3 to 7 ft (0.9 to 2.1 m) in length, one flexible coupling is adequate.

(b) At the ceiling of each story in multistory buildings.

(c) On one side of concrete or masonry walls within 3 ft (0.9 m) of the wall surface.

(d) At or near building expansion joints.

(e) At the top of drops to hose lines, rack sprinklers, and mezzanines, regardless of pipe size.

(f) At the top of drops exceeding 15 ft (4.6 m) in length to sprinklers or portions of systems, regardless of pipe size.

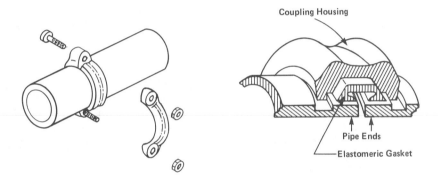

Figure 3.1. Flexible pipe coupling.

Figure 3.1 shows one type of flexible pipe coupling.

A-3-5.3.2 Strains on sprinkler piping can be greatly lessened and, in many cases, damage prevented by increasing the flexibility between major parts of the sprinkler system. One part of the piping should never be held rigidly and another part allowed to move freely without provision for relieving the strain. Flexibility can be provided by use of listed flexible couplings, by joining grooved end pipe at critical points, and by allowing clearances at walls and floors.

Tank or pump risers should be treated the same as sprinkler risers for their portion within a building. The discharge pipe of tanks on buildings should have a control valve above the roof line so any pipe break within the building can be controlled.

Piping 3 in. (76 mm) or smaller in size is pliable enough so that flexible couplings are not usually necessary. A flexible coupling is a mechanical coupling or fitting that permits some angular displacement, axial displacement, and rotation of the piping without failure of the pipe or fitting. "Rigid-type" mechanical couplings that do not permit movement at the grooved connections are not considered flexible couplings.

A-3-5.3.2(d)   A building expansion joint is usually a bituminous fiber strip used to separate blocks or units of concrete to prevent cracking due to expansion as a result of temperature changes. In this case, the flexible coupling required on one side by 3-5.3.2(d) will suffice.

For seismic separation joints, considerably more flexibility is needed, particularly for piping above the first floor. The following figure shows a method of providing additional flexibility through the use of swing joints.

3-5.3.3* Swing Joints.   Swing joints assembled with flexible fittings shall be installed where sprinkler piping, regardless of size, crosses building seismic joints.

A-3-5.3.3   Plan and elevation views of a swing joint assembled with flexible elbows are shown in Figure A-3-5.3.3.

A seismic swing joint is considered to be an assembly of fittings, pipe, and couplings, or an assembly of pipe and couplings that permits movement in all directions. The extent of permitted movement should be sufficient to accommodate calculated differential motions during earthquakes. In lieu of calculations, permitted movement can be made at least twice the actual separations, at right angles to the separation as well as parallel to it.

Figure A-3-5.3.3 is shown on page 96.

3-5.3.4* Clearance.   Clearance shall be provided around all piping extending through walls, floors, platforms, and foundations, including drains, fire department connections, and other auxiliary piping.

(a) Minimum clearance on all sides shall be not less than 1 in. (25 mm) for pipes 1 in. (25 mm) through 3½ in. (89 mm) and 2 in. (51 mm) for pipe sizes 4 in. (102 mm) and larger.

Exception No. 1:   When clearance is provided by a pipe sleeve, a nominal diameter 2 in. (51 mm) larger than the nominal diameter of the pipe is acceptable for pipe sizes 1 in. (25 mm) through 3½ in. (89 mm) and the clearance provided by a pipe sleeve of nominal diameter 4 in. (102 mm) larger than the nominal diameter of the pipe is acceptable for pipe sizes 4 in. (102 mm) and larger.

Exception No. 2:   No clearance is necessary for piping passing through gypsum board or equally frangible construction which is not required to have a fire-resistance rating.

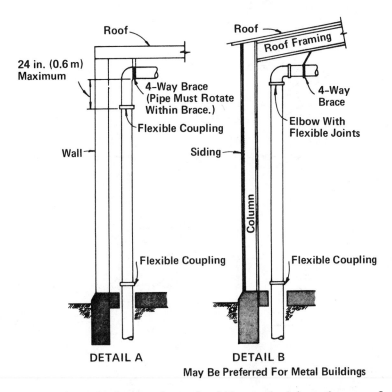

DETAIL A

DETAIL B
May Be Preferred For Metal Buildings

Note to Detail A: The four-way brace should be attached above the upper flexible coupling required for the riser, and preferably to the roof structure if suitable. The brace should not be attached directly to a plywood or metal deck.

Figure A-3-5.3.2(a)   Riser Details.

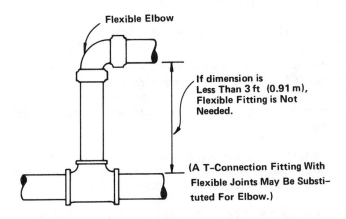

Figure A-3-5.3.2(b)   Detail at Short Riser.

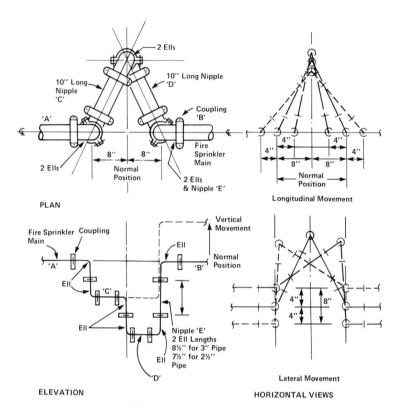

PLAN

Longitudinal Movement

ELEVATION

HORIZONTAL VIEWS

Lateral Movement

Metric Equivalent
1″ = 25.4 mm
1′ = 0.305 m

NOTE: The figure illustrates an 8-in. separation crossed by pipes up to 4 in. in nominal diameter. For other separation distances and pipe sizes, lengths and distances should be modified proportionally.

Figure A-3-5.3.3   Seismic Flexible Joint.

Exception No. 3:   No clearance is necessary if flexible couplings or swing joints are located within 1 ft (0.3 m) of each side of a wall.

(b) When required the clearance shall be filled with a flexible material such as mastic.

A-3-5.3.4 While clearances are necessary around the sprinkler piping to prevent breakage due to building movement, suitable provision should also be made to prevent passage of water, smoke, or fire.

Drains, fire department connections, and other auxiliary piping connected to risers should not be cemented into walls or floors; similarly, pipes that pass horizontally through walls or foundations should not be cemented solidly or strains will accumulate at such points.

When risers or lengths of pipe extend through suspended ceilings, they should not be fastened to the ceiling framing members.

Although B-3-1.1 recommends thimbles or sleeves for sprinkler piping passing through floors of concrete or waterproof construction to prevent floor leakage, it exempts areas subject to earthquakes. Adequate clearance must be provided to avoid damage to the sprinkler system in the event of an earthquake. Pipe sleeves are a means to that end but not the exclusive means.

3-5.3.5 Sway Bracing.

Sway bracing is provided for feed and cross mains, to hold them fairly rigid against a ceiling or roof assembly expected to move as a unit during an earthquake. The braces are required to withstand a horizontal force equal to 50 percent of the weight of the water-filled pipe (horizontal acceleration $A_H = 0.5$ g).

The hangers provided on branch lines will sufficiently restrict movement of that piping.

Lateral braces and longitudinal braces are both "two-way" braces in that they prevent piping from translating back and forth in one direction. "Four-way bracing" requires the simultaneous effect of lateral and longitudinal braces.

While the forces applied to the piping in the event of an earthquake might be both vertical and horizontal, it is expected that the normal system hangers can adequately handle the vertical loads. Consideration should nevertheless be given to the possibility of vertical loads acting upward on the piping.

The standard contains requirements for both lateral (perpendicular to the piping) and longitudinal (parallel to the piping) horizontal braces. The normal maximum spacing of lateral braces [50 ft (15.2 m)] is based on the strength of the piping as a beam under the uniform load of its expected horizontal "weight." Longitudinal braces are required at a maximum spacing of 80 ft (24 m). Unlike the requirements for flexible couplings, braces are required for feed and cross mains regardless of their size.

As with regular hangers, sway braces must be directly attached to the building structure. Fasteners and structural elements at the points of connection must be adequate to handle the intended loads.

Branch lines are not required to be braced except where movement could damage other equipment, since the smaller piping used for branches is capable of considerable movement without damage. However, a wraparound hanger or other approved means is required at the end of each branch line to prevent excessive movement.

3-5.3.5.1* Both lateral and longitudinal sway braces shall be sized and fastened such that the horizontal loads assigned to the braces in Table 3-5.3.5.1(1) do not exceed the allowable loads on the braces as shown in Table 3-5.3.5.1(2) and the allowable loads on fasteners as shown in Table 3-5.3.5.1(3). Sway bracing shall be tight and concentric. All parts and fittings of a brace shall lie in a straight line to avoid eccentric loadings on fittings and fasteners. For longitudinal braces only, the brace may be connected to a tab welded to the pipe in conformance with 3-7.2.

Exception: In lieu of using Table 3-5.3.5.1(1), horizontal loads for braces may be determined by analysis. Sway braces shall be designed to withstand a force in tension or compression equivalent to not less than half the weight of water-filled piping. For lateral braces, the load shall include all branch lines and mains within the zone of influence of the brace. For longitudinal braces, the load shall include all mains within the zone of influence of the brace. For individual braces the slenderness ratio l/r shall not exceed 200, where l is the length of the brace and r is the least radius of gyration, both in inches.

A-3-5.3.5.1 Sway Bracing.

Location of Bracing.   [*See Figure A-3-5.3.5.1(a).*]

Figure A-3-5.3.5.1(a) is shown on page 104.

Two-way braces are either longitudinal or lateral depending on their orientation with the axis of the piping. [*See Figures A-3-5.3.5.1(a), (b), (c), and (d)*]. The simplest form of two-way brace is a piece of steel pipe or angle. Because the brace must act in both compression and tension, it is necessary to size the brace to prevent buckling.

An important aspect of sway bracing is its location.

In Building 1, the relatively heavy main will pull on the branch lines when shaking occurs. If the branch lines are held rigidly to the roof or floor above, the fittings can fracture due to the induced stresses.

Bracing should be on the main as indicated at Location B. With shaking in the direction of the arrows, the light branch lines will be held at the fittings. When a branch line can pound against a piece of equipment, such as a space heater or a structural member, a lateral brace should be installed on the branch line to help prevent rupture.

A four-way brace is indicated at Location A. [*Also see Figure A-3-5.3.2(a).*] This keeps the riser and main lined up and also prevents the main from shifting.

Table 3-5.3.5.1(1) Assigned Load Table
(Based on half the weight of the water-filled pipe)

Spacing of Lateral Braces (ft)	Spacing of Longitudinal Braces** (ft)	Assigned Load for Pipe Size to be Braced (lb)						
		2	2½	3	4	5	6	8
10	20	380	395	410	435	470	655	915
20	40	760	785	815	870	940	1305	1830
25	50	950	980	1020	1090	1175	1630	2290
30	60	1140	1180	1225	1305	1410	1960	2745
40	80	1515	1570	1630	1740	1880	2610	3660
50*		1895	1965	2035	2175	2350	3260	4575

*Permitted only under Exception No. 4 to 3-5.3.5.4.
**If branch lines are provided with lateral bracing or hung with U-hooks bent out at least 10 degrees from vertical, half the assigned load may be used for longitudinal braces.

Table 3-5.3.5.1(2)

Shape and Size	Least Radius of Gyration	Maximum Length for l/r—200	Maximum Horizontal Load (lb)		
			30° Angle From Vertical	45° Angle From Vertical	60° Angle From Vertical
Pipe (Schedule 40)	$=\dfrac{\sqrt{r_0^2 + r_i^2}}{2}$				
1 in.	.42	7'0"	1767	2500	3061
1¼ in.	.54	9'0"	2393	3385	4145
1½ in.	.623	10'4"	2858	4043	4955
2 in.	.787	13'1"	3828	5414	6630
Pipe (Schedule 10)	$=\dfrac{\sqrt{r_0^2 + r_i^2}}{2}$				
1 in.	.43	7'2"	1477	2090	2559
1¼ in.	.55	9'2"	1900	2687	3291
1½ in.	.634	10'7"	2194	3103	3800
2 in.	.802	13'4"	2771	3926	4803
Angles					
1½ × 1½ × ¼	.292	4'10"	2461	3481	4263
2 × 2 × ¼	.391	6'6"	3356	4746	5813
2½ × 2 × ¼	.424	7'0"	3792	5363	6569
2½ × 2½ × ¼	.491	8'2"	4257	6021	7374
3 × 2½ × ¼	.528	8'10"	4687	6628	8118
3 × 3 × ¼	.592	9'10"	5152	7286	8923
Rods	$=\dfrac{r}{2}$				
⅜	.094	1'6"	395	559	685
½	.125	2'6"	702	993	1217
⅝	.156	2'7"	1087	1537	1883
¾	.188	3'1"	1580	2235	2737
⅞	.219	3'7"	2151	3043	3726
Flats	= 0.29 h (where h is smaller of two side dimensions)				
1½ × ¼	.0725	1'2"	1118	1581	1936
2 × ¼	.0725	1'2"	1789	2530	3098
2 × ⅜	.109	1'9"	2683	3795	4648

Table 3-5.3.5.1(3)
Maximum Loads for Various Types of Fasteners to Structure

NOTE: Loads (given in pounds) are keyed to vertical angles of braces and orientation of connecting surface. These values are based on concentric loadings of the fastener. Use figures to determine proper reference within table. For angles between those shown, use most restrictive case. Braces should not be attached to light structure members.

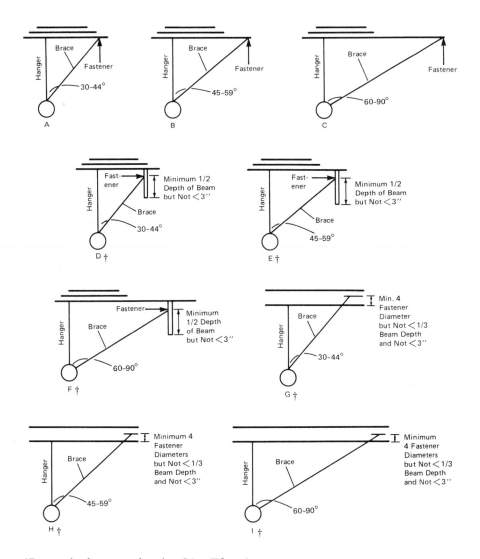

†For wooden beams not less than 3 in. (76 mm).

Table 3-5.3.5.1(3) (Continued)

Lag Screws in Wood (load perpendicular to grain—holes predrilled using good practice)
Shank Diameter of Lag (in.)

		3/8								1/2						
		A	B,E	C	D	F	G	H	I	A	B,E	C	D	F	G	H	I
Length	3	304	325	292	168	526	230	324	400	366	—	—	—	834	—	—	—
Under	4	392	354	317	183	678	250	352	435	473	509	456	264	818	360	507	626
Head	5	476	375	336	194	824	265	373	461	582	545	488	282	1008	385	542	670
(inches)	6	564	382	342	196	976	270	380	470	689	559	501	209	1192	395	556	687
	8	—	—	—	—	—	—	—	—	905	573	513	296	1586	405	570	704

		5/8								7/8						
		A	B,E	C	D	F	G	H	I	A	B,E	C	D	F	G	H	I
Length	3	410	—	—	—	716	—	—	—	487	—	—	—	843	—	—	—
Under	4	538	—	—	—	532	—	—	—	548	—	—	—	1122	—	—	—
Head	5	687	728	653	277	1154	515	725	896	813	—	—	—	1407	—	—	—
(inches)	6	791	778	697	403	1360	550	775	957	971	—	—	—	1630	—	—	—
	8	1044	806	723	416	1807	570	803	991	1297	1365	1223	685	2244	965	1359	1678

Through Bolts in Wood (load perpendicular to grain)
Diameter of Bolt (in.)

		3/8						1/2					
		ABCE	D	F	G	H	I	ABCE	D	F	G	H	I
Length	1½	300	173	519	150	211	261	340	197	589	170	239	296
of	2	370	214	641	185	261	322	420	243	727	210	296	365
Bolt	2½	460	266	796	230	324	400	550	318	952	275	387	478
in	3	480	277	831	240	338	417	630	364	1091	315	444	548
Timber	3⅝	460	268	797	230	324	400	720	416	1247	360	507	626
(in.)	5½	—	—	—	—	—	—	680	393	1177	340	479	591

		5/8						7/8					
		ABCE	D	F	G	H	I	ABCE	D	F	G	H	I
Length	1½	390	225	675	195	275	339	470	272	614	235	331	409
of	2	470	272	814	235	331	409	580	335	1004	290	408	504
Bolt	2½	620	358	1074	310	437	539	760	439	1316	380	535	661
in	3	710	410	1229	355	500	617	870	503	1506	435	613	757
Timber	3⅝	850	491	1472	425	599	739	1050	607	1818	525	739	913
(in.)	5½	1020	590	1766	510	718	887	1580	913	2736	790	1113	1374

For SI Units: 1 in. = 25.4 mm.

Table 3-5.3.5.1(3) (Continued)

Expansion Shields in Concrete
Diameter of Bolt (in.)

		3/8								1/2						
		A	B,E	C	D	F	G	H	I	A	B,E	C	D	F	G	H	I
Min. Depth of Hole (in.)	2½	498	962	1173	678	962	925	1303	1609	—	—	—	—	—	1638	2306	2848
	3¼	—	—	—	—	—	925	1303	1609	923	1782	2076	1200	1597	1638	2306	2848
	3¾	—	—	—	—	—	925	1303	1609	—	—	—	—	—	1638	2306	2848
	4½	—	—	—	—	—	925	1303	1609	—	—	—	—	—	1638	2306	2848

		5/8								7/8						
		A	B,E	C	D	F	G	H	I	A	B,E	C	D	F	G	H	I
Min. Depth of Hole (in.)	2½	—	—	—	—	—	2080	2930	3617	—	—	—	—	—	2970	4113	5078
	3¼	—	—	—	—	—	2080	2930	3617	—	—	—	—	—	2970	4113	5078
	3¾	1480	2857	2637	1524	2581	2080	2930	3617	—	—	—	—	—	2970	4113	5078
	4½	—	—	—	—	—	2080	2930	3617	3070	4130	3702	2139	5312	2970	4113	5078

Connections to Steel (values assume bolt perpendicular to mounting surface)
Diameter of Unfinished Steel Bolt (in.)

1/4								3/8							
A	B,E	C	D	F	G	H	I	A	B,E	C	D	F	G	H	I
400	500	600	300	650	325	458	565	900	1200	1400	800	1550	735	1035	1278

1/2								5/8							
A	B,E	C	D	F	G	H	I	A	B,E	C	D	F	G	H	I
1600	2050	2550	1450	2850	1300	1830	2260	2500	3300	3950	2250	4400	2045	2880	3557

For SI Units: 1 in. = 25.4 mm.

In Building 1, the branch lines are flexible in a direction parallel to the main, regardless of building movement. The heavy main cannot shift under the roof or floor, and it also steadies the branch lines.

While the main is braced, the flexible couplings on the riser allow the sprinkler system to move with the floor or roof above, relative to the floor below.

Figures A-3-5.3.5.1(b), (c), and (d) show typical locations of sway bracing for pipe schedule, gridded, and looped sprinkler systems.

See Figures A-3-5.3.5.1(b), (c), and (d) on pages 105 and 106.

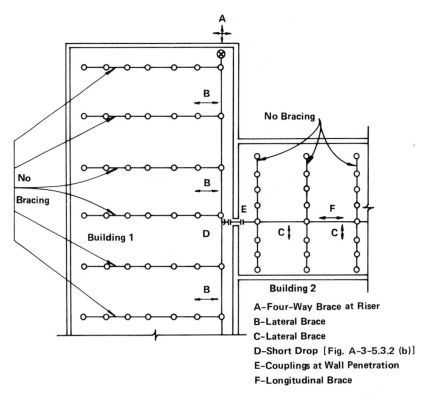

Figure A-3-5.3.5.1(a)   Earthquake Protection for Sprinkler Piping.

Listed devices permitting connection of braces to both the pipe and the building structure are available and are recommended. However, alternate means of attachment capable of handling the expected loads are acceptable.

Connection of the brace to the pipe can be made with a pipe clamp or U-bolt. One bolt of the pipe clamp can pass through a flattened end of pipe or one leg of an angle. (The other leg and filet of the angle can be cut away.) Pipe rings should be avoided because they result in a loose fit. Once the pipe is able to vibrate within a loose fitting, the bolts in the ring assembly can be fractured.

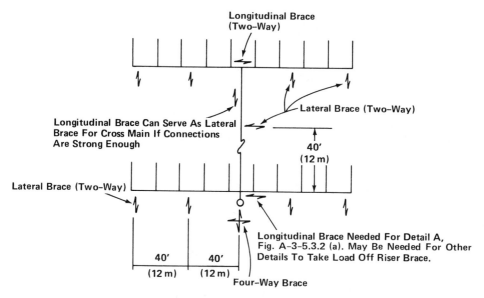

Figure A-3-5.3.5.1(b)   Typical Location of Bracing on a Pipe Schedule System.

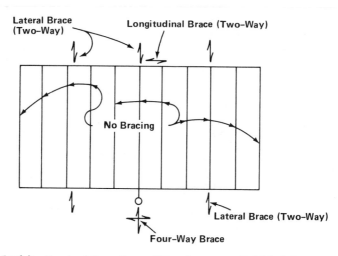

Figure A-3-5.3.5.1(c)   Typical Location of Bracing on a Gridded System.

The brace can be attached to the structural system directly through a leg of an angle or a flattened portion of pipe. Where dimensions are tight or some play must be allowed, a special fitting can be used. [*See Figure A-3-5.3.5.1(b)*]. This threads on an end of pipe. Rotation of the flat around the bolt allows play in the angle of the brace without sacrificing snugness.

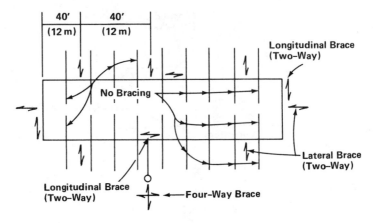

Figure A-3-5.3.5.1(d)   Typical Location of Bracing on a Looped System.

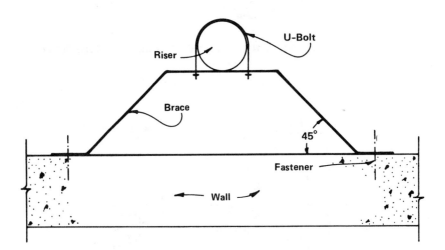

Figure A-3-5.3.5.1(e)   Detail of Four-Way Brace at Riser.

Some adjustment can be provided in a pipe brace by use of a left-hand/right-hand coupling. For all threaded connections, holes or other means should be provided to permit indication that sufficient thread is engaged.

To properly size and space braces, it is necessary to employ the following steps:

(a) Based on the distance of mains from the structural members that will support the braces, choose brace shapes and sizes from Table 3-5.3.5.1(2) such that the maximum slenderness ratios l/r do not exceed 200. The angle of the braces from the vertical should be at least 30 degrees, and preferably 45 degrees or more.

(b) Tentatively space lateral braces at 40 ft (12 m) maximum distances along mains and tentatively space longitudinal braces at 80 ft (24 m) maximum distances along mains. Lateral braces should meet the piping at right angles, and longitudinal braces should be aligned with the piping.

(c) Determine the total load tentatively applied to each brace in accordance with the examples shown in Figure A-3-5.3.5.1(h) and the following:

Figure A-3-5.3.5.1(h) is shown on page 110.

1. For the loads on lateral braces on cross mains, add one-half the weight of branch to one-half the weight of the portion of the cross main within the zone of influence of the brace. [*See examples 1, 3, 6, and 7 in Figure A-3-5.3.5.1(h).*]

2. For the loads on longitudinal braces on cross mains, consider only one-half the weight of the cross mains and feed mains within the zone of influence. [*See examples 2, 3, 5, and 8 in Figure A-3-5.3.5.1(h).*]

3. For the four-way brace at the riser, add the longitudinal and lateral loads within the zone of influence of the brace. [*See examples 2, 3, 4, 5, 7, and 8 in Figure A-3-5.3.5.1(h).*]

Use the information on weights of water-filled piping contained within Table A-3-5.3.5.1.

(d) If the total expected loads are less than the maximums permitted in Table 3-5.3.5.1(2) for the particular brace and orientation, go on to step (e). If not, add additional braces to reduce the zones of influence of overloaded braces.

(e) Check that fasteners connecting the braces to structural supporting members are adequate to support the expected loads on the braces in accordance with Table 3-5.3.5.1(3). If not, again add additional braces or additional means of support.

The ratio of length of brace to radius of gyration of brace cross section is kept as close to 200 as practical in order to provide dampening for system vibrations. The value of 200 is chosen as the traditional limit for determining the maximum allowable fiber-stress for secondary compression members. Above 200 there would be a concern for buckling and eventual failure of the brace.

3-5.3.5.2 Longitudinal sway bracing spaced at a maximum of 80 ft (24 m) on center shall be provided for feed and cross mains.

3-5.3.5.3* Tops of risers shall be secured against drifting in any direction, utilizing a four-way sway brace.

Table A-3-5.3.5.1
Piping Weights for Determining Horizontal Load

Schedule 40 Pipe	Weight of Water-Filled Pipe (lb per ft)	½ Weight of Water-Filled Pipe (lb per ft)
1	2.05	1.03
1¼	2.93	1.47
1½	3.61	1.81
2	5.13	2.57
2½	7.89	3.95
3	10.82	5.41
3½	13.48	6.74
4	16.40	8.20
5	23.47	11.74
6	31.69	15.85
8*	47.70	23.85
Schedule 10 Pipe		
1	1.81	0.91
1¼	2.52	1.26
1½	3.04	1.52
2	4.22	2.11
2½	5.89	2.95
3	7.94	3.97
3½	9.78	4.89
4	11.78	5.89
5	17.30	8.65
6	23.03	11.52
8	40.08	20.04

*Schedule 30

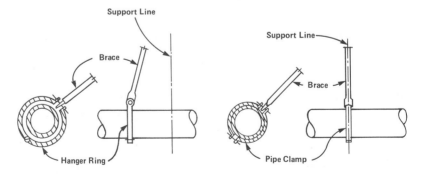

Figure 3.2.   Details of connection to brace pipe. Arrangement on left is not acceptable because hanger ring is not snug and brace is not perpendicular to pipe. Arrangement on right is acceptable.

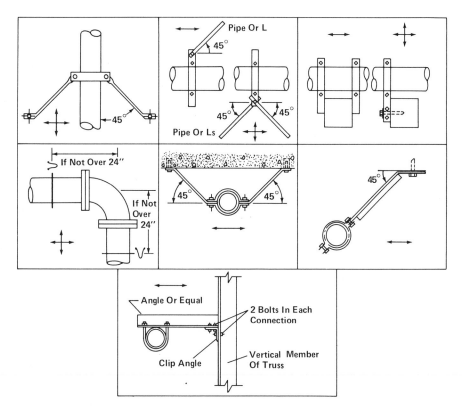

Figure A-3-5.3.5.1(f) Acceptable Types of Sway Bracing.

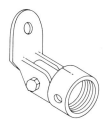

Figure A-3-5.3.5.1(g) Special Fitting.

The four-way brace at the top of the riser keeps the riser and main lined up and prevents the main from shifting. For multistory buildings, the intent is to require the four-way brace only at the highest floor. Where a combined sprinkler/standpipe riser extends through the ceiling to a roof outlet, the four-way brace located below the roof is sufficient.

A-3-5.3.5.3 The four-way brace provided at the riser may also provide longitudinal and lateral bracing for adjacent mains.

3-5.3.5.4 Lateral sway bracing spaced at a maximum of 40 ft (12 m) on center shall be provided for feed and cross mains.

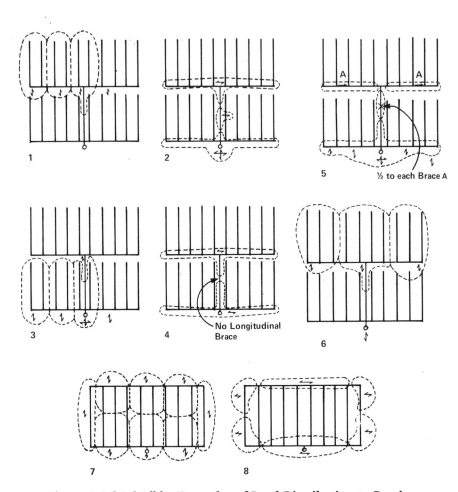

Figure A-3-5.3.5.1(h)   Examples of Load Distribution to Bracing.

Exception No. 1: Lateral sway bracing may be omitted on pipes individu- ally supported by rods less than 6 in. (152 mm) long.

Exception No. 2: U-type hangers used to support the mains may be used to satisfy the requirements for lateral sway bracing provided the legs are bent out at least 10 degrees from the vertical.

Exception No. 3: When flexible couplings are installed on mains other than as required in 3-5.3.2, a lateral brace shall be provided within 24 in. (610 mm) of every other coupling, but not more than 40 ft (12 m) on center.

Exception No. 4: When building primary structural members exceed 40 ft (12 m) on center, lateral braces may be spaced up to 50 ft (15.2 m) on center.

Short rods are expected to limit the lateral movement of mains, thus eliminating the need for lateral bracing. A hole in a beam which is used to support the pipe will give the same intended result as either Exception No. 1 or No. 2.

Formal Interpretation

Question: What is the maximum length at which a U-hook will satisfy sway bracing requirements as outlined in 3-5.3.5.4 Exception No. 2 of NFPA 13?

Answer: There is no maximum length restriction.

Formal Interpretation

Question 1: Is it the intent of 3-5.3.5 of NFPA 13 to require a four-way earthquake brace only at the top of risers in a multi-story building?

Answer: A four-way brace is required at the top of the riser in all buildings subject to earthquake damage. Four-way bracing is not required for intermediate floors in multi-story buildings.

Question 2: Is it the intent of 3-5.3.5 of NFPA 13 to require a four-way earthquake brace above the roof line when the sprinkler riser is used to supply 2½ in. hose outlets on each floor and above the roof?

Answer: No. The four-way brace just below the roof slab is adequate.

The intent of Exception No. 3 is to prevent excessive movement of the piping, possibly resulting in "bellows" or accordion effects. Additional bracing is not required where rigid type mechanical couplings are used.

3-5.3.5.5 Bracing shall be attached directly to feed and cross mains.

The intent is to disallow bracing of the feed and cross mains through the branch lines.

3-5.3.5.6 A length of pipe shall not be braced to sections of the building that will move differentially.

3-5.3.5.7 The last length of pipe at the end of a feed or cross main shall be provided with a lateral brace. Lateral braces may also act as longitudinal braces if they are within 24 in. (610 mm) of the center line of the piping braced longitudinally.

3-5.3.5.8   Sway bracing is not required for branch lines.

Exception No. 1:   The end sprinkler on a line shall be restrained against excessive movement by use of a wraparound U-hook (see Figure A-3-10.1) or by other approved means.

Exception No. 2:   Branch lines 2½ in. (64 mm) or larger shall be provided with lateral bracing in accordance with 3-5.3.5.4.

The wraparound U-hook or similar restraint at the end of each branch line is intended to prevent the branch line from whipping and possibly bouncing out of hangers such as standard U-hooks. Acceleration of larger branch lines [2½ in. (64 mm)] during seismic activity will warrant additional protection.

3-5.3.5.9   C-type clamps used to attach hangers to the building structure in areas subject to earthquakes shall be equipped with a retaining strap or other approved means to prevent movement. (*See Figure A-3-10.1.*)

3-5.3.5.10   C-type clamps, with or without retaining straps, shall not be used to attach braces to the building structure.

3-6 Drainage.

3-6.1 Pitching of Piping for Drainage.

3-6.1.1*   All sprinkler pipe and fittings shall be so installed that the system may be drained.

A-3-6.1.1   All piping should be arranged where practicable to drain to the main drain valve.

Ideally the entire system is pitched to drain at the main drain. Where this cannot be done, auxiliary drainage facilities as required in this section are sized and arranged with consideration of the type of system and the volume of trapped piping.

3-6.1.2   On wet-pipe systems, sprinkler pipes may be installed level. Trapped piping shall be drained in accordance with 3-6.3.

In pipe schedule systems the reduction in pipe sizes from the supply to the branch lines has the effect of a slight pitch when the piping is installed level because of the slope of the reducing fittings that results from reduction in pipe size.

3-6.1.3   On those portions of preaction systems subject to freezing and on dry-pipe systems, sprinkler pipe on branch lines shall be pitched at least ½-in. in 10 ft (4 mm/m) and the pipe of cross and feed mains shall be given a pitch of not less than ¼-in. in 10 ft (2 mm/m). A pitch of ¾ to 1-in. (19 to 25-mm) shall be provided for short branch lines and ½-in. in 10 ft (4 mm/m) for cross and feed mains in refrigerated areas and in buildings of light construction that may settle under heavy loads.

Dry-pipe systems are installed in areas subject to freezing. Proper drainage of such systems becomes very critical due to potential freezing of a pipe network after a system trip.

3-6.2 System, Main Drain, or Sectional Drain Connections.   [*See Figures 3-6.2 and A-3-4.1.1(b)*.]

Formal Interpretation

Question: Would the pressure gage arrangement shown in Figure 2 be considered an acceptable alternate to that shown in Figure 3-6.2 and Figure A-2-9.1?

Answer: No. The pressure gage location, as illustrated, will not give a true residual reading. It will indicate an excessive pressure drop.

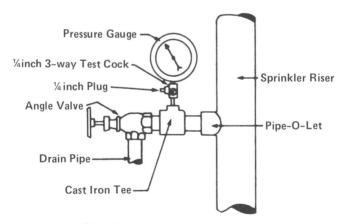

Drain Connection for Sprinkler Riser

Figure 2.

This section describes the facilities needed to drain any portion of a system in an efficient manner without undue risk of water damage. The arrangement of both main and auxiliary drains will expedite the manner in which a system may be taken out of service for maintenance or repairs.

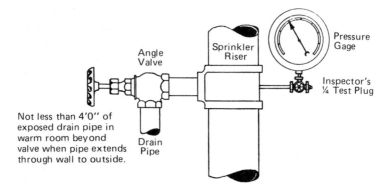

For SI Units: 1 in. = 25.4 mm; 1 ft = 0.3048 m.

Figure 3-6.2   Drain Connection for System Riser.

3-6.2.1   Provisions shall be made to properly drain all parts of the system.

3-6.2.2   Drain connections for systems supply risers and mains shall be sized as shown in Table 3-6.2.2.

Table 3-6.2.2

Riser or Main Size	Size of Drain Connection
Up to 2 in.	¾ in. or larger
2½ in., 3 in., 3½ in.	1¼ in. or larger
4 in. and larger	2 in. only

3-6.2.3   Each interior sectional control valve shall be provided with a drain connection sized as shown in Table 3-6.2.2 so as to drain that portion of the system controlled by the sectional valve. These drains shall discharge either outside or to a drain connection.

Piping controlled by sectional valves represents a significant segment of the sprinkler system. The drain valves provided with such sectional control valves serve the same function for that part of the system as the main drain does for the entire system. The use of express drains in high-rise buildings is one method of disposing of water from upper floors.

3-6.2.4   The test valves required by 2-9.1 may be used as main drain valves.

3-6.3 Auxiliary Drains.

Drainage facilities are necessary for trapped piping. The type of facility, its size, and its arrangement are dependent on the type of system and the volume of trapped piping.

3-6.3.1 Auxiliary drains shall be provided when a change in piping direction prevents drainage of sections of branch lines or mains through the main drain valve.

3-6.3.2 Auxiliary Drains for Wet-Pipe Systems.

3-6.3.2.1 When the capacity of trapped sections of pipes is 5 gal (18.9 L) or less, the auxiliary drain shall consist of a nipple and cap or brass plug not less than ¾ in. (19 mm) in size.

Exception: Auxiliary drains are not required for piping that can be drained by removing a single pendent sprinkler.

It is rarely necessary to drain a trapped section of piping in a wet system. Therefore, a nipple and cap or a plug is adequate when the volume of the trapped section is 5 gal (18.9 L) or less. A nipple and cap are preferable to a plug because of the comparative ease with which they can be removed. A plug must be brass in order to make it easier to remove.

The Exception recognizes that the small volume of water trapped in one pendent sprinkler can be drained as readily through the sprinkler connection as through a plug. It would be unrealistic to require a drain where pendent sprinklers installed below a ceiling are supplied directly from branch lines concealed above the ceiling.

3-6.3.2.2 When the capacity of isolated trapped sections of pipe is more than 5 gal (18.9 L) and less than 50 gal (189 L), the auxiliary drain shall consist of a valve not smaller than ¾ in. (19 mm) in size and a plug, at least one of which shall be brass. In lieu of a plug, a nipple and cap may be used.

A valved drain is required when the trapped section in a wet system exceeds 5 gal (18.9 L) so that the trapped piping may be emptied in a reasonable length of time without water damage.

3-6.3.2.3* When the capacity of isolated trapped sections of pipe is 50 gal (189 L) or more, the auxiliary drain shall consist of a valve not smaller than 1 in. (25 mm), piped to an accessible location.

A-3-6.3.2.3 An example of a suitable location would be a valve located approximately 7 ft (2 m) above the floor level to which a hose could be connected to discharge the water in an acceptable manner.

3-6.3.2.4 Tie-in drains are not required on wet-pipe systems.

Tie-in drains are cross connections of the ends of trapped branch lines that are piped to a single drain valve. They are not required on wet systems because such piping is very seldom drained.

3-6.3.3 Auxiliary Drains for Dry-Pipe Systems.

3-6.3.3.1 When capacity of trapped sections of pipe is 5 gal (18.9 L) or less, the auxiliary drain shall consist of a valve not smaller than ¾ in. (19 mm) and a plug, at least one of which shall be brass. In lieu of a plug, a nipple and cap may be used.

Exception: Auxiliary drains are not required for a drop nipple when installed in accordance with 5-2.2.

Formal Interpretation

Question: Is the 5-gal limit intended to apply to each individual trapped section of pipe?

Answer: Yes.

Because dry-pipe systems are subject to freezing, trapped areas of 5 gal (18.9 L) or less must be provided with drain valves to remove water that has entered the system either because of dry-pipe valve tripping or condensation of moisture from the pressurized air in the system. An auxiliary drain is not required when pendent sprinklers in a heated area are supplied by piping that is also in a heated area.

3-6.3.3.2 When capacity of isolated trapped sections of pipe is more than 5 gal (18.9 L), the auxiliary drain shall consist of two 1-in. (25-mm) valves, and one 2-in. by 12-in. (51-mm by 305-mm) condensate nipple or equivalent, accessibly located. (*See Figure 3-6.3.3.*)

Also see commentary to 3-6.3.2.4 and A-3-6.4.5.
The condensate nipple illustrated in Figure 3-6.3.3 or its equivalent is required for each section of trapped piping in a

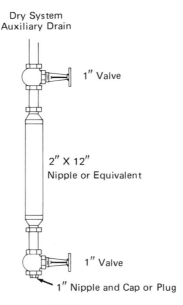

Dry System
Auxiliary Drain

1″ Valve

2″ X 12″
Nipple or Equivalent

1″ Valve

1″ Nipple and Cap or Plug

For SI Units: 1 in. = 25.4 mm; 1 ft = 0.3048 m.

Figure 3-6.3.3   Dry System Auxiliary Drain.

dry-pipe system with more than 5-gal (18.9-L) capacity. The condensate nipple is capable of collecting and removing moisture from the system while minimizing the potential for excessive loss of air pressure and possible unplanned operation of a dry-pipe valve. The upper valve is normally open, allowing moisture to enter the chamber, while the lower valve is closed and sealed to avoid leaks. To drain the chamber, the upper valve is closed to temporarily isolate it from the system and the lower valve is opened to remove the moisture. Where inside and outside temperatures are extreme, it may be desirable to isolate the exterior drainpipe from the interior drainpipe with a threaded fitting. This will have the effect of limiting moisture build-up by checking the transfer of a high exterior ambient temperature to the cooler interior temperature of the drainpipe.

3-6.3.3.3* Tie-in drains shall be provided for multiple adjacent trapped branch lines and shall be a minimum of 1 in. (25 mm). Tie-in drain lines shall be pitched a minimum of ½ in. in 10 ft (4 mm/m).

A-3-6.3.3.3 The size of tie-in drain lines should be increased as necessary to provide efficient removal of trapped water. Consideration should be given to the volume of water trapped, elevation head available, and the time required to discharge water. Water or condensation or both that remains after initial drain down should not be permitted to collect since freezing may cause failure of the drain/system piping.

When two or more adjacent branch lines are trapped in a dry-pipe or a preaction system, the ends of the lines must be piped together and run to a low-point drain of at least 1 in. (25 mm) in size that is equipped with a drain valve and nipple and cap or plug to facilitate removal of moisture from the system.

It is extremely important that tie-in drains be properly pitched and that condensate be removed from them prior to freezing weather. Typically, tie-in drains consist of smaller sized piping connected to a rather large volume of piping and, with these smaller drains, freeze-ups are possible.

3-6.3.4 Auxiliary Drains for Preaction Systems.

3-6.3.4.1   When trapped sections of pipe are in areas subject to freezing, auxiliary drains shall conform to 3-6.3.3.

3-6.3.4.2   When trapped sections of pipe are in areas not subject to freezing, auxiliary drains shall consist of a valve not smaller than ¾ in. (19 mm) and a plug, at least one of which shall be brass. In lieu of a plug, a nipple and cap may be used.

Exception: Auxiliary drains are not required for piping that can be drained by removing a single pendent sprinkler when capacity of the trapped sections of pipe is 5 gal (18.9 L) or less.

A drain valve is required for trapped sections of preaction system piping of 5-gal (18.9-L) capacity or less that are not subject to freezing because water would need to be removed in order not to interfere with the supervision of the piping provided by the low-pressure air maintained on most preaction systems.

3-6.4 Discharge of Drain Valves.

3-6.4.1*   Direct interconnections shall not be made between sewers and sprinkler drains of systems supplied by public water. The drain discharge shall be in conformity with any health or water department regulations.

A-3-6.4.1   When possible, the main sprinkler riser drain should discharge outside the building at a point free from the possibility of causing water damage. When it is not possible to discharge outside the building wall, the drain should be piped to a sump, which in turn should discharge by gravity or be pumped to a waste water drain or sewer. The main sprinkler riser drain connection should be of a size sufficient to carry off water from the fully open drain valve while it is discharging under normal water system pressures. When this is not possible, a supplementary drain of equal size should be provided for test purposes with free discharge, located at or above grade.

The restriction regarding connection of sprinkler drains and sewers is intended to prevent any harmful element from entering the sprinkler system by way of the drain connection and then the public water system by way of the sprinkler system. The valve between the sprinkler system and the drain connection, and the check valve between the sprinkler system and the public water supply provide some protection, but this additional regulation is considered necessary to provide better assurance.

3-6.4.2 When drain pipes are buried underground, approved corrosion-resistant pipe shall be used.

3-6.4.3 Drain pipes shall not terminate in blind spaces under the building.

The drain discharge must be piped in a manner that allows the operator to ascertain that the drain is not obstructed and that it is not causing water damage.

3-6.4.4 When exposed, drain pipes shall be fitted with a turned down elbow.

The downturn elbow discourages the use of the drain piping as a refuse receptacle and minimizes the possibility of damage to property or wetting of passers-by.

3-6.4.5* Drain pipes shall be arranged so as not to expose any part of the sprinkler system to freezing conditions.

A-3-6.4.5 When exterior ambient temperatures are subject to freezing [32°F (0°C) or less], at least 4 ft (1.2 m) of pipe should be installed beyond the valve, in a warm room.

The recommended 4 ft (1.2 m) of pipe beyond the valve provides a frost break that protects water in the system from freezing due to exterior cold acting on the pipe. Additionally, a good maintenance program is required to properly remove all

moisture from trapped unheated areas in dry-pipe and preaction systems. (*See NFPA 13A, Recommended Practice for the Installation, Testing and Maintenance of Sprinkler Systems, 4-8.2.*)

3-7 Joining of Pipe and Fittings.

3-7.1 Threaded Pipe and Fittings.

3-7.1.1 All threaded fittings and pipe shall have threads cut to ANSI/ASME standard B1.20.1. Care shall be taken that the pipe does not extend into the fitting sufficiently to reduce the waterway.

Poor workmanship can result in threads that allow pipe protrusions to partially obstruct fitting openings. If such joints are permitted they will seriously restrict the flow in the system, greatly increase the pressure lost to friction, and thereby impair the operation of the system.

3-7.1.2* Steel pipe with wall thicknesses less than Schedule 30 [in sizes 8 in. (203 mm) and larger] or Schedule 40 [in sizes less than 8 in. (203 mm)] shall not be joined by threaded fittings, unless a threaded assembly has been investigated for suitability in automatic sprinkler installations and listed for this service.

This section relates back to 3-1.1.3 and in effect permits an exception allowing for new technology to develop an innovative threaded assembly for use with thinner-walled pipe and to submit it to the testing laboratories for listing.

A-3-7.1.2 Some steel piping material having lesser wall thickness than specified in 3-7.1.2 has been listed for use in sprinkler systems when joined with threaded connections. The service life of such products may be significantly less than that of Schedule 40 steel pipe and it should be determined if this service life will be sufficient for the application intended.

All such threads should be checked by the installer using working ring gages conforming to the Basic Dimensions of Ring Gages for USA (American) Standard Taper Pipe Threads, NPT, as per ANSI/ASME B1.20.1, Table 8.

Threading of listed thin-wall pipe requires careful workmanship and good quality control. (*See 3-5.2.3 and 3-10.1.11 Exception for additional information.*)
To allay concerns associated with the life expectancy of listed thin-wall pipe, part of its evaluation includes an examination of its corrosion ratio with respect to Schedule 40 steel pipe. These results are available in the listing information for a given product.

3-7.1.3   Joint compound or tape shall be applied to the threads of the pipe and not in the fitting.

Joint compound or tape that is applied to a fitting rather than to the threads of the pipe forms a ridge inside the pipe when the joint is made, reducing the inside diameter of the pipe and thereby adversely affecting the flow rate. The operation of systems other than wet systems involves a high-velocity flow when the valve trips. The impact of the water tends to break off chunks of joint compound and carry them in a mass to the opened sprinklers, where they tend to obstruct the orifices.

3-7.2* Welded Piping.

See Figures A-3-7.2(a) and (b) on page 122.

3-7.2.1   Welding methods that comply with all of the requirements of AWS D10.9, *Specification for Qualification of Welding Procedures and Welders for Piping and Tubing*, Level AR-3, are acceptable means of joining fire protection piping.

Welding of any sprinkler piping is subject to strict quality control procedures. Failure to adhere to the procedures outlined in this section may result in rejection of a system.

Formal Interpretation

Question 1:   Do welding methods and weld inspection conforming to the requirements of ANSI B31.1-1978 meet the intent of 3-7.2.1?

Question 2:   Does qualification of welding procedures, welders, and welding operators to ANSI B31.1-1977, *Code for Power Piping,* and the requirements of the ASME *Boiler and Pressure Vessel Code* referenced therein meet the intent of 3-7.2.11?

Answer: Yes. The standard describes the minimum acceptable welding methods of procedure. Other standards requiring a higher level of weld quality, test procedures and welder qualification meet the intent of NFPA 13.

The AWS standard D10.9, Level AR-3, reflects the minimum welding requirements needed for the pressures found in sprinkler systems. As the above interpretation indicates, methods meeting higher standards are also acceptable.

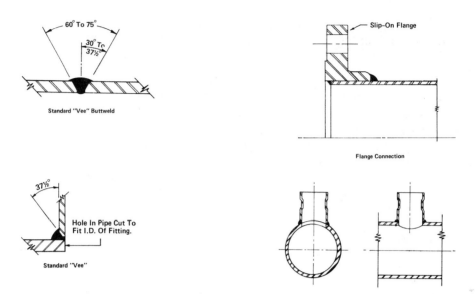

Figure A-3-7.2(a)   Acceptable Weld Joints.

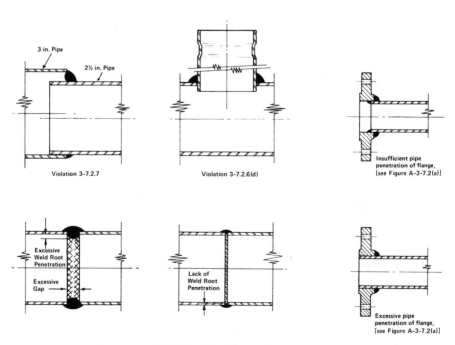

Figure A-3-7.2(b)   Unacceptable Weld Joints.

3-7.2.2* Welding sections of sprinkler piping in place inside the building shall not be permitted. Sections of branch lines, cross mains, or risers may be shop welded.

Exception: Welding sections of sprinkler piping in place inside new buildings under construction may be permitted only when the construction is noncombustible and no combustible contents are present and when the welding process is performed in accordance with NFPA 51B, Standard for Fire Prevention in Use of Cutting and Welding Processes.

Welding in place inside existing occupancies is not permitted. Welding inside new buildings may be permitted in an area free of combustible construction or combustible components.

Formal Interpretation

Question: **Can repair welding be done to an existing sprinkler system while in place?**

Answer: **No. 3-7.2 does not address itself to welding inside a building.**

The Committee interprets *shop welding* **as meaning either on the sprinkler contractor's own premises or in a specified area on the construction site sufficiently remote from the place of installation so as not to present any exposure hazard. On-site welding inside the building under construction is not permitted.**

A-3-7.2.2 As used in this standard, *shop* in the term *shop welded* means either:

(a) At the sprinkler contractor's or fabricator's premise.

(b) An approved welding area at the building site.

The definition of *shop* **is very broad and encompasses almost any location other than those places where the welding cannot be safely performed.**

3-7.2.3 Welding procedures, welders, and welding machine operators shall be qualified as required by 3-7.2.11.

3-7.2.4 Fittings used to join pipe shall be listed fabricated fittings, or manufactured in accordance with Table 3-8.1.1. Such fittings joined in conformance with a qualified welding procedure as set forth in this section are an acceptable product under this standard, provided that materials and wall thickness are compatible with other sections of this standard.

Exception: Fittings are not required when pipe ends are buttwelded.

The use of unlisted fabricated fittings is not permitted. A fabricated fitting will have a known equivalent length factor for expediting hydraulic calculations, and will have some inherent protection against misuse and thus an increased level of quality control.

3-7.2.5   No welding shall be performed if there is impingement of rain, snow, sleet, or high wind on the weld area of the pipe product.

Welding must not be performed under conditions that introduce a hazard to the mechanic. Welding performed in adverse conditions as described in this paragraph may also result in a structurally weak product.

3-7.2.6   When welding is performed:

(a)* Holes in piping for outlets shall be cut to the full inside diameter of fittings prior to welding in place of the fittings.

A-3-7.2.6(a)   Listed, shaped, contoured nipples meet the definition of fabricated fittings.

(b) Discs shall be retrieved.

(c) Openings cut into piping shall be smooth bore and all internal slag and welding residue shall be removed.

(d) Fittings shall not penetrate the internal diameter of the piping.

(e) Steel plates shall not be welded to the ends of piping or fittings.

(f) Fittings shall not be modified.

(g) Nuts, clips, eye rods, angle brackets, or other fasteners shall not be welded to pipe or fittings.

Exception: Tabs may be welded to pipe for longitudinal earthquake braces only. (See 3-5.3.5.1.)

In the past, quality control has been a problem with welded sprinkler piping. This section highlights abuses that must not be permitted.

When welded outlets are formed, some of the concerns addressed by these requirements are:

(a) That the area of flow must not be restricted;

(b) That holes have been cut with a machine, not a torch;

(c) That rough edges cause turbulence with resultant increased friction loss, and welding residue could obstruct opened sprinklers;

(d) That penetration beyond the internal diameter of the pipe would restrict flow and cause turbulence; and

(e) That some of these items are considered poor practice within the industry and must be discouraged.

Retrieval of the discs, mentioned in (b), is best handled by attaching the disc to the piping at the point at which it was cut.

Formal Interpretation

Question 1: Is it the intent of the Committee to accept the welding performed at those fabrication facilities whose operating procedures differ from those spelled out in paragraph 3-7.2.6, but have demonstrated that those procedures and methods will cut out, account for, and dispose of weld discs in a manner that has insured a full inside diameter opening for the weld fitting used?

Answer: Yes. Provided the hole is cut to the full inside diameter, the discharge characteristics are not adversely affected and subparagraphs (b) and (c) are complied with.

Question 2: Will tests conducted by a nationally recognized laboratory be accepted as satisfactory support for that fabrication's facility?

Answer: N/A. This is outside the scope of NFPA 13.

3-7.2.7 When reducing a pipe size in the run of a main, cross main, or branch, a reducing fitting designed for that purpose shall be used.

The use of unlisted fabricated fittings is not permitted. Unlisted fabricated reducing fittings will not have characteristics comparable to a fitting specifically designed for the purpose.

3-7.2.8 Torch cutting and welding shall not be permitted as a means of modifying or repairing sprinkler systems.

Torch cutting and welding are restricted to new installations and then only when they comply with the Exception to 3-7.2.2. This restriction bears in mind the possible introduction of ignition sources at a time when protection is impaired.

3-7.2.9 When welding is planned, a contractor shall specify the section to be shop welded on drawings and the type of fittings or formations to be used.

3-7.2.10 Sections of shop welded piping shall be joined by means of flanged or flexible gasketed joints or other approved fittings.

Exception: See 3-7.2.2.

When sections of piping are shop welded, they are usually joined to each other by means of flanged or flexible gasketed joints because of the ease of assembly.

3-7.2.11 Qualifications.

3-7.2.11.1 A welding procedure shall be prepared and qualified before any welding is done. Qualification of the welding procedure to be used and the performance of welders and welding operators is required and shall comply with the requirements of American Welding Society standard AWS D10.9, Level AR-3.

3-7.2.11.2 Contractors or fabricators shall be responsible for all welding they install. Each contractor or fabricator shall have an established written quality assurance procedure related to control of the requirements of 3-7.2.6, available to the authority having jurisdiction.

Contractors or fabricators are directly responsible for qualifying procedures and welders in accordance with this standard as well as for the quality of the welds performed by their employees.

3-7.2.11.3 Contractors or fabricators shall be responsible for qualifying any welding procedure that they intend to have used by personnel of their organization.

3-7.2.11.4 Contractors or fabricators shall be responsible for qualifying all of the welders and welding machine operators employed by them in compliance with the requirements of AWS D10.9, Level AR-3.

3-7.2.12 Records.

3-7.2.12.1 Welders or welding machine operators shall, upon completion of each weld, stamp an imprint of their identification into the side of the pipe adjacent to the weld.

3-7.2.12.2 Contractors or fabricators shall maintain certified records, which are available to the authority having jurisdiction, of the procedures used and the welders or welding machine operators employed by them along with their welding identification imprints. Records shall show the date and the results of procedure and performance qualifications.

3-7.3 Groove Joining Methods.

3-7.3.1   Pipe joined with mechanical grooved fittings shall be joined by a listed combination of fittings, gaskets, and grooves. When grooves are cut or rolled on the pipe they shall be dimensionally compatible with the fitting.

Exception: Steel pipe with wall thicknesses less than Schedule 30 [in sizes 8 in. (203 mm) and larger] or Schedule 40 [in sizes less than 8 in. (203 mm)] shall not be joined by fittings used with pipe having cut grooves.

Formal Interpretation

Question: Is it the intent of 3-7.3.1 to include rolled grooves in Schedule 40 pipe?

Answer: Yes. Grooves may be rolled on Schedule 40 pipe provided the instructions of the manufacturer of the groove-rolling equipment are followed.

Unlike threads, grooves are not fabricated to a standard. Therefore, the listing of a grooved fitting entails the combination of the fitting, the gasket, and the groove. (*See Figure 3.1.*)

3-7.3.2   Mechanical grooved couplings including gaskets used on dry-pipe systems shall be listed for dry-pipe service.

The gaskets on grooved couplings will have a limited fire endurance when used on dry-pipe systems. To compensate for this, the evaluation of the coupling includes its performance in a dry-pipe sprinkler system.

3-7.4* Brazed and Soldered Joints.   Joints for the connection of copper tube shall be brazed.

Exception No. 1:   Solder joints may be permitted for wet-pipe systems in Light Hazard Occupancies where the temperature classification of the installed sprinklers is Ordinary or Intermediate.

Exception No. 2:   Solder joints may be permitted for wet-pipe systems in Ordinary Hazard (Group 1) Occupancies where the piping is concealed.

A-3-7.4   The fire hazard of the brazing process should be suitably safeguarded.

Self-cleaning fluxes should not be used. Continued corrosive action after the soldering process is completed could result in leaks from the seats of sprinklers.

Paragraph 3-7.4 restricts the use of soldered joints to circumstances under which the system is water-filled and the exposure to heat will not be enough to compromise the integrity of the joint.

3-7.5 Other Types.   Other types of joints shall be made or installed in accordance with the requirements of the listing for this service.

This section encourages innovative technology by permitting listed joints not specifically addressed in the standard.

3-7.6 End Treatment.   After cutting, pipe ends shall have burrs and fins removed.

Piping must be reamed in order to avoid reduction of the inside diameter of the pipe and to remove any rough edges from the end of the pipe. The removal of irregular edges is particularly important when the fitting utilized has internal gaskets.

3-7.6.1   When using listed fittings, the pipe and its end treatment shall be in accordance with the manufacturer's installation instructions and the listing.

The end treatment required will vary depending on the fitting that is utilized. For example, some mechanical-type fittings require that the varnish be removed from the exterior wall of the pipe entering the joint.

3-8 Fittings.

3-8.1 Types of Fittings.

3-8.1.1   Fittings used in sprinkler systems shall be of the materials listed in Table 3-8.1.1 or in accordance with 3-8.1.2. The chemical properties, physical properties, and dimensions of the materials listed in Table 3-8.1.1 shall be at least equivalent to the standards cited in the table. Fittings used in sprinkler systems shall be designed to withstand the working pressures involved, but not less than 175 psi (12.1 bars) cold water [125 psi (8.6 bars) saturated steam] pressure.

Fittings of the types and materials indicated in Table 3-8.1.1 must be manufactured to the standards indicated or to stan-

Table 3-8.1.1   Fittings Materials and Dimensions

Material and Dimensions	Standard
Cast Iron	
Cast Iron Threaded Fittings, Class 125 and 250	ANSI B16.4
Cast Iron Pipe Flanges and Flanged Fittings	ANSI B16.1
Malleable Iron	
Malleable Iron Threaded Fittings, Class 150 and 300	ANSI B16.3
Steel	
Factory-made Wrought Steel Buttweld Fittings	ANSI B16.9
Buttwelding Ends for Pipe, Valves, Flanges and Fittings	ANSI B16.25
Spec. for Piping Fittings of Wrought Carbon Steel and Alloy Steel for Moderate and Elevated Temperatures	ASTM A234
Steel Pipe Flanges and Flanged Fittings	ANSI B16.5
Forged Steel Fittings, Socket Welded and Threaded	ANSI B16.11
Copper	
Wrought Copper and Bronze Solder-Joint Pressure Fittings	ANSI B16.22
Cast Bronze Solder-Joint Pressure Fittings	ANSI B16.18

dards meeting or exceeding the indicated standards. **Any fittings for use in a sprinkler system must be designed for a working pressure of at least 175 psi (12.1 bars).** (*See 3-8.1.3 for requirements in which pressure exceeds 175 psi.*)

3-8.1.2* Other types of fittings may be used, but only those investigated and listed for this service.

This section is related to 3-1.1.5, which gives permission for the use of alternative types of listed pipe and tube. Innovations in the technology of new fittings can be linked to new piping materials, as previously discussed. Any new fitting that has been listed for use in a sprinkler system can safely be incorporated into this standard, according to this section.

A-3-8.1.2 Rubber-gasketed pipe fittings and couplings should not be installed where ambient temperatures can be expected to exceed 150°F (66°C) unless listed for this service. If the manufacturer further limits a given gasket compound, those recommendations should be followed.

3-8.1.2.1* When unique characteristics of a fitting, such as a tendency to rotate, require support in addition to that required in Section 3-10, restraint shall be provided in accordance with its listing.

A-3-8.1.2.1 Unless properly restrained, gravitational forces on unsupported nonvertical branches can cause the pipe to rotate out of position.

There have been instances of unsupported branch outlets rotating out of position. Any fitting having the potential for such rotation must have as part of its listing appropriate measures to ensure against that circumstance.

3-8.1.3 Fittings used in sprinkler systems shall be extra-heavy pattern where pressures exceed 175 psi (12.1 bars).

Exception No. 1:   Standard weight pattern cast-iron fittings 2 in. (51 mm) in size and smaller may be used where pressures do not exceed 300 psi (20.7 bars).

Exception No. 2:   Standard weight pattern malleable iron fittings 6 in. (152 mm) in size and smaller may be used where pressures do not exceed 300 psi (20.7 bars).

Exception No. 3:   Fittings may be used for system pressures up to the limits specified in listings by a testing laboratory.

Formal Interpretation

Question: Is the interpretation correct that the use of standard weight cast-iron fittings is permissible in sizes up to and including 2 in. where normal pressure in the piping system exceeds 175 psi? In sizes 2½-in. and larger, cast-iron fittings shall be of the extra-heavy pattern where normal pressure exceeds 175 psi?

Answer: Yes.

This section allows cast-iron fittings 2 in. (51 mm) in size and smaller and malleable iron fittings 6 in. (152 mm) in size and smaller to be used up to 300 psi (20.7 bars), because the rupture strength of those fittings allows an adequate factor of safety.

3-8.1.4   When water pressures are 175 psi to 300 psi (12.1 to 20.7 bars), extra heavy valves shall be used in accordance with their pressure ratings.

Extra heavy valves must be utilized for working pressures in excess of 175 psi (12.1 bars). Flanges must be in compliance with 3-8.1.3.

3-8.1.5*   When individual floor/zone control valves are not provided, a flanged joint or mechanical coupling shall be used at the riser at each floor for connections to piping serving floor areas in excess of 5000 sq ft (465 m²).

The flanged joint or mechanical coupling at the riser provides a facility to isolate a section of the system, so that the remainder of the system may be returned to service while repairs or modifications are made. One popular form used to isolate the segmented areas is commonly referred to as a "frying pan" because of its characteristic shape.

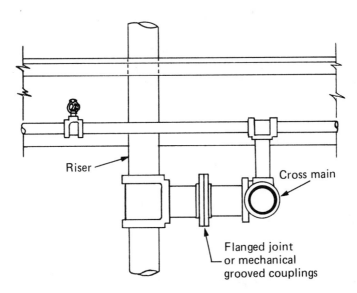

Figure A-3-8.1.5   One Arrangement of Flanged Joint at Sprinkler Riser.

3-8.2* **Couplings and Unions.**   Screwed unions shall not be used on pipe larger than 2 in. (51 mm). Couplings and unions of other than screwed-type shall be of types approved specifically for use in sprinkler systems. Unions, screwed or mechanical couplings, or flanges may be used to facilitate installation.

Screwed unions larger than 2 in. (51 mm) in size present a maintenance problem, due to their tendency to develop leaks. There are a variety of acceptable fittings available for use in connecting portions of systems.

A-3-8.2   Approved flexible connections are permissible and encouraged for sprinkler installations in racks to reduce the possibility of physical damage. When flexible tubing is used it should be located so that it will be protected against mechanical injury.

3-8.3 Reducers and Bushings.   A one-piece reducing fitting shall be used wherever a change is made in the size of the pipe.

Exception No. 1:   Hexagonal or face bushings may be used in reducing the size of openings of fittings when standard fittings of the required size are not available.

Exception No. 2:   Hexagonal bushings as permitted in 3-3.5.1 are acceptable.

For other than the circumstance cited in the Exception to 3-3.5.1, in which bushings are temporarily installed to facilitate an unrestricted flow to sprinklers to be installed below a future drop ceiling, bushings may be used only when reducing fittings are unavailable. They have poorer flow characteristics than reducing fittings and possess a greater tendency to leak.

Formal Interpretation

Question: Is it the intent of 3-8.3 to exclude the use of bushings when the required reducing fitting is manufactured?

Answer: Yes. The intent is to exclude the use of bushings when the one-piece reducing fitting is available on the market at the time the system is fabricated.

3-9 Valves.

3-9.1 Types of Valves to Be Used.

3-9.1.1   All valves on connections to water supplies and in supply pipes to sprinklers shall be listed indicating valves, unless a nonindicating valve, such as an underground gate valve with approved roadway box complete with T-wrench, is accepted by the authority having jurisdiction.

Such valves shall not close in less than 5 seconds when operated at maximum possible speed from the fully open position. This is to avoid damage to piping by water hammer.

The following may not incorporate indicating devices as part of the valve, but the valve assembly described shall qualify as an indicating valve:

(a) A listed underground gate valve equipped with a listed indicator post,

(b) A listed water control valve assembly with a reliable position indication connected to a remote supervisory station.

All valves controlling the flow of water to sprinklers must be listed and must incorporate a method of readily determining that the valve is open. The method of indicating position may be part of the valve itself, such as a rising stem, or a feature of the assembly as described in this section.

The one exception is the use of a roadway box, in which case the valve must be installed where nothing can extend above the surface of the box and the authority having jurisdiction accepts it in lieu of a valve pit.

3-9.1.2   Drain valves and test valves shall be of approved type of 175 psi (12.1 bars) cold water [125 psi (8.6 bars) saturated steam] pressure rating.

3-9.1.3   Check valves shall be listed and installed in a vertical or horizontal position in accordance with their listing.

In this section, _check valves_ refers to those in the sprinkler system waterway. It is not the intent of this standard to require the listing of check valves that do not affect system reliability, such as those used in the cross connection of drains. Not all valves are listed for both horizontal and vertical use. Therefore, care must be taken to ensure that valves are only installed in accordance with their listing.

3-9.1.4   When wafer type valves are used, the disc may extend beyond the end of the valve body. Valve discs shall not interfere with the operation of other system components.

The discs of some listed wafer type valves extend beyond the end of the valve body. There have been documented instances of such discs contacting other system components and partially obstructing the waterway.

3-9.2* Valves Controlling Sprinkler Systems.

3-9.2.1*   Each system shall be provided with a listed indicating valve so located as to control all sources of water supply except fire department connections.

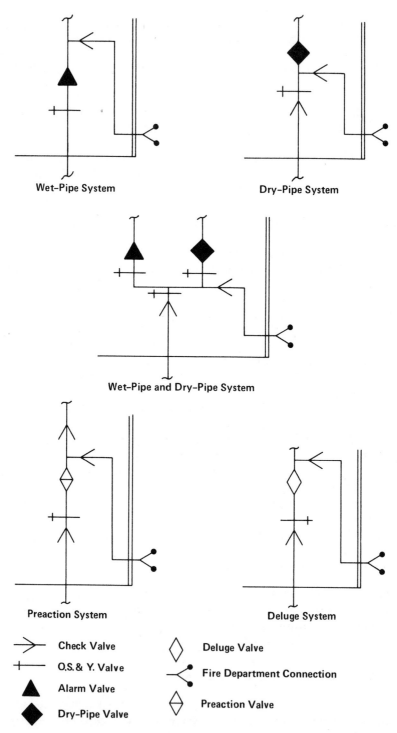

Wet–Pipe System

Dry–Pipe System

Wet–Pipe and Dry–Pipe System

Preaction System

Deluge System

Check Valve

O.S. & Y. Valve

Alarm Valve

Dry-Pipe Valve

Deluge Valve

Fire Department Connection

Preaction Valve

Figure A-3-9.2   Examples of Acceptable Valve Arrangements.

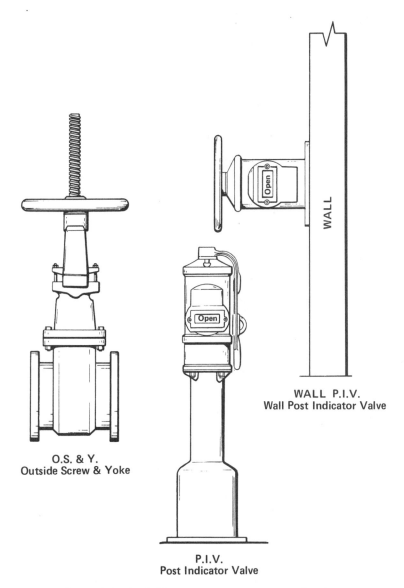

WALL P.I.V.
Wall Post Indicator Valve

O.S. & Y.
Outside Screw & Yoke

P.I.V.
Post Indicator Valve

**Figure 3.3. Sprinkler system water control valves
(Insurance Services Office).**

Formal Interpretation

Question: Is an approved indicating valve required on the system
side of an alarm valve, in addition to the approved indicating or
indicator post valve on the supply side of the alarm valve?

Answer: **No. It was never the intent of the Committee to require a control valve on the system side of the alarm valve, only on the supply side. Hence, only one control valve per sprinkler system.**

A-3-9.2.1 A water supply connection should not extend into or through a building unless such connection is under the control of an outside listed indicating valve or an inside listed indicating valve located near the outside wall of the building.

All valves controlling water supplies for sprinkler systems or portions thereof, including floor control valves, should be accessible to authorized persons during emergencies. Permanent ladders, clamped treads on risers, chain-operated hand wheels, or other accepted means should be provided when necessary.

Outside control valves are suggested in the following order of preference:

(a) Listed indicating valves at each connection into the building at least 40 ft (12.2 m) from buildings if space permits.

(b) Control valves installed in a cutoff stair tower or valve room accessible from outside.

(c) Valves located in risers with indicating posts arranged for outside operation.

(d) Key-operated valves in each connection into the building.

Each system must be provided with a valve that isolates that system from all automatic sources of supply. It is preferred, but not always possible, for the fire department connection to be on the system side of the valve so that, with the valve shut, the fire department can supply water to the system through the fire department connection. (*See 2-7.3.*)

3-9.2.2 At least one listed indicating valve shall be installed in each source of water supply except fire department connections.

Formal Interpretation

Question: **For a high-rise building with two risers supplying both sprinklers and 2½-in. hose outlets with a connection from each of the risers to a sprinkler loop main on each floor, where each connection from the riser to the loop contains an indicating-type pressure regulating control valve, waterflow switch, and O.S. & Y. control valve—is the valve arrangement indicated below accepta-ble?**

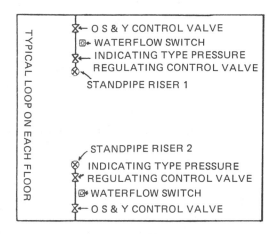

TYPICAL LOOP ON EACH FLOOR

O S & Y CONTROL VALVE
WATERFLOW SWITCH
INDICATING TYPE PRESSURE
REGULATING CONTROL VALVE
STANDPIPE RISER 1

STANDPIPE RISER 2
INDICATING TYPE PRESSURE
REGULATING CONTROL VALVE
WATERFLOW SWITCH
O S & Y CONTROL VALVE

Figure 3.

Answer: Yes. The Committee answer is with respect to the valve arrangement only and excludes all other aspects of the system.

Formal Interpretation

Question: According to 3-9.2.1 and 3-9.2.2, would a listed indicating valve be required at "A" in Figure 4?

Answer: A valve is not required at point "A" for the sketch provided. A more comprehensive answer would require consideration of the configuration of the outside underground piping and the control valves on it in addition to the location of the fire department

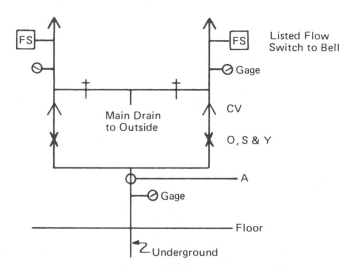

Figure 4. Sectional detail of automatic sprinkler risers.

connection. The pressure gage indicated below point "A" is not required.

3-9.2.3 Valves on connections to water supplies, sectional control valves, and other valves in supply pipes to sprinklers shall be supervised open by one of the following methods:

(a) Central station, proprietary, or remote station signaling service,

(b) Local signaling service that will cause the sounding of an audible signal at a constantly attended point,

(c) Locking valves open,

(d) Sealing of valves and approved weekly recorded inspection when valves are located within fenced enclosures under the control of the owner.

Exception No. 1:   Underground gate valves with roadway boxes need not be supervised.

Exception No. 2:   For floor control valves in high-rise buildings, see 3-12.6; for circulating closed-loop systems, see 5-6.1.6.

The four methods of supervision represented here will help to assure that a common mode of sprinkler system "failure" is reduced to a minimum. Approximately one-third of system failures are attributable to closed or partially closed sprinkler system control valves. Listed in descending order of preference, these methods are derived from NFPA 26, *Recommended Practice for the Supervision of Valves Controlling Water Supplies for Fire Protection.*

Central station, proprietary, or remote station supervision methods are preferred over the other methods listed. Supervisory signals received at such alarm facilities are usually forwarded to trouble or maintenance crews responsible for the protected property. This method also allows much quicker notification of fire departments which respond to fire alarm boxes at the protected premise. Knowing early on that a control valve has been closed for any reason may alter the fire department response and fire fighting tactics. These particular methods have not, however, been mandated due to the cost of such a system, particularly in existing large, spread-out plants.

3-9.2.4 When there is more than one source of water supply, a check valve shall be installed in each connection.

Exception: When cushion tanks are used with automatic fire pumps, no check valve is required in the cushion tank connection.

Each source of supply must have a check valve to isolate the supplies from each other. In a fire situation, water flows from the highest pressure source. For example, consider a system supplied by a gravity tank, a city connection, and a fire department connection, with the gravity tank being the highest pressure source. Initial flow would be from the gravity tank. If its pressure dropped below the city connection pressure, the tank's check valve would close and supply would be from the city. When the fire department connected onto the system, the check valves in both automatic supplies would close and supply would be through the fire department pumps via the fire department connection.

A cushion tank, when installed, is a part of the fire pump supply and, as such, is subject to the pump's check valve.

3-9.2.5* A check valve shall be installed in each water supply connection if there is a fire department connection on the system.

A-3-9.2.5 Pits for underground valves, except those located at the base of a tank riser, are described in NFPA 24, *Standard for Installation of Private Fire Service Mains and Their Appurtenances*. For pits protecting valves located at the base of a tank riser, refer to NFPA 22, *Standard for Water Tanks for Private Fire Protection*.

If a check valve were not installed in an automatic supply, pressure developed via the fire department connection would be dissipated into the supply. All sources, whether automatic or nonautomatic, must have check valves so that pressure from the highest pressure source is concentrated in the system. The presence of an alarm check valve installed upstream of the fire department connection, or a backflow prevention device in-

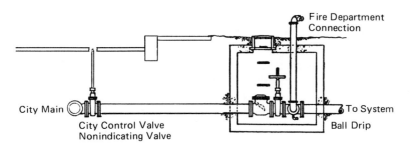

Figure A-3-9.2.5   Pit for Gate Valve, Check Valve, and Fire Department Connection.

stalled underground, would satisfy the condition expressed in this paragraph.

3-9.2.6* When a single wet-pipe sprinkler system is equipped with a fire department connection the alarm valve is considered a check valve and an additional check valve shall not be required.

A-3-9.2.6 When a system having only one dry-pipe valve is supplied with city water and fire department connection it will be satisfactory to install the main check valve in water supply connection immediately inside of the building; in case there is no outside control, the system indicating valve should be placed at the wall flange ahead of all fittings.

An alarm valve is a check valve designed to provide facilities to activate waterflow alarms. Another check valve would be redundant and would increase the friction loss characteristics of the system.

3-9.2.7 In a city connection serving as one source of supply, the city valve in the connection may serve as one of the required valves. A listed indicating valve or an indicator post valve shall be installed on the system side of the check valve. *(See Figure A-3-9.2.5.)*

Exception: When a wet-pipe sprinkler system is equipped with an (alarm) check valve, a gate valve is not required on the system side of the (alarm) check valve.

The intent of requiring gate valves on both sides of a check valve is to provide facilities for isolating the check valve for maintenance or repair while permitting the system to remain in service. If the system has only one automatic supply, a gate valve on the system side of the check valve is not required.

It should be recognized that the alarm valve is a specialized device and that control valves are not required on the system side of alarm check valves.

3-9.3 Pressure Reducing Valves.

There have been documented instances of pressure regulating valves intentionally being set to provide an inadequate outlet pressure. For this reason, 1-11.5 requires that each pressure reducing valve be tested upon completion of the installation. The reliability of these valves is also a matter for concern if they are not periodically exercised; thus, they should be tested annually. The intervals at which the valve should be tested will be governed by the manufacturer. Para-

graph 3-9.3.4 and its Exception call to attention that some pressure regulating valves incorporate an indicating shutoff valve lacking the tolerances needed for listing.

3-9.3.1 In portions of systems where all components are not listed for pressure greater than 175 psi (12.1 bar) and the potential exists for normal (nonfire condition) water pressure in excess of 175, a listed pressure reducing valve shall be installed and set for an outlet pressure not exceeding 165 psi (11.4 bar) at the maximum inlet pressure.

3-9.3.2 Pressure gages shall be installed on the inlet and outlet sides of each pressure reducing valve.

3-9.3.3 A relief valve of not less than ½ in. (13 mm) in size shall be provided on the discharge side of the pressure reducing valve set to operate at a pressure not exceeding 175 psi (12.1 bar).

3-9.3.4 A listed indicating valve shall be provided on the inlet side of each pressure reducing valve.

Exception: Where the pressure reducing valve meets the listing requirements for use as an indicating valve.

3-9.4* Identification of Valves. When there is more than one control valve, permanently marked identification signs indicating the portion of the system controlled by each valve shall be provided.

Embossed plastic tape, pencil, ink, crayon, etc., shall not be considered permanent markings. The sign shall be secured with noncorrosive wire, chain, or other means.

A-3-9.4 All control, drain, and test connections should be provided with identification signs.

Signs identifying control valves must be of a clearly legible type and installed in a manner that ensures their ability to remain in place. See Figure 3.4 on page 142.

3-10 Hangers.

3-10.1* General. Type of hangers and installation methods shall be in accordance with the requirements of Section 3-10.

Exception: Hangers and installation methods certified by a registered professional engineer for the following:

(a) Designed to support five times the weight of the water-filled pipe plus 250 lb (114 kg) at each point of piping support.

Figure 3.4. Typical sprinkler control valve identification sign, Style "A."

(b) These points of support are enough to support the sprinkler system.

(c) Ferrous materials are used for hanger components.

Detailed calculations shall be submitted, when required by the reviewing authority, showing stresses developed both in hangers and piping, and safety factors allowed.

This section provides the necessary parameters to ensure that sprinkler piping is reliably supported by the building structure. Established use of listed hangers, tables, and design methods are one means to that end.

For unusual applications or situations, permission is given to design hanger systems with alternative methods. These methods require the services of a professional engineer (preferably in the area of structural engineering) to certify that all three conditions, (a), (b), and (c), are satisfied.

As with any system design, consideration of safety factors is inherent in this alternative approach. The hanger must be capable of supporting five times the weight of the water-filled pipe. This is based upon the ultimate strength of the component. A 250 lb (114 kg) point load must be carried concurrently at each point of hanging. This load represents the "average" weight of a sprinkler fitter who grasps the pipe during a fall or slip from a ladder.

A-3-10.1 Branch line hangers under metal decking may be attached by drilling or punching vertical members and using through bolts. The distance from the bottom of the bolt hole to the bottom of the vertical member should be not less than ⅜ in. (9.5 mm).

To take care of the thrust in a steeply pitched roof branch line, a clamp should be installed on the pipe just above the lowest hanger.

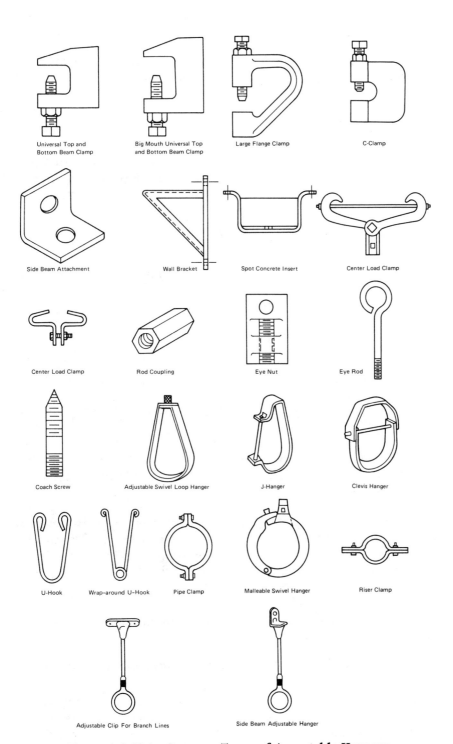

Figure A-3-10.1   Common Types of Acceptable Hangers.

3-10.1.1   Hangers and their components shall be ferrous.

Exception: Nonferrous components that have been proven by fire tests to be adequate for the hazard application, are listed for this purpose, and are in compliance with the other requirements of this section.

Formal Interpretation

Question: Does the Subsection 3-10.1 Exception mean that pipe supports need not be designed to accommodate a loading equal to five times the weight of water-filled pipe plus 250 lb at each support when the following are accomplished?

1.   A detailed piping stress and pipe support analysis which substantiates stresses, loads, and safety factors is performed. All calculations are retained as part of the permanent project records.

Type of pipe analysis performed:

(a)   2½-in. and larger piping — Verified Computer Analysis

(b)   2-in. and smaller piping — Computer Analysis or in accordance with a company approved design standard.

2.   Ferrous materials are utilized for piping and piping support components.

3.   Types of pipe supports and installations are in accordance with industry standards (MSS SP-58 and SP-69) and the requirements of 3-10 of NFPA 13.

Answer: No. The standard mandates all pipe supports must be designed to support five times the weight of the water-filled pipe plus 250 lbs at each point of piping support. This is the same requirement used by laboratories when reviewing hanger supports for listing purposes.

3-10.1.2   The components of hanger assemblies that directly attach to the pipe or to the building structure shall be listed.

Exception: Mild steel hangers formed from rods need not be listed.

Hanger assembly components must be listed. The Exception is granted for mild steel rod hangers because data has been submitted substantiating the fact that in the sizes required in this section the rods will support, at maximum permitted spacing, five times the weight of water-filled Schedule 40 pipe plus 250 lb (114 kg).

Formal Interpretation

Question: If we comply with 3-10, do we have any further burden of proof other than "support the added load of the water-filled pipe plus a minimum load of 250 lbs applied at the point of hanging"?

Answer: No. The Committee feels there has been a misinterpretation of the requirements of 3-10.1.4, which were quoted in part, as this applies only to the structure to which the hanger is attached. If the installation otherwise meets the requirement of Section 3-10, the requirements of 3-10.1 Exception as to certification by a registered professional engineer do not apply.

Formal Interpretation

Question: Does the method of hanging sprinkler piping as shown in Figure 5 comply with Section 3-10? It is capable of supporting five times the weight of water-filled piping plus 250 lb and has been certified by a registered professional engineer.

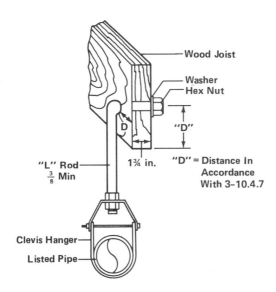

Figure 5.

Answer: Yes. Certification by a registered professional engineer also includes compliance with other requirements of Section 3-10.1 Exception, namely: "(b) These points of support are enough to support the sprinkler system. (c) Ferrous materials are used for hanger components."

In addition, detailed calculations shall be submitted, when required by the reviewing authority, showing stresses developed both in hangers and piping, and safety factors allowed.

3-10.1.3* Sprinkler piping or hangers shall not be used to support nonsystem components.

A-3-10.1.3 The rules covering the hanging of sprinkler piping take into consideration the weight of water-filled pipe plus a safety factor. No allowance has been made for the hanging of nonsystem components from sprinkler piping.

Wiring and conduit used as a part of the supplemental detection system for a deluge or preaction system are exempt from this paragraph.

3-10.1.4 Sprinkler piping shall be substantially supported from the building structure, which must support the added load of the water-filled pipe plus a minimum of 250 lb (114 kg) applied at the point of hanging.

Each individual hanger must support the weight of the water-filled pipe, plus 250 lb (114 kg), but it is not the intent to add 250 lb concurrently for each hanger when determining the minimum strength of the building structure. As mentioned in the commentary to 3-10.1, the 250 lb represents the average weight of a pipe fitter who grasps the pipe during a mishap.

Formal Interpretation

Question: Is it the intent of 3-10.1.4 to require all hangers (main, crossmain, and branch) to support the added load of the water-filled pipe plus a minimum of 250 lb applied at the point of hanging?

Answer: Yes.

3-10.1.5 Sprinkler piping shall be supported independently of the ceiling sheathing.

Exception: Toggle hangers shall be used only for the support of pipe 1½ in. (38 mm) or smaller in size under ceilings of hollow tile or metal lath and plaster.

Toggle hangers are not permitted for use with gypsum wallboard or other less substantial types of ceiling materials.

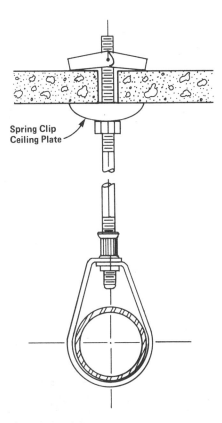

Spring Clip
Ceiling Plate

Figure 3.5. Toggle hanger for pipe 1½ in. or smaller.

Sprinkler piping must be supported from the building structure with the exception of 1½-in. (38-mm) and smaller piping,
which may be supported under ceilings of hollow tile or metal
lath and plaster with toggle hangers.

3-10.1.6 When sprinkler piping is installed below ductwork, piping shall
be substantially supported from the building structure or from the steel
angles supporting the ductwork provided the angles conform to Tables
3-10.1.7(a) and 3-10.1.7(b).

If the steel angles supporting the ductwork have dimensions
less than those required by Table 3-10.1.7, the pipe must be
supported by a hanger connected to the building structure, or
the duct hangers must be increased in size to satisfy the
requirements of the table.

3-10.1.7* For trapeze hangers, the minimum size of steel angle or pipe
span between purlins or joists shall be such that the available section

Table 3-10.1.7(a) Section Modulus Required for Trapeze Members (in.³)

Span of Trapeze	Pipe Size											
	1 in.	1¼ in.	1½ in.	2 in.	2½ in.	3 in.	3½ in.	4 in.	5 in.	6 in.	8 in.	10 in.
1 ft 6 in.	.08	.09	.09	.09	.10	.11	.12	.13	.15	.18	.24	.32
	.08	.09	.09	.10	.11	.12	.13	.15	.18	.22	.30	.41
2 ft 0 in.	.11	.12	.12	.13	.13	.15	.16	.17	.20	.24	.32	.43
	.11	.12	.12	.13	.15	.16	.18	.20	.24	.29	.40	.55
2 ft 6 in.	.14	.14	.15	.16	.17	.18	.20	.21	.25	.30	.40	.54
	.14	.15	.15	.16	.18	.21	.22	.25	.30	.36	.50	.68
3 ft 0 in.	.17	.17	.18	.19	.20	.22	.24	.26	.31	.36	.48	.65
	.17	.18	.18	.20	.22	.25	.27	.30	.36	.43	.60	.82
4 ft 0 in.	.22	.23	.24	.25	.27	.29	.32	.34	.41	.48	.64	.87
	.22	.24	.24	.26	.29	.33	.36	.40	.48	.58	.80	1.09
5 ft 0 in.	.28	.29	.30	.31	.34	.37	.40	.43	.51	.59	.80	1.08
	.28	.29	.30	.33	.37	.41	.45	.49	.60	.72	1.00	1.37
6 ft 0 in.	.33	.35	.36	.38	.41	.44	.48	.51	.61	.71	.97	1.30
	.34	.35	.36	.39	.44	.49	.54	.59	.72	.87	1.20	1.64
7 ft 0 in.	.39	.40	.41	.44	.47	.52	.55	.60	.71	.83	1.13	1.52
	.39	.41	.43	.46	.51	.58	.63	.69	.84	1.01	1.41	1.92
8 ft 0 in.	.44	.46	.47	.50	.54	.59	.63	.68	.81	.95	1.29	1.73
	.45	.47	.49	.52	.59	.66	.72	.79	.96	1.16	1.61	2.19
9 ft 0 in.	.50	.52	.53	.56	.61	.66	.71	.77	.92	1.07	1.45	1.95
	.50	.53	.55	.59	.66	.74	.81	.89	1.08	1.30	1.81	2.46
10 ft 0 in.	.56	.58	.59	.63	.68	.74	.79	.85	1.02	1.19	1.61	2.17
	.56	.59	.61	.65	.74	.82	.90	.99	1.20	1.44	2.01	2.74

For SI units: 1 in. = 25.4 mm; 1 ft = 0.3048m.
Top values are for Schedule 10 pipe; bottom values are for Schedule 40 Pipe.
Note: The table is based on a maximum allowable bending stress of 15 KSI and a midspan concentrated load from 15 ft of water-filled pipe, plus 250 lb.

Table 3-10.1.7(b)   Available Section Moduli
of Common Trapeze Hangers

Pipe	Modulus	Angles					Modulus
Schedule 10							
1 in.	.12	1½	X	1½	X	³⁄₁₆	.10
1¼ in.	.19	2	X	2	X	⅛	.13
1½ in.	.26	2	X	1½	X	³⁄₁₆	.18
2 in.	.42	2	X	2	X	³⁄₁₆	.19
2½ in.	.69	2	X	2	X	¼	.25
3 in.	1.04	2½	X	1½	X	³⁄₁₆	.28
3½ in.	1.38	2½	X	2	X	³⁄₁₆	.29
4 in.	1.76	2	X	2	X	⁵⁄₁₆	.30
5 in.	3.03	2½	X	2½	X	³⁄₁₆	.30
6 in.	4.35	2	X	2	X	⅜	.35
		2½	X	2½	X	¼	.39
Schedule 40		3	X	2	X	³⁄₁₆	.41
		3	X	2½	X	³⁄₁₆	.43
1 in.	.13	3	X	3	X	³⁄₁₆	.44
1¼ in.	.23	2½	X	2½	X	⁵⁄₁₆	.48
1½ in.	.33	3	X	2	X	¼	.54
2 in.	.56	2½	X·	2	X	⅜	.55
2½ in.	1.06	2½	X	2½	X	⅜	.57
3 in.	1.72	3	X	3	X	¼	.58
3½ in.	2.39	3	X	3	X	⁵⁄₁₆	.71
4 in.	3.21	2½	X	2½	X	½	.72
5 in.	5.45	3½	X	2½	X	¼	.75
6 in.	8.50	3	X	2½	X	⅜	.81
		3	X	3	X	⅜	.83
		3½	X	2½	X	⁵⁄₁₆	.93
		3	X	3	X	⁷⁄₁₆	.95
		4	X	4	X	¼	1.05
		3	X	3	X	½	1.07
		4	X	3	X	⁵⁄₁₆	1.23
		4	X	4	X	⁵⁄₁₆	1.29
		4	X	3	X	⅜	1.46
		4	X	4	X	⅜	1.52
		5	X	3½	X	⁵⁄₁₆	1.94
		4	X	4	X	½	1.97
		4	X	4	X	⅝	2.40
		4	X	4	X	¾	2.81
		6	X	4	X	⅜	3.32
		6	X	4	X	½	4.33
		6	X	4	X	¾	6.25
		6	X	6	X	1	8.57

For SI Units: 1 in. = 25.4 mm; 1 ft. = 0.3048 m.

modulus of the trapeze member from Table 3-10.1.7(b) equals or exceeds the section modulus required in Table 3-10.1.7(a).

Any other sizes or shapes giving equal or greater section modulus will be acceptable. All angles are to be used with the longer leg vertical. The trapeze member shall be secured to prevent slippage. When a pipe is suspended from a pipe trapeze, ring, strap, or clevis, hangers of the size corresponding to the suspended pipe shall be used on both ends.

The top numbers in Table 3-10.1.7(a) provide the required section modulus for support of Schedule 10 pipe, while the lower numbers show the required section modulus for support of Schedule 40 pipe. Table 3-10.1.7(b) provides the available section modulus for various trapeze bars. The materials given in the table are for Schedule 10 and Schedule 40 pipe and angle iron. The section modulus for a material is denoted as "Z."

The following example shows the intended use of these tables.

Situation: Installation involves the support of Schedule 10 pipe with a diameter of 4 in. (102 mm). The span (horizontal distance between members) is given as 6 ft (1.9 m).

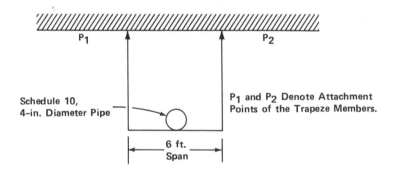

Determine: Acceptable materials to satisfy a correct trapeze installation.

Solution: From Table 3-10.1.7(a), the required section modulus is 0.51 in.[3] From Table 3-10.1.7(b), the following materials provide a section modulus of 0.51 in.[3]:

(a)   Schedule 10 pipe with a 2½ in. diameter: $Z = 0.69$ in.[3]

(b)   Schedule 40 pipe with a 2 in. diameter: $Z = 0.56$ in.[3]

(c)   An angle iron with dimensions of 3 in. × 2 in. × ¼ in.: $Z = 0.54$ in.[3]

Shapes or materials which meet or exceed the predetermined Z of 0.51 in.[3] are also acceptable for this problem.

For SI Units: 1 in. = 25.4 mm; 1 ft = 0.3048 m.

A-3-10.1.7   Table 3-10.1.7(a) assumes that the load from 15 ft (5 m) of water-filled pipe, plus 250 lb (114 kg), is located at the midpoint of the span of the trapeze member, with a maximum allowable bending stress of 15 KSI (111 kg). If the load is applied at other than the midpoint, for the purpose of sizing the trapeze member, an equivalent length of trapeze may be used, derived from the formula

$L = \dfrac{4ab}{a+b}$ where "L" is the equivalent length, "a" is the distance from one support to the load, and "b" is the distance from the other support to the load.

When multiple mains are to be supported or multiple trapeze hangers are provided in parallel, the required or available section modulus may be added.

> **This method allows an equivalent length, "L," to be calculated for other than symetrically loaded trapeze hangers.**

3-10.1.8   The size of hanger rods and fasteners required to support the steel angle iron or pipe indicated in Table 3-10.1.7(a) shall comply with 3-10.4.

3-10.1.9   Eye rods and ring hangers shall be secured with necessary lock washers to prevent lateral motion at the point of support.

3-10.1.10   Holes through concrete beams may also be considered as a substitute for hangers for the support of pipes.

> **Holes through building materials other than concrete may not be considered as a substitute for hangers. In some wood materials, drilling of holes through the material may tend to weaken the structural integrity of the member.**

3-10.1.11 Maximum Distance Between Hangers.

3-10.1.11.1*   For steel pipe sizes 1½ in. (38 mm) and larger, the maximum distance between hangers shall be 15 ft (4.5 m). For steel pipe sizes less than 1½ in. (38 mm), the maximum distance between hangers shall be 12 ft (3.6 m).

Exception No. 1:   Threaded lightweight steel pipe shall have a maximum distance between hangers not exceeding 12 ft (3.6 m) for pipe sizes 3 in. (76 mm) or less.

Exception No. 2:   The maximum distance between hangers may be modified in accordance with other paragraphs of Section 3-10.

Except as specified in 3-10.6, where cross mains receive additional support from hangers on the branch lines, hanger spacing for steel pipe cannot exceed 15 ft (4.5 m) for 1½-in. (38-mm) and larger pipe nor 12 ft (3.6 m) for 1¼-in. (33-mm) and smaller pipe.

In instances where pipe or tubing does not have the beam strength anticipated by the spacing specifications of 3-10.1.11 or 3-10.1.12, additional hangers will probably be necessary.

Because of the lack of material between the inside diameter and the root diameter of the thread, hangers must be spaced closer for threaded thin-wall pipe.

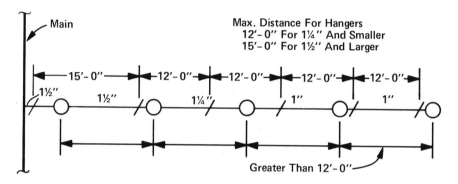

For SI Units: 1 in. = 25.4 mm; 1 ft = 0.3048 m.

Figure A-3-10.1.11.1   Distance Between Hangers with Steel Pipe.

3-10.1.11.2*   For copper tubes as specified in Table 3-1.1.1 the maximum distance between hangers shall not exceed that in Table 3-10.1.11.2.

Table 3-10.1.11.2   Hanger Spacing for Copper Tube

Tube Size	Maximum Hanger Spacing
¾ in.—1 in.	8 ft
1¼ in.—1½ in.	10 ft
2 in.—3 in.	12 ft
3½ in.—8 in.	15 ft

For SI units: 1 in. = 25.4 mm; 1 ft = 0.3048 m.

A-3-10.1.11.2   When copper tube is to be installed in moist areas or other environments conducive to galvanic corrosion, copper hangers or ferrous hangers with an insulating material should be used.

3-10.1.12 When sprinkler piping is installed in storage racks as defined in NFPA 231C, *Standard for Rack Storage of Materials*, piping shall be substantially supported from the storage rack structure or building in accordance with all applicable provisions of Section 3-10.

NFPA 231C, *Standard for Rack Storage of Materials*, stipulates requirements for sprinkler location, sprinkler spacing, and water supplies for rack storage areas. However, everything not modified by NFPA 231C should be installed in accordance with the requirements of this standard.

3-10.2 Hangers in Concrete.

3-10.2.1 Listed inserts set in concrete may be installed for the support of hangers. Wood plugs shall not be used.

Inserts listed for use in concrete have been tested for such use. Wood plugs used in concrete do not have the required retention to be used for the support of piping.

3-10.2.2 Listed expansion shields for supporting pipes under concrete construction may be used in a horizontal position in the sides of beams. In concrete having gravel or crushed stone aggregate, expansion shields may be used in the vertical position to support pipes 4 in. (102 mm) or less in diameter.

Expansion shields may be installed either horizontally or vertically. Shields installed horizontally are not subject to a direct downward pull and are preferable. Pipe 4 in. (102 mm) and smaller may be supported vertically by expansion shields at the spacing indicated in 3-10.1.11. For larger piping, see 3-10.2.3.

3-10.2.3 For the support of pipes 5 in. (127 mm) and larger, expansion shields if used in the vertical position shall alternate with hangers connected directly to the structural members such as trusses and girders, or to the sides of concrete beams. In the absence of convenient structural members, pipes 5 in. (127 mm) and larger may be supported entirely by expansion shields in the vertical position, but spaced not over 10 ft (3 m) apart.

Due to stress on expansion shields installed in the vertical position for pipe 5 in. (127 mm) or larger, shields must either be alternated with hangers providing other means of support or their maximum spacing reduced from 15 ft (4.5 m) to 10 ft (3 m) apart.

3-10.2.4 Expansion shields shall not be used in ceilings of gypsum or similar soft material. In cinder concrete, expansion shields shall not be used except on branch lines where they shall alternate with through bolts or hangers attached to beams.

The effectiveness of expansion shields depends on the gripping power of the material with which they are used. They will pull out of soft materials such as gypsum.

3-10.2.5 When expansion shields are used in the vertical position, the holes shall be drilled to provide uniform contact with the shield over its entire circumference. Depth of the hole shall be not less than specified for the type of shield used.

The listings of expansion shields, as with any other hanger, assume that they are properly installed.

3-10.2.6 Holes for expansion shields in the side of concrete beams shall be above the center line of the beam or above the bottom reinforcement steel rods.

Shields must be located where they will provide maximum strength. If a reinforcing rod is contacted when drilling for a shield, a new hole above the bottom reinforcing rod must be made to the required depth.

3-10.3 Powder-Driven Studs and Welding Studs.

Formal Interpretation

Question 1: Do the design requirements set forth in 3-10.1 apply to powder driven studs?

Answer: Yes.

Question 2: Does 3-10.1 apply to powder driven studs when used in a manner other than that listed (such as limited penetration)?

Answer: No. 3-10.3.2 is imposed on powder driven studs and they must be used per their listing.

Question 3: Does Section 3-10.1.4 relate directly to 3-10.3.2 in that "Point of Hanging" equates to the "Ability of concrete to hold the stud?"

Answer: Yes.

Question 4: Are the minimum loads given in 3-10.3.2 test loads as opposed to actual working loads?

Answer: Yes.

Question 5: With due consideration for industry standards, is it safe to assume that a stud that passes a tension pull test (for example, of 750 lbs for 2 in. pipe) is safe to carry a working load of 326.62 lbs or 383.1 lbs, whichever is applicable?

Answer: Yes.

Question 6: Does Section 3-10.3.1 limit the loads (working) put on studs to that given in their listing?

Answer: No. The loads applied to the stud must meet the listing *and* NFPA 13 requirements.

Question 7: If "Yes" to Question 6, would that limit the engineer to the listed load?

Answer: N/A.

Question 8: If "No" to Question 6, is approval then subject only to the acceptance of the authority having jurisdiction?

Answer: No. Must comply with Section 3-10.3.

Question 9: Is the bottom line here the intent to require the hanger and its parts to meet at least 3-10.1 and that, if a powder driven stud in concrete or steel is used, that it test under tension to carry a load equal to the requirement of 3-10.1.4 except that it shall not pull out at less than 750 lbs tension for 2 in. or smaller etc? Test under tension as used here would mean a load inclusive of the manufacturer's safety factor.

Answer: No. It is not intended to require safety factors beyond those in 3-10.3.2.

3-10.3.1* Powder-driven studs, welding studs, and the tools used for installing these devices shall be listed by a testing laboratory and installed within the limits of pipe size, installation position, and construction material into which they are installed as expressed in individual listings or approvals.

A-3-10.3.1 Powder-driven studs should not be used in steel of less than ³⁄₁₆ in. (4.8 mm) total thickness.

The limitations placed on powder-actuated tools and stud welders vary depending on the individual tool. Limitations of each device and the components of its installation unit are based on its individual listing.

3-10.3.2 The ability of concrete to hold the studs varies widely according to type of aggregate and quality of concrete, and it shall be established in each case by testing concrete on the job to determine that the studs will hold a minimum load of 750 lb (341 kg) for 2-in. (51 mm) or smaller pipe, 1000 lb (454 kg) for 2½-, 3-, or 3½-in. (64-, 76-, or 89-mm) pipe, and 1200 lb (545 kg) for 4- or 5-in. (102- or 127-mm) pipe.

The concrete in the individual occupancy must be tested to ascertain that powder-driven studs will have not less than the indicated holding power. It is the intent of NFPA 13 that the safety factor imposed in 3-10.1.4 be increased as specified in 3-10.3.2 when powder-driven studs are installed in concrete but that no additional safety factors beyond this be prescribed.

3-10.3.3 When increaser couplings are used, they shall be attached directly to the powder-driven stud or welding stud.

Listings commonly permit studs with a smaller diameter than required by Table 3-10.4.1. Rods in compliance with the table must be connected directly to the studs with increaser couplings.

3-10.3.4 Welded studs or other hanger parts shall not be attached by welding to steel less than U.S. Standard, 12 gage.

Steel less than U.S. Standard, 12 gage, lacks the necessary strength for welding.

3-10.4 Rods and "U" Hooks.

3-10.4.1 Hanger rod size shall be the same as that approved for use with the hanger assembly and the size of rods shall not be less than that given in Table 3-10.4.1.

Exception: Rods of smaller diameter may be used when the hanger assembly has been tested and listed by a testing laboratory and installed within the limits of pipe sizes expressed in individual listings or approvals. For rolled threads, the rod size shall not be less than the root diameter of the thread.

Table 3-10.4.1 indicates the smallest rod sizes that may be used as part of a hanger assembly. The Exception allows for

Table 3-10.4.1   Hanger Rod Sizes

Pipe Size	Dia. of Rod in.	Dia. of Rod mm	Pipe Size	Dia. of Rod in.	Dia. of Rod mm
Up to and including 4 in.	³⁄₈	9.5	5, 6, and 8 in.	½	12.7
			10 and 12 in.	⁵⁄₈	15.9

For SI Units: 1 in. = 25.4 mm.

smaller diameters if improved technology results in such rods securing a listing.

3-10.4.2 U-Hooks.   The size of the rod material of U-hooks shall not be less than that given in Table 3-10.4.2. Drive screws shall be used only in a horizontal position as in the side of a beam in conjunction with U-hangers only.

Table 3-10.4.2   U-Hook Rod Sizes

Pipe Size	Hook Material Diameter in.	Hook Material Diameter mm
Up to 2 in.	⁵⁄₁₆	7.9
2½ in. to 6 in.	³⁄₈	9.5
8 in.	½	12.7

For SI Units: 1 in. = 25.4 mm.

3-10.4.3   The size of the rod material for eye rods shall not be less than specified in Table 3-10.4.3.

Table 3-10.4.3   Eye Rod Sizes

Pipe Size	With Bent Eye in.	With Bent Eye mm	With Welded Eye in.	With Welded Eye mm
Up to 4 in.	³⁄₈	9.5	³⁄₈	9.5
5-6 in.	½	12.7	½	12.7
8 in.	¾	19.1	½	12.7

For SI Units: 1 in. = 25.4 mm.

3-10.4.4   Threaded sections of rods shall not be formed or bent.

Bending on the threaded section of hanger rods could result in cracks at the thread root. Such cracks will affect the capability of the member structurally and may lead to failure of the rod.

3-10.4.5 Screws.   For ceiling flanges and U-hooks, screw dimensions shall not be less than those given in Table 3-10.4.5.

Exception: When the thickness of planking and thickness of flange do not permit the use of screws 2 in. (51 mm) long, screws 1¾ in. (44 mm) long may be permitted with hangers spaced not over 10 ft (3 m) apart. When the thickness of beams or joists does not permit the use of screws 2½ in. (64 mm) long, screws 2 in. (51 mm) long may be permitted with hangers spaced not over 10 ft (3 m) apart.

The screw dimensions in Table 3-10.4.5 are intended for use with hangers spaced to the maximums permitted by 3-10.1.11. The Exception recognizes that it is not always possible to use the screws specified in the table and allows for use of shorter screws with closer hanger spacing.

Table 3-10.4.5   Screw Dimension for Ceiling Flanges and U-Hooks

Pipe Size	2 Screw Flanges
Up to 2 in.	Wood Screw No. 18 × 1½ in.

Pipe Size	3 Screw Flanges
Up to 2 in.	Wood Screw No. 18 × 1½ in.
2½ in., 3 in., 3½ in.	Lag Screw ⅜ in. × 2 in.
4 in., 5 in., 6 in.	Lag Screw ½ in. × 2 in.
8 in.	Lag Screw ⅝ in. × 2 in.

Pipe Size	4 Screw Flanges
Up to 2 in.	Wood Screw No. 18 × 1½ in.
2½ in., 3 in., 3½ in.	Lag Screw ⅜ in. × 1½ in.
4 in., 5 in., 6 in.	Lag Screw ½ in. × 2 in.
8 in.	Lag Screw ⅝ in. × 2 in.

Pipe Size	U-Hooks
Up to 2 in.	Drive Screw No. 16 × 2 in.
2½ in., 3 in., 3½ in.	Lag Screw ⅜ in. × 2½ in.
4 in., 5 in., 6 in.	Lag Screw ½ in. × 3 in.
8 in.	Lag Screw ⅝ in. × 3 in.

For SI Units: 1 in. = 25.4 mm.

Drive screws, which are intended to be installed with a hammer, do not have the holding power of wood screws. They can only be installed horizontally in conjunction with U-hooks to support 2-in. (51-mm) and smaller pipe.

3-10.4.6   The size bolt or lag (coach) screw used with an eye rod or flange on the side of the beam shall not be less than specified in Table 3-10.4.6.

Exception: When the thickness of beams or joists does not permit the use of screws 2½ in. (64 mm), screws 2 in. (51 mm) may be permitted with hangers spaced not over 10 ft (3 m) apart.

Table 3-10.4.6 Minimum Bolt or Lag Screw Sizes

Size of Pipe	Size of Bolt or Lag Screw		Length of Lag Screw Used with Wood Beams	
	in.	mm	in.	mm
Up to and including 2 in.	⅜	9.5	2½	64
2½ to 6 in. (inclusive)	½	12.7	3	76
8 in.	⅝	15.9	3	76

3-10.4.7 Wood screws shall not be driven. Nails are not acceptable for fastening hangers.

3-10.4.8 Screws in the side of a timber or joist shall be not less than 2½ in. (64 mm) from the lower edge when supporting branch lines, and not less than 3 in. (76 mm) when supporting main lines. This shall not apply to 2-in. (51-mm) or thicker nailing strips resting on top of steel beams.

Screws must be located correctly to ensure that the weight of the water-filled pipe is adequately supported in accordance with 3-10.1.2.

3-10.4.9 The minimum plank thickness and the minimum width of the lower face of beams or joists in which lag screw rods are used shall be as given in Table 3-10.4.9.

Table 3-10.4.9
Minimum Plank Thicknesses and Beam or Joist Widths

Pipe Size	Nominal Plank Thickness		Nominal Width of Beam or Joist Face	
	in.	mm	in.	mm
Up to 2 in.	3	76	2	51
2½ in. to 3½ in.	4	102	2	51
4 in. and 5 in.	4	102	3	76
6 in.	4	102	4	102

3-10.4.10 Lag screw rods shall not be used for support of pipes larger than 6 in. (152 mm). All holes for lag screw rods shall be predrilled ⅛ in. (3.2 mm) less in diameter than the root diameter of the lag screw thread.

The thickness required in Table 3-10.4.9 is needed for compliance with 3-10.1 Exception. Predrilling holes ⅛ in. (3.2 mm) smaller than the root diameter of the lag screw thread combines ease of installation with assurance of the needed gripping of the lag into the wood.

3-10.5 Location of Hangers on Branch Lines.   This subsection applies to the support of steel pipe or copper tube as specified in 3-1.1.1, subject to the provisions of 3-10.1.11.

3-10.5.1   On branch lines, there shall be not less than one hanger for each length of pipe.

Exception: Hangers may be located as provided in 3-10.5.2 to 3-10.5.5 inclusive.

Because of unique characteristics of their fittings, some manufacturers may recommend hanger locations exceeding the requirements of this section. Under such circumstances the manufacturer's instructions should be followed.

3-10.5.2*   When sprinklers are spaced less than 6 ft (1.8 m) apart, hangers may be spaced up to a maximum of 12 ft (3.7 m). (*See Figure A-3-10.5.2.*)

Unless unique characteristics of the fitting utilized require support at each fitting, spacing hangers up to 12 ft (3.7 m) apart will adequately support the branch line piping when sprinklers are less than 6 ft (1.8 m) apart. Under this section, consideration should be given to the flexibility of the joining methods being used at the individual sprinklers in judging omission of hangers at each sprinkler.

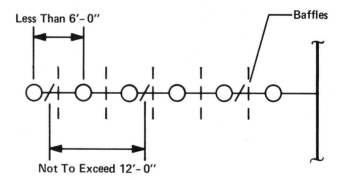

For SI Units: 1 in. = 25.4 mm; 1 ft = 0.3048 m.

Figure A-3-10.5.2   Distance Between Hangers.

3-10.5.3   Starter lengths less than 6 ft (1.8 m) do not require a hanger, except on the end line of a side-feed system or where an intermediate cross main hanger has been omitted.

3-10.5.4   The distance between a hanger and the center line of an upright sprinkler shall not be less than 3 in. (76 mm).

Guidance for positioning the deflector with respect to the hanger will insure that the hanger does not become an obstruction to discharge.

3-10.5.5* The unsupported length between the end sprinkler and the last hanger on the line shall not be greater than 36 in. (914 mm) for 1-in. (25-mm) pipe or 48 in. (1219 mm) for 1¼-in. (33-mm) or larger pipe. When either of these limits is exceeded, the pipe shall be extended beyond the end sprinkler and supported by an additional hanger. (*See 3-8.1.2.1.*)

Exception: When the maximum pressure at the sprinkler exceeds 100 psi (6.9 bars), and a branch line above a ceiling supplies sprinklers in a pendent position below the ceiling the hanger assembly supporting the pipe supplying an end sprinkler in a pendent position shall be of a type that prevents upward movement of the pipe. [See Figure A-3-10.5.5(a).]

The unsupported length between the end sprinkler in a pendent position or drop nipple and the last hanger on the branch line shall not be greater than 12 in. (305 mm) for steel pipe or 6 in. (152 mm) for copper pipe. When this limit is exceeded, the pipe shall be extended beyond the end sprinkler and supported by an additional hanger. The hanger closest to the sprinkler shall be of a type that clamps to and prevents upward movement of the piping. [See Figure A-3-10.5.5(a).]

As Figure A-3-10.5.5 illustrates, excessive overhang is overcome by extending the branch line to the next structural member. The 36-in. (914-mm) overhang limitation is also applicable to ¾-in. (19-mm) copper pipe. Also see Figure A-3-10.5.5 Exception on page 162.

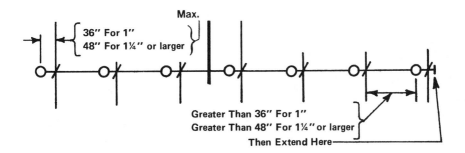

For SI Units: 1 in. = 25.4 mm; 1 ft = 0.3048 m.

NOTE: For pendent sprinklers below a ceiling and exposed to maximum pressure greater than 100 psi, see Figure A-3-10.5.5 Exception.

Figure A-3-10.5.5   Distance from Sprinkler to Hanger.

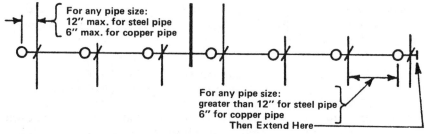

For any pipe size:
12" max. for steel pipe
6" max. for copper pipe

For any pipe size:
greater than 12" for steel pipe
6" for copper pipe
Then Extend Here

For SI Units: 1 in. = 25.4 mm; 1 ft = 0.3048 m.

NOTE 1:   The pendent sprinkler may be installed either directly in the fitting at the end of the line, or in a fitting at the bottom of a drop nipple.

NOTE 2:   Hanger closest to sprinkler shall be of a type that clamps to and prevents upward movement of the pipe.

Figure A-3-10.5.5   Exception. Distance from sprinkler to hanger where maximum pressure exceeds 100 psi (6.9 bars) and a branch line above a ceiling supplies pendent sprinklers below the ceiling.

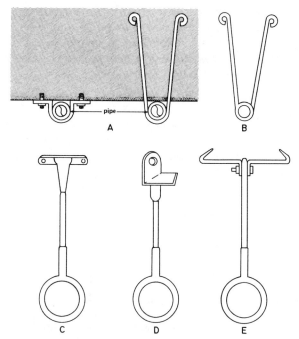

A U-type hangers for branch lines
B Wraparound U-hook
C Adjustable clip for branch lines
D Side beam adjustable hanger
E Adjustable coach screw clip for branch lines

Figure A-3-10.5.5(a)   Examples of Acceptable Hangers for End of Line (or Armover) Pendent Sprinklers.

3-10.5.6* The length of an unsupported armover to a sprinkler shall not exceed 24 in. (610 mm) for steel pipe or 12 in. (305 mm) for copper tube. *(See 3-8.1.2.1.)*

Exception: When the maximum pressure at the sprinkler exceeds 100 psi and a branch line above a ceiling supplies sprinklers in a pendent position below the ceiling, the length of an unsupported armover to a sprinkler or drop nipple shall not exceed 12 in. (305 mm) for steel pipe and 6 in. (152 mm) for copper tube.

When the limits of unsupported armover lengths of 3-10.5.6 or this Exception are exceeded, the hanger closest to the sprinkler shall be of a type that prevents upward movement of the piping. [See Figure A-3-10.5.5(a).]

Tests have shown that unrestrained pendent sprinklers operating at high pressures create a thrust that can lift the branch line piping. At the end sprinkler and at armovers the amount of thrust is sufficient to lift a pendent sprinkler out of the ceiling tile and up into the concealed space.

The requirements relevant to unsupported armovers must be used in combination with 3-8.1.2.1. Fittings having a tendency to rotate unless restrained will either incorporate a mechanism to make rotation impossible or will as part of their listing have

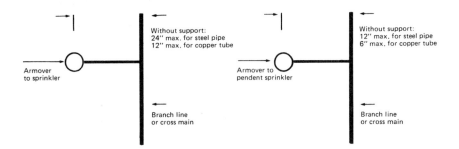

NOTE: For a pendent sprinkler below a ceiling and exposed to maximum pressure greater than 100 psi, see Figure A-3-10.5.6 Exception.

Figure A-3-10.5.6. Maximum Length for Unsupported Armover.

NOTE: The pendent sprinkler may be installed either directly in the fitting at the end of the armover or in a fitting at the bottom of a drop nipple.

Figure A-3-10.5.6 Exception. Maximum length of unsupported armover when the maximum pressure exceeds 100 psi (6.9 bars) and a branch line above a ceiling supplies pendent sprinklers below the ceiling.

a requirement for a hanger on any arm regardless of length. Incorporation of these requirements will maintain system effectiveness at unusually high operating pressures.

3-10.5.7 Wall mounted sidewall sprinklers shall be restrained to prevent movement.

3-10.6 Location of Hangers on Cross Mains. This subsection applies to the support of steel pipe only as specified in 3-1.1.1, subject to the provisions of 3-10.1.11. Intermediate hangers shall not be omitted for copper tube.

3-10.6.1* On cross mains, there shall be at least one hanger between each two branch lines.

Exception No. 1: In bays having two branch lines, the intermediate hanger may be omitted provided that a hanger attached to a purlin is installed on each branch line located as near to the cross main as the location of the purlin permits. [See Figure A-3-10.6.1(a).] Remaining branch line hangers shall be installed in accordance with 3-10.5.

Exception No. 2: In bays having three or more branch lines, either side or centerfeed, one (only) intermediate hanger may be omitted provided that a hanger attached to a purlin is installed on each branch line located as near to the cross main as the location of the purlin permits. [See Figures A-3-10.6.1(b) and A-3-10.6.1(c).] Remaining branch line hangers shall be installed in accordance with 3-10.5.

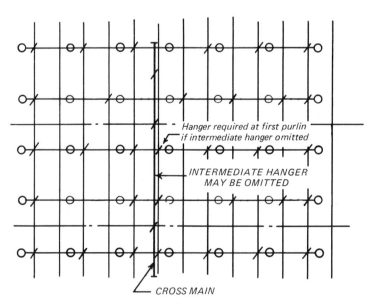

Figure A-3-10.6.1(a)   Hangers on Cross Main.

Copper tube cross mains must have at least one hanger between each two branch lines. With steel pipe cross mains, the designer can either install a hanger between each two branch lines or install hangers on each branch line as near as possible to the cross main and omit one intermediate cross main hanger in each bay.

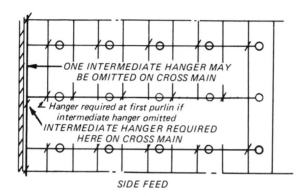

Figure A-3-10.6.1(b)   Hanger Omission on Side Feed System.

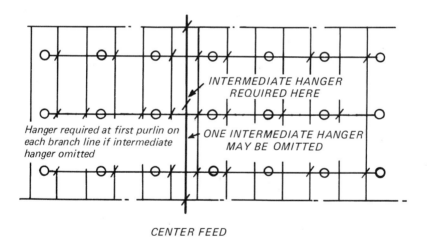

Figure A-3-10.6.1(c)   Hangers on Cross Main—Center Feed System.

3-10.6.2   At the end of the cross main, intermediate trapeze hangers shall be installed unless the cross main is extended to the next framing member with an ordinary hanger installed at this point, in which event an intermediate hanger may be omitted in accordance with 3-10.6.1, Exceptions No. 1 and No. 2.

The option to omit the intermediate cross main hanger in accordance with the Exceptions to 3-10.6.1 is applicable to the last piece of cross main only if the main is extended to the next framing member and a hanger is also installed at that point.

3-10.7 Support of Risers.

3-10.7.1 Risers shall be supported by attachments directly to the riser or by hangers located on the horizontal connections close to the riser.

Risers must be adequately supported to avoid excess strain on fittings and joints. Listed supports exist for risers and must be used to comply with this section.

3-10.7.2 In multistory buildings, riser supports shall be provided at the lowest level, at each alternate level above, above and below offsets, and at the top of the riser. Supports above the lowest level shall also restrain the pipe to prevent movement by an upward thrust when flexible fittings are used. Where risers are supported from the ground, the ground support constitutes the first level or riser support. Where risers are offset or do not rise from the ground, the first ceiling level above the offset constitutes the first level of riser support.

Restraints are required at alternate levels and at the points most subject to stress.

3-10.7.3 Sprinkler and tank risers in vertical shafts, or in buildings with ceilings over 25 ft (7.6 m) high, shall have at least one support for each riser pipe section.

3-10.7.4 Clamps supporting pipe by means of set screws shall not be used.

To adequately support a riser, the full surface of a clamp must bear against the surface of the pipe.

B-3 System Components.

B-3-1 Sleeves for Pipe Risers.   (*See Figure B-3-1.*)

B-3-1.1 Sprinkler piping passing through floors of concrete or waterproof construction should have properly designed substantial thimbles or sleeves projecting 3 to 6 in. (76 to 152 mm) above the floor to prevent possible floor leakage, except in areas subject to earthquakes. (*See A-3-5.3.1.*) The space between the pipe and sleeve should be caulked with oakum or equivalent material. If floors are of cinder concrete, thimbles or sleeves should extend all the way through to protect the piping against corrosion.

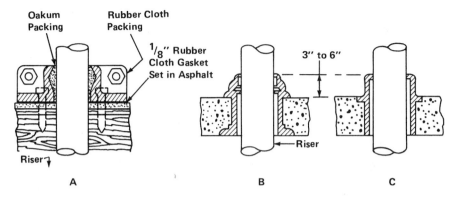

A–For Wood or Concrete Floors; B and C–For Concrete Floors.

Figure B-3-1   Watertight Riser Sleeves.

B-3-1.2   Ordinary floors through which pipes pass should be made reasonably tight around the risers, except in areas subject to earthquakes. (*See A-3-5.3.1.*)

B-3-1.3   The time required to remove water from a trapped section of the system is important. In extreme cases the time required to drain the system may allow water to freeze.

3-11 Sprinklers.

Currently, no less than 20 different types and variations of automatic sprinklers exist on the market. This section describes the correct uses of some of these sprinklers.

3-11.1 Types of Sprinklers.   Some of the commonly used sprinklers are as follows:

(a)   *Upright Sprinklers.*   Sprinklers designed to be installed in such a way that the water spray is directed upwards against the deflector.

(b)   *Pendent Sprinklers.*   Sprinklers designed to be installed in such a way that the water stream is directed downward against the deflector.

(c)   *Sidewall Sprinklers.*   Sprinklers having special deflectors that are designed to discharge most of the water away from the nearby wall in a pattern resembling one quarter of a sphere, with a small portion of the discharge directed at the wall behind the sprinkler.

(d)   *Extended Coverage Sidewall Sprinklers.*   Sprinklers with special extended, directional, discharge patterns.

Figure 3.6. An approved and listed sprinkler showing upright (left) and pendent (right) models of the same issue. Sprinklers shown are Reliable Model G.

Figure 3.7. Star sidewall sprinkler.

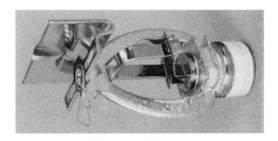

Figure 3.8. Grinnell Model FR-1 extended coverage horizontal sidewall sprinkler.

(e) *Open Sprinklers.* Sprinklers from which the actuating elements have been removed.

Figure 3.9. Grinnell open sprinkler.

(f) *Corrosion-Resistant Sprinklers.* Sprinklers with special coatings or platings to be used in an atmosphere that would corrode an uncoated sprinkler.

Corrosion-resistant sprinklers are usually standard sprinklers with a corrosion-resistant coating, such as wax or lead. Any coating of this type can be applied only by the manufacturer.

Figure 3.10. Grinnell automatic Protectospray nozzle.

(g) *Nozzles.* Devices for use in applications requiring special discharge patterns, directional spray, fine spray, or other unusual discharge characteristics.

(h)* *Dry-Pendent Sprinklers.* Sprinklers for use in a pendent position in a dry-pipe system or a wet-pipe system with the seal in a heated area.

(i)* *Dry Upright Sprinklers.* Sprinklers that are designed to be installed in an upright position, on a wet-pipe system, to extend into an unheated area with a seal in a heated area.

A-3-11.1(h) and (i) Under certain ambient conditions, wet-pipe systems having dry-pendent (or upright) sprinklers may freeze due to heat loss by conduction. Therefore, due consideration should be given to the amount of heat maintained in the heated space, the length of the nipple in the heated space, and other relevant factors.

The dry upright sprinkler is similar to the dry-pendent except that the sprinkler uses an upright deflector.

(j) *Ornamental Sprinklers.* Sprinklers that have been painted or plated by the manufacturer.

Some ornamental sprinklers may be nothing more than a corrosion-resistant sprinkler with a special coating that may be similar to many interior building finish products.

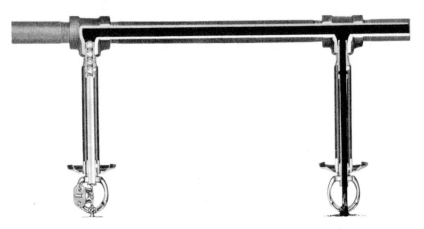

Figure 3.11. Representative dry-pendent sprinkler. When the solder holding the fusible link together melts, the levers and link parts are ejected away from the sprinkler. The inner tube, which also serves as an orifice, drops to a predetermined position, allowing the sealing elements to pass through the tube and away from the sprinkler. This, in turn, allows water to flow through the tube and strike the deflector, which distributes it in a spray pattern comparable to a ½-in. standard sprinkler. Shown is Reliable's Model C dry-pendent.

(k) *Flush Sprinklers.* Sprinklers in which all or part of the body, including the shank thread, is mounted above the lower plane of the ceiling.

Figure 3.12. Reliable Model A flush type automatic sprinkler before operation (left) and after operation (right).

(l) *Recessed Sprinklers.* Sprinklers in which all or part of the body, other than the shank thread, is mounted within a recessed housing.

(m) *Concealed Sprinklers.* Recessed sprinklers with cover plates.

(n) *Old-Style Sprinklers.* Sprinklers that direct only from 40 to 60 percent of the total water initially in a downward direction and that are designed to be installed with the deflector either upright or pendent.

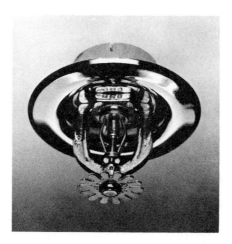

Figure 3.13. A Central Model "A" recessed sprinkler.

(o) *Residential Sprinklers.* Sprinklers that have been specifically listed for use in residential occupancies.

(p) *Intermediate Level Sprinklers.* Sprinklers equipped with integral shields to protect their operating elements from the discharge of sprinklers installed at higher elevations.

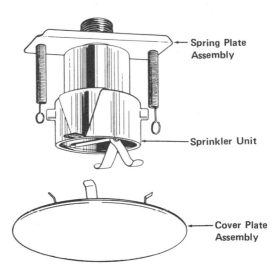

Spring Plate
Assembly

Sprinkler Unit

Cover Plate
Assembly

Figure 3.14. Stargard Model G concealed ceiling sprinkler. Cover plate drops away when heat is applied to bottom side of plate.

Figure 3.15. Grinnell Model F954 residential sprinkler.

(q) *Special Sprinklers.*   Sprinklers that have been tested and listed as prescribed in 4-1.1.3.

Special sprinklers are listed for installations with protection areas or distances between sprinklers different from those specified in this standard. Such listings are intended to be based on tests specifically designed to evaluate the sprinkler's performance under the special listing criteria. Extended coverage and extra large orifice sprinklers are types of special sprinklers.

Figure 3.16. Star Model LD intermediate level sprinkler.

(r) *Quick-Response Sprinklers.*   A type of special sprinkler incorporating a fast-actuating heat-responsive element.

Quick-Response Sprinkler (QRS) technology is a derivative of the residential sprinkler technology. Quick-response sprinklers are tested by the same standard as standard response sprinklers, specifically UL 199, *Standard for Automatic Sprinklers for Fire Protection Service.* Many people incorrectly interchange the terms residential, fast response, and quick-response. See commentary to 3-11.2.

(s) *Large-Drop Sprinklers.*   Listed large-drop sprinklers are characterized by a K factor between 11.0 and 11.5, and proven ability to meet prescribed penetration, cooling, and distribution criteria prescribed in large-drop sprinkler examination requirements. The deflector/discharge characteristics of large-drop sprinklers generate drops of such size and velocity as to enable effective penetration of the high-velocity fire plume.

Figure 3.17. Viking Model A large-drop sprinkler.

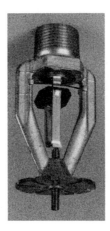

Figure 3.18. Grinnell Model ESFR-1 early suppression fast response sprinkler.

The upright, the pendent, and the sidewall are the basic sprinklers. The remaining sprinklers defined in this section, with the exception of the nozzles and the old-style sprinklers, are variations of the basic sprinklers modified to address specific needs. The primary difference between the old-style sprinkler, which is designed for installation in either the upright or pendent position, and the current upright and pendent sprinklers is the design of the deflector, which for the old-style sprinkler directed approximately 40 percent of the water up against the ceiling with the remainder of the water flowing to the fire. Nozzles are special application devices that have unique characteristics designed to meet specific needs.

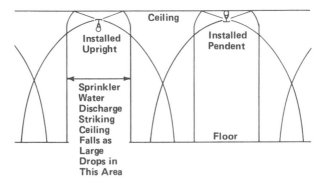

Figure 3.19. Principal distribution pattern of water from old-style sprinklers (previous to 1953).

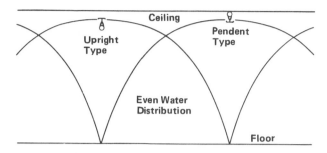

Figure 3.20. Principal distribution pattern of water from standard sprinklers (in use since 1953).

3-11.2 Use of Sprinklers.

To describe the recent developments in sprinkler link technology, the following table provides a summary of various sprinkler criteria.

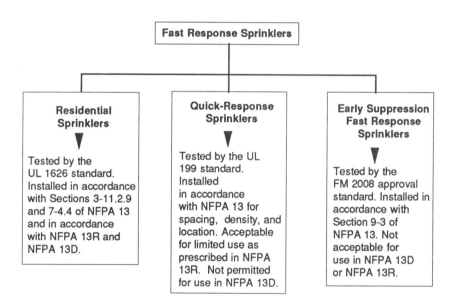

Each of these sprinklers will respond sooner to a fire than a conventional standard response sprinkler. The Response Time Index (RTI) is used to comparatively describe the sensitivity of the sprinkler link for any given sprinkler. The RTI for the group of fast response sprinklers shown in the table will on average range from 40 ft$^{1/2}$ s$^{1/2}$ - 60 ft$^{1/2}$ s$^{1/2}$ (22 s$^{1/2}$ m$^{1/2}$ - 33 s$^{1/2}$ m$^{1/2}$). RTI values for standard response sprinklers will typically be in the range of 150 ft$^{1/2}$ s$^{1/2}$ - 200 ft$^{1/2}$ s$^{1/2}$ (83 s$^{1/2}$ m$^{1/2}$ - 110 s$^{1/2}$ m$^{1/2}$).

The confusion surrounding the fast response sprinkler group may be alleviated with the adoption of a convention of terms as suggested by this table. Stating that all three types of sprinklers in the table contain a fast response link will allow one to discriminate against these three basic types which have been placed in the table.

The supplement contained at the end of this book provides a summary of the fast response sprinkler technology and data.

3-11.2.1* Only listed sprinklers shall be used and shall be installed in accordance with their listing.

Exception: When construction features or other special situations require unusual water distribution, listed sprinklers may be installed in other positions than anticipated by their listing to achieve specific results.

Any limitations placed on a sprinkler, such as the hazard classifications where it may be used or limitations placed on its temperature rating, are included as part of its listing.

A-3-11.2.1 Upright sprinklers should be installed with the frame parallel to the branch line pipe to reduce to minimum the obstruction of the discharge pattern.

Formal Interpretation

Question: Does 3-11.2.1 provide for the listing of sprinklers designed specifically for smaller protection areas or distances between sprinklers than specified elsewhere in NFPA 13?

Answer: No.

3-11.2.2 Sprinklers shall not be altered in any respect or have any type of ornamentation or coating applied after shipment from the place of manufacture.

Any finish applied to a sprinkler can only be applied by the manufacturer. This includes both decorative finishes and any finish applied to protect the sprinkler from a corrosive environment.

3-11.2.3 Sprinklers shall not be used for system working pressures exceeding 175 psi (12.1 bars).

Exception: Higher design pressures may be used when sprinklers are listed for those pressures.

Currently, all sprinklers are designed for a maximum working pressure of 175 psi (12.1 bars). The Exception allows for the future listing of higher pressure designs. See also the Exception to 1-8.1.3.

3-11.2.4 Old-style sprinklers shall not be used in a new installation.

Exception No. 1:   For installation under piers and wharves where construction features require an upward discharge to wet the underside of decks and structural members supporting the decks, a sprinkler that projects water upward to wet the overhead shall be used. This can be accomplished by using standard pendent sprinklers installed in an upright position or by the use of old-style sprinklers. See NFPA 307, Standard for the Construction and Fire Protection of Marine Terminals, Piers, and Wharves.

Exception No. 2:   Old-style sprinklers shall be installed in fur storage vaults. See 4-4.16.3. Also see NFPA 81, Standard for Fur Storage, Fumigation and Cleaning.

Exception No. 3:   Listed old-style sprinklers may be used when construction features or other special situations require unique water distribution.

The discharge (floor-level wetting) characteristics of old-style sprinklers are inferior to those of upright and pendent sprinklers. Upright and pendent sprinklers provide a more even distribution of water over a greater area with a more consistent droplet size. Fires originating under a pier or wharf are best controlled by old-style sprinklers because of their upward discharge characteristics. Fire testing done on fur storage vaults exclusively used old-style sprinklers, and thus the edict for their use remains some 40 years later.

3-11.2.5 Sidewall sprinklers shall be installed only in Light Hazard Occupancies.

Exception: Sidewall sprinklers specifically listed for use in Ordinary Hazard Occupancies.

The discharge characteristics of sidewall sprinklers are inferior to those of upright or pendent sprinklers. Accordingly, they are confined to use in Light Hazard Occupancies, unless the sprinkler has been specifically listed for use in Ordinary Hazard Occupancies. Many horizontal sidewall sprinklers are listed for use in Ordinary Hazard Occupancies.

3-11.2.6 Extended coverage sidewall sprinklers shall be installed only in accordance with their listing.

The characteristics of standard sidewall sprinklers are such that the rules covering their installation found in Section 4-5 of this standard are applicable regardless of manufacturer. Extended coverage (EC) sidewall sprinklers are limited to use in Light Hazard Occupancies that have smooth ceilings. Because of the allowance for greater areas of coverage than are typically designated for horizontal sidewall sprinklers, higher operating pressures are usually necessary for EC sidewall sprinklers.

3-11.2.7 Open sprinklers may be used to protect special hazards, for protection against exposures, or in other special locations.

Open sprinklers referenced in this section are standard sprinklers with the operating elements removed. For information on open window or cornice sprinklers designed for outside use for protection against exposure fires, see Chapter 6. Open sprinklers are also used in deluge systems as discussed in Chapter 5. On a larger scale, NFPA 409, *Standard on Aircraft Hangars*, requires foam-water deluge systems (which utilize open sprinklers) as the predominant means of fire protection.

3-11.2.8 Escutcheon Plates.

3-11.2.8.1 When nonmetallic ceiling plates (escutcheons) are used they shall be listed.

3-11.2.8.2 Escutcheon plates used to create a recessed or flush type sprinkler shall be part of a listed sprinkler assembly.

Components for escutcheon plates are integral to suitable operation of the sprinkler. The listing of these components assures that they will not significantly retard the operating time of the sprinkler and that they fall completely free of the deflector and allow an unobstructed spray pattern.

There have been instances of actual installations in which the dimensions of an unlisted recessed cup and a sprinkler were such that the deflector was not clear of the cup and thereby had a seriously impaired spray pattern.

3-11.2.9 Residential Sprinklers.

As presented in Table 3.1 earlier in this chapter, residential sprinklers are tested and listed in accordance with UL 1626.

Since residential sprinklers are tested for listing using a residential fire scenario they are permitted in residential portions of all occupancies. In other portions of such occupancies, sprinklers must be listed for general usage and installed in accordance with the requirements of this standard.

NFPA 13R, *Standard for the Installation of Sprinkler Systems in Residential Occupancies up to Four Stories in Height,* covers sprinkler system design in certain low-rise residential facilities.

Formal Interpretation

Question: Is NFPA 13D appropriate for use in multiple (three or more) attached dwellings under any condition?

Answer: No. NFPA 13D is appropriate for use only in one- and two-family dwellings and mobile homes. Buildings which contain more than two dwelling units shall be protected in accordance with NFPA 13. Section 3-11.2.9 of NFPA 13 permits residential sprinklers to be used in residential portions of other buildings provided all other requirements of NFPA 13, including water supplies, are satisfied.

Note: Building codes may contain requirements such as 2-hour fire separations which would permit adjacent dwellings to be considered unattached.

3-11.2.9.1* Residential sprinklers may be used in dwelling units located in any occupancy provided they are installed in conformance with their listing and the positioning requirements of NFPA 13D, *Standard for the Installation of Sprinkler Systems in One- and Two-Family Dwellings and Mobile Homes.* One-half inch or larger orifice residential sprinklers may be used in dry-pipe systems when the design area is in compliance with Chapter 2.

A-3-11.2.9.1 The response and water distribution pattern of listed residential sprinklers have been shown by extensive fire testing to provide better control than conventional sprinklers in residential occupancies. These sprinklers are intended to prevent flashover in the room of fire origin, thus improving the chance for occupants to escape or be evacuated.

3-11.2.9.2 When residential sprinklers are installed within a compartment as defined in 7-4.4, all sprinklers shall be from the same manufacturer and have the same heat-response element including temperature rating.

Depending upon the velocity of airflow, the sensitivity of differing residential sprinklers varies. To avoid the possibility of reverse order or spot operation, all residential sprinklers in a compartment must be of the same manufacturer and have the same heat-responsive element.

3-11.3 Replacement of Sprinklers.

3-11.3.1   When sprinklers are replaced, the replacement sprinkler shall be of the same type, orifice, and temperature rating unless conditions require a different type sprinkler be installed. The replacement sprinkler shall then be of a type, orifice, and temperature rating to suit the new conditions.

In replacing sprinklers, in addition to matching the type (upright, pendent, etc.), the orifice size, and the temperature rating, care must be taken to ensure that the replacement sprinkler is installed in accordance with its listing as required by 3-11.2.1. It would be very easy, for example, to replace a sidewall sprinkler in an Ordinary Hazard Occupancy with a sprinkler that is listed for Light Hazard use only.

3-11.3.2   Old-style sprinklers may be replaced with old-style sprinklers, or with the appropriate pendent or upright sprinkler.

3-11.3.3   Old-style sprinklers shall not be used to replace pendent or upright sprinklers.

Because of their better discharge characteristics, greater coverage area per sprinkler is permitted for upright and pendent sprinklers than was permitted for the old-style sprinklers.

3-11.3.4   Extreme care shall be exercised when replacing horizontal sidewall and extended coverage sidewall sprinklers to assure the correct replacement sprinkler is installed.

3-11.3.5   Sprinklers that have been painted or coated, except by the manufacturer, shall be replaced and shall not be cleaned by use of chemicals, abrasives, or other means. (*See 3-11.9.2.*)

One instance has been documented in which an installer, in order to save expenses, attempted to provide chrome plating on sprinklers in the shop. The fusible link and sprinkler seal had both been coated, and one of the largest product recalls in the sprinkler industry was undertaken.

3-11.4 Corrosion-Resistant, Wax-Coated, or Similar Sprinklers.

Sprinklers with special coatings or platings are specifically listed for use in atmospheres that would corrode an uncoated sprinkler. Attempts by parties other than the manufacturer to protect sprinklers would result in ineffective protection of the sprinkler and/or serious impairment of its operation. The listing of sprinklers with protective or special coatings requires that the coatings be applied at the manufacturer's facility. Field application of these platings and coatings is not allowed.

3-11.4.1* Listed corrosion-resistant or special coated sprinklers shall be installed in locations where chemicals, moisture, or other corrosive vapors sufficient to cause corrosion of such devices exist.

A-3-11.4.1 Examples of such locations are paper mills, packing houses, tanneries, alkali plants, organic fertilizer plants, foundries, forge shops, fumigation, pickle and vinegar works, stables, storage battery rooms, electroplating rooms, galvanizing rooms, steam rooms of all descriptions, including moist vapor dry kilns, salt storage rooms, locomotive sheds or houses, driveways, areas exposed to outside weather such as piers and wharves exposed to salt air, areas under sidewalks, around bleaching equipment in flour mills, all portions of cold storage buildings where a direct ammonia expansion system is used, and portions of any plant where corrosive vapors prevail.

3-11.4.2 Care shall be taken in the handling and installation of wax-coated or similar sprinklers to avoid damaging the coating.

3-11.4.3 Corrosion-resistant coatings shall be applied only by the manufacturer of the sprinkler.

Exception: Any damage to the protective coating occurring at the time of installation shall be repaired at once using only the coating of the manufacturer of the sprinkler in the approved manner so that none of the sprinkler will be exposed after installation has been completed.

3-11.5* Sprinkler Discharge Characteristics and Identification.

A-3-11.5 The following Table A-3-11.5 shows the nominal discharge capacities of approved sprinklers having a nominal ½-in. (13-mm) orifice at various pressures up to 100 psi (6.9 bars).

Table A-3-11.5   Nominal Discharge Capacities

Pressure at Sprinkler psi	Discharge gpm	Pressure at Sprinkler psi	Discharge gpm
10	18	35	34
15	22	50	41
20	25	75	50
25	28	100	58

For SI Units: 1 gpm = 3.785 L/min; 1 psi = 0.0689 bar.

3-11.5.1 Table 3-11.5 shows the K factor, relative discharge, and identification for sprinklers having different orifice sizes.

Exception: Special listed sprinklers may have pipe threads different from those shown in Table 3-11.5.

Table 3-11.5   Sprinkler Discharge Characteristics Identification

Nominal Orifice Size (in.)[1]	Orifice Type	K Factor[2]	Percent of Nominal ½ in. Discharge	Thread Type	Pintle	Nominal Orifice Size Marked On Frame
¼	Small	1.3-1.5	25	½ in. NPT	Yes	Yes
⁵⁄₁₆	Small	1.8-2.0	33.3	½ in. NPT	Yes	Yes
⅜	Small	2.6-2.9	50	½ in. NPT	Yes	Yes
⁷⁄₁₆	Small	4.0-4.4	75	½ in. NPT	Yes	Yes
½	Standard	5.3-5.8	100	½ in. NPT	No	No
¹⁷⁄₃₂	Large	7.4-8.2	140	¾ in. NPT	No	No
				or ½ in. NPT	Yes	Yes
⅝	Extra Large	11.0-11.5	200	¾ in. NPT	Yes	Special Deflector

[1]See A-3-11.5.2

[2]K factor is the constant in the formula $Q = K \sqrt{P}$

Where Q = Flow in gpm
P = Pressure in psi

For SI Units: $Qm = Km \sqrt{Pm}$

Where Qm = Flow in L/min
Pm = Pressure in bars
Km = 14 K

The orifice sizes are indicated as nominal in Table 3-11.5 because the various sprinklers differ slightly, plus or minus, from the norm. The exact orifice size of a sprinkler is indicated by its "K" factor. The K factor is used in determining the sprinkler's flow rate, which is of major consequence in hydraulically designed systems. (*See Chapter 7.*) In most instances, sprinklers with ½-in. (12.7-mm) orifices and ½-in. pipe threads or sprinklers with ¹⁷⁄₃₂-in. (13-mm) orifices and ¾-in. (19-mm) pipe threads are installed. Sprinklers of other than those combinations are indicated by a pintle, which is a metal rod extending beyond the sprinkler's deflector.

3-11.5.2 For Light Hazard Occupancies not requiring as much water as is discharged by a nominal ½-in. (12.7-mm) orifice sprinkler operating at 7 psi (0.5 bar), sprinklers having a smaller orifice may be used subject to the following restrictions:

The reference to a 7 psi (0.5 bar) minimum clarifies that small orifice sprinklers should only be permitted when the need for water is less than that provided by a ½-in. (12.7-mm) sprinkler operating at minimum pressure provides. The discharge pattern of small orifice sprinklers becomes distorted at high pressures.

Small orifice sprinklers are restricted to Light Hazard Occupancies because they are not as efficient in fire control as standard and large orifice sprinklers. They are restricted to wet systems due to the danger of the orifice being obstructed by scale and foreign material carried at the head of the high-velocity water flowing in other types of systems. Figure 3.21 on page 184 is one example of a small orifice sprinkler.

(a) The system shall be hydraulically calculated. (*See Section 8-1.*)

(b) Small orifice sprinklers shall not be used on dry-pipe, preaction, or combined dry-pipe and preaction systems.

Exception: Outside sprinklers for protection from exposure fires installed in conformance with Chapter 6.

On systems that normally are filled with air rather than water, there is a tendency to form more internal scale; therefore, small orifice sprinklers might become clogged.

(c) An approved strainer shall be provided in the riser or feed main that supplies sprinklers having orifices smaller than ⅜ in. (9.5 mm).

3-11.5.3 For locations or conditions requiring more water than is discharged by a nominal ½-in. (12.7-mm) orifice sprinkler, a sprinkler having a larger orifice may be used.

3-11.5.4 Sprinklers having orifice sizes exceeding ½ in. (12.7 mm) and having ½ in. NPT shall not be installed in new sprinkler systems.

Large orifice sprinklers with ½-in. (12.7-mm) national pipe threads are intended solely for use where conditions merit replacing existing ½-in. orifice sprinklers with sprinklers having ¹⁷/₃₂-in. orifices. Some sprinklers having larger orifices have been developed and listed with special conditions of use and are limited by those conditions. See Figure 3.22 on page 184.

3-11.6* Temperature Ratings, Classifications, and Color Coding.

Figure 3.21. Viking small orifice sprinkler.

Figure 3.22. Viking large orifice sprinkler. This model is fitted with ¾-in. threads.

Temperature selection of a particular sprinkler is important in order to achieve optimum fire control. The selection criteria for the temperature rating of a given sprinkler is a function of the occupancy classification and expected ambient ceiling temperatures in the vicinity of the sprinkler.

If sprinklers with differing response characteristics are installed in a compartment, there is a distinct possibility of sprinklers remote from a fire operating prior to sprinklers in the immediate vicinity of the fire.

A-3-11.6 Information regarding the highest temperature that may be encountered in any location in a particular installation may be obtained by use of a thermometer that will register the highest temperature encountered, which should be hung for several days in the questionable location, with the plant in operation.

When an occupancy hazard normally may be expected to produce a fast-developing fire or a rapid rate of heat release, the use of sprinklers of high temperature classification, as a means of limiting the total number of sprinklers that might open in a fire, is recommended. Since the number of sprinklers that might be expected to open will be reduced where the water pressure effective in first operating sprinklers is at least 75 psi (5.2 bars) without the disadvantage of a potential increase in fire damage, this alternative should be given first consideration.

NOTE: Fire tests have shown that the number of sprinklers that might be expected to open, particularly under conditions where fast-developing fires may be expected, can be limited by the use of sprinklers of High Temperature Classification. This may be of advantage in reducing the number of sprinklers that would otherwise open outside the area directly involved in a fire and decrease the overall water demand. However, some increase in fire damage and fire temperatures may be expected when sprinklers of Intermediate or High Temperature Classification are used.

Some occupancies employ high-temperature fumigation processes requiring consideration in the selection of sprinkler temperature ratings.

3-11.6.1   The standard temperature ratings of automatic sprinklers are shown in Table 3-11.6.1. Automatic sprinklers shall have their frame arms colored in accordance with the color code designated in Table 3-11.6.1, with the following Exceptions:

Exception No. 1:   The color identification for corrosion-resistant sprinklers may be a dot on the top of the deflector, the color of the coating material, or colored frame arms.

Exception No. 2:   Color identification is not required for ornamental sprinklers such as factory plated or factory painted sprinklers, or for recessed, flush, or concealed sprinklers.

Table 3-11.6.1   Temperature Ratings, Classifications, and Color Codings

Max. Ceiling Temp.		Temperature Rating		Temperature Classification	Color Code	Glass Bulb Colors
°F	°C	°F	°C			
100	38	135 to 170	57 to 77	Ordinary	Uncolored or Black	Orange or Red
150	66	175 to 225	79 to 107	Intermediate	White	Yellow or Green
225	107	250 to 300	121 to 149	High	Blue	Blue
300	149	325 to 375	163 to 191	Extra High	Red	Purple
375	191	400 to 475	204 to 246	Very Extra High	Green	Black
475	246	500 to 575	260 to 302	Ultra High	Orange	Black
625	329	650	343	Ultra High	Orange	Black

Sprinklers are color coded in accordance with 3-11.6.1 to provide a ready means of establishing the temperature classifications of the their operating elements. Table 3-11.6.1 indicates the range of temperatures for sprinklers in each classification and the maximum ceiling temperatures for which each classification may be installed. Exception No. 2 recognizes that traditional color codings are not applicable to specially coated sprinklers, such as decorative or ornamental sprinklers.

3-11.6.2 Ordinary temperature rated sprinklers shall be used throughout buildings.

Exception No. 1: Where maximum ceiling temperatures exceed 100°F (38°C), sprinklers with temperature ratings in accordance with the maximum ceiling temperatures of Table 3-11.6.1 shall be used.

Exception No. 2: Intermediate and high-temperature sprinklers may be used throughout Ordinary and Extra Hazard Occupancies.

Exception No. 3: Sprinklers of intermediate and high temperature classifications shall be installed in specific locations as required by 3-11.6.3 and 5-6.1.4.1.

Exception No. 4: When permitted or required by other NFPA standards.

Higher temperature classification sprinklers are preferable for some types of fires where ordinary temperature classification sprinklers would tend to operate beyond the fire area and thereby reduce the water discharge density available over the fire. In some high-heat-release fires with high thermal updraft discharge, water from sprinklers is carried back toward the ceiling as steam and, when it condenses on ordinary temperature sprinklers, will cause them to operate. This is the phenomenon considered to be responsible for sprinklers operating beyond the fire area.

Extra high, very extra high, and ultra high temperature sprinklers may be specified in areas immediately above industrial ovens, furnaces, and boilers.

3-11.6.3 The following practices shall be observed to provide sprinklers of other than ordinary temperature classification unless maximum expected temperatures are otherwise determined, or unless high-temperature sprinklers are used throughout [*see Tables 3-11.6.3(a) and 3-11.6.3(b) and Figure 3-11.6.3(a)*]:

(a) Sprinklers near unit heaters. Sprinklers in the heater zone shall be high and sprinklers in the danger zone intermediate temperature classification.

(b) Sprinklers located within 12 in. (305 mm) to one side or 30 in. (762 mm) above an uncovered steam main, heating coil, or radiator shall be intermediate temperature classification.

(c) Sprinklers within 7 ft (2.1 m) of a low-pressure blowoff valve that discharges free in a large room shall be high temperature classification.

(d) Sprinklers under glass or plastic skylights exposed to the direct rays of the sun shall be intermediate temperature classification.

(e) Sprinklers in an unventilated concealed space under an uninsulated roof, or in an unventilated attic, shall be of intermediate temperature classification.

(f) Sprinklers in unventilated show windows having high-powered electric lights near the ceiling shall be intermediate temperature classification.

(g) For sprinklers protecting commercial-type cooking equipment and ventilation systems, temperature classifications of intermediate, high, or extra high shall be provided as determined by use of a temperature measuring device. (*See 4-4.17.2.*)

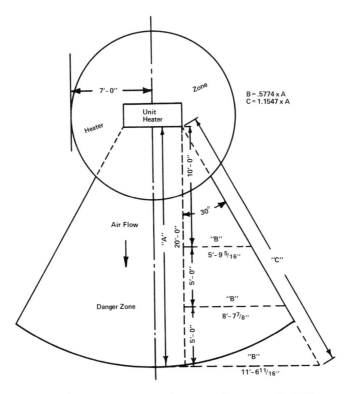

Figure 3-11.6.3(a)   Heater and Danger Zones at Unit Heaters.

Table 3-11.6.3(a)   Temperature Ratings of Sprinklers Based on Distance from Heat Sources

Type of Heat Condition	Ordinary Degree Rating	Intermediate Degree Rating	High Degree Rating
1. Heating Ducts (a) Above	More than 2 ft 6 in.	2 ft 6 in. or less	—
(b) Side and Below	More than 1 ft 0 in.	1 ft 0 in. or less	—
(c) Diffuser Downward Discharge Horizontal Discharge	Any distance except as shown under Intermediate	*Downward:* Cylinder with 1 ft 0 in. radius from edge, extending 1 ft 0 in. below and 2 ft 6 in. above *Horizontal:* Semi-cylinder with 2 ft 6 in. radius in direction of flow, extending 1 ft 0 in. below and 2 ft 6 in. above	—
2. Unit Heater (a) Horizontal Discharge	—	*Discharge Side:* 7 ft 0 in. to 20 ft 0 in. radius pie-shaped cylinder [see Figure 3-11.6.3(a)] extending 7 ft 0 in. above and 2 ft 0 in. below heater; also 7 ft 0 in. radius cylinder more than 7 ft 0 in. above unit heater	7 ft 0 in. radius cylinder extending 7 ft 0 in. above and 2 ft 0 in. below unit heater
(b) Vertical Downward Discharge [Note: For sprinklers below unit heater, see Figure 3-11.6.3(a).]	—	7 ft 0 in. radius cylinder extending upward from an elevation 7 ft 0 in. above unit heater	7 ft 0 in. radius cylinder extending from the top of the unit heater to an elevation 7 ft 0 in. above unit heater
3. Steam Mains (Uncovered) (a) Above	More than 2 ft 6 in.	2 ft 6 in. or less	—
(b) Side and Below	More than 1 ft 0 in.	1 ft 0 in. or less	—
(c) Blowoff Valve	More than 7 ft 0 in.	—	7 ft 0 in. or less

For SI Units: 1 in. = 25.4 mm; 1 ft = 0.3048 m.

Table 3-11.6.3(b) Ratings of Sprinklers in Specified Locations

Location	Ordinary Degree Rating	Intermediate Degree Rating	High Degree Rating
Skylights	—	Glass or plastic	—
Attics	Ventilated	Unventilated	—
Peaked Roof: Metal or thin boards; concealed or not concealed; insulated or uninsulated	Ventilated	Unventilated	—
Flat Roof: Metal, not concealed; insulated or uninsulated	Ventilated or unventilated	Note: For uninsulated roof, climate and occupancy may require Intermediate sprinklers. Check on job.	—
Flat Roof: Metal; concealed; insulated or uninsulated	Ventilated	Unventilated	—
Show Windows	Ventilated	Unventilated	—

Note: A check of job condition by means of thermometers may be necessary.

3-11.6.4 In case of change of occupancy involving temperature change, the sprinklers shall be changed accordingly.

3-11.7* Stock of Spare Sprinklers.

A-3-11.7 For equipment aboard vessels or in isolated locations, a greater number of sprinklers should be provided to permit equipment to be put back into service promptly after a fire. When a great number of sprinklers are likely to be opened by a flash fire, a greater number of sprinklers should be provided.

3-11.7.1 There shall be maintained on the premises a supply of spare sprinklers (never less than 6) so that any sprinklers that have operated or been damaged in any way may be promptly replaced. These sprinklers shall correspond to the types and temperature ratings of the sprinklers in the property. The sprinklers shall be kept in a cabinet located where the temperature to which they are subjected will at no time exceed 100°F (38°C).

3-11.7.2 A special sprinkler wrench shall also be provided and kept in the cabinet, to be used in the removal and installation of sprinklers.

3-11.7.3 The stock of spare sprinklers shall be as follows:

(a) For equipments not over 300 sprinklers, not less than 6 sprinklers.

(b) For equipments 300 to 1,000 sprinklers, not less than 12 sprinklers.

(c) For equipments above 1,000 sprinklers, not less than 24 sprinklers.

(d) Stock of spare sprinklers shall include all types and ratings installed.

The stock of spare sprinklers required is a minimum. Spare sprinklers of all types and ratings installed must be available. For an occupancy with a variety of types and ratings of sprinklers, the amount of spare sprinklers should be increased above the minimum.

3-11.8* Guards and Shields.   Sprinklers that are so located as to be subject to mechanical injury (in either the upright or the pendent position) shall be protected with listed guards.

A-3-11.8 Sprinklers under open gratings should be provided with shields. Shields over automatic sprinklers should not be less, in least dimension, than four times the distance between the shield and fusible element, except special sprinklers incorporating a built-in shield need not comply with this recommendation if listed for the particular application.

Any sprinkler suffering physical damage that could affect its efficiency, such as a bent deflector, must be replaced.

3-11.9 Painting and Ornamental Finishes.

3-11.9.1* When the sprinkler piping is given any kind of coating, such as whitewash or paint, care shall be exercised to see that no automatic sprinklers are coated.

A-3-11.9.1 When painting sprinkler piping or painting in areas near sprinklers, the sprinklers may be protected by covering them with a bag that should be removed immediately after the painting has been finished.

3-11.9.2* Sprinklers shall not be painted and any sprinklers that have been painted shall be replaced with new listed sprinklers of the same characteristics.

Exception: Factory-applied coatings to sprinkler frames for identifying sprinklers of different temperature ratings in accordance with 3-11.6.1.

A-3-11.9.2 Painting of sprinklers may retard the thermal response of the heat-responsive element, may interfere with the free movement of parts, and may render the sprinkler inoperative. Moreover, painting may invite the application of subsequent coatings, thus increasing the possibility of a malfunction of the sprinkler.

Painting is the primary, but not the only, example of a problem known as loading, in which a build-up on the sprinkler would delay or prevent proper response. Sprinklers in textile mills are subject to loading by lint type material, which is a by-product of the cloth manufacturing process. Any sprinkler subject to loading that cannot be readily dusted or blown away must be replaced.

3-11.9.3 Ornamental finishes shall not be applied to sprinklers by anyone other than the sprinkler manufacturer and only sprinklers listed with such finishes shall be used.

Any coatings for identification or ornamentation purposes must be installed by the manufacturer in accordance with the sprinkler's listing.

3-12 Sprinkler Alarms.

3-12.1 Definition. A local alarm unit is an assembly of apparatus approved for the service and so constructed and installed that any flow of water from a sprinkler system equal to or greater than that from a single automatic sprinkler of the smallest orifice size installed on the system will result in an audible alarm on the premises within 5 minutes after such flow begins. For remote sprinkler waterflow alarm transmission see 3-12.7.1.

This standard does not mandate the activation of a building fire alarm system when the sprinkler system is activated. If this feature is desired and specifically required by a building code, then details (electrical circuit arrangement, time for alarm annunciation, and acceptable components) on such an alarm system are contained in NFPA 72A, *Standard for the Installation, Maintenance, and Use of Local Protective Signaling Systems.*

3-12.2* Where Required. Local waterflow alarms shall be provided on all sprinkler systems having more than 20 sprinklers.

A-3-12.2 Central station, auxiliary, remote station, or proprietary protective signaling systems are a highly desirable supplement to local alarms, especially from a safety to life standpoint. (*See 3-12.6.*)

Identification Signs. Approved identification signs should be provided for outside alarm devices. The sign should be located near the device in a conspicuous position and should be worded as follows:

SPRINKLER FIRE ALARM—WHEN BELL RINGS CALL FIRE DEPARTMENT OR POLICE. (*See Figure A-3-12.2.*)

Figure A-3-12.2 Identification Sign.

This standard requires facilities to sound an audible water-flow alarm on the premises for all systems having more than 20 sprinklers. It recommends, but does not require, locally sounding alarms for fewer than 20 sprinklers. It does not require supplemental alarm systems. It requires alarm supervision of valves controlling sprinkler systems with nonfire protection connections (*see 5-6.1.6*) and recommends, but does not require, such supervision in other instances (*see 3-9.2.3*). It should be noted that the sprinkler alarms are, as indicated, waterflow alarms and are not considered to be building evacuation alarms. However, in some applications they may be utilized as such. The purpose for which the alarm is intended should be considered in the overall design of both actuating mechanism and audible alarm devices.

3-12.3 Waterflow Detecting Devices.

3-12.3.1 Wet-Pipe Systems.   The alarm apparatus for a wet-pipe system shall consist of a listed alarm check valve or other listed waterflow detecting alarm device with the necessary attachments required to give an alarm.

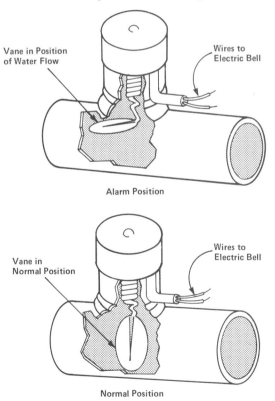

Figure 3.23. Waterflow detector.

On wet-pipe systems, alarm check valves may be preferred to other listed waterflow detecting alarm devices, particularly where fluctuating pressure water supplies are provided. Fluctuating pressure can cause false alarms when other types of waterflow detecting devices are used. On larger systems an alarm check valve makes it possible to introduce an excess pressure higher than the supply on the system side of the alarm check valve. Waterflow is then detected by a drop in pressure on the system side of the alarm check. In other instances, the alarm check valve serves to prevent waterflow alarms on risers not involved in an actual waterflow condition when fire pumps start or similar large fluctuations occur.

Formal Interpretation

Question: Would the use of a listed water flow switch and bell as the alarm apparatus meet the intent of 3-12.3.1?

Answer: Yes.

3-12.3.2 Dry-Pipe Systems. The alarm apparatus for a dry-pipe system shall consist of listed alarm attachments to the dry-pipe valve. When a dry-pipe valve is located on the system side of an alarm valve, the actuating device of the alarms for the dry-pipe valve may be connected to the alarms on the wet-pipe system.

In some instances, small dry-pipe systems may be necessary in areas of otherwise heated buildings, such as outside canopies, loading docks, unheated penthouses, etc. In these cases, it may be more economical to install an auxiliary dry-pipe system supplied from the building's wet system. Under these circumstances, the waterflow alarm from the dry-pipe system may be connected into the wet-pipe system alarm sounding or annunciating device.

3-12.3.3* Preaction and Deluge Systems. In addition to the waterflow alarms required for systems having more than 20 sprinklers, all deluge and preaction systems shall be provided with listed alarm attachments actuated by the detection system.

A-3-12.3.3 A mechanical alarm (water motor gong) may also be required.

Inasmuch as preaction and deluge valves are not actuated by the operation of an automatic sprinkler, a separate detection system is required to actuate the valves. This detection system gives an indication of fire prior to flow of water in the system. It may also be utilized to initiate other fire-related activities.

3-12.3.4* Waterflow alarm indicators (paddle-type) shall not be installed in dry-pipe, preaction, or deluge systems.

A-3-12.3.4 The surge of water when the valve trips may seriously damage the device.

In addition to damaging the device itself, the high-velocity flow could totally disengage the paddle and carry it downstream until it lodged in the piping, obstructing it. This restriction is also applicable to a situation in which a wet-pipe system acts as the supply source for a preaction, deluge, or dry-pipe system.

3-12.4 Attachments—General.

3-12.4.1* An alarm unit shall include a listed mechanical alarm, horn, or siren, or a listed weatherproof electric gong, bell, horn, or siren.

A-3-12.4.1 Audible alarms are normally located on the outside of the building. Listed electric gongs, bells, horns, or sirens inside the building or a combination inside and outside are sometimes advisable.

The required audible local alarm may be either mechanically or electrically operated. The standard does not stipulate whether its location be indoors or outdoors, as this must be determined by individual circumstances. The location and number of audible alarms will normally be dictated by the purpose of the alarms (waterflow or evacuation). They will also be determined by normal operations within the protected premises and location at which the alarm would be expected to be heard.

3-12.4.2* Outdoor mechanical or electrically operated bells shall be of weatherproof and guarded type.

A-3-12.4.2 All alarm apparatus should be so located and installed that all parts are accessible for inspection, removal, and repair, and should be substantially supported.

Formal Interpretation

Question: Is it the intent of 3-12.4.2 to require guards against mechanical injury on all outdoor water motor gongs?

Answer: No. It is the intent of the Committee that the word "guarded" relates to the protection against birds, vermin, and debris.

3-12.4.3   On each alarm check valve used under conditions of variable water pressure, a retarding device shall be installed. Valves shall be provided in the connections to retarding devices, to permit repair or removal without shutting off sprinklers; these valves shall be so arranged that they may be locked or sealed in the open position.

Retarding chambers are required when alarm devices con-nected to alarm valves would otherwise be subject to nuisance operation caused by surges from variable pressure supplies, such as city water or fire pumps. When the supply is a constant pressure source, such as a gravity tank or pressure tank, retarding chambers are not required.

3-12.4.4   Alarm valves, dry-pipe, preaction, and deluge valves shall be fitted with an alarm bypass test connection for an electric alarm switch, water motor gong, or both. This pipe connection shall be made on the water supply side of the system and provided with a control valve and drain for the alarm piping. A check valve shall be installed in the pipe connection from the intermediate chamber of a dry-pipe valve.

Local waterflow alarms should be tested at least quarterly. (See NFPA 13A, Recommended Practice for the Installation, Test-ing and Maintenance of Sprinkler Systems, Sections 4-5 and 4-6.1.) The required bypass from the water side of the system provides the facility to conduct these tests without tripping the system which, in addition to being costly and time consuming, would tend to introduce foreign material and scale into the piping.

3-12.4.5   A control valve shall be installed in connection with a pressure-type contactor or water-motor-operated alarm devices, and such valves shall be of the type that will clearly indicate whether they are open or closed and be so constructed that they may be locked or sealed in the open position. The control valve for the retarding chamber on alarm check valves of wet-pipe systems may be accepted as complying with this paragraph.

The valve being discussed is located in the piping between the connection from the alarm valve, a dry-pipe valve, a preaction valve, or a deluge valve and the local waterflow alarm device(s). This piping is normally subject to only atmospheric air pres-sure and, lacking an indicating type valve, there would be no means of visually ascertaining if the valve were open and that water could reach the alarm actuators in the event of fire.

3-12.5* **Attachments—Mechanically Operated.**   For all types of sprinkler systems employing water-motor-operated alarms, an approved ¾-in. (19-mm) strainer shall be installed at the alarm outlet of the waterflow detecting device except that when a retarding chamber is used in connection with an alarm valve, the strainer shall be located at the outlet of the retarding chamber unless the retarding chamber is provided with an approved integral strainer in its outlet. Water-motor-operated devices shall be protected from the weather, and shall be properly aligned and so installed as not to get out of adjustment. All piping to these devices shall be galvanized or brass or other approved corrosion-resistant material of a size not less than ¾ in. (19 mm).

A-3-12.5 Water-motor-operated devices should be located as near as practicable to the alarm valve, dry-pipe valve, or other waterflow detecting device. The total length of the pipe to these devices should not exceed 75 ft (22.9 m) nor should the water-motor-operated device be located over 20 ft (6.1 m) above the alarm device or dry-pipe valve.

Formal Interpretation

Question: Is it the intent of 3-12.5 to require water-motor-operated alarm devices drain piping to be of corrosion-resistant material?

Answer: No.

The strainer and the pipe of corrosion-resistant material are required to protect against obstruction of the small orifice through which water enters the water motor.

3-12.6 Alarm Attachments—High-rise Buildings. When a fire must be fought internally due to the height of a building, the following additional alarm supervision must be provided:

(a) When each sprinkler system on each floor is equipped with a separate waterflow device, it shall be connected to an alarm system in such a manner that operation of one sprinkler will actuate the alarm system, and the location of the operated flow device shall be indicated on an annunciator and/or register. The annunciator or register shall be located at grade level at the normal point of fire department access, at a constantly attended building security control center, or both locations.

Exception: When the location within the protected buildings where supervisory or alarm signals are received is not under constant supervision by qualified personnel in the employ of the owner, a connection shall be provided to transmit a signal to a remote or central station.

(b) A distinct trouble signal shall be provided to indicate a condition that will impair the satisfactory operation of the sprinkler system. This shall include but not be limited to monitoring control valves, building temperatures, fire pump power supplies and running conditions, and water tank levels and temperatures. Pressure supervision shall also be provided on pressure tanks.

To improve the reliability of the automatic sprinkler systems in high-rise buildings, supervision of any portion of the system that would impair its operation is required.

A sprinkler waterflow alarm annunciated by floor is not required, as some systems may take on configurations that would not lend themselves to this type of zoning. On the other hand, where sprinkler waterflow alarms are provided on each floor, it is the intent that they be annunciated at a point to allow rapid identification of the fire location by the fire department upon arrival.

The overall reliability of the system is further improved by requiring remote monitoring of supervisory signals in the numerous cases where 24-hour-per-day surveillance of on-site supervisory equipment is not provided.

3-12.7 Attachments—Electrically Operated.

3-12.7.1   Electrically operated alarm attachments forming part of an auxiliary, central station, proprietary, or remote station signaling system shall be installed in accordance with the following applicable NFPA standards:

(a) NFPA 71, *Standard for the Installation, Maintenance, and Use of Signaling Systems for Central Station Service,*

(b) NFPA 72B, *Standard for the Installation, Maintenance, and Use of Auxiliary Protective Signaling Systems for Fire Alarm Service,*

(c) NFPA 72C, *Standard for the Installation, Maintenance, and Use of Remote Station Protective Signaling Systems,*

(d) NFPA 72D, *Standard for the Installation, Maintenance, and Use of Proprietary Protective Signaling Systems.*

3-12.7.2*   The circuits of electrical alarm attachments forming part of a local sprinkler waterflow alarm system need not be supervised.

Exception: If the local sprinkler waterflow alarm system is part of a required local fire alarm system, it shall be installed in accordance with NFPA 72A, Standard for the Installation, Maintenance, and Use of Local Protective Signaling Systems.

A-3-12.7.2   Switches that will silence electric alarm sounding devices by interruption of electrical current are not desirable; however, if such means are provided, then the electrical alarm sounding device circuit should be arranged so that when the sounding device is electrically silenced, that fact shall be indicated by means of a conspicuous light located in the vicinity of the riser or alarm control panel. This light shall remain in operation during the entire period of the electrical circuit interruption.

Electrically operated waterflow alarm attachments must be installed in accordance with NFPA 72A except as modified in this section.

3-12.7.3 Waterflow detecting devices, including the associated alarm circuits, shall be tested by an actual waterflow through use of a test connection. (*See 3-12.8.*)

3-12.7.4 Outdoor electric alarm devices shall be of a type specifically listed for outdoor use, and the outdoor wiring shall be in approved conduit, properly protected from the entrance of water in addition to the requirements of 3-12.7.1 and 3-12.7.2.

3-12.8 Drains. Drains from alarm devices shall be so arranged that there will be no danger of freezing, and so that there will be no overflowing at the alarm apparatus at domestic connections or elsewhere with the sprinkler drains wide open and under system pressure. (*See 3-6.4.*)

REFERENCES CITED IN COMMENTARY

The following publications are available from the National Fire Protection Association, Batterymarch Park, Quincy, MA 02269.

NFPA 13A-1987, *Recommended Practice for the Inspection, Testing and Maintenance of Sprinkler Systems*
NFPA 13D-1989, *Standard for the Installation of Sprinkler Systems in One- and Two-Family Dwellings and Mobile Homes*
NFPA 13R-1989, *Standard for the Installation of Sprinkler Systems in Residential Occupancies up to Four Stories in Height*
NFPA 14-1986, *Standard for the Installation of Standpipe and Hose Systems*
NFPA 26-1988, *Recommended Practice for the Supervision of Valves Controlling Water Supplies for Fire Protection*
NFPA 72A-1987, *Standard for the Installation, Maintenance, and Use of Local Protective Signaling Systems for Guard's Tour, Fire Alarm, and Supervisory Service*
NFPA 231C-1986, *Standard for Rack Storage of Materials*
NFPA 409-1985, *Standard on Aircraft Hangars.*

The following publication is available from American Welding Society, 2501 N.W. 7th Street, Miami, FL 33125.

AWS D10.9-1980, *Specification for Qualification of Welding Procedures and Welders for Piping and Tubing.*

The following publication is available from Factory Mutual Research Corporation, 1151 Boston-Providence Turnpike, Norwood, MA 02062.

FM 2008-1988, *Approval for Class #2008.*

The following publications are available from Underwriters Laboratories Inc., 333 Pfingsten Road, Northbrook, IL 60062.

UL 199-1989, *Standard for Automatic Sprinklers for Fire-Protection Service*
UL 1626-1988, *Standard for Residential Sprinklers for Fire-Protection Service.*

4

Spacing, Location, and Position of Sprinklers

4-1 General Information.

4-1.1* Basic Requirements.

A-4-1.1 The installation requirements are specific for the normal arrangement of structural members. There will be arrangements of structural members not specifically detailed by the requirements. By applying the basic principles, layouts for such construction can vary from specific illustrations, provided the maximum specified for the spacing of sprinklers (Section 4-2) and position of sprinklers (Section 4-3) are not exceeded.

All needless ceiling sheathing, hollow siding, tops of high shelving, partitions, or decks should be removed. Sheathing of paper and similar light flammable materials is particularly objectionable.

4-1.1.1* The basic requirements for spacing, location, and position of sprinklers are specified in this chapter and are based on the following principles:

(a) Sprinklers installed throughout the premises,

(b) Sprinklers located so as not to exceed maximum protection area per sprinkler,

(c) Sprinklers positioned and located so as to optimize performance with respect to activation time and distribution,

(d) And as specified herein.

Exception No. 1: See 4-4.3, 4-4.4, and 4-4.19.2 for locations from which sprinklers may be omitted.

Exception No. 2: Special sprinklers may be installed in accordance with 4-1.1.3.

Exception No. 3: When sprinklers are specifically tested and test results prove that deviations from clearance requirements to structural members offer no obstruction to spray discharge, they may be positioned and located accordingly.

Exception No. 4:   Clearance between sprinklers and ceilings may exceed the maximum specified in Sections 4-3 and 4-5.4 provided that, for the conditions of occupancy protected, tests or calculations show comparable sensitivity and performance of the sprinklers to those installed in confor-mance with Sections 4-3 and 4-5.4.

Automatic sprinkler system designs are based on a variety of parameters which relate to water supply and components, among others. This chapter deals exclusively with another of those parameters, namely the positioning and location of sprinklers. In practical terms, sprinklers are required through-out the premises, with selective exclusions permitted such as concealed noncombustible spaces (*see 4-4.4.1*), and hotel bath-rooms and clothes closets (*see 4-4.19*). They could also include areas in noncombustible buildings where introduction of com-bustible contents is precluded by building usage. The installa-tion of sprinklers throughout the building is necessary to ensure that the purpose of a particular design will not be defeated by fire originating in an unsprinklered area and growing beyond system ability to control it.

The positioning of sprinklers with respect to their spacing and spacing of sprinklers with respect to ceiling members allows for prompt operation of the fusible element of the sprinkler and minimizes obstruction to water discharge pat-terns. By limiting distances between sprinklers, the sprinklers will activate faster, provide more efficient discharge patterns, and result in superior fire control.

Formal Interpretation

Background: Construction—Structural steel frame with con-crete floors. The floor to floor height is approximately 15½ ft. Approximately 9 ft above each floor, a gypsum deck has been poured to form an "interstitial space." Noncombustible ductwork, plumb-ing, electrical conduit, sprinkler piping, and open cable trays run in the interstitial space. Access to the space is by five stairways, with access doors each being approximately 3 ft × 3 ft.

Question: Is it the intent of Section 4-1 of NFPA 13 that sprinkler protection of the interstitial space described above be provided to consider the building completely protected by automatic sprinklers?

Answer: No. The space described is essentially noncombustible with no occupancy and no combustible services [with the possible exception of the cable trays, which could be protected in accordance with 4-4.4.4(b) if necessary] and so could be treated as a noncombustible concealed space. Use of the space for storage or the introduction of combustibles would require provision of sprinklers to maintain classification as completely protected by automatic sprinklers.

Formal Interpretation

Question: Is it the intent of 4-1.1.1 to require installation of a sprinkler in every room in a building including (a) Shower Rooms, (b) Clothes Closets?

Answer: Yes. The intent is to require sprinklers in every area except where specifically excluded (*for example, see 4-4.4.1*), or except where the authority having jurisdiction permits omission of the sprinklers.

Formal Interpretation

Question: Is it the intent of 4-1.1.1 to require automatic sprinklers or equivalent automatic protection in rooms designated for the specific use of electrical equipment including buss ducts and circuit breaker panels?

Answer: Yes.

Formal Interpretation

Question: Does 4-1.1.1 require that a fire pump room containing a fire pump for a sprinkler system for a completely sprinklered building be protected by automatic sprinklers even if the fire pump is in a separate building?

Answer: Yes. Wherever the fire pump is located, it would be considered a part of the building and therefore required to be protected by automatic sprinklers.

A-4-1.1.1 This standard contemplates full sprinkler protection for all areas. Other NFPA standards that mandate sprinkler installation may not require sprinklers in certain areas. The requirements of this standard should be used insofar as they are applicable. The authority having jurisdiction should be consulted in each case.

The specific placement of automatic sprinklers is necessary to achieve effective control of the fire and to assure that sprinklers will operate in a timely manner. In accomplishing this goal, obstructions to discharge must be minimized to the extent practicable.

4-1.1.2   Residential sprinklers shall be installed in conformance with their listing and the positioning requirements of NFPA 13D, *Standard for the Installation of Sprinkler Systems for One- and Two-Family Dwellings and Mobile Homes.*

The test procedure under which residential sprinklers have been listed differs from the procedure followed in listing sprinklers for this standard. Therefore, their use is confined to residential portions of occupancies and their spacing and location must be in accordance with NFPA 13D, *Standard for the Installation of Sprinkler Systems in One- and Two-Family Dwellings and Mobile Homes.* Residential sprinklers are also governed by paragraphs 3-11.2.9 and 7-4.4 of this standard.

4-1.1.3   Special sprinklers may be installed with protection areas, locations, and distances between sprinklers differing from those specified in Sections 4-2 and 4-5 when found suitable for such use based on: fire tests related to the hazard category; tests to evaluate distribution, wetting of floors and walls, interference to distribution by structural elements; and tests to characterize response sensitivity, when installed in accordance with any special sprinkler listing limitation.

Exception No. 1:   No sprinkler shall be installed to protect an area greater than 400 sq ft (36 m²).

Exception No. 2:   Maximum area of coverage for individual extended coverage pendent and upright sprinklers shall be limited to areas having coverage with equal-sided dimensions.

An example is the extended coverage horizontal sidewall sprinkler. The intent is to permit the testing laboratories to list new products with spacings greater than permitted in this chapter provided they can achieve the same degree of fire control.

However, because of concerns for proliferation of unlimited pressure/area combinations with no guidance to the testing laboratories, and because of concerns that existing test methods may not appropriately evaluate the resulting throw distances for other than equal-sided maximum areas of coverage or various ceiling configurations, the two Exceptions place an upper limit on the area of coverage [400 sq ft (36 m²)] and mandate that the areas have equal-sided dimensions.

The position and clearance rules are specific but may be modified if it can be demonstrated by test that there is no impairment of fire control capability. In some cases, this may imply no impairment of water discharge. Some tests have indicated comparable fire control capability even with minor water discharge interference.

Formal Interpretation

Question: May a horizontal sidewall sprinkler be installed at distances greater than 12 in. below the ceiling, when that sprinkler has been tested and listed by Underwriters Laboratories Inc. for distances greater than 12 in. below the ceiling, and the sprinkler is installed in accordance with that listing?

Answer: Yes. This is reflected in 4-1.1.1 Exception No. 4 of NFPA 13 which, while not specifically addressing sidewall sprinklers, was intended to apply to all sprinklers.

4-1.2* When partial sprinkler systems are installed, the requirements of this standard shall be used insofar as they are applicable. The authority having jurisdiction shall be consulted in each case.

> The standard recognizes that, although it is intended to apply to fully sprinklered buildings, some codes and ordinances require that only sprinklers in certain areas or occupancies be installed according to this standard. As an example, NFPA *101*®, *Life Safety Code*®, allows one such condition for apartment buildings, namely sprinkler protection in common areas such as corridors.
>
> It is expected that even in these circumstances, most if not all of the same components prescribed by this standard could be utilized. This would include the piping, hangers, and sprinklers as described in Chapter 3. In these circumstances, the authority having jurisdiction must be consulted to determine what additional requirements may be necessary to compensate for the lack of sprinklers in the remainder of the building. These may include fire separations between sprinklered and unsprinklered areas and an increased water supply.

A-4-1.2   Installation of sprinklers throughout the premises is necessary for protection of life and property. In some cases partial sprinkler installations covering hazardous sections and other areas are specified in codes or standards or are required by authorities having jurisdiction for minimum protection to property or to provide opportunity for safe exit from the building.

When buildings or portions of buildings are of combustible construction or contain combustible material, standard fire barriers should be provided to separate the areas that are sprinkler protected from adjoining unsprinklered areas. All openings should be protected in accordance with applicable standards and no sprinkler piping should be placed in an unsprinklered area unless the area is permitted to be unsprinklered by this standard.

Water supplies for partial systems should be adequate and designed with due consideration to the fact that in a partial system more sprinklers may be opened in a fire that originates in an unprotected area and spreads to the sprinklered area than would be the case in a completely protected building. Fire originating in a nonsprinklered area may overpower the partial sprinkler system.

When sprinklers are installed in corridors only, sprinklers should be spaced up to the maximum of 15 ft (4.5 m) along the corridor, with one sprinkler opposite the center of any door or pair of adjacent doors opening onto the corridor, and with an additional sprinkler spaced inside each adjacent room above the door opening. When the sprinkler in the adjacent room provides full protection for that space, an additional sprinkler is not required in the corridor adjacent to the door.

Tests conducted at the National Bureau of Standards support this type of partial installation.

4-1.3 Definitions.

It should be noted that these definitions are used to describe particular construction features which in turn can be used to establish the placement of sprinklers with respect to those features.

4-1.3.1 Smooth Ceiling Construction.   The term *smooth ceiling construction* as used in this standard includes:

(a) Flat slab, pan-type reinforced concrete, concrete joist less than 3 ft (0.9 m) on centers.

(b) Continuous smooth bays formed by wood, concrete, or steel beams spaced more than 7½ ft (2.3 m) on centers — beams supported by columns, girders, or trusses.

(c) Smooth roof or floor decks supported directly on girders or trusses spaced more than 7½ ft (2.3 m) on centers.

(d) Smooth monolithic ceilings of at least ¾ in. (19 mm) of plaster on metal lath or a combination of materials of equivalent fire-resistive rating attached to the underside of wood joists, wood trusses, and bar joists.

(e) Open web-type steel beams, regardless of spacing.

(f) Smooth shell-type roofs, such as folded plates, hyperbolic paraboloids, saddles, domes, and long barrel shells.

(g) In (b) through (f) above, the roof and floor decks may be noncombustible or combustible. Item (b) would include standard mill construction.

(h) Suspended ceilings of noncombustible construction.

(i) Suspended ceilings of combustible construction where there is a full complement of sprinklers in the space immediately above such a ceiling and the space is unfloored and unoccupied.

(j) Smooth monolithic ceilings with fire resistance less than that specified under item (d) attached to the underside of wood joists, wood trusses, and bar joists.

(k) Combustible suspended ceilings arranged other than as specified under item (i).

In general, smooth ceiling construction is such that it does not incorporate supporting members that are less than 7½ ft (2.3 m) on center or have members that would interfere with the distribution of water from sprinklers. The 7½-ft dimension is selected as the demarcation point. This is the maximum dimension at which a sprinkler may be placed from a wall for Light Hazard and Ordinary Hazard Occupancies. Sprinklers may be placed up to 9 ft (2.7 m) from the wall under conditions imposed by 4-4.19.

4-1.3.2 Beam and Girder Construction. The term *beam and girder construction* as used in this standard includes noncombustible and combustible roof or floor decks supported by wood beams of 4 in. (102 mm) or greater nominal thickness, or concrete or steel beams spaced 3 to 7½ ft (0.9 to 2.3 m) on centers and either supported on or framed into girders. [When supporting a wood plank deck, this includes semi-mill and panel construction, and when supporting (with steel framing) gypsum plank, steel deck, concrete, tile, or similar material, this would include much of the so-called noncombustible construction.]

Beam and girder construction is similar to smooth ceiling, but with supporting members at 3- to 7½-ft (0.9- to 2.3-m) centers which could interfere with the distribution of water from sprinklers.

4-1.3.3* Bar Joist Construction. The term *bar joist construction* refers to construction employing joists consisting of steel truss-shaped members. Wood truss-shaped members which consist of wood top and bottom chord members not exceeding 4 in. (102 mm) in depth with steel tube or bar webs are also defined as bar joists. Bar joist includes noncombustible or combustible roof or floor decks on bar joist construction.

Also known as open-web steel joist construction, these members may be provided with a wood chord or a steel chord. This construction offers little resistance to heat traveling under a ceiling, unlike solid beams. Obstruction to water distribution is minimal (*see 4-2.4.4 and 4-2.4.5*), and the protection area per sprinkler is 200 sq ft (18.6 m²) for pipe schedule systems and 225 sq ft (20.9 m²) for hydraulically designed systems (*see 4-2.2.2.1*). Figures A-4-1.3.3(a) and (b) show two of the several types of bar joist construction.

A-4-1.3.3   See Figures A-4-1.3.3(a) and (b) for examples of bar joist construction.

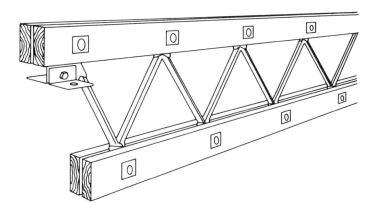

Figure A-4-1.3.3(a)   Wood Bar Joist Construction.

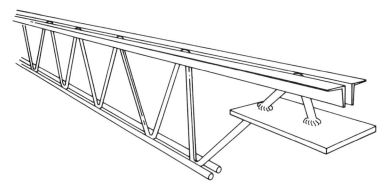

Figure A-4-1.3.3(b)   Open-Web Bar Joist Construction.

4-1.3.4 Panel Construction.   The term *panel construction* as used in this standard includes ceiling panels formed by members capable of trapping heat to aid the operation of sprinklers and limited to a maximum of 300 sq ft (27.9 m²) in area. Beams spaced more than 7½ ft (2.3 m) apart and framed into girders qualify for panel construction provided the 300 sq ft (27.9 m²) area limitation is met.

This is a version of beam and girder construction in which panels of up to a maximum area of 300 sq ft (27.9 m²) are formed. This construction is advantageous from the deflector position standpoint since the defined volume limitations of such construction allow for greater deflector distances from the ceiling.

4-1.3.5 Standard Mill Construction.   The term *standard mill construction* as used in this standard refers to heavy timber construction as defined in NFPA 220, *Standard on Types of Building Construction*.

See Figures 4.1 and 4.2 on pages 210 and 211.

4-1.3.6 Semi-Mill Construction.   The term *semi-mill construction* as used in this standard refers to a modified standard mill construction, where greater column spacing is used and beams rest on girders.

4-1.3.7 Wood Joist Construction.   The term *wood joist construction* refers to solid wood members of rectangular cross section, which may vary from 2 to 4 in. (51 to 102 mm) nominal width and up to 14 in. (356 mm) nominal depth, spaced up to 3 ft (0.9 m) on centers, and spanning up to 40 ft (12 m) between supports, supporting a floor or roof deck. Solid wood members less than 4 in. (102 mm) nominal thickness and up to 14 in. (356 mm) nominal depth, spaced more than 3 ft (0.9 m) on centers are also considered as wood joist construction.

The common type of light combustible construction utilizes nominal 2-in. (51-mm) wide "beams" on edge to support a wood deck, without sheathing. Sheathed joist is classified as smooth ceiling [*see 4-1.3.1(j)*]. With the exception of wood truss construction (*see 4-1.3.9*), other construction types utilizing light wood beams less than 4 in. (102 mm) thick are classified as wood joist.

4-1.3.8* Composite Wood Joist Construction.   The term *composite wood joist construction* refers to wood beams of I cross section constructed of wood flanges and solid wood web, supporting a floor or roof deck. Composite wood joists may vary in depth up to 48 in. (1.2 m), may be spaced up to 48 in. (1.2 m) on centers, and may span up to 60 ft (18 m) between supports. Joist channels shall be fire-stopped to the full depth of the joists with material equivalent to the web construction so that individual channel areas do not exceed 300 sq ft (27.9 m²).

The fire-stopping between the composite wood joists to form the 300 sq ft (27.9 m²) pocket is to trap the heat and accelerate sprinkler operation. See Figure A-4-1.3.8 on page 210.

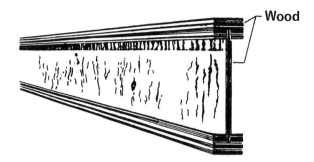

Figure A-4-1.3.8   Typical Composite Wood Joist Construction.

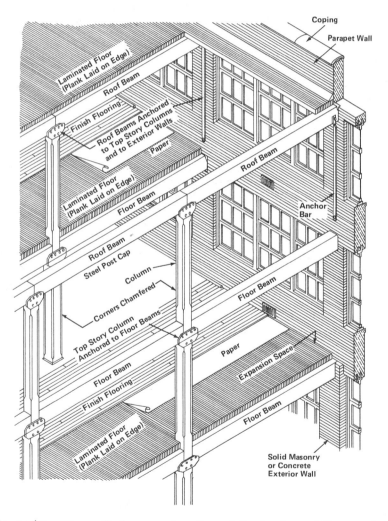

Figure 4.1.   Heavy timber construction of the laminated floor and beam type. (National Forest Products Association).

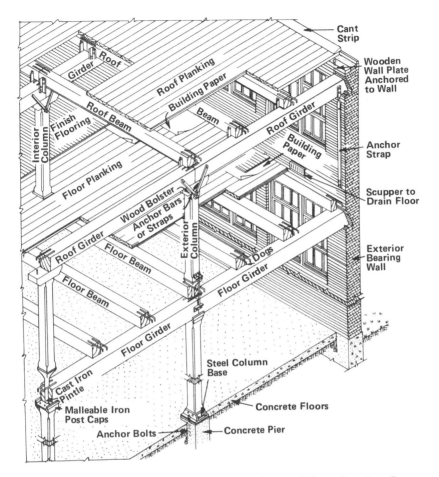

Figure 4.2. Components of a heavy timber building showing floor framing and identifying components of a type known as semi-mill.

4-1.3.9 Wood Truss Construction. The term *wood truss construction* refers to parallel or pitched wood chord members connected by open wood members (webbing), supporting a roof or floor deck. Trusses with steel webbing, similar to bar joist construction, having top and bottom wood chords exceeding 4 in. (102 mm) in depth, shall also be considered wood truss construction.

Several editions of this standard had included wood truss construction in the same category as wood joist construction.

This caused some confusion; thus, the new, separate definition was included in the 1989 edition. Truss types discussed by this standard include both wood web and steel web member types.

4-1.3.10 High-Piled Storage.   The term *high-piled storage* refers to solid piled, palletized, rack storage, bin box, and shelf storage in excess of 12 ft (3.7 m) in height. See 10-1.2 for availability of information for sprinkler protection of high-piled storage.

Storage heights exceeding 12 ft (3.7 m) are qualified by this standard as high-piled. Special provisions as contained in NFPA 231, *General Storage*, or NFPA 231C, *Rack Storage*, will provide the necessary design parameters, while many of the detail items will be contained in this standard.

4-2 Spacing and Location of Upright and Pendent Sprinklers.   (*See also Sections 4-3 and 4-4.*)

The provisions of this section apply to upright and pendent sprinklers. The special provisions applicable to sidewall sprinklers are given in Section 4-5.

B-4 Spacing, Location, and Position of Sprinklers.

B-4-1.1   Cutting holes through partitions, either solid or slatted, to allow sprinklers on one side thereof to distribute water to the other side is not effective.

B-4-1.2   When wood cornices on masonry buildings face an exposure they should be replaced with a parapet, or the projecting woodwork should be cut away and metal flashing extended to cover the exposed edge of planking, or suitable sprinkler protection should be provided.

4-2.1 Distance Between Sprinklers, on the Branch Lines and Between the Branch Lines.

The following rules limit the distances between adjacent sprinklers on the same branch line and on subsequent branch lines.

4-2.1.1   For Light Hazard Occupancies, the distance between sprinklers, either on branch lines or between branch lines, shall not exceed 15 ft (4.6 m).

4-2.1.2*   For Ordinary Hazard Occupancies, the distance between sprinklers, either on branch lines or between branch lines, shall not exceed 15 ft (4.6 m).

A-4-2.1.2 For examples of sprinkler layouts under smooth ceiling construction, refer to Figures A-4-2.1.2(a) and (b).

These two figures illustrate the spacing rules. Figure
A-4-2.1.2(b) is located on page 214.

Flat Slab or Pan-Type Reinforced Concrete

Maximum Spacing: 130 sq ft per Sprinkler
L × S = 130 or less

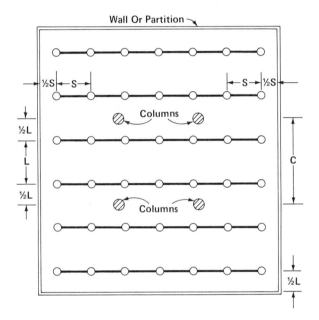

KEY

C = Column spacing.
L = Distance between branch lines, limit 15 ft.
S = Distance between sprinklers on branch lines, limit 15 ft.

Examples

C	L	S (Max)	C	L	S (Max)
21 ft 8 in.	10 ft 10 in.	12 ft 0 in.	21 ft 6 in.	10 ft 9 in.	12 ft 1 in.
24 ft 2 in.	12 ft 1 in.	10 ft 9 in.			

For SI Units: 1 in. = 25.4 mm; 1 ft = 0.3048 m; 1 ft² = 0.0929 m².

**Figure A-4-2.1.2(a) Layout of Sprinklers under Smooth Ceiling
Construction — Ordinary Hazard Occupancy.**

Continuous Smooth Bays with Beams Supported on Columns

Maximum Spacing: 130 sq ft per Sprinkler
L × S = 130 or less

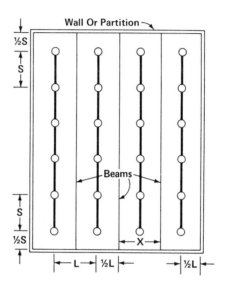

KEY

L = Distance between branch lines, limit 15 ft.

S = Distance between sprinklers on branch lines, limit 15 ft.

X = Width of bay.

Examples

X	L	S (Max)	X	L	S (Max)
10 ft 10 in.	10 ft 10 in.	12 ft 0 in.	10 ft 9 in.	10 ft 9 in.	12 ft 1 in.
12 ft 1 in.	12 ft 1 in.	10 ft 9 in.			

For SI Units: 1 in. = 25.4 mm; 1 ft = 0.3048 m; 1 ft² = 0.0929 m².

Figure A-4-2.1.2(b)   Layout of Sprinklers under Smooth Ceiling Construction — Ordinary Hazard Occupancy.

4-2.1.3   For Extra Hazard Occupancies, the distance between sprinklers, either on branch lines or between branch lines, shall not exceed 12 ft (3.7 m).

4-2.1.4   In areas used for high-piled storage (as defined in 4-1.3.10), the distance between sprinklers shall not exceed 12 ft (3.7 m).

Exception No. 1:   In bays 25 ft (7.6 m) wide, a spacing of 12 ft 6 in. (3.8 m) between sprinklers is permitted.

Exception No. 2:   For systems hydraulically designed for densities below 0.25 gpm per sq ft [(10.2 L/min)/m²)] spacing between sprinklers to 15 ft (4.6 m) is permitted.

The effect of high-piled storage is twofold. First, increased fuel loads are anticipated, and second, more obstructions to the sprinkler spray pattern are anticipated due to the increase vertically in the fuel geometry. The maximum distance of 15 ft (4.6 m) between sprinklers on branch lines or between branch lines is reduced to 12 ft (3.7 m) in buildings used for high-piled storage. If the protection area per sprinkler is permitted, by Exception No. 1 to 4-2.2.5, to exceed 100 sq ft (9.3 m²) [but not be greater than 130 sq ft (12.1 m²)], then the reduction from 15 to 12 ft need not be applied.

Formal Interpretation

Question 1:   Is it the intent of 4-2.1.2 to permit 15-ft spacing for high-piled storage for systems designed in accordance with NFPA 231 for a density below 0.25 gpm/sq ft as noted in the Exception of 4-2.2.5?

Question 2:   Does the Exception for hydraulically designed systems with densities below 0.25 gpm noted in 4-2.2.5 of NFPA 13 allow designers to exceed 12 ft-0 in. spacing for high-piled stock in buildings with bays other than 25 ft-0 in. as noted in 4-2.1.2 of NFPA 13?

Answer: Yes. When sprinkler systems are hydraulically designed for densities below 0.25 gpm/sq ft, the distance between the branch lines or between sprinklers on branch lines shall not exceed 15 ft provided the sprinkler area coverage does not exceed 130 sq ft.

4-2.1.5 Distance from Walls.

4-2.1.5.1   The distance from walls to sprinklers shall not exceed one-half of the allowable distance between sprinklers.

Exception: For small rooms, see 4-4.19.

The allowable distance between sprinklers on the branch lines is determined by the actual distance between the branch lines and the permissible protection area per sprinkler. (*See 4-2.2.*) To minimize the amount of piping used, branch lines are usually spaced as far apart as possible while still maintaining even spacing in the bay or room. If the spacing is not even, then the greatest distance between lines should be used to determine the allowable distance between sprinklers on the lines. If the distance of a line from a wall exceeds one-half of the distance between lines, then twice the distance of the line from the wall should be used in determining the allowable distance between sprinklers on the lines. For example, if the distance from the branch line to the wall was 7 ft (2.1 m) and the distance between branch lines was 13 ft (4.0 m), twice the distance from the branch line to the wall would be 14 ft (4.3 m), which should be used to determine the distance between sprinklers on the branch lines on the first line parallel to the wall. Other lines could use the 13- or the 14-ft dimension, depending on whether symmetry was desired. If an Ordinary Hazard system with a protection area of 130 sq ft (12.1 m^2) were being installed, the distance between sprinklers on the branch lines would be: 130 sq ft ÷ 14 ft = 9.3 ft (2.8 m). [*See 7-4.3.1.2(a).*]

4-2.1.5.2   Sprinklers shall be located a minimum of 4 in. (102 mm) from a wall.

Dead air spaces in corners can affect a sprinkler's operation time. The 4-in. (102-mm) limitation ensures that the sprinkler will operate properly. NFPA 72E, *Standard on Automatic Fire Detectors*, provides more discussion on this phenomenon.

4-2.2* Protection Area Limitations.

A-4-2.2   The protection area per sprinkler should be determined as follows:

1. Along Branch Lines. Determine distance to next sprinkler (or to wall in case of end sprinkler on branch line) upstream and downstream. Choose the larger of either twice the distance to the wall or distance to the next sprinkler. Call this "S".

2. Between Branch Lines. Determine perpendicular distance to sprinkler on branch lines (or to wall in the case of the branch line) on each side of the branch line on which the subject sprinkler is positioned. Choose the larger of (1) the larger distance to the sprinklers on the next branch line, or (2) in the case of the last branch line, twice the distance to the wall. Call this "L".

3. Protection area of the sprinkler = S × L.

4. This does not apply to any sprinkler located more than 7.5 ft (2.3 m) from any single wall in a small room. (*See 4-4.19.*)

4-2.2.1 System Areas. The maximum floor area on any one floor to be protected by sprinklers supplied by any one sprinkler system riser or combined system riser shall be as follows:

Light Hazard — 52,000 sq ft (4831 m²)

Ordinary Hazard — 52,000 sq ft (4831 m²)

Extra Hazard — Pipe Schedule — 25,000 sq ft (2323 m²)

— Hydraulically Calculated — 40,000 sq ft (3716 m²)

Storage — High-piled storage (as defined in 4-1.3.10) and storage covered by other NFPA standards — 40,000 sq ft (3716 m²).

Exception: When single systems protect Extra Hazard, high-piled storage, or storage covered by other NFPA standards and Ordinary or Light Hazard areas, the Extra Hazard or storage area coverage shall not exceed the floor area specified for that hazard and the total area coverage shall not exceed 52,000 sq ft (4831 m²).

A single system protecting both Ordinary or Light Hazard areas and solid-piled, palletized, or rack storage in excess of 12 ft (3.7 m) high, or Extra Hazard areas, may have a coverage area of up to 52,000 sq ft (4831 m²), but not more than 40,000 sq ft (3716 m²) of that coverage area may be high-piled storage or hydraulically designed for Extra Hazard Occupancies. The 40,000 sq ft maximum Extra Hazard coverage area for hydraulically designed systems is consistent with the requirements of the storage standards for similar fire loading.

Formal Interpretation

Question: **Would the 52,000 sq ft maximum area on one floor apply to several buildings being supplied by one riser assembly? Can the 52,000 sq ft be exceeded if the individual buildings are not in excess of 52,000 sq ft and are separated by a clear space of 20 feet or more?**

Answer: **No. The 52,000 sq ft area limitation and the 20 ft separating distance are not relevant to the problem. Each building should have its individual system riser.**

Implicit in this Formal Interpretation is the definition of *a* sprinkler system (*see 1-3*) which is intended to be installed in *a* building.

The specified maximum coverage areas are the maximum floor area per system on any one floor. There is no limit on the number of floors, in that each floor constitutes a separate fire area. Openings are assumed to be protected as outlined in 4-4.7. Thus, in a single-story 312,000-sq ft (28,986-m²) Ordinary or Light Hazard Occupancy at least six systems would be required. However, if the occupancy was six stories high and had 52,000 sq ft (4831 m²) on each floor, it could be protected by a single system. A maximum of 40,000 sq ft (3716 m²) of solid-piled, palletized, or rack storage may be protected by a single system as may an Extra Hazard area with hydraulically designed sprinkler protection. However, that system may also protect Ordinary Hazard areas provided the total coverage area does not exceed 52,000 sq ft (4831 m²). This standard formerly limited system size by the number of sprinklers on a system; for example, 400 sprinklers for Ordinary Hazard areas. The 400 sprinklers times 130 sq ft maximum spacing developed the 52,000 sq ft (4831 m²) limitation. These area limitations are not related to the hydraulics of the system, nor are they related to its operating characteristics. Rather, the area limitations are judgmental factors as to the maximum area within a single, vertical fire division that it is felt should be protected by a single system (or that could be out of service at any one time).

This standard should be used for solid-piled storage or rack storage heights of 12 ft (3.7 m) or less. For heights of more than 12 ft (3.7 m), storage documents such as the following should be used: NFPA 231, *General Storage*; NFPA 231C, *Rack Storage of Materials*; NFPA 231D, *Storage of Rubber Tires*; NFPA 231E, *Storage of Baled Cotton*; and NFPA 231F, *Storage of Roll Paper*.

When multiple risers are necessary to adhere to these area limitations, consideration should be given to the use of a manifold riser arrangement. Figure 4.3 depicts one such arrangement. For a Light Hazard or Ordinary Hazard Occupancy building, each of the three risers could protect up to 52,000 sq ft in a single-story 156,000 sq ft (14,470 m²) building.

4-2.2.2 Light Hazard Occupancy.

4-2.2.2.1   Under smooth ceiling, beam and girder, and bar joist construction (as defined in 4-1.3.1, 4-1.3.2, and 4-1.3.3) the protection area per sprinkler shall not exceed 225 sq ft (20.9 m²).

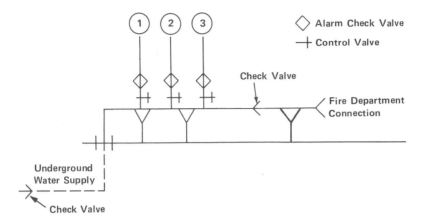

Figure 4.3. Manifold riser.

Exception: The protection area per sprinkler for pipe schedule systems shall not exceed 200 sq ft (18.6 m²).

The standard recognizes the more efficient designs which result with hydraulic calculation techniques. For Light Hazard, smooth ceiling construction, an area of coverage increase to 225 sq ft (20.9 m²) is permitted when hydraulic calculation methods are used.

4-2.2.2.2* Under exposed wood joist construction (as defined in 4-1.3.7) the protection area per sprinkler shall not exceed 130 sq ft (12.1 m²).

Because of the nature of open wood joist construction (numerous small pockets, tendency to impede the flow of heat and cause more obstruction to the sprinkler discharge pattern, and because it is readily subject to fire damage), the protection area is reduced to 130 sq ft (12.1 m²).

Figure A-4-2.2.2.2 (*see page 220*) is an illustration of the spacing rules for open wood joist construction, which are the same for both Light and Ordinary Hazard Occupancies.

4-2.2.2.3 Under exposed composite wood joist and exposed parallel chord wood truss construction (as defined in 4-1.3.8 and 4-1.3.9 respectively) of 16 in. (406 mm) nominal depth or less and exposed pitch chord wood truss construction of any depth, the protection area per sprinkler shall not exceed 130 sq ft (12.1 m²).

Exception: For composite wood joist construction and exposed parallel wood chord construction with members spaced more than 3 ft (0.9 m) on centers, see 4-2.2.2.4.

**Joists Above Girders or Framed into Girders;
Branch Lines Uniformly Spaced Between Girders**

Maximum Spacing: 130 sq ft per Sprinkler
L × S = 130 or less

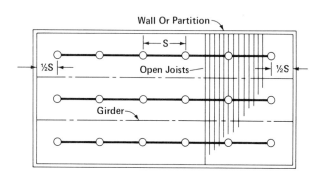

KEY
L = Distance between branch lines, limit 15 ft.
S = Distance between sprinklers on branch lines, limit 15 ft.
Y = Maximum distance between girders.

Examples

Y	L	S (Max)	Y	L	S (Max)
10 ft 9 in.	10 ft 9 in.	12 ft 1 in.	10 ft 10 in. 12 ft 1 in.	10 ft 10 in. 12 ft 1 in.	12 ft 0 in. 10 ft 9 in.

For SI Units: 1 in. = 25.4 mm; 1 ft = 0.3048 m; 1 ft² = 0.0929 m².

**Figure A-4-2.2.2.2   Layout of Sprinklers under Open Wood Joist
Construction — Light and Ordinary Hazard Occupancies.**

The nature of closely spaced composite wood joist members and wood truss construction presents situations in which the obstructions must be minimized while at the same time providing adequate protection to the floor area. If those goals are not satisfied, it is possible that the construction itself may become a part of the fuel package.

4-2.2.2.4   For other types of construction the protection area per sprinkler shall not exceed 168 sq ft (15.6 m²).

The tables on pages 221-223 are intended to explain the spacing rules as they apply to these common types of construction:

Bar Joist Construction
(Defined in 4-1.3.3)

	Sprinkler Deflector Position	Maximum Protection Area Per Sprinkler
Exposed Bar Joists	Deflector maximum 10 in. below combustible roof or deck or 12 in. below noncombustible roof or deck. (See 4-2.4.5 and 4-3.3.)	Regular NFPA 13 (225 sq ft hydraulically calculated—See 4-2.2.2.1.)
Bar Joists with Ceiling Attached	Required in concealed space if enclosed at least partly by combustible construction and more than 6 in. between the roof or floor deck and ceiling. (See 4-4.4.1 Exc. 2).	Regular NFPA 13 below smooth ceiling. Regular NFPA 13 Light Hazard in inaccessible concealed space if space is sprinklered. (See 4-4.4.2.)

Wood Joist Construction
(Defined in 4-1.3.7)

	Sprinkler Deflector Position	Maximum Protection Area Per Sprinkler
Exposed Wood Joists (Solid) up to 3 ft on Center	1 to 6 in. below bottom of joists. (See 4-3.5.)	130 sq ft (See 4-2.2.2.2.)
Exposed Wood Joists (Solid) more than 3 ft on Center	In accordance with smooth ceiling or beam and girder construction needs, as appropriate. (See 4-3.5 and 4-3.1 or 4-3.2.)	130 sq ft (See 4-2.2.2.2.)
Wood Joists (Solid) with Ceiling	Sprinklers required in concealed space if ceiling more than 6 in. below bottom of joist (See 4-4.4.1 Exc. 3.)	Regular NFPA 13 below smooth ceiling. 130 sq ft in concealed space if required.

Composite Wood Joist Construction
(Defined in 4-1.3.8)

	Sprinkler Deflector Position	Maximum Protection Area Per Sprinkler
Exposed Composite Wood Joists up to 16 in. Depth up to 3 ft on Center	1 to 6 in. below bottom of joists. (See 4-3.6.)	130 sq ft (See 4-2.2.2.3.)
Exposed Composite Wood Joists up to 16 in. Depth more than 3 ft on Center	1 to 6 in. below bottom (See 4-3.6.)	168 sq ft (See 4-2.2.2.3 Exception and 4-2.2.2.4.)
Composite Wood Joist with Ceiling	Sprinklers required in concealed space unless ceiling directly attached to underside and joist channels fire stopped into 160 cu ft volumes. (See 4-4.4.1 Exc. 4.)	Regular NFPA 13 spacings below smooth ceiling. 130 sq ft in concealed space if required, except 168 sq ft if more than 3 ft on center.
Exposed Composite Wood Joists over 16 in. Depth	Not addressed by NFPA 13.	Not addressed by NFPA 13.

Wood Truss Construction
(Defined in 4-1.3.9)

	Sprinkler Deflector Position	Maximum Protection Area Per Sprinkler
Exposed Wood Trusses up to 3 ft on Center	Within trusses, conforming to 4-2.4.1, 4-2.4.5, and 4-3.1 except that sprinklers may be 1 to 6 in. below bottom of trusses up to 16 in. nominal depth. (See 4-3.7.)	130 sq ft (See 4-2.2.2.3.)

Wood Truss Construction (Continued)

	Sprinkler Deflector Position	Maximum Protection Area Per Sprinkler
Exposed Wood Trusses more than 3 ft on Center	Within trusses, conforming to 4-2.4.1, 4-2.4.5, and 4-3.1 except that sprinklers may be 1 to 6 in. below bottom of trusses up to 16 in. nominal depth. (See 4-3.7.)	168 sq ft (See 4-2.2.2.3 Exception and 4-2.2.2.4.)
Wood Trusses with Ceiling	Required in concealed space where more than 6 in. clear depth between inside edges of top and bottom chords. (See 4-4.4.1 Exc. 1.) Smooth ceiling sprinkler positions in concealed space, i.e., sprinkler deflectors 1 to 10 in. below combustible ceiling. (See 4-3.7.)	Regular NFPA 13 below smooth ceiling. 130 sq ft maximum in concealed space, except 168 sq ft if more than 3 ft on center. (See 4-2.2.2.3.)

Current requirements in NFPA 13 are not valid when composite wood joist construction exceeds 16 in. (406 mm) in depth. In designs utilizing this construction type, such spaces should be filled with noncombustible insulation. Without this approach, sprinklers within each channel may be necessary. Until full-scale testing is completed on the behavior of heat travel and sprinkler operation, no specific rules can be established by the Technical Committee on Automatic Sprinklers.

Formal Interpretation

Question: Series L, M, H, 60, and 60 open web joists as manufactured by the "Trus Joist Corporation" are trusses with structural wood cord members less than 4 in. nominal thickness and tubular steel web members.

If this type of truss is spaced more than 3 ft on centers in a combustible attic area, not accessible for storage (light hazard), would the protection area per sprinkler be limited to 168 sq ft as referenced in 4-2.2.2.4?

Answer: Yes.

4-2.2.3 Ordinary Hazard Occupancy. For all types of construction the protection area per sprinkler shall not exceed 130 sq ft (12.1 m²).

4-2.2.4 Extra Hazard Occupancy. The protection area per sprinkler shall not exceed 100 sq ft (9.3 m²) for any type of building construction. Sprinkler spacing may exceed 100 sq ft (9.3 m²) but shall not exceed 130 sq ft (12.1 m²) for densities below 0.25 gpm per sq ft [(10.2 L/min)/m²].

Exception: The protection area per sprinkler for pipe schedule systems shall not exceed 90 sq ft (8.4 m²).

Extra Hazard Occupancies have a high fuel load characteristic which is accompanied by great potential for rapidly developing fires. Decreasing allowable areas of coverage to 90 sq ft (8.4 m²) for pipe schedule systems and 100 sq ft (9.3 m²) for hydraulically designed systems is intended to compensate for the potential larger fires which are typical of Extra Hazard Occupancies.

To align this standard with certain provisions of NFPA 30, *Flammable and Combustible Liquids Code*; NFPA 231, *General Storage*; and NFPA 231C, *Rack Storage*, areas of coverage for hydraulically designed systems at or below densities of 0.25 gpm/sq ft [(10.2 L/min)/m²] are permitted to extend as high as 130 sq ft (12.1 m²) per sprinkler.

4-2.2.5 High-Piled Storage. In areas used for high-piled storage (as defined in 4-1.3.10) or for storage covered by other NFPA standards the protection area per sprinkler shall not exceed 100 sq ft (9.3 m²).

Exception No. 1: Sprinkler protection areas may exceed 100 sq ft (9.3 m²) but shall not exceed 130 sq ft (12.1 m²) in systems hydraulically designed in accordance with NFPA 231, Standard for General Storage, and 231C, Standard for Rack Storage of Materials, for densities below 0.25 gpm per sq ft [(10.2 L/min)/m²].

Exception No. 2: Where protection areas are specifically indicated in the design criteria of other NFPA standards.

Tests conducted on various commodities may indicate specific protection areas per sprinkler. Any such criteria should be followed per the appropriate NFPA standard. In addition, note that these protection areas may be greater or smaller than the limitations of 4-2.2.5. (*See 4-2.2.6.*)

4-2.2.6 Special Conditions. When other NFPA standards have developed more stringent sprinkler system spacing criteria, they shall take precedence when supported by valid fire tests.

4-2.3* Location of Sprinklers and Branch Lines with Respect to Structural Members.

A-4-2.3 The arrangement of branch lines depends upon such construction features as the distance between girders or trusses, columns of mushroom-type reinforced concrete, and beams of standard mill construction. Each space or bay should usually be treated as a unit, installing the same number of branch lines uniformly in each space. When single branch lines will suffice, they should be placed midway in each bay or space. The arrangement of branch lines also depends upon the structural members available and suitable for the attachment of hangers, and upon the need for properly locating sprinkler deflectors in accordance with 4-2.4 and Section 4-3.

The direction in which branch lines are usually run in common types of ceiling construction and framing is shown in Table A-4-2.3.

Table A-4-2.3 is shown on page 226.

4-2.3.1 Sprinklers may be located under beams, in bays, or both, in combination, but the locations must meet the provisions outlined in Sections 4-2.4 and 4-3.

The rules that follow give permissible locations for sprinklers either under beams or under a roof or ceiling. A combination of these two positions is permissible provided the clearance and position rules are followed.

4-2.3.2 Where there are two sets of joists under a roof or ceiling and there is no flooring over the lower set, sprinklers shall be installed above and below the lower set of joists where there is a clearance of 6 in. (152 mm) or more between the top of the lower joist and bottom of the upper joist. (*See Figure 4-2.3.2.*) Sprinklers may be omitted from below the lower set of joists where at least 18 in. (457 mm) is maintained between sprinkler deflectors and tops of lower joists.

Formal Interpretation

Question: Under 4-2.3.2 and its respective diagram, in spacing sprinklers above the lower set of joists, is it not the intent to measure the distance from the point where the space between joists (and/or wood trusses) is 6 in. and greater and not from the wall or where the top and bottom chords of the truss intersect?

Table A-4-2.3   Common Branch Line Run Directions

Type of Ceiling	Location of Branch Lines
Smooth Continuous:	
Concrete mushroom.	Either direction
Concrete pan-type or flat	Either direction
Sheathed (ceiling attached to bottom of beams, wood joists, or bar joists):	
Girders beneath sheathing	Across the beam or joists
No girders beneath sheathing.	Whichever direction facilitates easy and proper hanging
Bays more than 7½ ft (2.3 m) wide:	
Formed by beams supported on columns.	Parallel to beams
Formed by beams supported on girders or trusses . .	Either across beams or parallel to beams in the bays above girders or trusses
Supported directly on girders.	Parallel to girders
Supported directly on trusses.	Either direction, parallel to or through trusses
Beam and Girder:	
Wood or steel beams spaced 3 to 7½ ft (0.9 to 2.3 m) apart . . .	Across beams
Open Bar Joist.	Across the joists or trusses (either through or under them)
Open Joist (wood, steel, or concrete)	Across joists

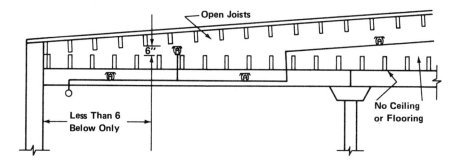

For SI Units: 1 in. = 25.4 mm.

Figure 4-2.3.2   Arrangement of Sprinklers under Two Sets of Open Joists — No Sheathing on Lower Joists.

Answer: No. This paragraph addresses only the special conditions stated in the beginning of the paragraph, and does not apply to wood trusses. Under the conditions stated in 4-2.3.2, the first sprinkler between the two sets of joists should be located at the point where the clear space between the two sets of joists is 6 in. In wood truss construction, sprinklers should be spaced from the end of the truss.

This is a special rule for a special condition. The first sprinkler above the lower set of joists should be located where the clear space is 6 in. (152 mm). The sprinklers below the lower set of joists should be spaced evenly in the space from the wall to the point where the clearance from the sprinkler deflector above the lower joist to the top of the lower joists is at least 18 in. (457 mm). Sprinklers are not required below the lower set of joists beyond the point where this clearance exceeds 18 in. (457 mm).

4-2.4 Clearance Between Sprinklers and Structural Members.

4-2.4.1 Trusses.

4-2.4.1.1   Sprinklers installed between the top and bottom chord members shall be located at least 2 ft (0.6 m) laterally from truss members (web or chord) that are more than 4 in. (102 mm) wide, and at least 1 ft (0.3 m) laterally from truss members 4 in. (102 mm) or less in width. For trusses with steel webbing similar to bar joists, clearance from webs shall be in accordance with 4-2.4.5.

Exception: Sprinklers running through or above the trusses may be located on the center line of a truss provided chord members are not more than 8 in. (203 mm) wide, and the deflector is at least 6 in. (152 mm) above the chord member.

4-2.4.1.2 Where sprinklers are located laterally beside the chord members, clearances between the chord members and the sprinkler deflectors shall be in accordance with 4-2.4.6.

> **When installing sprinklers within a truss system it is critical to provide adequate horizontal and vertical clearances between the deflector and the truss member.**

4-2.4.2 Girders. When sprinkler lines are located perpendicular to and above girders, sprinklers shall be located at least 3 ft 9 in. (1.14 m) from girders except that they may be located directly above girders with the top flange not more than 8 in. (203 mm) wide, in which case the deflectors shall be at least 6 in. (152 mm) above the top of the girder.

4-2.4.3 When sprinkler deflectors are in accordance with Table 4-2.4.6, the girders may be disregarded in the spacing of the branch lines.

4-2.4.4 Open Web-Type Steel Beams. *(See Figure 4-2.4.4.)* When branch lines are run across and through openings of open web-type steel beams, sprinklers may be spaced "bay and beam" provided:

(a) The distance between sprinklers and between branch lines conforms to 4-2.1,

(b) Sprinklers in the beam openings are located within 1 in. (25 mm) horizontally of the opening center line,

(c) The branch line is located within 1 in. (25 mm) horizontally of the opening center line, and

(d) Sprinklers on alternate lines are staggered.

4-2.4.5 Bar Joists. Sprinklers shall be at least 3 in. (76 mm) laterally from web members of open bar joists which do not exceed ½ in. (13 mm) and at least 6 in. (152 mm) laterally from web members which do not exceed 1 in. (25 mm). When the dimensions of the web member exceed 1 in. (25 mm), see 4-2.4.1.

4-2.4.6* Beams. Deflectors of sprinklers in bays shall be at sufficient distances from the beams, as shown in Table 4-2.4.6 and Figure 4-2.4.6, to avoid obstruction to the sprinkler discharge pattern. Otherwise the spacing of sprinklers on opposite sides of the beams shall be measured from the center line of the beam and the distance shall not exceed one-half of the allowable distance between sprinklers.

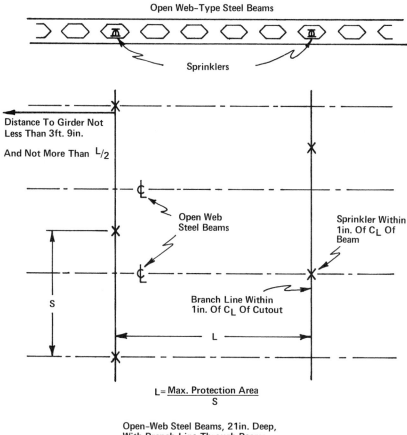

Open Web-Type Steel Beams

Sprinklers

Distance To Girder Not Less Than 3ft. 9in.

And Not More Than $L/2$

Open Web Steel Beams

Sprinkler Within 1in. Of C_L Of Beam

Branch Line Within 1in. Of C_L Of Cutout

S

L

$L = \dfrac{\text{Max. Protection Area}}{S}$

Open-Web Steel Beams, 21in. Deep, With Branch Line Through Beams

S= Spacing Of Sprinklers On Branch Lines
L= Distance Between Branch Lines
Lx S= Maximum Protected Area Per Sprinkler

For SI Units: 1 in. = 25.4 mm.

Figure 4-2.4.4   Location of Branch Lines and Sprinklers.

The provision addressed by this section is commonly referred to as the beam rule. The maximum spacing of branch lines and of sprinklers on branch lines must be followed. Therefore, the distance between sprinklers on opposite sides of the beam is measured from the center line of the beam. It may be possible to exceed the distances above the ceiling obstruction, tabulated in Table 4-2.4.6 (*see page 230*), if the trajectory and resultant discharge pattern of a particular sprinkler have been verified by test.

Table 4-2.4.6   Position of Deflector when Located above Bottom of Beam

Distance from Sprinkler to Side of Beam	Maximum Allowable Distance Deflector above Bottom of Beam
Less than 1 ft	0 in.
1 ft to less than 2 ft	1 in.
2 ft to less than 2 ft 6 in.	2 in.
2 ft 6 in. to less than 3 ft	3 in.
3 ft to less than 3 ft 6 in.	4 in.
3 ft 6 in. to less than 4 ft	6 in.
4 ft to less than 4 ft 6 in.	7 in.
4 ft 6 in. to less than 5 ft	9 in.
5 ft to less than 5 ft 6 in.	11 in.
5 ft 6 in. to less than 6 ft	14 in.

For SI Units: 1 in. = 25.4 mm; 1 ft = 0.3048 m.

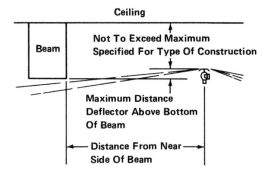

Figure 4-2.4.6   Position of Deflector, Upright, or Pendent Sprinkler When Located above Bottom of Beam.

This rule is also useful for evaluation of spray patterns with respect to architectural features that may cause obstructions, such as soffits, light fixtures, and similar features.

Formal Interpretation

Question: Does Table 4-2.4.6 apply to the layout shown in Figure 1?

Answer: No. 4-2.4.6 offers the option as an alternate to Table 4-2.4.6 of spacing the sprinklers from the beam at a distance of not more than ½ the allowable distance between sprinklers. Under those conditions, the sprinklers may be positioned other than as specified in Table 4-2.4.6.

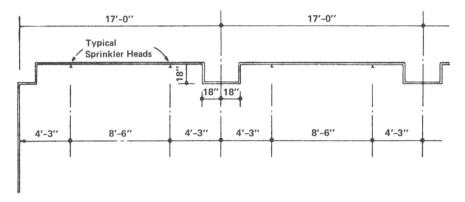

Figure 1. Typical ceiling cross section.

Formal Interpretation

Question 1:   Does Table 4-2.4.6 apply to valances at stock fixtures approximately 3 ft from a wall and up to 5 ft deep from the ceiling?

Answer: No.

Question 2:   If the answer to Question 1 is "no," what sections of NFPA 13 would apply?

Answer: 4-4.10 and 4-4.12 more closely approximate the situation referred to.

A-4-2.4.6 Narrow Pocket.   Girders, beams, or trusses forming narrow pockets of combustible construction along walls when of a depth that will obstruct the spray discharge pattern may require additional sprinklers.

4-2.4.7* Position of Deflectors.   Deflectors of sprinklers shall be parallel to ceilings, roofs, or the incline of stairs, but when installed in the peak of a pitched roof they shall be horizontal. Low-pitched roofs having slopes not greater than 1 in. per ft (83 mm/m) may be considered level in the application of this rule, and sprinklers may be installed with deflectors horizontal.

> **Maintaining the deflector parallel to the ceiling will result in minimum obstructions to discharge and a somewhat superior discharge pattern.**

A-4-2.4.7   On sprinkler lines larger than 2 in. (51 mm), consideration should be given to the distribution interference caused by the pipe, which can be minimized by installing sprinklers on riser nipples or installing sprinklers in the pendent position.

4-2.5 Clear Space Below Sprinklers.

4-2.5.1 A minimum of 18 in. (457 mm) clearance shall be maintained between top of storage and ceiling sprinkler deflectors. For in-rack sprinklers, the clear space shall be in accordance with NFPA 231C, *Standard for Rack Storage of Materials*.

> The discharge pattern of upright and pendent sprinklers is a half-paraboloid filled with spray. In order to permit the distribution of water over the area that the sprinkler has been designed to protect, there must be no obstruction to the spray pattern. The rack storage fire tests and other tests with solid-piled storage have shown that sprinklers are effective with an 18-in. (457-mm) clearance.

4-2.5.2* The distance from sprinklers to privacy curtains, free-standing partitions, or room dividers shall be sufficient, as shown in Table 4-2.5.2 and Figure 4-2.5.2, to avoid obstruction to the sprinkler discharge pattern.

Table 4-2.5.2   Horizontal and Minimum Vertical Distances for Sprinklers

Horizontal Distance	Minimum Vertical Distance Below Deflector
6 in. or less	3 in.
More than 6 in. up to 9 in.	4 in.
More than 9 in. up to 12 in.	6 in.
More than 12 in. up to 15 in.	8 in.
More than 15 in. up to 18 in.	9½ in.
More than 18 in. up to 24 in.	12½ in.
More than 24 in. up to 30 in.	15½ in.
More than 30 in.	18 in.

For SI Units: 1 in. = 25.4 mm.

A-4-2.5.2 The distances given in Table 4-2.5.2 were determined through tests in which privacy curtains with either a solid fabric or close mesh ¼ in. (6.4 mm) top panel were installed. For broader-mesh top panels [e.g., ½ in. (13 mm)] the obstruction of the sprinkler spray is not likely to be severe and the authority having jurisdiction may not need to apply the requirements in 4-2.5.2.

> The distances specified in Table 4-2.5.2 were derived from NBS Report NBSIR 80-2097, *Full-Scale Fire Tests with Automatic Sprinklers in a Patient Room.* This rule should not be applied beyond Light Hazard Occupancies since the testing only evaluated sprinkler performance in a Light Hazard environment.

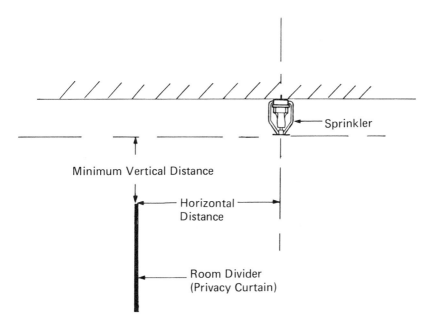

Figure 4-2.5.2   Standard Sprinkler Installed Near Privacy Curtains, Free-standing Partitions, or Room Dividers.

4-3 Position of Upright and Pendent Sprinklers.

This section establishes the rules for the location of sprinklers with respect to a specific construction type. These positions are measured from the sprinkler deflector to the particular ceiling configuration. In all cases, a minimum 1-in. (25-mm) clearance exists in order to allow easy replacement of upright sprinklers. Maximum distances vary and are dependent on the ability of a particular construction type to aid in the collection of heat, thus speeding the operating time of the sprinkler.

Deflector positioning requirements are given for conditions under ceilings, beams, and combinations thereof. These requirements typically allow a greater distance between the ceiling and deflector if the construction type is capable of trapping the heat in an enclosed volume, thus resulting in a faster sprinkler operating time.

4-3.1 Smooth Ceiling Construction (*as defined in 4-1.3.1*).   Deflectors of sprinklers shall be located 1 to 10 in. (25 to 254 mm) below combustible ceilings or 1 to 12 in. (25 to 305 mm) below noncombustible ceilings. The operating elements of sprinklers shall be located below the ceiling.

Exception No. 1:   Deflectors of sprinklers under beams shall be located 1 to 4 in. (25 to 102 mm) below beams, and not more than 14 in.

(356 mm) below combustible ceilings or not more than 16 in. (406 mm) below noncombustible ceilings.

Exception No. 2: Special ceiling-type pendent sprinklers (concealed, recessed, and flush-types) may have the operating element above the ceiling and the deflector located nearer to the ceiling when installed in accordance with their listing.

Formal Interpretation

Question 1: Is it the intent of 4-3.1 Exception No. 1 to require sprinklers to be installed in (4 ft × 8 ft) skylights having 55 flame spread, when they do not support combustion? Additionally, the skylights would have to be heated over 600°F to start melting out.

Answer: No. It is not the intent of the standard to require the installation of sprinklers in pockets formed by 4 ft × 8 ft skylights as such areas would not significantly retard the operation of sprinklers.

Question 2: Is it the intent of 4-3.1 Exception No. 1 to require sprinklers to be installed in (4 ft × 8 ft) plastic skylights fitted with aluminum ventilation louvers approximately one foot high supporting the skylight?

Answer: No. It is not the intent of the standard to require the installation of sprinklers in pockets formed by plastic skylights as such areas would not significantly retard the operation of the sprinklers.

These two interpretations point to but one of numerous conditions in which certain ceiling configurations, usually the result of a desired architectural feature, require a great deal of judgment to decide at what point a given volume becomes "too much." Void spaces created by an arrangement of soffits or beams may allow large quantities of heat to build up while at the same time delaying operation of sprinklers below the open space.

The argument here is not that areas below such a void cannot be adequately covered by sprinklers installed below this void, but that a significant delay in the operation of those sprinklers may result since the void would first have to fill with heat that would presumably bank down from the high ceiling, spill into surrounding areas, and then cause the sprinklers to operate. Figure 4.4 illustrates this point.

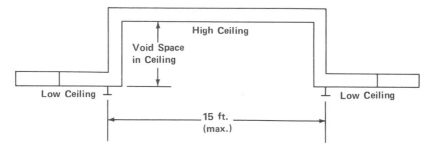

Figure 4.4. Ceiling cross section.

4-3.2 Beam and Girder Construction *(as defined in 4-1.3.2)*.

4-3.2.1 Deflectors of sprinklers in bays shall be located 1 to 16 in. (25 to 406 mm) below combustible or noncombustible roof or floor decks.

This section applies when the sprinkler is not located under a beam.

4-3.2.2 Deflectors of sprinklers under beams shall be located 1 to 4 in. (25 to 102 mm) below beams and not more than 20 in. (508 mm) below combustible or noncombustible roof or floor decks.

This section applies when the sprinkler is located under a beam.

4-3.2.3 Deflectors of sprinklers under concrete tee construction with stems spaced less than 7½ ft (2.3 m) but more than 3 ft (0.9 m) on centers shall, regardless of the depth of the tee, be located at or above a plane 1 in. (25 mm) below the level of the bottom of the stems of the tees and comply with Table 4-2.4.6.

The sprinkler is to be located as high as possible, taking into account the obstruction that could be offered by the leg of the concrete tee. This section is based on sensitivity tests conducted by the sprinkler industry.

4-3.3 Open Bar Joist Construction *(as defined in 4-1.3.3)*. Deflectors of sprinklers shall be located 1 to 10 in. (25 to 254 mm) below combustible roof or floor decks or not more than 12 in. (305 mm) below noncombustible roof or floor decks.

The open nature of bar joist construction allows it to be treated in the same manner as smooth ceiling construction with respect to deflector positioning.

4-3.4 Panel Construction (*as defined in 4-1.3.4*).

Formal Interpretation

Question: Under so-called Berkley Construction with a plywood deck on 2 in. by 6 in. wood stiffeners at 2 ft-0 in. centers framed into Glu-lam beams, with fiberglass insulation batts between the stiffeners with aluminum foil stapled to the lower edge of the stiffeners supporting the insulation, should sprinkler deflectors be positioned as for open wood joist construction or for panel construction with the aluminum sheathing considered as the "ceiling?"

Answer: The sheathing could be considered the ceiling in panel construction.

4-3.4.1 Deflectors of sprinklers in bays formed by members, such as beams framed into girders, resulting in panels up to 300 sq ft (27.9 m²) shall be located 1 to 18 in. (25 to 457 mm) below combustible or noncombustible roof or floor decks.

The distances given in this section recognize the ability of panel construction to trap the heat and operate sprinklers in a quicker manner than other types of construction. Since panel construction has a specified maximum square footage, an additional 2 in. (51 mm) beyond that specified by the beam and girder rule is permitted.

4-3.4.2 Deflectors of sprinklers under the members, such as under beams framed into girders, forming panels up to 300 sq ft (27.9 m²) shall be located 1 to 4 in. (25 to 102 mm) below such members and not more than 22 in. (559 mm) below combustible or noncombustible roof or floor decks.

This section applies when the sprinkler is located under a beam.

4-3.5 Wood Joist Construction (*as defined in 4-1.3.7*). In exposed wood joists spaced 3 ft (0.9 m) or less on centers, deflectors of sprinklers shall be located 1 to 6 in. (25 to 152 mm) below the bottom of the joist. If exposed joists are spaced more than 3 ft (0.9 m) on centers, deflectors of sprinklers shall be located in accordance with 4-3.1 or 4-3.2.

Heat travel across joists is impaired by the numerous narrow pockets which tend to direct the heat toward the joist channels, so sprinkler deflector distance is limited to 6 in. (152 mm) to account for sensitivity.

4-3.6 Composite Wood Joist Construction (*as defined in 4-1.3.8*). In exposed composite wood joist construction of 16 in. (406 mm) nominal depth or less, deflectors of sprinklers shall be located 1 to 6 in. (25 to 152 mm) below the bottom chord.

Specific requirements for composite wood joists in excess of 16 in. (406 mm) do not exist. See the commentary to 4-2.2.2.4.

4-3.7 Wood Truss Construction (*as defined in 4-1.3.9*). In exposed wood truss construction, sprinklers shall be located within the trusses and shall conform to 4-2.4.1, 4-2.4.5, and 4-3.1.

Exception: For wood truss construction with members 16 in. (406 mm) nominal depth or less, deflectors of sprinklers may be located 1 to 6 in. (25 to 152 mm) below the bottom chord.

4-3.8 Location Under Sheathed or Suspended Ceiling Under Any Type of Construction. The position of sprinklers under sheathed or suspended ceilings with any type of construction shall be the same as for smooth-ceiling construction. (*See 4-3.1.*)

4-4* Locations or Conditions Involving Special Consideration.

The rules in this section have been developed to deal with a number of special conditions.

A-4-4 Special Occupancy Considerations.

(a) Subject to the approval of the authority having jurisdiction, sprinklers may be omitted in rooms or areas where they are considered undesirable because of the nature of the contents, or in rooms or areas of noncombustible construction with wholly noncombustible contents and that are not exposed by other areas. Sprinklers should not be omitted from any room merely because it is damp or of fire-resistive construction.

(b) It is not advisable to install sprinklers when the application of water, or of flame and water, to a room or area's contents may constitute a serious life or fire hazard, as in the manufacture or storage of quantities of aluminum powder, calcium carbide, calcium phosphide, metallic sodium and potassium, quicklime, magnesium powder, and sodium peroxide. The manufacture and storage of such materials should be confined to specially cut-off, unsprinklered rooms, or buildings of fire-resistive construction.

For further general discussion of special occupancy considerations, refer to Section B-4-2, which follows the commentary to 4-4.19.3.

Formal Interpretation

Question: Is it the intent of Section 4-4 to require sprinkler protection in walk-in type coolers and freezers in fully-sprinklered buildings?

Answer: Yes.

Formal Interpretation

Question: Are automatic sprinklers required in the combustible concealed space that is developed in the installation of a Mansard Roof on a wood frame structure that requires automatic sprinklers because of occupancy, i.e., nursing homes?

Answer: Yes. Automatic sprinklers are required in Mansard Roofs as described in the material submitted to the Subcommittee.

4-4.1 Combustible Form Board. When roof or floor decks consist of poured gypsum or concrete on combustible form board supported on steel supports, the position of sprinkler deflectors shall be the same as for noncombustible construction as stated in Section 4-3. When combustible form board is located above suspended ceilings or in concealed spaces, see 4-4.4.1.

4-4.2 Metal Roof Decks. When roof decks are metal with combustible adhesives or vapor seal, the position of sprinklers shall be the same as for combustible construction.

The subject of combustible metal roof decks is dealt with in detail in the NFPA *Fire Protection Handbook*, 16th Edition, Chapter 7.

4-4.3 Spaces Under Ground Floors. Sprinklers shall be installed in all spaces below combustible ground floors except that, by permission of the authority having jurisdiction, sprinklers may be omitted when all of the following conditions prevail:

(a) The space is not accessible for storage purposes or entrance of unauthorized persons and is protected against accumulation of wind-borne debris;

(b) The space contains no equipment such as steam pipes, electric wiring, shafting, or conveyors;

(c) The floor over the space is tight;

(d) No combustible or flammable liquids or materials that under fire conditions may convert into combustible or flammable liquids are processed, handled, or stored on the floor above.

This is an exception to the general rule that all combustible spaces must be sprinklered, but it is subject to jurisdictional approval and recognizes that there can be maintenance problems, such as lack of access and danger of freezing, with piping under a floor. The conditions indicated are intended to eliminate sources of ignition from the concealed space and to limit combustibles to the floor only.

4-4.4 Concealed Spaces.

Formal Interpretation

Question 1: In a typical franchise restaurant having a roof system of a plywood deck on open web joist at 32-in. centers having steel web members and wood top and bottom chords with a ceiling sheathing attached to the bottom chord, is it the intent of 4-4.4 of NFPA 13 to permit 200 sq ft spacing of sprinklers in the blind space and under the ceiling in the restaurant seating area, the piping being sized by the pipe schedules for Light Hazard Occupancy Tables 8-2.2 and 8-2.3?

Answer: No. The spacing of sprinklers in the blind space is governed by 4-2.2.2.4 and is limited to 168 sq ft per sprinkler. If the ceiling below conforms to 4-2.2.2.1 then 200 sq ft spacing would apply. Pipe schedule Tables 8-2.2 and 8-2.3 would be applicable.

Question 2: In the restaurant above, is it intended to permit the sprinklers in the kitchen, mechanical and storage areas which would be spaced at 130 sq ft maximum per sprinkler to be fed by the piping sized by Tables 8-2.2 and 8-2.3?

Answer: No. The pipe schedule for these sprinklers should be sized by Table 8-3.2(a), 8-3.2(b), or 8-3.3 as these areas are considered to be Ordinary Hazard Occupancies.

Sprinklers are to be installed below close-spaced [less than 3 ft (0.9 m) on center] open wood trusses due to the high degree of obstruction, but within open wood trusses installed at greater spacings. When a ceiling is attached to the bottom of the wood trusses and sprinklers are required in the concealed space, the maximum sprinkler coverage of 168 sq ft (15.5 m²) (regarding sprinklers within the trusses) prevails regardless of the truss spacing.

Formal Interpretation

Question: Is it the intent of 4-4.4 of NFPA 13 to require sprinkler heads in the exterior canopy space of a shopping center, the interior of which is fully sprinklered and which is separated from the exterior canopy space by a rated partition in compliance with the *National Building Code?* The soffit of said canopy space is accessible by the removal of noncombustible 2 ft by 4 ft panels in a lay-in system with hold-down clips.

Answer: Yes. The only exception is where the space is less than 50 sq ft in area as covered in 4-4.4.1 Exception No. 7.

Formal Interpretation

Question 1: 4-4.4 of NFPA 13 deals with the installation of sprinklers in concealed spaces. Does 4-4.4 require the installation of sprinklers in a concealed space formed between a noncombustible roof or floor assembly and a suspended noncombustible lay-in tile ceiling system with enclosing end walls of masonry construction if the only combustibles present are fire-retardant treated wood studs and top caps forming the top of one-hour rated interior partitions? The partitions consist of gypsumboard and noncombustible insulation fill terminating immediately above the ceiling and the studs with top caps extending another 4 to 6 in. above the gypsumboard but not to the structure above. The entire building is provided with an automatic sprinkler system below the ceiling.

Answer: No. The Committee intent is to permit limited combustibles if the exposed surfaces have been demonstrated not to propagate fire in the form in which they are installed in the space.

Question 2: Does the response to No. 1 depend on whether or not the floor or roof/ceiling design has an hourly fire rating?

Answer: No.

4-4.4.1* All concealed spaces enclosed wholly or partly by exposed combustible construction shall be protected by sprinklers.

This section applies to those portions of a building which have construction materials of a combustible nature or combustible finish materials, or are used for the storage of combustible materials, all of which would be found in a concealed space. If none of the three prescribed conditions exists, with respect to combustible objects, the space is defined as a concealed, noncombustible space and requires no additional sprinkler protection.

This section recognizes that some minor quantities of combustible materials, such as communication wiring, will be present in some concealed spaces but should not typically be viewed as requiring sprinklers.

Formal Interpretation

Question: Would partitions constructed of 2- by 4-wood studs with ⅝-in. sheet rock on each side extending from floor through a 1-hour rated drop ceiling and up to a concrete slab above be considered "combustible construction" as contemplated under 4-4.4.1?

Answer: No. It is the opinion of the Committee that where a combustible member is protected on both sides by noncombustible material such as sheet rock or metal-lath and plaster, the partition is not considered "exposed combustible construction" in the intent of 4-4.4.1.

Exception No. 1: Spaces formed by studs or joists with less than 6 in. (152 mm) between the inside or near edges of the studs or joists. (See Figure 4-2.3.2.)

Exception No. 2: Spaces formed by bar joists with less than 6 in. (152 mm) between the roof or floor deck and ceiling.

Exception No. 3: Spaces formed by ceilings attached directly to or to within 6 in. (152 mm) of wood joist construction.

Exception No. 3 applies to sheathed joist construction or similar narrow spaces formed by attachment of a ceiling to or within 6 in. (152 mm) of beams. It is the intent that this Exception apply only to joists with no openings in the members and up to a nominal depth of 14 in. (356 mm).

Exception No. 4: Spaces formed by ceilings attached directly to the underside of composite wood joist construction, provided the joist channels are fire-stopped into volumes each not exceeding 160 cu ft (4.53 m³) using materials equivalent to the joists.

Exception No. 5: Spaces entirely filled with noncombustible insulation.

In some cases it might be economically advantageous to fill an unsprinklered combustible concealed space with noncombustible insulation, rather than to install sprinklers as suggested by Exception No. 5.

Exception No. 6:   In wood joist construction and composite wood joist construction with noncombustible insulation filling the space from the ceiling up to the bottom edge of the joist of the roof or floor deck, provided that in composite wood joist construction, the joist channels are fire-stopped into volumes each not exceeding 160 cu ft (4.53 m³) using materials equivalent to the joists.

Exception No. 7:   Small spaces over rooms not exceeding 50 sq ft (4.6 m²) in area.

Exception No. 8:   When the exposed surfaces have a flame spread rating of 25 or less and the materials have been demonstrated not to propagate fire in the form in which they are installed in the space.

Exception No. 8 allows the use of paper-coated insulation material.

Exception No. 9:   When the Btu content of the facing and substrate of insulation material does not exceed 1000 Btu per sq ft (11 356 kJ/m²).

Exception No. 9 is intended to permit fire-retardant treated lumber or materials classed as limited-combustible.

A-4-4.4.1   Exceptions No. 1, 2, and 3 do not require sprinkler protection because it is not physically practical to install sprinklers in these spaces. To prevent the possibility of uncontrolled fire spread, consideration should be given in these unsprinklered concealed space situations by using other means such as Exceptions No. 5, 8, and 9.

4-4.4.2   Sprinklers in concealed spaces having no access that will allow storage or other use may be installed on the basis of Light Hazard Occupancy.

The existence of an access to a concealed space could lead to the space being used for storage or other purposes. However, if the size or arrangement of the access is such that it is determined to not allow any use of the space, then sprinklers can be installed according to the requirements for a Light Hazard Occupancy.

4-4.4.3   When heat-producing devices such as furnaces or process equipment are located in the joist channels above a ceiling attached directly to the underside of composite wood joist construction that would not otherwise require sprinkler protection of the spaces, the joist channel containing the heat-producing devices shall be sprinklered by installing two sprinklers in each joist channel, one on each side, adjacent to the heat-producing device. The temperature rating of the sprinklers shall be as prescribed in Table 3-11.6.1 and Figure 3-11.6.3(a).

4-4.4.4*   In concealed spaces having exposed combustible construction, or containing exposed combustibles, in localized areas, the combustibles shall be protected as follows:

(a) If the exposed combustibles are in the vertical partitions or walls around all or a portion of the enclosure, a single row of sprinklers spaced not over 12 ft (3.7 m) apart nor more than 6 ft (1.8 m) from the inside of the partition may be installed to protect the surface. The first and last sprinklers in such a row shall not be over 5 ft (1.5 m) from the ends of the partitions.

(b) If the exposed combustibles are in the horizontal plane, permission may be given to protect the area of the combustibles on a light hazard spacing and add a row of sprinklers not over 6 ft (1.8 m) outside the outline of the area and not over 12 ft (3.7 m) on center along the outline. When the outline returns to a wall or other obstruction, the last sprinkler shall not be over 6 ft (1.8 m) from a wall or obstruction.

The presence of localized combustibles can be abated when the requirements of this paragraph are followed. The presence of significant concentrated quantities of exposed computer cable above a ceiling, is one candidate for this special protection.

Formal Interpretation

Background: **A roof construction consisting of 2½-in. permadeck on 2-by-4 wood nailers on top chord of open web steel joist, with a noncombustible suspended ceiling below. The only combustible material is the 2-by-4 wood nailer and the only exposed surface is the 1-¾ face on the side of the nailer.**

Question 1:   **Does this type of construction qualify as noncombustible construction?**

Answer: **No. Permadeck is classed as a limited-combustible construction material under NFPA 220, *Standard on Types of Building Material.***

Question 2:   **Is it the intent of the Committee on Automatic Sprinklers to strictly classify all blind spaces with wood construction as combustible construction, even if the amount of combustible material is very restricted and less than the "exposed combustibles in localized areas" mentioned in 4-4.4.4 of NFPA 13?**

Answer: No. The intent of the Committee is to permit fire-retardant treated lumber in unrestricted amounts, or minor amounts of untreated wood.

Question 3: Are upright sprinklers required in the concealed spaces above noncombustible suspended ceiling in construction as described above?

Answer: No.

A-4-4.4.4 When there is a limited amount of combustibles available to burn and a limited prospect of fire propagation, sprinklers may not be required.

4-4.5 Spacing of Sprinklers Under Pitched Roofs.

Formal Interpretation

Question 1: Is it the intent of 4-4.5 to:

(a) Define the area of protection covered by a sprinkler head as the distance between sprinklers and branch lines as measured on the slope, or

(b) To establish the location of the sprinkler head with respect to the peak of a pitched roof building with the area of protection covered being determined by the spacing being projected to the horizontal plane of the floor?

Answer: The intent is expressed by (a).

Question 2: If the answer to the above is (a), why would it be allowed by 4-4.6 to project the spacing to the horizontal plane of the floor for curved roof buildings?

Answer: 4-4.6 does not permit the spacing of sprinklers under a curved roof to be determined by their projected distance on the floor. It does however, in 4-4.6.1, make an exception with respect to the sprinkler nearest a side wall when the roof curves down to the floor line. The provisions of 4-4.6.3 are also an exception with respect to extra hazard spacing, permitting the area at the roof to be increased to that permitted for ordinary hazard provided the projected spacing on the floor does not exceed that for extra hazard [*see* 7-4.3.1.2(a)]. The general requirement is that spacing be determined on the plane of the roof.

4-4.5.1 Branch lines parallel to peaks of pitched roofs and sprinklers on lines perpendicular to peaks shall be spaced throughout the distance measured along the slope. This will place a row of sprinklers either in the peak or one-half the spacing down the slope from the peak.

> The spacing of sprinklers is to be measured on the slope of the roof, and the maximum distances and areas given in Section 4-2 are applied on this slope. Note, however, that for hydraulically calculated systems the density is still calculated on the basis of floor area.

4-4.5.2 Under saw-toothed roofs, the row of sprinklers at the highest elevation shall not be more than 3 ft (0.9 m) down the slope from the peak.

Formal Interpretation

Question 1: Please define "saw-toothed" roofs.

Answer: Saw-toothed roofs have regularly spaced monitors of saw-tooth shape, with the nearly vertical side glazed and usually arranged for venting.

Question 2: Why is the row of sprinklers limited to maximum of 3 ft down the slope from the peak?

Answer: Sprinklers are limited to a maximum of 3 ft down the slope from the peak because if they are further removed from the venting they may not operate.

Question 3: (a) Where a peaked roof has a corridor parallel with the peak (center line of corridor is center line of peak) and the walls of the corridor are extended above the ceiling to the roof, thus subdividing the peaked roof into a peak with adjoining right triangles, could this "shed roof" (two right triangle areas) ever be considered a "saw-toothed" roof?

(b) Would the 3-ft maximum distance from the peak apply here?

(c) If so, why?

Answer: (a) No. The construction described does not include the glazing and venting features of saw-toothed roofs.

(b) No. 4-4.5.1 would apply.

(c) Does not apply.

4-4.5.3 In 4-4.5.1 or 4-4.5.2, sprinklers in or near the peak shall have deflectors located not more than 3 ft (0.9 m) vertically down from the peak. [*See Figures 4-4.5.3(a) and 4-4.5.3(b).*]

Exception: In a steeply pitched roof the distance from the peak to the deflectors may be increased to maintain a horizontal clearance of not less than 2 ft (0.6 m) from other structural members. [See Figure 4-4.5.3(c).]

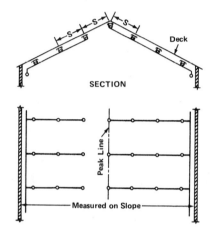

For SI Units: 1 in. = 25.4 mm; 1 ft = 0.3048 m.

Figure 4-4.5.3(a)   Sprinklers at Pitched Roofs; Branch Lines Run Up the Slope.

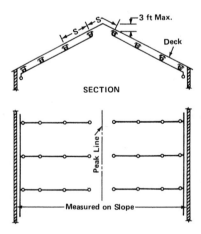

For SI Units: 1 in. = 25.4 mm; 1 ft = 0.3048 m.

Figure 4-4.5.3(b)   Sprinklers at Pitched Roofs; Branch Lines Run Up the Slope.

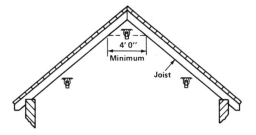

For SI Units: 1 in. = 25.4 mm; 1 ft = 0.3048 m.

Figure 4-4.5.3(c)   Desirable Horizontal Clearance for Sprinklers at Peak of Pitched Roof.

Sprinklers are required to be within 3 ft (0.9 m) vertically of the peak but must also meet the position requirements of Section 4-3. The 3-ft limitation reflects concern for sensitivity.

If the roof is steeply pitched, sensitivity is compromised by distribution considerations.

4-4.6 Spacing of Sprinklers Under Curved Roof Buildings.

4-4.6.1 When roofs are curved down to the floor line, the horizontal distance measured at the floor level from the sidewall or roof construction to the nearest sprinklers shall not be greater than one-half the allowable distance between sprinklers in the same direction.

When the slope of the curved roof becomes steep near the side of the building, the consideration of spacing changes from on the plane of the roof to that of the floor for the location of the first sprinkler from the sidewall.

4-4.6.2 Deflectors of sprinklers shall be parallel with the curve of the roof or tilted slightly toward the peak of the roof. Deflectors of sprinklers shall be located as described for beam and girder construction or for the closest comparable type of ceiling construction.

4-4.6.3 When Extra Hazard Occupancy spacing of sprinklers is used under curved ceilings of other than fire-resistive construction, as in aircraft storage or servicing areas, the spacing as projected on the floor shall be not wider than that required for Extra Hazard Occupancies, but in no case shall the spacing on the roof or ceiling be wider than that required for Ordinary Hazard Occupancies.

The designer must consider the spacing under the curved ceiling on the slope and also the projected area on the floor.

4-4.7 Elevators, Stairs, and Floor Openings.

4-4.7.1 Vertical Shafts.

4-4.7.1.1 One sprinkler shall be installed at the top of all shafts.

Sprinklers are to be provided at the top of all shafts used for elevators or stairs, or other shafts open to more than one floor. Concealed combustible shafts must be sprinklered; concealed shafts of noncombustible construction and contents in a suitably rated enclosure do not require sprinklers.

Codes that cover elevator design do not permit water discharge in elevator shafts until electrical power to the elevator cab has been secured. This necessitates some special arrangement, such as a preaction system, to protect such shafts.

4-4.7.1.2* When vertical shafts have combustible sides, one sprinkler shall be installed at each alternate floor level. When a shaft having combustible surfaces is trapped, an additional sprinkler shall be installed at the top of each trapped section.

The additional sprinklers for shafts with combustible sides must be placed to effectively wet the combustible surfaces. Some elevator and stair shafts are equipped with floor level trap doors; in this case, sprinklers are required at the ceiling of each level.

A-4-4.7.1.2 When practicable, sprinklers should be staggered at the alternate floor levels, particularly when only one sprinkler is installed at each floor level.

4-4.7.1.3 When accessible shafts have noncombustible surfaces, one sprinkler shall be installed near the bottom.

Formal Interpretation

Question: Is it the intention of 4-4.7.1.3 to require protection at the bottom of a noncombustible elevator shaft, not having direct access?

Answer: Yes.

Refuse has a tendency to collect at the bottom of shafts. A properly located sprinkler will control a fire in such material.

4-4.7.1.4 When vertical openings are not protected by standard enclosures, sprinklers shall be so placed as to fully cover them. This necessitates placing sprinklers close to such openings at each floor level.

By placing sprinklers close to a ceiling opening, the floor area under the opening may be protected. Note that this is discussed under the heading of Vertical Shafts, implying small openings. Large openings are addressed in 4-4.7.2.3.

4-4.7.2* Stairways.

A-4-4.7.2 Floor or wall openings tending to create vertical or horizontal drafts, or other structural conditions that would delay the prompt operation of automatic sprinklers by preventing the banking up of the heated air from the fire, should be properly stopped in order to permit control of fire at any point by local sprinklers.

4-4.7.2.1 Stairways of combustible construction shall be sprinklered underneath, whether risers are open or not.

4-4.7.2.2 Stairways of noncombustible construction with combustible storage beneath shall be sprinklered.

4-4.7.2.3* When moving stairways, staircases, or similar floor openings are unenclosed, the floor openings involved shall be protected by draft stops in combination with closely spaced sprinklers.

The draft stops shall be located immediately adjacent to the opening, shall be at least 18 in. (457 mm) deep, and shall be of substantially noncombustible material that will stay in place before and during sprinkler operation. Sprinklers shall be spaced not more than 6 ft (1.8 m) apart and placed 6 to 12 in. (152 to 305 mm) from the draft stop on the side away from the opening to form a water curtain. Sprinklers in this water curtain shall be hydraulically designed to provide a discharge of 3 gpm per lineal ft [(37 L/min)/m] of water curtain, with no sprinklers discharging less than 15 gpm (56.8 L/min). The number of sprinklers calculated in this water curtain shall be the number in the length corresponding to the length parallel to the branch lines in the design area determined by 7-4.3.1. The water supply for these sprinklers shall be added to the water supply required for the area of operation in hydraulically designed systems or to the water supply required as determined in accordance with Table 2-2.1.1(a). Supplies shall be balanced to the higher pressure demand in either case. Sprinklers shall be nominal ½ in., 7⁄16 in., or 3⁄8 in. orifice. When sprinklers are closer than 6 ft (1.8 m), cross baffles shall be provided in accordance with 4-4.18. When sprinklers in the normal pattern are closer than 6 ft (1.8 m) from the water curtain, it may be preferable to locate the water curtain sprinklers in recessed baffle pockets.

Exception: Closely spaced sprinklers are not required around large openings such as those found in shopping malls, atrium buildings, and similar structures where all adjoining levels and spaces are protected by automatic sprinklers in accordance with this standard, when the openings have all horizontal dimensions between opposite edges of 20 ft (6 m) or greater, and an area of 1000 sq ft (93 m²) or greater.

This section limits itself to openings that do not meet the definition of an atrium. These smaller openings tend to behave in a similar manner as a chimney, as they tend to allow for rapid vertical movement of the hot gases from the fire. The closely spaced sprinklers in conjuction with draft stops have proven to be an effective method of gaining control of the fire in these smaller sized openings.

When the features of the opening warrant sprinklers spaced at intervals of less than 6 ft (1.8 m), the installation of baffles between adjacent sprinklers will prevent the occurrence of the "cold solder" effect. The installation of such devices is discussed in 4-4.18.

Formal Interpretation

Question: Is it the intent of 4-4.7.2.3 that the close spaced sprinklers used in combination with draft stops at moving stairways or large monumental staircases of similar unenclosed floor openings be closed sprinklers?

Answer: No. Close spaced sprinklers used in combination with draft stops and moving stairways, large monumental staircases or similar unenclosed floor openings may be either closed sprinklers or open sprinklers actuated by fixed temperature detection devices.

Use of deluge water curtains has become quite rare since the early 1960s. Privately conducted tests using closed sprinklers indicated their effectiveness. Accidental discharge of deluge-type water curtains has resulted in considerable water damage as well as personal injury to persons on escalators when such false actuation has occurred.

4-4.7.2.4* In noncombustible stair shafts, sprinklers shall be installed at the top and under the first landing above the lowest level. When the stair shaft serves two or more separate fire sections, sprinklers shall also be installed at each floor landing.

When a noncombustible stair shaft serves two fire-separated buildings or fire sections of one building as shown in Figure A-4-4.7.2.4(a) (that is, when the stair landing serves as a horizontal exit), sprinklers are required at each floor landing. If the stair serves only one fire section, then sprinklers are

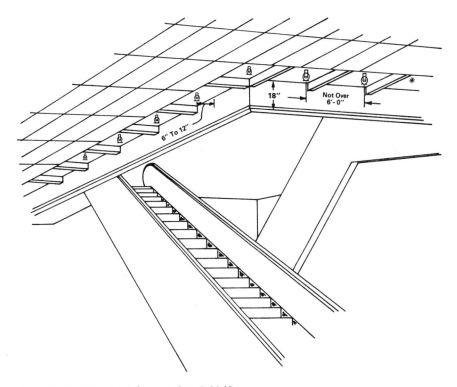

For SI Units: 1 in. = 25.4 mm; 1 ft = 0.3048 m.
*Baffle (see 4-4.18).

Figure A-4-4.7.2.3   Sprinklers Around Escalators.

required only at the roof and under the lowest landing. See Figures A-4-4.7.2.4(a) and A-4-4.7.2.4(b) on page 252.

4-4.8* Building Service Chutes.   Building service chutes (linen, rubbish, etc.) shall be protected internally by automatic sprinklers. A sprinkler shall be provided above the top service opening of the chute, above the lowest service opening, and above service openings at alternate levels in buildings over two stories in height. The room or area into which the chute discharges shall also be protected by automatic sprinklers.

A-4-4.8   The installation of sprinklers at floor levels should be so arranged as to protect the sprinklers from mechanical injury, from falling materials, and not cause obstruction within the chute. This can usually be accomplished by recessing the sprinkler in the wall of the chute or by providing a protective deflector canopy over the sprinkler. Sprinklers should be placed so that there will be minimum interference of the discharge therefrom. (*See also 4-1.2.*) Sprinklers with special directional discharge characteristics may be advantageous.

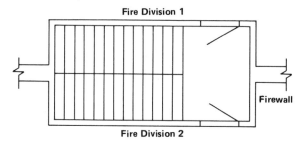

Figure A-4-4.7.2.4(a)   Noncombustible Stair Shaft Serving Two Fire Sections.

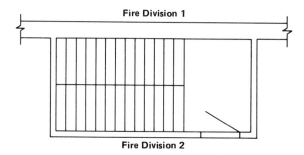

Figure A-4-4.7.2.4(b)   Noncombustible Stair Shaft Serving One Fire Section.

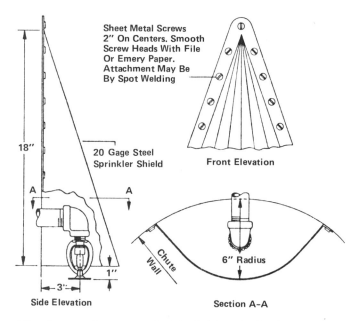

Figure 4.5.   Canopy for protecting sprinklers in building service chutes.

4-4.9* Exterior Canopies, Docks, and Platforms.

A-4-4.9 Small loading docks, covered platforms, ducts, or similar small unheated areas may be protected by dry-pendent sprinklers extending through the wall from wet sprinkler piping in an adjacent heated area, as shown in Figure A-4-4.9.

Where possible, the dry-pendent sprinkler should extend down at a 45-degree angle. The width of the area to be protected should not exceed 7½ ft (2.3 m). Sprinklers should be spaced not over 12 ft (3.7 m) apart.

4-4.9.1 Sprinklers shall be installed under roofs or canopies over outside-loading platforms, docks, or other areas where combustibles are stored or handled.

Figure A-4-4.9 shows one method of protecting under roofs or canopies up to 7½ ft (2.3 m) wide. Sprinklers are required under all such coverings where combustible goods are stored or handled.

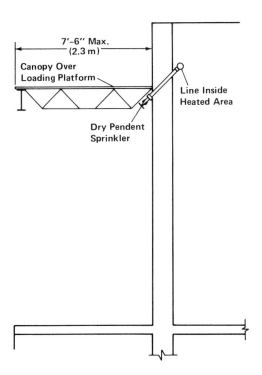

Figure A-4-4.9   Dry-Pendent Sprinklers for Protection of Covered Platforms, Shipping Docks, and Similar Areas.

4-4.9.2 Sprinklers shall be installed under exterior combustible roofs or canopies exceeding 4 ft (1.2 m) in width.

Exception: Sprinklers may be omitted where construction is noncombustible and areas under the roofs or canopies are not used for storage or handling of combustibles.

If of combustible construction, exterior canopies exceeding 4 ft (1.2 m) in width are to be sprinklered whether or not goods are stored or handled underneath. If construction is noncombustible and the area underneath is essentially restricted to pedestrian use, sprinklers may be omitted. The roof canopy typically found on strip shopping malls is one example of this condition, where the area under the canopy is limited to pedestrians. Automobiles stopping briefly to pick up or drop off passengers would not be considered storage.

4-4.9.3 Sprinklers shall be installed under exterior docks and platforms of combustible construction unless such space is closed off and protected against accumulation of debris.

This is an exception to the general requirement that sprinklers be installed in all areas of combustible construction but does not apply if wind-borne debris can accumulate and enhance the chance of a fire developing.

4-4.10* Decks. Sprinklers shall be installed under decks and galleries over 4 ft (1.2 m) wide. Slatting of decks, walkways, or the use of open gratings as a substitute for such sprinklers is not acceptable. Sprinklers installed under open gratings shall be of the listed intermediate level type or shielded from the discharge of overhead sprinklers.

Several sections of this chapter deal with positioning sprinklers in order to avoid obstructions. For some situations, the obstruction cannot be avoided and additional sprinklers are necessary to compensate for areas under the obstruction which would not receive adequate coverage.

The point at which obstructions become too large to ignore is typically thought of as 4 ft (1.2 m). Openings in grated decks are not adequate to compensate for obstructions to the sprinkler spray pattern; thus, supplemental sprinklers under the deck become necessary. Because gratings or slatted constructions are frequently covered with goods in storage or by a light surface dust stop, sprinklers are required under such gratings and walkways.

A-4-4.10 Frequently, additional sprinkler equipment can be avoided by reducing the width of decks or galleries and providing proper clearances. Slatting of decks or walkways or the use of open grating as a substitute for automatic sprinklers thereunder is not acceptable. The use of cloth or paper dust tops for rooms forms obstruction to water distribution. If employed, the area below should be sprinklered.

Formal Interpretation

Question: Is 4-4.10 intended to apply to steel racking with combustible materials stored on combustible shelves presenting a barrier (8 ft wide and 12 ft high, usually 2 or 3 shelves and of any length) to the proper operation of the sprinkler system?

Answer: No. It was not the intent of 4-4.10, when originally written, to apply to steel racking with combustible material stored on combustible shelves presenting a barrier to the proper operation of the sprinkler system.

Although it is not a mandatory requirement, the Committee wishes to call attention to B-4-2.4.8, Stock Fixtures, which states, "Sprinklers should be installed in all stock fixtures that exceed 5 ft (1.5 m) in width, also in those that are less than 5 ft (1.5 m) but more than 2½ ft (0.8 m) in width unless bulkheaded with tight partitions. Sprinklers should be installed in any compartments that are larger than 5 ft (1.5 m) deep, 8 ft (2.4 m) long, and 3 ft (0.9 m) high."

4-4.11 Library Stack Rooms. Sprinklers shall be installed in every aisle and at every tier of stacks with distance between sprinklers along aisles not to exceed 12 ft (3.6 m). [*See Figure 4-4.11(a).*]

Exception No. 1: When vertical shelf dividers are incomplete and allow water distribution to adjacent aisles, sprinklers may be omitted in alternate aisles on each tier. When ventilation openings are also provided in tier floors, sprinklers shall be staggered vertically. [See Figure 4-4.11(b).]

Exception No. 2: Install sprinklers without regard to aisles when there is 18 in. (457 mm) or more clearance between sprinkler deflectors and tops of racks.

Library stack rooms are generally thought of as high density book storage on shelves and in some ways may be thought of as a type of rack storage. To overcome the obstruction caused by

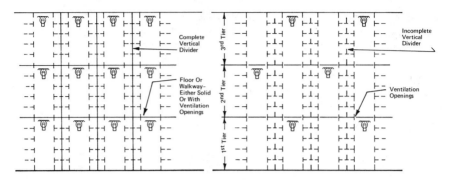

Figures 4-4.11(a) and (b)   Sprinklers in Multitier Library Bookstacks.

the bookshelves, sprinklers are to be installed at alternating tiers unless the Exceptions can be satisfied.

4-4.12* Ducts. Sprinklers shall be installed beneath ducts over 4 ft (1.2 m) wide unless ceiling sprinklers can be spaced in accordance with Table 4-2.4.6.

A-4-4.12 For ducts less than 4 ft (1.2 m) wide that obstruct distribution from ceiling sprinklers, see B-4-2.3.

Ducts are similar to decks *(see 4-4.10)* because they can obstruct sprinkler discharge. The size limit is set at 4 ft (1.2 m) unless the spacing of the ceiling sprinklers can conform to the distances shown in Table 4-2.4.6, in which case they need not be installed under the duct. For ducts that are less than 4 ft in width, the provisions of the beam rule as discussed in 4-2.4.6 should also be considered.

Formal Interpretation

Question 1: Is it the intent of 4-4.12 (Ducts), of NFPA 13, to require sprinkler heads to be located below ducts greater than 4 ft wide in addition to the ceiling sprinkler heads located above the ducts when spacing requirements in accordance with Table 4-2.4.6 cannot be satisfied?

Answer: Yes.

Question 2: If the answer to the question above is "yes," could the ceiling sprinkler heads above the obstruction be omitted when the area of coverage above the obstruction protected by the omitted ceiling heads is noncombustible? Assume that sprinkler heads are located below the obstruction.

Answer: The Committee feels that this is a question that should properly be answered by the authority having jurisdiction who can properly assess all the circumstances.

4-4.13 Electrical Equipment. When sprinkler protection is provided in generator and transformer rooms, hoods or shields installed to protect important electrical equipment from water shall be noncombustible.

Sprinkler protection in these types of rooms is covered by 4-1.1.1(a) of this standard. To prevent direct contact between the electrical equipment and the discharging water, water shields of noncombustible material are necessary. This will minimize any shock hazard and will allow for containment of fires in these areas. Frequently, electrical equipment rooms become storage rooms for combustible materials.

4-4.14* Open-Grid Ceilings. The following requirements are applicable to open-grid ceilings in which the openings are ¼ in. (6.4 mm) or larger in the least dimension, when the thickness or depth of the material does not exceed the least dimension of the openings, and when such openings constitute at least 70 percent of the area of the ceiling material. Other types of open-grid ceilings shall not be installed beneath sprinklers unless they are listed by a testing laboratory and are installed in accordance with the instructions contained in each package of the ceiling material. Ceilings made of highly flammable material may spread fire faster than sprinklers can control.

(a) In Light Hazard Occupancies when spacing of either standard or old-style sprinklers is not wider than 10 by 10 ft (3 by 3 m), a minimum clearance of at least 18 in. (457 mm) shall be provided between the sprinkler deflectors and the upper surface of the open-grid ceiling. When spacing is wider than 10 by 10 ft (3 by 3 m) but not wider than 10 by 12 ft (3 by 3.7 m), a clearance of at least 24 in. (610 mm) shall be provided from standard sprinklers and at least 36 in. (914 mm) from old-style sprinklers. When spacing is wider than 10 by 12 ft (3 by 3.7 m), a clearance of at least 48 in. (1219 mm) shall be provided.

(b) In Ordinary Hazard Occupancies, open-grid ceilings may be installed beneath sprinklers only where such use is approved by the authority having jurisdiction, and shall be installed beneath standard sprinklers only. When sprinkler spacing is not wider than 10 by 10 ft (3 by 3 m), a minimum clearance of at least 24 in. (610 mm) shall be provided between the sprinkler deflectors and the upper surface of the open-grid ceiling. When spacing is wider than 10 by 10 ft (3 by 3 m), a clearance of at least 36 in. (914 mm) shall be provided.

These requirements for clearance between the sprinkler deflectors and the top of the grid ceiling are to ensure that sprinkler discharge is not too severely impaired, for the grid ceiling will obstruct the discharge pattern. In addition to the height above the grid ceiling, other obstructions such as pipes or ducts must be considered.

A-4-4.14   The installation of open-grid egg crate, louver, or honeycomb ceilings beneath sprinklers restricts the sideways travel of the sprinkler discharge and may change the character of discharge.

4-4.15 Drop-out Ceilings.

4-4.15.1   Drop-out ceilings may be installed beneath sprinklers when ceilings are listed for that service and are installed in accordance with their listing. The authority having jurisdiction shall be consulted in all cases.

Exception: Special sprinklers shall not be installed above drop-out ceilings unless specifically listed for this purpose.

Drop-out ceilings are designed to shrivel and fall when heated by a fire, permitting the sprinkler that had been concealed above to come into play. Extreme care must be exercised in using only panels that are listed and tested for this use. When these ceilings are opaque, thus making it difficult to determine if they are painted or have material stored above them, their use should be discouraged. When any appreciable delay in sprinkler operation is undesirable, these ceilings become undesirable also.

The response time for quick-response and some extended coverage sprinklers installed above drop-out ceiling panels may cause the sprinklers to operate before all the ceiling panels drop out. Since both types of sprinklers have fast acting heat-responsive elements, they should not be used above ceiling panels unless they have been specifically investigated for such use.

4-4.15.2   Drop-out ceilings shall not be considered ceilings within the context of this standard.

Because the ceiling falls in the early stages of a fire, it need not be considered with respect to the position of sprinklers.

4-4.15.3*   Piping installed above drop-out ceilings shall not be considered concealed piping. *(See 3-7.4, Exception No. 2.)*

Since piping is exposed to a fire, solder joints are not permitted above drop-out ceilings in Ordinary Hazard Occupancies.

A-4-4.15.3   Drop-out ceilings do not provide the required protection for soft-soldered copper joints or other piping that requires protection.

4-4.15.4*   Sprinklers shall not be installed beneath drop-out ceilings.

A-4-4.15.4   The ceiling tiles may drop before sprinkler operation. Delayed operation may occur because heat must then bank down from the deck above before sprinklers will operate.

Additionally, the danger exists for the drop-out ceiling tiles to hang up on a sprinkler and interfere with that sprinkler's proper operation.

4-4.16 Fur Vaults.

4-4.16.1   Sprinklers in fur storage vaults shall be located centrally over the aisles between racks and shall be spaced not over 5 ft (1.5 m) apart along the aisles.

Baffles are not required because old-style sprinklers *(see 4-4.16.3)* can be installed as close as 5 ft (1.5 m) apart without wetting the adjacent sprinklers.

4-4.16.2   When sprinklers are spaced 5 ft (1.5 m) apart along the sprinkler branch lines, pipe sizes may be in accordance with the following schedule:

1 in.	4 sprinklers	2 in.	.20 sprinklers
1¼ in.	6 sprinklers	2½ in.	.40 sprinklers
1½ in. . . . . .10 sprinklers		3 in.	.80 sprinklers

This schedule applies to piping running in the same direction as the aisles.

4-4.16.3   Sprinklers shall be listed old-style having orifice sizes selected to provide as closely as possible but not less than 20 gal per min (76 L/min) per sprinkler, for four sprinklers, based on the water pressure available.

NOTE: See NFPA 81, *Standard on Fur Storage, Fumigation and Cleaning.* For tests of sprinkler performance in fur vaults see Fact Finding Report on Automatic Sprinkler Protection for Fur Storage Vaults of Underwriters Laboratories Inc., dated November 25, 1947.

This is the only section of the standard that requires old-style sprinklers to be installed. Full-scale testing, conducted in 1947 on fur storage vaults, found that the optimum system design included operation of four old-style sprinklers in this scenario.

4-4.17 Commercial-type Cooking Equipment and Ventilation.

The original requirements with respect to protection of cooking equipment and ventilation systems were developed as a result of test work done by a task group of the National Fire Sprinkler Association and were introduced in 1968. Automatic sprinklers are effective for extinguishing fires in greases and cooking oils because the fine droplets of the water spray lower the temperatures to below the point at which the fire can sustain itself. These current requirements have been revised based on research and input from both the cooking equipment ventilation industry and the sprinkler manufacturing industry.

In order to avoid a large and sudden increase in the volume of fire resulting from large droplets of water sprayed into cooking oil (boil-over), sprinklers listed for this purpose must be used when protecting deep fat fryers. (*See 4-4.17.8.2.*)

4-4.17.1 In cooking areas protected by automatic sprinklers, additional sprinklers or automatic spray nozzles shall be provided to protect commercial-type cooking equipment and ventilation systems that are designed to carry away grease-laden vapors unless otherwise protected. (*See NFPA 96, Standard for the Installation of Equipment for the Removal of Smoke and Grease-Laden Vapors from Commercial Cooking Equipment.*)

4-4.17.2 Standard sprinklers or automatic spray nozzles shall be so located as to provide for the protection of exhaust ducts, hood exhaust duct collars, and hood exhaust plenum chambers.

Exception: Sprinklers or automatic spray nozzles in ducts, duct collars, and plenum chambers may be omitted when all cooking equipment is served by listed grease extractors.

Sprinklers are an effective way of controlling fires in cooking equipment, filters, and ducts if they are located to cover the cooking surfaces and all areas of an exhaust system to which fires could spread. Subject to acceptance by the authority having jurisdiction, if all cooking equipment is served by listed grease extractors, the sprinkler protection may be limited to the cooking surfaces. Some manufacturers of exhaust systems incorporating listed grease extractors provide listed built-in water spray fire protection for cooking surfaces in a preengineered package ready for connection to the sprinkler system.

4-4.17.3 Exhaust ducts shall have one sprinkler or automatic spray nozzle located at the top of each vertical riser and at the midpoint of each offset. The first sprinkler or automatic spray nozzle in a horizontal duct shall be installed at the duct entrance. Horizontal exhaust ducts shall have such devices located on 10-ft (3-m) centers beginning no more than 5 ft (1.5 m) from the duct entrance. Sprinkler(s) or automatic spray nozzle(s) in exhaust ducts subject to freezing shall be properly protected against freezing by approved means. (*See 3-5.1.*)

Exception: Sprinklers or automatic spray nozzles may be omitted from a vertical riser located outside of a building provided the riser does not expose combustible material or the interior of a building and the horizontal distance between the hood outlet and the vertical riser is at least 25 ft (7.6 m).

4-4.17.4* Each hood exhaust duct collar shall have one sprinkler or automatic spray nozzle located 1 in. minimum to 12 in. maximum (25.4 mm min. to 305 mm max.) above the point of duct collar connection in hood plenum. Hoods that have listed fire dampers located in the duct collar shall be protected with a sprinkler or automatic spray nozzle located on the discharge side of the damper and be so positioned as not to interfere with damper operation.

See Figure A-4-4.17.4 on page 262.

4-4.17.5 Hood exhaust plenum chambers shall have one sprinkler or automatic spray nozzle centered in each chamber not exceeding 10 ft (3 m) in length. Plenum chambers greater than 10 ft (3 m) in length shall have two sprinklers or automatic spray nozzles evenly spaced with the maximum distance between the two sprinklers not to exceed 10 ft (3 m).

4-4.17.6 Sprinklers or automatic spray nozzles being used in duct, duct collar, and plenum areas shall be of the temperature classification extra high [325 to 375°F (163 to 191°C)] and have orifice sizes no less than ¼ in. (6.4 mm) and no more than ½ in. (13 mm).

Exception: When use of a temperature measuring device indicates temperatures above 300°F (149°C) a sprinkler or automatic spray nozzle of higher classification shall be used.

It is important that the sprinklers over the cooking surface operate first, so that there is no chance of water from the sprinklers in the plenum cooling them and preventing operation.

4-4.17.7 Access must be provided to all sprinklers or automatic spray nozzles for examination and replacement.

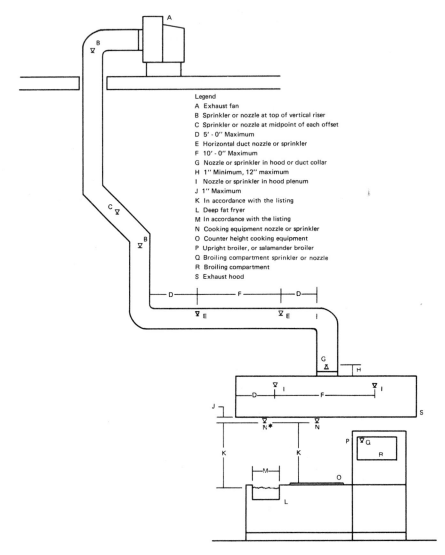

Legend
A Exhaust fan
B Sprinkler or nozzle at top of vertical riser
C Sprinkler or nozzle at midpoint of each offset
D 5' - 0" Maximum
E Horizontal duct nozzle or sprinkler
F 10' - 0" Maximum
G Nozzle or sprinkler in hood or duct collar
H 1" Minimum, 12" maximum
I Nozzle or sprinkler in hood plenum
J 1" Maximum
K In accordance with the listing
L Deep fat fryer
M In accordance with the listing
N Cooking equipment nozzle or sprinkler
O Counter height cooking equipment
P Upright broiler, or salamander broiler
Q Broiling compartment sprinkler or nozzle
R Broiling compartment
S Exhaust hood

* Listed for deep fat fryer protection.

Figure A-4-4.17.4   Typical Installation Showing Automatic Sprinklers or Automatic Spray Nozzles Being Used for the Protection of Commercial Cooking Equipment and Ventilation Systems.

The "access" should not jeopardize the integrity of the hood or duct. Refer to NFPA 96, *Standard for the Installation of Equipment for the Removal of Smoke and Grease-Laden Vapors from Commercial Cooking Equipment.*

4-4.17.8 Cooking Equipment.

4-4.17.8.1 Cooking equipment (such as deep fat fryers, ranges, griddles, and broilers) that may be a source of ignition shall be protected in accordance with the provisions of 4-4.17.1.

4-4.17.8.2 A sprinkler or automatic spray nozzle used for protection of deep fat fryers shall be listed for that application. The position, arrangement, location, and water supply for each sprinkler or automatic spray nozzle shall be in accordance with its listing.

4-4.17.8.3 The operation of any cooking equipment sprinkler or automatic spray nozzle shall automatically shut off all sources of fuel and heat to all equipment requiring protection. Any gas appliance not requiring protection but located under ventilating equipment shall also be shut off. All shutdown devices shall be of the type that requires manual resetting prior to fuel or power being restored.

4-4.17.9* A listed indicating valve shall be installed in the water supply line to the sprinklers and spray nozzles protecting the cooking and ventilating system.

A-4-4.17.9 The indicating control valve should be readily accessible and properly identified. (*See 3-9.4.*)

   This indicating valve should be separate, readily accessible, and properly identified.

4-4.17.10 An approved line strainer shall be installed in the main water supply preceding sprinklers or automatic spray nozzles having orifices smaller than ⅜ in. (9.5 mm).

4-4.17.11 A system test connection shall be provided to verify proper operation of equipment specified in 4-4.17.8.3.

4-4.17.12 Sprinklers and automatic spray nozzles used for protecting commercial-type cooking equipment and ventilating systems shall be replaced annually.

Exception: When automatic bulb-type sprinklers or spray nozzles are used and annual examination shows no build-up of grease or other material on the sprinklers or spray nozzles.

   The effect of grease and elevated temperature may seal the sprinkler orifice cap to the frame so that it will not release when the fusible element operates. Grease deposits may insulate and retard sprinkler operation. Because of the migration of solder components found in 360°F (182°C) solder, the Exception applies only to bulb-type sprinklers.

4-4.18 Baffles. Baffles (*except for in-rack sprinklers, see NFPA 231C, Standard for Rack Storage of Materials*) shall be installed whenever sprinklers are less than 6 ft (1.8 m) apart to prevent the sprinkler first opening from wetting adjacent sprinklers, thus delaying their operation. Baffles shall be located midway between sprinklers and arranged to baffle the actuating elements. Baffles may be of sheet metal about 8 in. (203 mm) wide and 6 in. (152 mm) high. The top of baffles shall extend 2 to 3 in. (51 to 76 mm) above the deflectors of upright sprinklers; and the bottom of baffles shall extend downward to a level at least even with the deflectors of pendent sprinklers.

Exception: Old-style sprinklers protecting fur storage vaults.

The baffles are to prevent water spray from striking the adjacent sprinkler fusible links and preventing the operation of sprinklers that otherwise would be activated.

4-4.19 Small Rooms. Rooms with smooth ceilings, and having floor areas not exceeding 800 sq ft (74.3 m²), of Light Hazard Occupancy classification.

Formal Interpretation

Question 1: **Is it permissible to apply this rule to a hospital patient room?**

Answer: **Yes. This rule applies to any room of Light Hazard Occupancy that has a smooth ceiling and does not exceed 800 sq ft.**

Question 2: **Is it permissible to use multiple sprinklers in conjunction with the referenced sections in a pipe schedule system as long as the area per sprinkler does not exceed 200 sq ft?**

Answer: **Yes. This rule may be applied to either pipe schedule or hydraulically designed systems.**

4-4.19.1* Within small rooms, sprinklers may be located not more than 9 ft (2.7 m) from any single wall; however, sprinkler spacing limitations of 4-2.1.1 and area limitations of 4-2.2.2.1 shall not be exceeded.

A-4-4.19.1 Examples of Sprinkler Spacing within Small Rooms. An example of sprinklers in small rooms for hydraulically designed and pipe schedule systems is shown in Figure A-4-4.19.1(a), and examples for hydraulically designed systems only are shown in Figures A-4-4.19.1(b), (c), and (d).

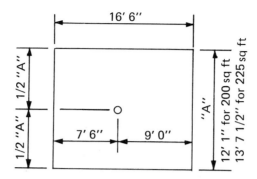

For SI Units: 1 in. = 25.4 mm; 1 ft = 0.3048 m.

Figure A-4-4.19.1(a).

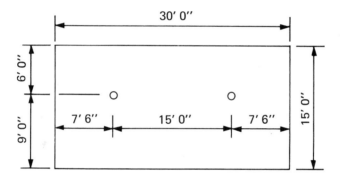

For SI Units: 1 in. = 25.4 mm; 1 ft = 0.3048 m.

Figure A-4-4.19.1(b).

To accommodate ceiling patterns, lighting fixtures, diffusers, etc., in Light Hazard Occupancies with smooth ceilings, it is permissible to move the sprinklers a maximum of 1 ft 6 in. (0.5 m) in one direction beyond the usual rules so that the maximum distance from one wall is 9 ft (2.7 m). In small rooms where other fixtures prevent the location of a sprinkler within the usual spacing rules, this will prevent having to add additional sprinklers. This rule contemplates that only those sprinklers in the room would operate, the result being a higher available operating pressure and thus an increased throw length of the discharge pattern.

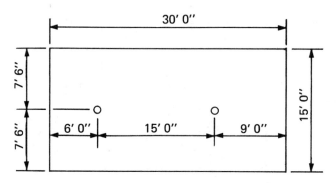

For SI Units: 1 in. = 25.4 mm; 1 ft = 0.3048 m.

Figure A-4-4.19.1(c).

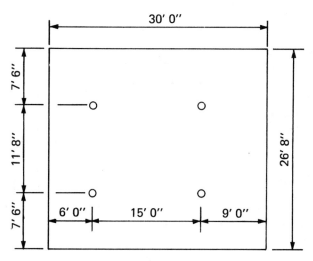

For SI Units: 1 in. = 25.4 mm; 1 ft = 0.3048 m.

Figure A-4-4.19.1(d).

4-4.19.2 In hotels, sprinklers may be omitted from bathrooms not exceeding 55 sq ft (5.1 m²) with noncombustible plumbing fixtures and with walls and ceilings surfaced with noncombustible materials.

Bathrooms, which usually are surfaced with tile or other noncombustible material and contain very few combustible articles, do not present a significant fire hazard.

4-4.19.3 In hotel guest rooms, sprinkler installation may be omitted in clothes closets where the area does not exceed 24 sq ft (2.2 m²), the least dimension does not exceed 3 ft (0.9 m), and the walls and ceilings are surfaced with noncombustible or limited-combustible materials.

Hotel clothes closets do not represent a substantial fuel load, so omission of sprinklers from such a small area is not considered detrimental to the system design or performance.

B-4-2 Locations or Conditions Involving Special Consideration.

B-4-2.1 Overhead Doors. When overhead doors form an obstruction to water distribution from sprinklers above, additional sprinkler protection may be required. When piping can be attached to the door structural framing, locate and space sprinklers under the doors in accordance with the rules for Ordinary Hazard Occupancy. When piping cannot be attached to the door structural framing, space sprinklers not more than 12 ft (3.1 m) apart around the perimeter of the three accessible sides of the doors and at least 12 in. (305 mm) in from the edges of the doors. Deflectors should not be more than 10 in. (254 mm) below the doors in the open position. Sidewall sprinklers may be used when their distribution would be more effective than that from standard sprinklers. When doors are predominantly glass construction and when those doors, in an open position, will merely be over a traffic aisle, sprinkler protection is not necessarily required.

B-4-2.2 Tables.

B-4-2.2.1 Sprinklers should be installed under cutting, pressing, sewing machine, and other work tables over 4 ft (1.2 m) wide. Sprinklers may be omitted under tables less than 5½ ft (1.7 m) but wider than 4 ft (1.2 m) if the tables are of temporary or semipermanent nature, as determined by the authority having jurisdiction, and tight vertical partitions of galvanized iron or other noncombustible material are provided not more than 10 ft (3.0 m) apart.

B-4-2.2.2 Partitions should be the full width of the table, extend from the underside of the table to the floor and from the front edge to back edge of the table; should be substantially fastened to the underside of the table and to the floor; and should be reinforced with angle or channel iron uprights.

B-4-2.2.3 The outer edges of each partition should be smoothly finished (rounded if of metal) so as to prevent injury to employees.

B-4-2.2.4 Instructions should be obtained relative to the installation of "stops" under tables of unusual construction.

B-4-2.3 Obstructions. Timbers, uprights, hangers, piping, lighting fixtures, ducts, etc., are likely to interfere with the proper distribution of water from sprinklers. Therefore, sprinklers should be so located or spaced that any interference is held to a minimum. The required clearance between such members and sprinklers is dependent upon the size of the obstruction to water distribution. The clearances should not be less than those specified between sprinklers and truss members in 4-2.4.1 and 4-2.4.5. (*See also 4-2.4.6.*)

B-4-2.4 Enclosures.

B-4-2.4.1 Sprinklers should be installed in enclosed equipment where combustible materials are processed or where combustible wastes or deposits may accumulate. Examples of such locations may include ovens, driers, dust collectors, conveyors, large ducts, spray booths, paper machine hoods, paper machine economizers, parts of textile preparatory machines, and similar enclosures.

B-4-2.4.2 Sprinklers should be installed in small enclosed structures of combustible construction or containing combustible material. Examples of such locations may include penthouses, passageways, small offices, stock rooms, closets, vaults, or similar enclosures.

B-4-2.4.3 For small enclosures, pipes may be run outside the enclosures and sprinklers installed in approved dome-shaped covers about 10 in. (254 mm) in diameter. Where sprinklers can be nippled into the enclosure without forming an obstruction this should be done and dome-shaped covers omitted.

B-4-2.4.4 Sprinkler piping may be run above hoods over paper machines and similar equipment where dripping of condensation from sprinkler piping must be avoided and the sprinklers nippled through. The lower sprinklers under the hoods should be located outside of the line of the cylinders or rolls.

B-4-2.4.5 When enclosures are subject to freezing temperatures, special types of sprinkler protection should be provided. Manually operated systems should not be used.

B-4-2.4.6 The provision of other approved extinguishing systems does not permit the omission of sprinklers except in special situations, such as high value records or museum displays in vaults, commercial-type cooking equipment, or computer or other electrical equipment enclosures. The authority having jurisdiction should be consulted.

B-4-2.4.7 Safe deposit or other vaults of fire-resistive construction will not ordinarily require sprinkler protection when used for the storage of records, files, and other documents, when stored in metal cabinets.

B-4-2.4.8 Stock Fixtures. Sprinklers should be installed in all stock fixtures that exceed 5 ft (1.5 m) in width, also in those that are less than 5 ft (1.5 m) but more than 2½ ft (0.8 m) in width unless bulkheaded with tight partitions. Sprinklers should be installed in any compartments that are larger than 5 ft (1.5 m) deep, 8 ft (2.4 m) long, and 3 ft (0.9 m) high.

B-4-2.4.9 Lighting fixtures of the pendent or surface-mounted type may offer obstruction to discharge from sprinklers unless the clearances specified in Table 4-2.4.6 are provided.

4-4.20 Theater Stages. Sprinklers shall be installed under the roof at the ceiling, in spaces under the stage either containing combustible materials or constructed of combustible materials; in all adjacent spaces and dressing rooms, storerooms, and workshops. When proscenium opening protection is required a deluge system shall be provided within 3 ft (0.9 m) of the stage side of the proscenium arch, with open sprinklers spaced up to a maximum of 6 ft (1.8 m) on center and designed to provide a discharge of 3 gpm/lineal ft [(37 L/min)/m] of water curtain, with no sprinkler discharging less than 15 gpm (56.8 L/min).

4-5* Spacing, Location, and Position of Sidewall Sprinklers. *(See 3-11.2.5.)*

Rules for horizontal sidewall, pendent sidewall, and upright sidewall sprinklers are covered by this section. Extended coverage (EC) sidewall sprinklers require specific tests, well beyond what horizontal sidewall sprinklers undergo. Extended coverage sidewall sprinklers must be installed in accordance with their listing, which includes maximum dimensions as well as minimum pressure and flow requirements.

A-4-5 The installation of sidewall sprinklers other than beneath smooth ceilings will require special consideration. Beams or other ceiling obstructions interfere with proper distribution and, when present, sidewall sprinklers should be spaced with regard to such obstructions.

4-5.1 Sidewall sprinklers shall only be installed along walls, lintels, or soffits where the distance from the ceiling to the bottom of the lintel or soffit is at least 2 in. (51 mm) greater than the distances from the ceiling to sidewall sprinkler deflector. Sidewall sprinklers shall not be installed back to back without being separated by a lintel or soffit.

The last sentence of this paragraph allows sidewall sprinklers to be installed back to back along the center line of a room provided the conditions for a lintel or soffit are satisfied.

4-5.2* Distance Between Sprinklers on Branch Lines.

A-4-5.2 The protection area per sprinkler should be determined using the S × L = Protection Area rule as follows:

1. "S" — Determine distance to the next sprinkler (or to the wall, in case of an end sprinkler on a branch line) upstream and downstream. Choose the larger of either twice the distance to the wall or the distance to the next sprinkler.

2. "L" — The distance to the opposite side of the room will be "L". Where sprinklers are provided on both sides of the room, "L" should be half the distance between the walls.

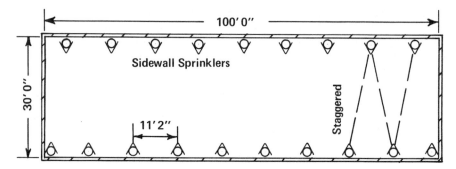

For SI Units: 1 in. = 25.4 mm; 1 ft = 0.3048 m.

Figure 4-5.2.1 Spacing of Sidewall Sprinklers under Combustible Smooth Ceilings.

4-5.2.1 Sidewall sprinklers shall be installed along the length of a single wall of rooms or bays not exceeding the width dimension specified in Table 4-5.2.1.

Exception: Sidewall sprinklers may be installed in Light Hazard Occupancies with smooth ceilings or in bays up to 30 ft (9 m) in width. When sidewall sprinklers are installed on two opposite walls or sides of bays, the spacing shall be as required in Section 4-5 with sprinklers regularly staggered. (See Figure 4-5.2.1.)

The standard limits the maximum room width to be protected by a single row of sidewall sprinklers to 14 ft (4.2 m). As shown by Figure 4-5.2.1, room widths of up to 30 ft (9 m) may be protected by two rows of sprinklers with the sprinklers arranged in a stagger mode. This design allows an extra two feet of credit.

4-5.3 Protection Area Limitations for Light Hazard Occupancy.

Protection areas for sidewall sprinklers are different than for upright and pendent sprinklers. This is necessary to help compensate for the characteristic "horizontal" or "sideways" travel which results in shadow areas or obstructions by furniture or similar objects which stay in place on the floor.

4-5.3.1 With noncombustible smooth ceilings the protection area per sprinkler shall not exceed 196 sq ft (18.2 m²).

4-5.3.2 With combustible ceiling construction sheathed with plasterboard, metal, or wood lath and plaster forming a smooth ceiling, the

Table 4-5.2.1 Dimensions for Sidewall Sprinkler Installation

	Light Hazard Occupancy			Ordinary Hazard Occupancy	
	Combustible Construction with Combustible Sheathing	Combustible Construction Sheathed with Plaster-board, Metal, or Wood Lath and Plaster	Noncombustible Construction with Noncombustible Sheathing	Combustible Construction with Combustible Sheathing*	Noncombustible Smooth Ceiling
Maximum distance between sprinklers on branch line	14	14	14	10	10
Maximum room width for single branch line along wall (ft)	12	12	14	10	10
Maximum area coverage (ft²)	120	168	196	80	100

For SI Units: 1 ft = 0.3048 m.
*See 4-5.4.2

protection area per sprinkler shall not exceed 168 sq ft (15.6 m²). *(See Table 4-5.2.1.)* When sheathing is combustible, such as wood, fiberboard, or other combustible material, the protection area per sprinkler shall not exceed 120 sq ft (11.1 m²).

Exception: Noncombustible smooth ceiling spacing is permitted beneath a noncombustible smooth ceiling located beneath any combustible concealed space. (See 4-4.4.1.)

4-5.4 Protection Area Limitations for Ordinary Hazard Occupancy.

4-5.4.1 With noncombustible smooth ceilings the protection area per sprinkler shall not exceed 100 sq ft (9.3 m²).

4-5.4.2 With combustible ceiling construction sheathed with plasterboard, metal, wood lath and plaster, wood fiberboard, or other combustible material forming a smooth ceiling, the protection area per sprinkler shall not exceed 80 sq ft (7.4 m²).

Exception: Noncombustible smooth ceiling spacing is permitted beneath a noncombustible smooth ceiling located beneath any combustible concealed space. (See 4-4.4.1.)

4-5.5 Position of Sidewall Sprinklers.

4-5.5.1* Sprinkler deflectors shall be at a distance from walls and ceilings not more than 6 in. (152 mm) or less than 4 in. (102 mm), unless special construction arrangements make a different position advisable for prompt operation and effective distribution.

The dead air space described in the commentary to 4-2.1.5.2 can also have a detrimental effect on sidewall sprinklers.

Exception No. 1: Horizontal sidewall sprinklers may be positioned 6 to 12 in. (152 to 305 mm) below noncombustible ceilings when listed for these positions.

Exception No. 2: Horizontal sidewall sprinklers may be positioned with their deflectors less than 4 in. (102 mm) from the wall on which they are mounted.

These Exceptions allow deflectors to be spaced up to 12 in. (305 mm) below the ceiling when they have been tested and listed for such installation, and installed less than 4 in. (102 mm) from the wall if circumstances do not permit other arrangements.

A-4-5.5.1 Sidewall sprinklers should be placed to receive heat from a fire and at the same time most effectively distribute the water discharged by them. This is likely to be particularly important when heavy decorative molding is encountered near the junction of walls and ceilings.

4-5.5.2   Where soffits are used for the installation of sidewall sprinklers, they shall not exceed an 8 in. (203 mm) projection from the wall.

The evaluation of a sidewall sprinkler includes testing of its ability to project water to the backwall area of a soffit and wall. This ability decreases as the depth of the soffit becomes greater; thus, the 8 in. (203 mm) maximum soffit is viewed as a maximum practical width.

REFERENCES CITED IN COMMENTARY

The following publications are available from the National Fire Protection Association, Batterymarch Park, Quincy, MA 02269.

NFPA 13D-1989, *Standard for the Installation of Sprinkler Systems in One- and Two-Family Dwellings and Mobile Homes*
NFPA 30-1987, *Flammable and Combustible Liquids Code*
NFPA 72E-1987, *Standard on Automatic Fire Detectors*
NFPA 96-1987, *Standard for the Installation of Equipment for the Removal of Smoke and Grease-Laden Vapors from Commercial Cooking Equipment*
NFPA *101*®-1988, *Life Safety Code*®
NFPA 231-1987, *Standard for General Storage*
NFPA 231C-1986, *Standard for Rack Storage of Materials*
NFPA 231D-1986, *Standard for Storage of Rubber Tires*
NFPA 231E-1989, *Recommended Practice for the Storage of Baled Cotton*
NFPA 231F-1987, *Standard for the Storage of Roll Paper*
Fire Protection Handbook, 16th Edition.

The following publication is available from the National Institute of Standards and Technology, Gaithersburg, MD 20899.

NBSIR 80-2097, *Full-Scale Fire Tests with Automatic Sprinklers in a Patient Room.*

5

Types of Systems

There are several types of sprinkler systems. (*See Chapter 1.*) The intended building use will generally dictate the type of system to be installed.

5-1 Wet-Pipe Systems.

5-1.1* Definition. A system employing automatic sprinklers attached to a piping system containing water and connected to a water supply so that water discharges immediately from sprinklers opened by a fire.

Wet-pipe systems are the most reliable and simple of all sprinkler systems since no equipment other than the sprinklers themselves need operate. Only those sprinklers that have been activated by heat from the fire will discharge water. All other types of systems included in this section utilize additional equipment for some specific purpose. As the number of events necessary for the system to discharge water increases, some increase in the probability of system failure may also occur.

A-5-1.1 A dry-pipe, preaction, or deluge system may be supplied from a larger wet-pipe system, provided the water supply is adequate.

These auxiliary systems are connected to wet-pipe systems beyond the main control valve. They are considered a part of the wet-pipe system and must satisfy all rules regarding total sprinklers, capacity, hydraulic calculations, etc., for that system. Such combined system risers must also adhere to the restrictions imposed by 4-2.2.1. Each auxiliary system will require its own separate control valve, drains, etc. Alarms on the wet-pipe system may not be adequate for the auxiliary system and additional alarms for the auxiliary system may be needed.

5-1.2 Pressure Gages. An approved pressure gage conforming to 2-9.2.2 shall be installed in each system riser. Pressure gages shall be installed above and below each alarm check valve when such devices are present.

275

At least one pressure gage is required in each system riser. When an alarm check valve is utilized, two gages are required — one on the supply side of the alarm check valve, and one on the system side. These gages may be used as a quick reference to establish water pressure quality at a given time. They are also used during the two-inch drain test recommended in NFPA 13A, Section 2-6.

5-1.3 Relief Valves.   A gridded wet-pipe system shall be provided with a relief valve not less than ¼ in. (6.4 mm) in size set to operate at pressure not greater than 175 psi (12.1 bars).

Exception No. 1:   When the maximum system pressure exceeds 165 psi (11.4 bars), the relief valve shall operate at 10 psi (0.7 bar) in excess of the maximum system pressure.

Exception No. 2:   When auxiliary air reservoirs are installed to absorb pressure increases.

Several cases have been reported where the pressure build-up in gridded wet-pipe systems has exceeded the working pressure of the system components. Pressures in excess of 200 psi (13.8 bars), due to temperature differentials in a building as a result of solar heating, have caused failure of fittings or other components. Expansion of water in the system causes compression of the entrapped air, resulting in extremely high pressures.

5-2 Dry-Pipe Systems.

5-2.1* Definition.   A system employing automatic sprinklers attached to a piping system containing air or nitrogen under pressure, the release of which (as from the opening of a sprinkler) permits the water pressure to open a valve known as a dry-pipe valve. The water then flows into the piping system and out the opened sprinklers.

Dry-pipe systems are installed in lieu of wet-pipe systems where piping is subject to freezing. Paragraph 3-5.1.1 requires dry-pipe systems to be installed when the building environment cannot be maintained at or above 40°F (4°C). Dry-pipe systems should not be used for the purpose of reducing water damage from pipe breakage or leakage since they operate too quickly to be of value for this purpose.

A-5-2.1   A dry-pipe system should be installed only where heat is not adequate to prevent freezing of water in all or sections of the system. Dry-pipe systems should be converted to wet-pipe systems when they become unnecessary because adequate heat is provided. Sprinklers should not be shut off in cold weather.

When two or more dry-pipe valves are used, systems should preferably be divided horizontally to prevent simultaneous operation of more than one system and resultant increased time delay in filling systems and discharging water, plus receipt of more than one waterflow alarm signal.

When adequate heat is present in sections of the dry-pipe system, consideration should be given to dividing the system into a separate wet-pipe system and dry-pipe system. Minimized use of dry-pipe systems is desirable where speed of operation is of particular concern.

Formal Interpretation

Question: **Is it the intent of Section A-5-2.1 to recommend the use of a wet-pipe sprinkler system in a building in a cold climate where the heating system is turned off during non-working hours?**

Answer: **No. This section merely points out that wet-pipe systems are impractical in unheated spaces. It is not the intent of this section to recommend providing heat in any space to accommodate the sprinkler system nor to replace dry-pipe systems with wet-pipe systems.**

B-5-3 Dry-Pipe Systems.

B-5-3.1 Eight-Inch Systems.   Where an 8-in. (203-mm) riser is employed in connection with a dry-pipe system, a 6-in. (152-mm) dry-pipe valve and a 6-in. (152-mm) gate valve between taper reducers may be used.

B-5-3.2 Dry-Pipe System Serving Several Remote Unheated Areas.   Where a single dry-pipe valve is used to supply piping and sprinklers located in several small unheated areas that are remote from each other, the dry-pipe valve and riser may be sized according to the number of sprinklers in the largest area. (*Also see 5-2.3.1.*)

5-2.2 Dry-Pendent Sprinklers.   Automatic sprinklers installed in the pendent position shall be of the listed dry-pendent type if installed in an area subject to freezing.

Exception: Pendent sprinklers installed on return bends are permitted when both the sprinklers and the return bends are located in a heated area.

Dry-pendent sprinklers are specially designed to prevent water from entering the drop pipe between the sprinkler supply pipe and the operating mechanism of the sprinkler.

These sprinklers may also be used on wet-pipe systems where individual sprinklers are extended into spaces subject to freezing. Extension of sprinklers into a freezer unit in a protected building is one example of this application. Special applications utilizing dry-pendent sprinklers in upright and other positions should be in accordance with their listing.

5-2.3* Size of Systems.

A-5-2.3   The capacities of the various sizes of pipe given in Table A-5-2.3 are for convenience in calculating the capacity of a system.

Table A-5-2.3   Capacity of One Foot of Pipe
(Based on actual internal pipe diameters)

Nominal Diameter	Gal Sch 40	Sch 10	Nominal Diameter	Gal Sch 40	Sch 10
¾ in.	0.028	—	3 in.	0.383	0.433
1 in.	0.045	0.049	3½ in.	0.513	0.576
1¼ in.	0.078	0.085	4 in.	0.660	0.740
1½ in.	0.106	0.115	5 in.	1.040	1.144
2 in.	0.174	0.190	6 in.	1.501	1.649[1]
2½ in.	0.248	0.283	8 in.	2.66[3]	2.776[2]

For SI Units: 1 in. = 25.4 mm; 1 ft = 0.3048 m; 1 gal = 3.785 L.

[1]0.134 Wall Pipe
[2]0.188 Wall Pipe
[3]Schedule 30

5-2.3.1 Volume Limitations.   Not more than 750-gal (2839 L) system capacity shall be controlled by one dry-pipe valve. Gridded dry-pipe systems shall not be installed.

Exception: Piping volume may exceed 750 gal (2839 L) for nongridded systems if the system design is such that water is delivered to the system test connection in not more than 60 seconds, starting at the normal air pressure on the system and at the time of fully opened inspection test connection.

Gridded dry-pipe systems are prone to excessive delays before water arrives at the first operating sprinkler. A survey conducted on gridded dry-pipe systems found that in more than 90 percent of the installations, the 60-second time limitation was being exceeded.

The size of dry-pipe systems is limited by volume of piping as well as the area limitations of 4-2.2.1. The primary limitation is the 750-gal (2839-L) capacity for nongridded systems. This volume limitation may be exceeded if delivery of water to the inspector's test connection can be achieved in 60 seconds or less. There is no 60-second limitation where these pipe volumes are not exceeded. Volume limitations, however, are imposed with the intent that water will be delivered to a sprinkler in a dry-pipe system as quickly as possible after it operates.

Without an established method for predicting water delivery time, the designer must exercise caution when the piping volume exceeds 750 gal (2839 L), as water delivery time in excess of 60 seconds may be cause for disapproval of the system by the authority having jurisdiction.

5-2.4 Quick-Opening Devices.

Quick-opening devices consist primarily of accelerators and exhausters. Accelerators cause the dry-pipe valve to operate more quickly, thus allowing water pressure to expel air in the system at a faster rate and speed water movement to the open sprinklers. Exhausters cause air to be more quickly evacuated from sprinkler piping, thus shortening the time from sprinkler opening to dry-pipe valve operation and allowing more rapid movement of water to operating sprinklers after the dry-pipe valve has operated.

5-2.4.1 Dry-pipe valves shall be provided with a listed quick-opening device when system capacity exceeds 500 gal (1893 L).

Exception: A quick-opening device shall not be required if the requirements of 5-2.3.1 Exception can be met without such a device.

Even though quick-opening devices are not required where systems are of smaller capacity, they may be used whenever speed of operation is important, such as with extended piping systems, and where life safety is an important factor. The Exception merely allows omission of such devices when water delivery occurs at the inspector's test connection in 60 seconds or less. Omitting this device and failing to stay within established time constraints on larger systems causes some risk for system rejection.

5-2.4.2* The quick-opening device shall be located as close as practical to the dry-pipe valve. To protect the restriction orifice and other operating parts of the quick-opening device against submergence, the connection to the riser shall be above the point at which water (priming water and back drainage) is expected when the dry-pipe valve and quick-opening device are set, except where design features of the particular quick-opening device make these requirements unnecessary.

Special circumstances may require a quick-opening device, such as an exhauster, to be located at a point remote from the dry-pipe valve. In these instances, consideration should be given to protection against freezing of the device or its components. Remote devices should be accessible for servicing.

A-5-2.4.2 In the case of dry-pipe valves having relatively small priming chambers and in which the normal quantity of priming water fills, or nearly fills, the entire priming chamber, the objective contemplated by this rule will be met by requiring connection of the quick-opening device at a point on the riser above the dry-pipe valve, which will provide a capacity measure between the normal priming level of the air chamber and the connection of 1½, 2, and 3 gal (5.7, 7.6, and 11.4 L) for 4-, 5-, and 6-in. (102-, 127-, and 152-mm) risers, respectively. Making the connection 24 in. (610 mm) above the normal priming water level will ordinarily provide this capacity.

5-2.4.3 A soft disc globe or angle valve shall be installed in the connection between the dry-pipe sprinkler riser and the quick-opening device.

The shutoff valve between the dry-pipe sprinkler riser and the quick-opening device is necessary since these devices may require separate maintenance. It is highly undesirable to have the entire system out of service because of a problem with the quick-opening device. As a consequence of this ability to separately shut off these devices, they are frequently left out of service. Since system design has been based on proper operation of the quick-opening device, it should be maintained in operating condition at all times, and any necessary repairs should be expeditiously completed.

5-2.4.4 A check valve shall be installed between the quick-opening device and the intermediate chamber of the dry-pipe valve. If the quick-opening device requires pressure feedback from the intermediate chamber, a valve of the type that will clearly indicate whether it is opened or closed may be installed in place of that check valve. This valve shall be constructed so that it may be locked or sealed in the open position.

The shutoff valve for this type of quick-opening device is much the same as the shutoff valve at the connection to the sprinkler riser and should always be maintained open.

5-2.4.5 An approved antiflooding device shall be installed in the connection between the dry-pipe sprinkler riser and the quick-opening device, unless the particular quick-opening device has built-in antiflooding design features.

Quick-opening devices and their accessory equipment, such as antiflooding devices, have in general had a poor service history. It is important that manufacturer's instructions be carefully followed when resetting each dry-pipe valve and quick-opening device. Many of the problems have been caused by not properly draining the equipment while resetting the dry-pipe valve after it has operated.

5-2.5* Location and Protection of Dry-Pipe Valve.

A-5-2.5 The dry-pipe valve should be located in an accessible place near the sprinkler system it controls.

When exposed to cold, the dry-pipe valve should be located in an approved valve room or enclosure and, where this is not possible, in an underground pit acceptable to the authority having jurisdiction. The room should be of sufficient size to give at least 2½ ft (0.8 m) of free space at the sides and in front of, above and below, the dry-pipe valve or valves, and this room, if feasible, should not be built until the valve is in position.

Size of enclosure should be governed by the number and arrangement of dry-pipe valves, so as to give ready access to these devices.

This serves to reduce the amount of piping in a dry-pipe system, and to help limit system capacity. Long bulk mains from dry-pipe valves to system piping may use a large proportion of the system capacity and thus cause the actual sprinkler system to be quite small. Reduction of bulk pipe can be accomplished in large buildings by extending underground piping below floors to the dry-pipe valve. However, this type of arrangement introduces the problem of underground piping leaks under buildings and may dictate an alternate solution.

5-2.5.1 The dry-pipe valve and supply pipe shall be protected against freezing and mechanical injury.

5-2.5.2 Valve rooms shall be lighted and heated. The source of heat shall be of a permanently installed type. Heat tape shall not be used in lieu of heated valve enclosures to protect the dry-pipe valve and supply pipe against freezing.

Heat tape is not considered an acceptable type of permanently installed heat source.

5-2.5.3 The supply for the sprinkler in the dry-pipe valve enclosure shall be from the dry side of the system.

5-2.5.4 Protection against accumulation of water above the clapper shall be provided for a low differential dry-pipe valve. This may be an automatic high water level signaling device or an automatic drain device.

5-2.6* Cold Storage Rooms.

Dry-pipe systems installed in cold storage rooms require additional care and consideration since temperatures are below freezing and frequently in subzero ranges. If a system trips accidentally, allowing the piping to fill with water, the nature of the occupancy may preclude heating this space to allow melting of ice and repair of the system. It may be necessary in such systems to move piping to a warm area where it would be thawed, emptied, and then reinstalled. Such operations are costly and result in long periods in which systems are out of service. Where conditions are particularly severe, it may be desirable to consider preaction or other types of special systems. It is very important in cold storage plants for air supply to be dry enough to prevent accumulation of moisture in piping. This can be accomplished by taking air supply from the coldest freezer area, installing air dryers, or using a moisture-free gas such as nitrogen.

A-5-2.6 Careful installation and maintenance, and some special arrangements of piping and devices as outlined in this section, are needed to avoid the formation of ice and frost inside piping in cold storage rooms that will be maintained at or below 32°F (0°C). Conditions are particularly favorable to condensation where pipes enter cold rooms from rooms having temperatures above freezing.

Whenever the opportunity offers, fittings such as those specified in 5-2.6.1 and illustrated in Figures 5-2.6.1(A) and 5-2.6.1(B), as well as flushing connections specified in 3-3.2, should be provided in existing systems.

When possible, risers should be located in stair towers or other locations outside of refrigerated areas. This would reduce the probabilities of ice or frost formation within the riser (supply) pipe.

Cross mains should be connected to risers or feed mains with flanges. In general, flanged fittings should be installed at points that would allow easy dismantling of the system. Split ring or other easily removable types of hangers will facilitate the dismantling.

Because it is not practical to allow water to flow into sprinkler piping in spaces that may be constantly subject to freezing, or where temperatures must be maintained at or below 40°F (4°C), it is important that means be provided at the time of system installation to conduct trip tests on dry-pipe valves which service such systems. NFPA 13A contains guidance in this matter.

5-2.6.1 Fittings for Inspection Purposes.

5-2.6.1.1 Fittings for inspection purposes shall be provided whenever a cross main connects to a riser or feed main. This may be accomplished by a blind flange on a fitting (tee or cross) in the riser or cross main or a flanged removable section 24 in. (610 mm) long in the feed main as shown in Figure 5-2.6.1(A). Such fittings in conjunction with the flushing connections specified in 3-3.2 would permit examination of the entire lengths of the cross mains. Branch lines may be examined by backing the pipe out of fittings.

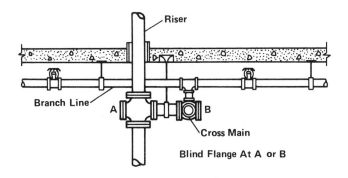

(a) Elevation At Riser And Cross Main

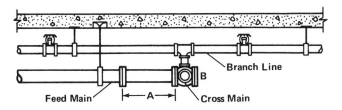

(b) Elevation At Feed Main And Cross Main

For SI Units: 1 in. = 25.4 mm.

Figure 5-2.6.1(A)   Fittings to Facilitate Examination of Feed Mains, Risers, and Cross Mains in Freezing Areas.

5-2.6.1.2 Whenever feed mains change direction, facilities shall be provided for direct observation of every length of feed main within the refrigerated area. This may be accomplished by means of 2-in. (51-mm) capped nipples or blind flanges on fittings.

5-2.6.1.3 Fittings for inspection purposes shall be provided whenever a riser or feed main passes through a wall or floor from a warm room to a cold room. This may be accomplished at floor penetrations by a tee with a blind flange in the cold room and at wall penetrations by a 24-in. (610-mm) flanged removable section in the warm room as shown in Figure 5-2.6.1(B).

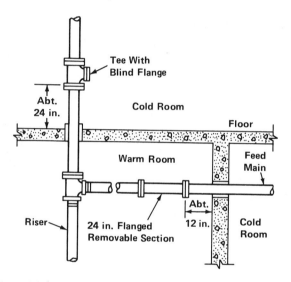

For SI Units: 1 in. = 25.4 mm.

Figure 5-2.6.1(B)   Fittings in Feed Main or Riser Passing through Wall or Floor from Warm Room to Cold Room.

5-2.6.2 A local low air-pressure alarm shall be installed on sprinkler systems supplying freezer sections.

5-2.6.3 Piping in cold storage rooms shall be installed with pitch, as outlined in 3-6.1.3.

5-2.6.4 The air supply for dry-pipe systems in cold storage plants shall be taken from the freezers of lowest temperature or through a chemical dehydrator. Compressed nitrogen gas from cylinders may be used in place of air in dry-pipe systems to eliminate introducing moisture.

5-2.7 Air Pressure and Supply.

Proper air pressure should be maintained on systems at all times. It is important to follow manufacturer's instructions regarding the range of air pressures to be maintained. Low air pressures could result in accidental operation of the dry-pipe valve. High air pressures result in slower operation because this air must be exhausted from the system before water can be delivered to open sprinklers.

5-2.7.1 Maintenance of Air Pressure.   Air or nitrogen pressure shall be maintained on dry-pipe systems throughout the year.

5-2.7.2* Air Supply.   The compressed air supply shall be from a source available at all times and having a capacity capable of restoring normal air pressure in the system within 30 minutes, except for low differential dry-pipe systems where this time may be 60 minutes. Where low differential dry-pipe valves are used, the air supply shall be maintained automatically.

Standard dry-pipe valves normally have a differential in water pressure to air pressure at trip point of approximately 5.5:1. A low differential dry-pipe valve usually is an alarm check valve that has been converted to a dry-pipe valve. The differential in water pressure and air pressure at the trip point of these valves is approximately 1.1:1.

A-5-2.7.2   The compressor should draw its air supply from a place where the air is dry and not too warm. Moisture from condensation may cause trouble in the system.

5-2.7.3 Air Filling Connection.   The connection pipe from the air compressor shall not be less than ½ in. (13 mm) and shall enter the system above the priming water level of the dry-pipe valve. A check valve shall be installed in this air line and a shutoff valve of renewable disc type shall be installed on the supply side of this check valve and shall remain closed unless filling the system.

5-2.7.4 Relief Valve.   An approved relief valve shall be provided between the compressor and controlling valve, set to relieve at a pressure 5 psi (0.3 bars) in excess of maximum air pressure that should be carried in the system.

5-2.7.5 Shop Air Supply.   When the air supply is taken from a shop system having a normal pressure greater than that required for dry-pipe systems and an automatic air maintenance device is not used, the relief valve shall be installed between two control valves in the air line, and a small air cock, which is normally left open, shall be installed in the fitting below the relief valve.

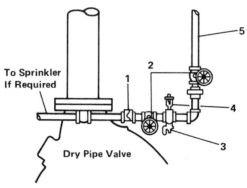

To Sprinkler If Required

Dry Pipe Valve

1. Check Valve
2. Control Valve (Renewable Disc Type)
3. Small Air Cock (Normally Open)
4. Relief Valve
5. Air Supply

Figure 5-2.7.5   Air Supply from Shop System.

The relief valve on systems that use shop air supply should be set to prevent overpressurizing of the sprinkler system. As described above, high pressures are undesirable since they result in slower operation of the system.

5-2.7.6 Automatic Air Compressor.   When a dry-pipe system is supplied by an automatic air compressor or plant air system, any device or apparatus used for automatic maintenance of air pressure shall be of a type specifically approved for such service and capable of maintaining the required air pressure on the dry-pipe system. Automatic air supply to more than one dry-pipe system shall be connected to enable individual maintenance of air pressure in each system. A check valve or other positive backflow prevention device shall be installed in the air supply to each system to prevent air or water flow from one system to another.

Pumping directly from an automatic air compressor through a fully opened supply pipe into the sprinkler system is not permitted. An approved device must be installed between the automatic air supply and the dry-pipe valve. The restricted opening in air maintenance devices prevents the air supply system from adding air at too fast a rate, thus preventing or slowing operation of the dry-pipe valve. Automatic air compressors may require an air holding tank to prevent damage to the air compressor.

5-2.7.7 Air Pressure to Be Carried.   The air pressure to be carried shall be in accordance with the instruction sheet furnished with the dry-pipe valve, when available, or 20 psi (1.4 bars) in excess of the calculated trip pressure of the dry-pipe valve, based on the highest normal water pressure of the system supply. The permitted rate of air leakage shall be as specified in 1-11.3.2.

5-2.7.8   When used, nitrogen shall be introduced through a pressure regulator set to maintain system pressure in accordance with 5-2.7.7.

5-2.8 Pressure Gages.   Approved pressure gages conforming to 2-9.2.2 shall be connected:

(a) On the water side and air side of the dry-pipe valve,

(b) At the air pump supplying the air receiver,

(c) At the air receiver,

(d) In each independent pipe from air supply to dry-pipe system, and

(e) At exhausters and accelerators.

5-3 Preaction and Deluge Systems.

These systems are more technical in nature and require specialized knowledge and experience in design and installation. Manufacturer's specifications and listings must be strictly adhered to. Since the systems are somewhat more complex, failure to maintain them properly will seriously reduce their reliability. Initial installation costs as well as long-term maintenance costs of these systems should be considered. The supplemental equipment necessary to operate these systems also results in increased economic concerns which must be considered.

5-3.1 Definitions.

Preaction System.   A system employing automatic sprinklers attached to a piping system containing air that may or may not be under pressure, with a supplemental fire detection system installed in the same areas as the sprinklers. Actuation of the fire detection system (as from a fire) opens a valve that permits water to flow into the sprinkler piping system and to be discharged from any sprinklers that may be open.

Deluge System.   A system employing open sprinklers attached to a piping system connected to a water supply through a valve that is opened by the operation of a fire detection system installed in the same areas as the sprinklers. When this valve opens water flows into the piping system and discharges from all sprinklers attached thereto.

The primary difference between preaction and deluge systems is that sprinklers in preaction systems are closed and sprinklers in deluge systems are open. Operation of the detection system in a preaction system only fills the pipe with water until a sprinkler opens. However, operation of the detection system in a deluge system results in flow from all sprinklers on the system. Deluge systems are normally used for high hazard areas requiring an immediate application of water over the entire hazard. Aircraft hangars are usually protected with deluge systems that discharge a foam solution. Preaction systems are normally used to protect properties when concern with accidental water discharge on property or equipment is paramount. Even though accidental or premature discharge of sprinkler systems is extremely rare, many sprinkler system "consumers" still prefer these systems in electronic equipment areas.

5-3.2* Description.   Preaction and deluge systems are normally without water in the system piping. The water supply is controlled by an automatic valve operated by means of fire detection devices and provided with manual means for operation that are independent of the sprinklers. Systems may have equipment of the types described in (a) through (f) below. (*See 5-3.6.2.*)

(a) Automatic sprinklers with both sprinkler piping and fire detection devices automatically supervised.

(b) Automatic sprinklers with sprinkler piping and fire detection devices not automatically supervised.

(c) Open sprinklers with only fire detection devices automatically supervised.

(d) Open sprinklers with fire detection devices not automatically supervised.

(e) Combination of open and automatic sprinklers with fire detection devices automatically supervised.

(f) Combination of open and automatic sprinklers with fire detection devices not automatically supervised.

The term "supervised" in this section refers to a method of insuring the overall integrity of the system. Supervision of the sprinkler piping ensures that a closed system exists. Pressurized air is usually introduced to the preaction system pipe network. The "normal" status would indicate that the system is tight or, in other words, no leaks exist.

Supervision of the associated electronic detection devices is very important since loss or failure of the detection system to an open circuit or ground fault may negate the benefit of automatic operation of the system.

A-5-3.2 Preaction and deluge systems may also have outside sprinklers for protection against exposure fire.

B-5 Types of Systems.

B-5-1 Preaction and Deluge Systems—Valves.

B-5-1.1 In hazardous locations, an approved indicating-type valve or manual means for operation of preaction or deluge valve should be installed in a location where access to the control valves is not likely to be prevented under fire emergency conditions.

B-5-1.2 With deluge systems, the deluge valve should be located as close as possible to the hazard protected, outside any fire or explosion hazard area.

5-3.3* General.

A-5-3.3 Conditions of occupancy or special hazards may require quick application of large quantities of water and in such cases deluge systems may be needed.

Fire detection devices should be selected to assure operation, yet guard against premature operation of sprinklers, based on normal room temperatures and draft conditions.

In locations where ambient temperature at the ceiling is high, from heat sources other than fire conditions, heat-responsive devices that operate at higher than ordinary temperature and are capable of withstanding the normal high temperature for long periods of time should be selected.

When corrosive conditions exist, materials or protective coatings that resist corrosion should be used.

To help avoid ice formation in piping due to accidental tripping of dry-pipe valves in cold storage rooms, a deluge automatic water control valve may be used on the supply side of the dry-pipe valve. When this combination is employed:

(a) Dry systems may be manifolded to a deluge valve, the protected area not exceeding 40,000 sq ft (3716 m²). The distance·between valves should be as short as possible to minimize water hammer.

(b) The dry-pipe valves should be pressurized to 50 psi (3.4 bars) to reduce the possibility of dry-pipe valve operation from water hammer.

5-3.3.1 A supply of spare fusible elements for heat-responsive devices, not less than two of each temperature rating, shall be maintained on the premises for replacement purposes.

5-3.3.2 When hydraulic release systems are used, it is possible to water column the deluge valve or deluge-valve actuator if the heat-actuated devices (fixed temperature or rate-of-rise) are located at extreme heights above the valve. Refer to the manufacturer for height limitations of a specific deluge valve or deluge-valve actuator.

5-3.3.3 All new preaction or deluge systems shall be tested hydrostatically as specified in 1-11.2.1. In testing deluge systems, plugs shall be installed in fittings and replaced with open sprinklers after the test is completed, or automatic sprinklers may be installed and the operating parts removed after the test is completed.

5-3.4 Location and Protection of Preaction and Deluge Systems.

As with dry-pipe systems, preaction and deluge systems need protection from freezing and mechanical injury.

5-3.4.1 The preaction and deluge system water control valves and supply pipes shall be protected against freezing and mechanical injury.

5-3.4.2 Valve rooms shall be lighted and heated. The source of heat shall be of a permanently installed type.

5-3.4.3 Heat tape shall not be used in lieu of heated valve enclosure rooms to protect preaction and deluge valves and supply pipe against freezing.

5-3.5 Location and Spacing of Fire Detection Devices. Spacing of fire detection devices other than automatic sprinklers shall be in accordance with their listing by testing laboratories or in accordance with manufacturer's specifications. When automatic sprinklers are used as detectors, the distance between detectors and the area per detector shall not exceed the maximum permitted for suppression sprinklers as specified in 4-2.1 and 4-2.2; they shall be positioned in accordance with Section 4-3, but need not conform with the clearance requirements of 4-2.4. (*See NFPA 72E, Standard on Automatic Fire Detectors.*)

It should be noted that a multitude of acceptable methods for activating a deluge or preaction system exist. Among these are pneumatic and hydraulic detection systems, smoke detectors, heat detectors, and detectors that operate on ultraviolet and infrared principles. Also note that this paragraph recognizes an automatic sprinkler system as being a type of detection system.

Even though the definition of sprinkler system (*see Section* 1-3) states that these systems are "usually activated by heat," this should not be construed as being a prohibition on smoke detection devices or similar non-heat sensing devices. Smoke, as well as ultraviolet and infrared radiation, are both by-products of heat.

As precisely stated in this paragraph, detection systems are to be installed in accordance with their listing or as directed by the manufacturer of the detection device.

5-3.6 Preaction Systems.

5-3.6.1 All components of pneumatic, hydraulic, or electrical preaction systems shall be compatible.

5-3.6.2 Size of Systems.   Not more than 1,000 closed sprinklers shall be controlled by any one preaction valve.

5-3.6.3 Supervision.   Sprinkler piping and fire detection devices shall be automatically supervised when there are more than 20 sprinklers on the system.

Formal Interpretation

Question 1: Is it the intent of 5-3.6.3, NFPA 13 (Supervision), to limit methods of supervision to pressurized air monitoring devices when there are more than 20 sprinklers on the system?

Answer: No.

Question 2: If the answer to Question 1 above is "no," would a downstream pressure sensing device for water flow be adequate to satisfy the supervision requirements of 5-3.6.3, NFPA 13 (Supervision)?

Answer: No. Supervision (electrical or mechanical) in this section refers to constant monitoring of piping and detection equipment to assure integrity of the system.

Specifics on supervision of fire detection system circuits, now referred to as "monitoring the integrity of installation conductors," can be found in NFPA 72E, *Standard on Automatic Fire Detectors.*

5-3.6.4   For pipe schedules see Chapter 7 and Sections 8-2, 8-3, 8-4, and 8-5.

5-3.6.5 Pendent Sprinklers.   Automatic sprinklers on preaction systems installed in the pendent position shall be of the listed dry-pendent type if installed in an area subject to freezing.

Exception: Pendent sprinklers installed on return bends are permitted when both the sprinklers and the return bends are located in a heated area.

5-3.7* Deluge Systems.   The fire detection devices or systems shall be automatically supervised when there are more than 20 sprinklers on the system.

A-5-3.7   Deluge systems are usually applied to severe conditions of occupancy. In designing piping systems the pipe sizes should be calculated in accordance with the standard for hydraulically designed sprinkler systems as given in Chapter 7.

When 8-in. (203-mm) piping is employed to reduce friction losses in a system operated by fire detection devices, a 6-in. (152-mm) preaction or deluge valve and 6-in. (152-mm) gate valve between taper reducers may be used.

5-3.8 Devices for Test Purposes and Testing Apparatus.

5-3.8.1   When fire detection devices installed in circuits are located where not readily accessible, an additional fire detection device shall be provided on each circuit for test purposes at an accessible location and shall be connected to the circuit at a point that will assure a proper test of the circuit.

5-3.8.2   Testing apparatus capable of producing the heat or impulse necessary to operate any normal fire detection device shall be furnished to the owner of the property with each installation. Where explosive vapors or materials are present, hot water, steam, or other methods of testing not involving an ignition source shall be used.

5-3.8.3 Pressure Gages.   Approved pressure gages conforming to 2-9.2.2 shall be installed as follows:

(a) Above and below preaction valve and below deluge valve.

(b) On air supply to preaction and deluge valves.

5-4 Combined Dry-Pipe and Preaction Systems.

5-4.1 General.

5-4.1.1* Definition. A combined dry-pipe and preaction sprinkler system is one employing automatic sprinklers attached to a piping system containing air under pressure with a supplemental fire detection system installed in the same areas as the sprinklers; operation of the fire detection system, as from a fire, actuates tripping devices that open dry-pipe valves simultaneously and without loss of air pressure in the system. Operation of the fire detection system also opens approved air exhaust valves at the end of the feed main, which facilitates the filling of the system with water, which usually precedes the opening of sprinklers. The fire detection system also serves as an automatic fire alarm system.

A-5-4.1.1 When Installed. Combined dry-pipe and preaction systems may be installed when wet-pipe systems are impractical. They are intended for use in but not limited to structures where a number of dry-pipe valves would be required if a dry-pipe system were installed.

Combined dry-pipe and preaction systems are employed primarily where more than one dry-pipe system is required by reason of area or capacity and it is not possible to install supply piping to each dry-pipe valve through a heated or protected space. An example of this use is on piers, wharves, and in large cold storage warehouses. These are special application systems and should not be used merely for large areas in an ordinary building.

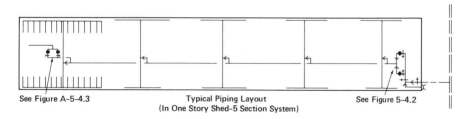

See Figure A-5-4.3   Typical Piping Layout   See Figure 5-4.2
                    (In One Story Shed-5 Section System)

Figure A-5-4.1.1   Typical Piping Layout for Combined Dry-Pipe and Preaction Sprinkler System.

5-4.1.2 Combined automatic dry-pipe and preaction systems shall be so constructed that failure of the fire detection system shall not prevent the system from functioning as a conventional automatic dry-pipe system.

Conventional automatic operation of the dry-pipe valve provides a fail-safe back-up to the fire detection system. Since the primary purpose of the fire detection system is to cause piping

to start filling sooner, loss of the detection system will only result in a much slower operating mode. Although somewhat impaired, the system will be reasonably fail-safe.

5-4.1.3 Combined automatic dry-pipe and preaction systems shall be so constructed that failure of the dry-pipe system of automatic sprinklers shall not prevent the fire detection system from properly functioning as an automatic fire alarm system.

Quite the opposite effect from that of 5-4.1.2 results when the preaction system operates but the dry-pipe valve does not. With this type of failure there would be no water to the piping system and sprinklers, only an alarm.

5-4.1.4 Provisions shall be made for the manual operation of the fire detection system at locations requiring not more than 200 ft (61.0 m) of travel.

Manual operation of the fire detection system provides rapid means of filling the system with water from remote locations throughout the protected area. This method can be used in special circumstances where there is danger of an incident and the system can be converted to a wet-pipe system prior to a fire.

5-4.1.5 Except as indicated in 5-2.2, automatic sprinklers installed in the pendent position shall be of the approved dry-pendent type.

5-4.2 Dry-Pipe Valves in Combined Systems.

5-4.2.1 Where the system consists of more than 600 sprinklers or has more than 275 sprinklers in any fire area, the entire system shall be controlled through two 6-in. (152-mm) dry-pipe valves connected in parallel and shall feed into a common feed main. These valves shall be checked against each other. (*See Figure 5-4.2.*)

The primary purpose of providing two valves is to maintain service on the system while servicing the other dry-pipe valve. The second dry-pipe valve in parallel also provides an additional degree of safety, since one valve would serve as a back-up to the other.

5-4.2.2 Each dry-pipe valve shall be provided with an approved tripping device actuated by the fire detection system. Dry-pipe valves shall be cross connected through a 1-in. (25.4-mm) pipe connection to permit simultaneous tripping of both dry-pipe valves. This 1-in. (25.4-mm) pipe connection shall be equipped with a gate valve so that either dry-pipe valve can be shut off and worked on while the other remains in service.

Although each dry-pipe valve serves as a back-up valve, simultaneous operation of both valves is desirable to increase flow characteristics of the system.

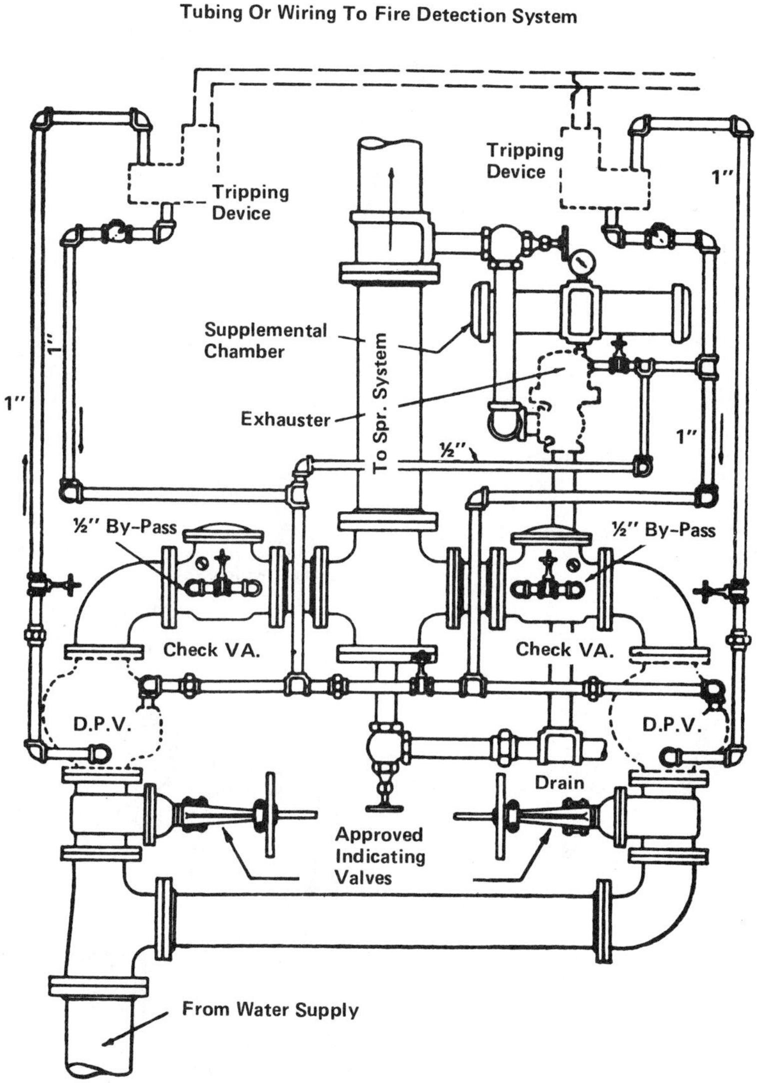

Tubing Or Wiring To Fire Detection System

Tripping Device

Tripping Device

1"

Tripping Device

Supplemental Chamber

To Spr. System

Exhauster

1"

1"

1"

½"

½" By-Pass

½" By-Pass

Check VA.

Check VA.

D.P.V.

D.P.V.

Drain

Approved Indicating Valves

From Water Supply

For SI Units: 1 in. = 25.4 mm.

Figure 5-4.2   Header for Combined Dry-Pipe and Preaction Sprinkler System, Standard Trimmings not Shown.

5-4.2.3 The check valves between the dry-pipe valves and the common feed main shall be equipped with ½-in. (13-mm) bypasses so that a loss of air from leakage in the trimmings of a dry-pipe valve will not cause the valve to trip until the pressure in the feed main is reduced to the tripping point. A gate valve shall be installed in each of these bypasses so that either dry-pipe valve can be completely isolated from the main riser or feed main and from the other.

5-4.2.4 Each combined dry-pipe and preaction system shall be provided with approved quick-opening devices at the dry-pipe valves.

5-4.3* Air Exhaust Valves. One or more approved air exhaust valves of 2-in. (51-mm) or larger size controlled by operation of a fire detection system shall be installed at the end of the common feed main. (*See Figure A-5-4.3.*) These air exhaust valves shall have soft-seated globe or angle valves in their intakes; also, approved strainers shall be installed between these globe valves and the air exhaust valves.

These systems have a very large air capacity. The exhaust valves serve to evacuate air at remote ends of the piping and allow water to fill the system more rapidly. The air exhaust valves should be located where they may be readily tested and maintained.

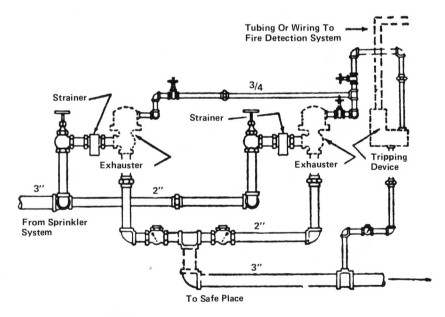

For SI Units: 1 in. = 25.4 mm.

Figure A-5-4.3 Arrangement of Air Exhaust Valves for Combined Dry-Pipe and Preaction Sprinkler System.

5-4.4 Subdivision of System Using Check Valves.

5-4.4.1 Where more than 275 sprinklers are required in a single fire area, the system shall be divided into sections of 275 sprinklers or less by means of check valves. If the system is installed in more than one fire area or story, not more than 600 sprinklers shall be supplied through any one check valve. Each section shall have a 1¼-in. (33-mm) drain on the system side of each check valve supplemented by a drum drip.

> **This section provides a means to increase the size of systems to effect a cost savings. Its use is usually determined by economic considerations.**

5-4.4.2 Section drain lines and drum drips shall be located in heated areas or inside of thermostatically controlled electrically heated cabinets of sufficient size to enclose drain valves and drum drips for each section. Drum drips shall also be provided for all low points except that heated cabinets need not be required for systems of 20 sprinklers or less.

5-4.4.3 Air exhaust valves at the end of a feed main and associated check valves shall be protected against freezing.

5-4.5 Time Limitation. The sprinkler system shall be so constructed and the number of sprinkler heads controlled shall be so limited that water shall reach the farthest sprinkler within a period of time not exceeding 1 minute for each 400 ft (122 m) of common feed main from the time the heat-responsive system operates. Maximum time permitted shall not exceed 3 minutes.

> **Since these systems are a compromise for economic reasons, time restrictions should not be exceeded. Excessive time for delivery of water should be reviewed with the authority having jurisdiction.**

5-4.6 System Test Connection. The end section shall have a system test connection as required for dry-pipe systems.

5-5 Antifreeze Systems.

5-5.1 Definition. An antifreeze system is one employing automatic sprinklers attached to a piping system containing an antifreeze solution and connected to a water supply. The antifreeze solution, followed by water, discharges immediately from sprinklers opened by a fire.

5-5.2* Where Used. The use of antifreeze solutions shall be in conformity with any state or local health regulations.

State or local plumbing and health regulations may not allow the introduction of foreign materials to piping systems connected to public water. Where these regulations are in effect, the use of antifreeze and the type of solution permitted should be checked with local authorities. It may be permitted where piping arrangements make contamination of public water virtually impossible. In many instances antifreeze systems are the only practical solution to a fire protection problem in a small area of a building and deviations from codes will be allowed. In other cases, requirements for a reduced pressure backflow preventer may be imposed.

A-5-5.2 Antifreeze solutions may be used for maintaining automatic sprinkler protection in small unheated areas. Antifreeze solutions are recommended only for systems not exceeding 40 gallons (151 L).

Because of the cost of refilling the system or replenishing small leaks, it is advisable to use small dry valves where more than 40 gallons (151 L) are to be supplied.

Propylene glycol or other suitable material may be used as a substitute for priming water, to prevent evaporation of the priming fluid, and thus reduce ice formation within the system.

In addition to the cost of the antifreeze configuration, one other reason for limiting the system size concerns the effect of antifreeze concentrate mixed with water. As the freezing point of the water is lowered by increasing the quantity of the antifreeze concentrate, the boiling point of the new solution is raised beyond 212°F (100°C) and is not as effective an extinguishing agent as pure water.

5-5.3 Antifreeze Solutions.

5-5.3.1 When sprinkler systems are supplied by public water connections the use of antifreeze solutions other than water solutions of pure glycerine (C.P. or U.S.P. 96.5 percent grade) or propylene glycol shall not be permitted. Suitable glycerine-water and propylene glycol-water mixtures are shown in Table 5-5.3.1.

5-5.3.2 If public water is not connected to sprinklers, the commercially available materials indicated in Table 5-5.3.2 are suitable for use in antifreeze solutions.

5-5.3.3* An antifreeze solution shall be prepared with a freezing point below the expected minimum temperature for the locality. The specific gravity of the prepared solution shall be checked by a hydrometer with suitable scale. [*See Figures 5-5.3.3(a) and (b).*]

Table 5-5.3.1 Antifreeze Solutions
to be Used if Public Water is Connected to Sprinklers

Material	Solution (by Volume)	Specific Gravity at 60°F (15.6°C)	Freezing Point °F	°C
Glycerine	50% Water	1.133	−15	−26.1
C.P. or U.S.P. Grade*	40% Water	1.151	−22	−30.0
	30% Water	1.165	−40	−40.0
Hydrometer Scale 1.000 to 1.200				
Propylene Glycol	70% Water	1.027	+ 9	−12.8
	60% Water	1.034	− 6	−21.1
	50% Water	1.041	−26	−32.2
	40% Water	1.045	−60	−51.1
Hydrometer Scale 1.000 to 1.200 (Subdivisions 0.002)				

*C.P. —Chemically Pure.
U.S.P.—United States Pharmacopoeia 96.5%.

Table 5-5.3.2 Antifreeze Solutions
to be Used if Public Water is not Connected to Sprinklers

Material	Solution (by Volume)	Specific Gravity at 60°F (15.6°C)	Freezing Point °F	°C
Glycerine	If glycerine is used, see Table 5-5.3.1			
Diethylene Glycol	50% Water	1.078	−13	−25.0
	45% Water	1.081	−27	−32.8
	40% Water	1.086	−42	−41.1
Hydrometer Scale 1.000 to 1.120 (Subdivisions 0.002)				
Ethylene Glycol	61% Water	1.056	−10	−23.3
	56% Water	1.063	−20	−28.9
	51% Water	1.069	−30	−34.4
	47% Water	1.073	−40	−40.0
Hydrometer Scale 1.000 to 1.120 (Subdivisions 0.002)				
Propylene Glycol	If propylene glycol is used, see Table 5-5.3.1			
Calcium Chloride 80% "Flake"	Lb CaCl₂ per gal of Water			
Fire Protection Grade*	2.83	1.183	0	−17.8
Add corrosion inhibitor	3.38	1.212	−10	−23.3
of sodium bichromate	3.89	1.237	−20	−28.9
¼ oz per gal water	4.37	1.258	−30	−34.4
	4.73	1.274	−40	−40.0
	4.93	1.283	−50	−45.6

*Free from magnesium chloride and other impurities.

Verification of the hydrometer scale is very important. Commonly available hydrometers, such as those used for automobile service stations, are often not adequate.

Figures 5-5.3.3(a) and (b) show how the densities of ethylene and propylene glycol can vary as the temperature at which the pipe is exposed varies. Extreme changes in temperature between summer and winter months may result in higher densities and thus increased pressure to the system components.

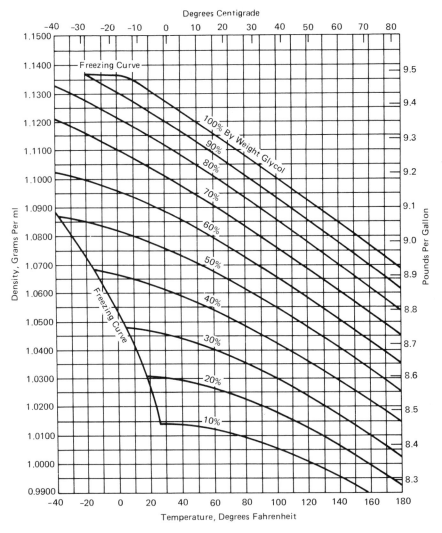

Figure 5-5.3.3(a)   Densities of Aqueous Ethylene Glycol Solutions (percent by weight).

A-5-5.3.3 Beyond certain limits, increased proportion of antifreeze does not lower the freezing point of solution. (*See Figure A-5-5.3.3.*) Glycerine, diethylene glycol, ethylene glycol, and propylene glycol should never be used without mixing with water in proper proportions, because these materials tend to thicken near 32°F (0°C).

See Figure A-5-5.3.3 on page 302.

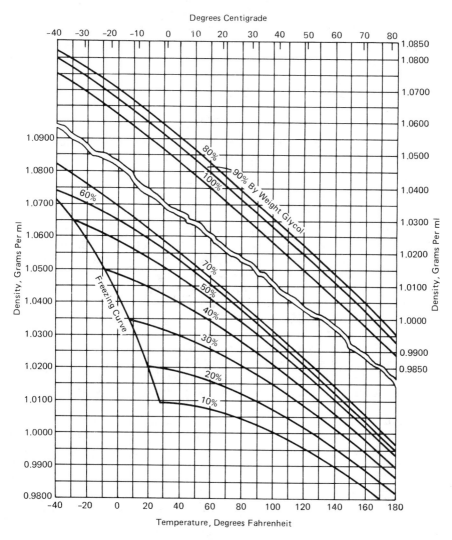

Figure 5-5.3.3(b) Densities of Aqueous Propylene Glycol Solutions (percent by weight).

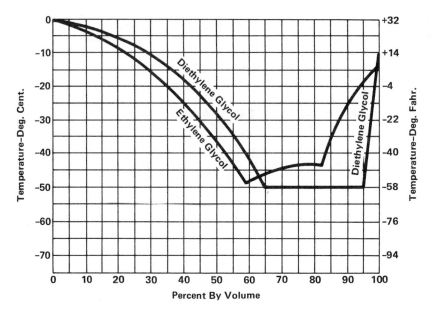

**Figure A-5-5.3.3   Freezing Points of Water Solutions of Ethylene Glycol
and Diethylene Glycol.**

5-5.4* Arrangement of Supply Piping and Valves.   All permitted antifreeze solutions are heavier than water. At the point of contact (interface) the heavier liquid will be below the lighter liquid in order to prevent diffusion of water into the unheated areas. In most cases, this necessitates the use of a 5-ft (1.5-m) drop pipe or U-loop as illustrated in Figure 5-5.4. The preferred arrangement is to have the sprinklers below the interface between the water and the antifreeze solution.

If sprinklers are above the interface, a check valve with a ⅛2-in. (0.8-mm) hole in the clapper shall be provided in the U-loop. A water control valve and two small solution test valves shall be provided as illustrated in Figure 5-5.4. An acceptable arrangement of a filling cup is also shown.

Exception: When the connection between the antifreeze system and the wet-pipe system incorporates a backflow prevention device, an expansion chamber shall be provided to compensate for the expansion of the antifreeze solution.

Figure 5-5.4 shows the most common arrangement of supply piping and valves. Actual field conditions may require modification of this arrangement. For example, if piping on the antifreeze system is at a higher elevation than the water supply, it is not practical to install a gravity-type fill cup, as shown in the

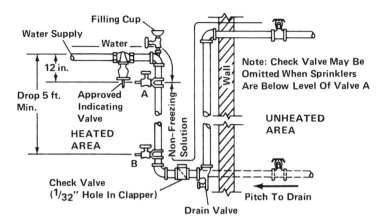

For SI Units: 1 in. = 25.4 mm; 1 ft = 0.3048 m.

NOTE: The ⅟₃₂-in. (0.8-mm) hole in the check valve clapper is needed to allow for expansion of the solution during a temperature rise and thus prevent damage to sprinkler heads.

Figure 5-5.4   Arrangement of Supply Piping and Valves.

drawing, at the supply location. It is generally relocated to the high point of the piping system. In a few instances, it may be necessary to use the valves at A or B to pump antifreeze into the system if the fill cup is impractical.

B-5-2 Filling with Antifreeze Solutions.   With the water supply valve closed and the system drained, fill the piping through the filling cup, using a suitable antifreeze solution of the proper concentration. Vent the air at the end sprinklers. Back out all sprinklers slightly until the liquid appears so that the piping will be completely filled and all air expelled. If the filling cup is not above the highest sprinklers, the piping may be filled through valve B by means of a small pump or through a filling cup installed at the highest branch sprinkler line. If the last-named method is used, the drop pipe should be filled through the filling cup shown in Figure 5-5.4. Then tighten the sprinkler heads and open valve A until the 12-in. (305-mm) section of pipe above this valve is empty and the level of the antifreeze solution in the drop pipe is at valve A. Close valve A. Close the filling connection valve and slowly open the supply valve wide.

A-5-5.4   To avoid leakage, the materials and workmanship should be excellent, the threads clean and sharp, and the joints tight. Use only metal-faced valves.

5-5.5* Testing. Before freezing weather each year, the solution in the entire system shall be emptied into convenient containers and brought to the proper specific gravity by adding concentrated liquid as needed. The resulting solution may be used to refill the system.

If emptying of the antifreeze piping is impractical, tests may be conducted as described in A-5-5.5. In order to facilitate this yearly testing, it is recommended that a permanent record be kept near the antifreeze fill connection which notes the type of antifreeze used, the temperature for which the system is designed, and the date of all tests and charging of the system.

A-5-5.5 Tests should be made by drawing a sample of the solution from valve B two or three times during the freezing season, especially if it has been necessary to drain the building sprinkler system for repairs, changes, etc. A small hydrometer should be used so that a small sample will be sufficient. When water appears at valve B or when the test sample indicates that the solution has become weakened, empty the entire system and recharge as previously described.

5-6 Automatic Sprinkler Systems with Nonfire Protection Connections.

5-6.1 Circulating Closed-Loop Systems.

It has been considered feasible for many years to make use of the sprinkler piping, which normally stands idle, for a more active purpose.

This section allows sprinkler piping to be used for additional functions other than fire protection, such as heating and air conditioning of a building. A basic criterion used in the development of this section of the standard is that the auxiliary functions be added in such a manner and with sufficient controls to ensure that there would be no reduction in overall fire protection as a result of adding these auxiliary uses. It was determined that this could be controlled more adequately with totally closed systems. The limitation of a closed system is viewed as a first step.

An inherent feature of these systems is their self-supervisory feature. Problems with the system would be readily obvious to building occupants if they were uncomfortable because the building temperature was too hot or too cold. Such a condition would result in maintenance personnel being dispatched immediately to facilitate repairs while at the same time making sure the sprinkler system remained in service.

5-6.1.1 Definition. A circulating closed loop is one with nonfire protection connections to automatic sprinkler systems in a closed-loop piping arrangement for the purpose of utilizing sprinkler piping to conduct water for heating or cooling. Water is not removed or used from the system, but only circulated through the piping system.

These systems are frequently heat pump systems in which water is circulated through heating and air conditioning equipment utilizing sprinkler pipe as the primary conductors. Heat is added to, or rejected from, the circulating water as dictated by requirements of the system. Water is neither added to nor removed from the piping system.

5-6.1.2 System Components.

5-6.1.2.1 Basic Principle. A circulating closed-loop system is primarily a sprinkler system, and all provisions of this standard such as control valves, area limitation of a system, alarms, fire department connections, sprinkler spacing, etc., are to be satisfied.

Exception: Items as specifically detailed within 5-6.1.

5-6.1.2.2 Piping, fittings, valves, and pipe hangers shall meet requirements specified in Chapter 3.

5-6.1.2.3 A dielectric fitting shall be installed in the junction where dissimilar piping materials are joined, e.g., copper to steel.

Exception: Dielectric fittings are not required in the junction where sprinklers are connected to piping.

5-6.1.2.4 It is not required that other auxiliary devices be listed for sprinkler service; however, these devices, such as pumps, circulating pumps, heat exchangers, radiators, and luminaires shall be pressure rated at 175 or 300 psi (12.1 or 20.7 bars) (rupture pressure of 5 X rated water working pressure), to match the required rating of sprinkler system components.

Although this section does not require testing and listing of the auxiliary equipment or components, they must be of materials acceptable in Chapter 3 and of sufficient strength to be equal to all other sprinkler system components (e.g., durability, suitability, and strength).

5-6.1.2.5 Auxiliary devices shall incorporate materials of construction and be so constructed that they will maintain their physical integrity under fire conditions to avoid impairment to the fire protection system.

5-6.1.2.6 Auxiliary devices where hung from the building structure shall be supported independently from the sprinkler portion of the system, following recognized engineering practices.

5-6.1.3 Hydraulic Characteristics. Piping systems for attached heating and cooling equipment shall have auxiliary pumps or an arrangement made to return water to the piping system in order to assure the following:

(a)* Water for sprinklers shall not be required to pass through heating or cooling equipment. At least one direct path shall exist for water flow from the sprinkler water supply to every sprinkler. Pipe sizing in the direct path shall be in accordance with design requirements of this standard.

A-5-6.1.3(a) Outlets should be provided at critical points on sprinkler system piping to accommodate attachment of pressure gages for test purposes.

(b) No portions of the sprinkler piping shall have less than the sprinkler system design pressure regardless of the mode of operation of the attached heating or cooling equipment.

(c) There shall be no loss or outflow of water from the system due to or resulting from the operation of heating or cooling equipment.

(d) Shutoff valves and a means of drainage shall be provided on piping to heating or cooling equipment at all points of connection to sprinkler piping and shall be installed in such a manner as to make possible repair or removal of any auxiliary component without impairing the serviceability and response to the sprinkler system. All auxiliary components, including the strainer, shall be installed on the auxiliary equipment side of the shutoff valves.

This section is intended to ensure that there is no reduction in flow to sprinklers regardless of operation or presence of the auxiliary equipment. There may be an enhancement of flow to the sprinkler system, but no reduction.

5-6.1.4 Water Temperature.

5-6.1.4.1 Maximum. In no case shall maximum water temperature flowing through the sprinkler portion of the system exceed 120°F (49°C). Protective control devices listed for this purpose shall be installed to shut down heating or cooling systems when temperature of water flowing through the sprinkler portion of the system exceeds 120°F (49°C). When water temperature exceeds 100°F (37.8°C), intermediate or higher temperature rated sprinklers shall be used.

5-6.1.4.2 Minimum. Precaution shall be taken to ensure that temperatures below 40°F (4°C) will not be permitted.

5-6.1.5 Obstruction to Discharge. Automatic sprinklers shall not be obstructed by auxiliary devices, piping, insulation, etc., from detecting fire or from proper distribution of water.

5-6.1.6 Valve Supervision. Position of all valves controlling sprinkler system (post indicator, main gate, sectional control) shall be supervised open by one of the following methods:

(a) Central station, proprietary, or remote station alarm service.

(b) Local alarm service, which will cause the sounding of an audible signal at a constantly attended point.

Supervision of all control valves is required in this type of system in consideration of the combined use characteristics of the system. Some repairs may be made by those not familiar with sprinkler systems and supervision lessens the chance of improper valve closure.

5-6.1.7 Signs. Caution signs shall be attached to all controlling sprinkler valves. The caution sign shall be worded as follows:

"This valve controls fire protection equipment. Do not close until after fire has been extinguished. Use auxiliary valves when necessary to shut supply to auxiliary equipment. CAUTION: Automatic alarm will be sounded if this valve is closed."

5-6.1.8 Water Additives. Materials added to water shall not adversely affect the fire-fighting properties of the water and shall be in conformity with any state or local health regulations. Due care and caution shall be given to the use of additives that may remove or suspend scale from older piping systems. When additives are necessary for proper system operation, due care shall be taken to ensure that additives are replenished after alarm testing or whenever water is removed from the system.

These additives are usually necessary to extend the life of heating units, chillers, or companion equipment which is associated with the HVAC systems. It is not the intent of 1-11.2.4 to prohibit the types of materials discussed in this chapter.

As a courtesy to fire suppression personnel, local fire departments with primary jurisdiction over the property should be made aware of the presence of and type of water additives that are included in the system.

5-6.1.9 Waterflow Detection. The supply of water from sprinkler piping through auxiliary devices, circulatory piping, and pumps shall not under any condition or operation, transient or static, cause false sprinkler waterflow signals.

5-6.1.9.1   A sprinkler waterflow signal shall not be impaired when water is discharged through an opened sprinkler or through the system test connection while auxiliary equipment is in any mode of operation (on, off, transient, stable).

5-6.1.10* Working Plans.   Working plans shall be prepared and submitted in accordance with Section 1-9. Special symbols shall be used and ex-

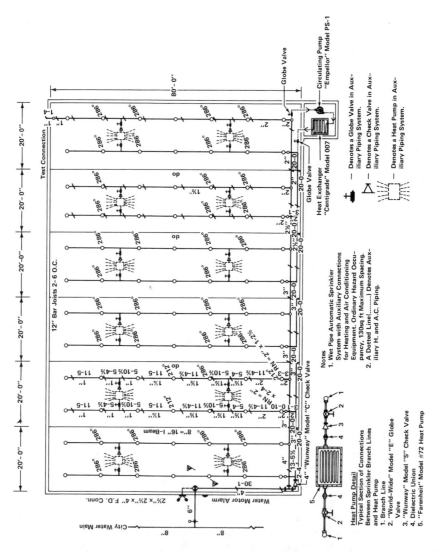

Figure A-5-6.1.10(a)   Working Plans for Circulating Closed-Loop Systems.

plained for auxiliary piping, pumps, heat exchangers, valves, strainers, and the like, clearly distinguishing those devices and piping runs from those of the sprinkler system. Model number, type, and manufacturer's name shall be identified for each piece of auxiliary equipment.

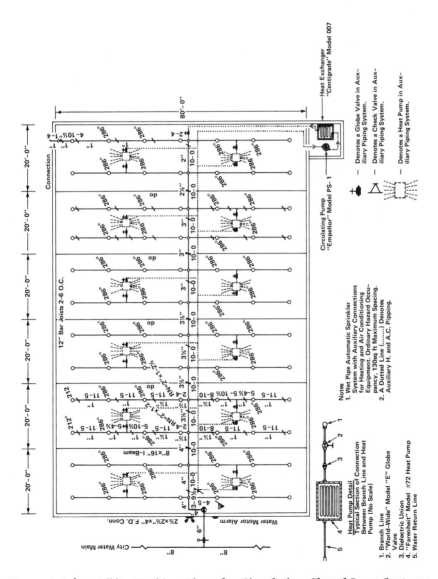

Figure A-5-6.1.10(b)   Working Plans for Circulating Closed-Loop Systems.

5-6.1.11 Testing.

5-6.1.11.1   All sprinkler system and auxiliary system components shall be hydrostatically tested in accordance with 1-11.3.

5-6.1.11.2   Sprinkler system discharge tests shall be conducted using system test connections described in 3-4.1. Pressure gages shall be installed at critical points and readings taken under various modes of auxiliary equipment operation. Waterflow alarm signals shall be responsive to discharge of water through system test pipes while auxiliary equipment is in each of the possible modes of operation.

5-6.1.12 Contractor's Material and Test Certificate.   Additional information shall be appended to the Contractor's Material and Test Certificate as shown in Figures 1-10.1(a) and (b), as follows:

(a) Certification that all auxiliary devices, such as heat pumps, circulating pumps, heat exchangers, radiators, and luminaires have a pressure rating of 175 or 300 psi (12.1 or 20.7 bars).

(b) All components of sprinkler system and auxiliary system have been pressure tested as a composite system in accordance with 1-11.2, Hydrostatic Tests.

(c) Waterflow tests have been conducted and waterflow alarms have operated while auxiliary equipment is in each of the possible modes of operation.

(d) With auxiliary equipment tested in each possible mode of operation and with no flow from sprinklers or test connection, waterflow alarm signals did not operate.

(e) Excess temperature controls for shutting down the auxiliary system have been properly field tested.

REFERENCES CITED IN COMMENTARY

The following publications are available from the National Fire Protection Association, Batterymarch Park, Quincy, MA 02269.

NFPA 13A-1987, *Recommended Practice for the Inspection, Testing and Maintenance of Sprinkler Systems*
NFPA 72E-1987, *Standard on Automatic Fire Detectors.*

6

Outside Sprinklers for Protection Against Exposure Fires

There are two purposes for the installation of sprinklers around the outer perimeter of a building that may be exposed to fire from an adjacent building or structure. The first is to prevent radiated or convected heat from an exposure fire from entering the building through windows, doors, or other openings in the exposed walls and igniting combustibles inside the building. The second is to deter ignition of or heat damage to combustible sheathing, eaves, cornices, or other exposed combustible surfaces.

Rapidity of water application should be ensured by automatic actuation of the outside sprinkler system. Delayed application of water to unruptured, heated glass surfaces may cause breakage from thermal shock. Delayed application of water could also result in the ignition of exterior combustible surfaces or damage to heat-sensitive materials, leading to unnecessary damage to these surfaces.

These special systems may be required to compensate for lack of physical separation between adjacent structures. NFPA 80A, *Recommended Practice for Protection of Buildings from Exterior Fire Exposures*, allows the reduction of separation distances between buildings when proper exposure protection from sprinklers is provided.

6-1 Water Supply and Control.

6-1.1 Water Supply.

6-1.1.1* Sprinklers installed for protection against exposure fires shall be supplied from a standard water supply as outlined in Chapter 2.

Exception: When approved, other supplies such as manual valves or pumps, or fire department connections, may be used.

The Exception permits other than the automatic water supply required in Chapter 2. The authority having jurisdiction must determine if this method of water supply, which includes fire department apparatus pumping into the fire department connection, is acceptable for the circumstances.

A-6-1.1.1 The water supply should be capable of furnishing the total demand for all exposure sprinklers operating simultaneously for protection against the exposure fire under consideration for a duration of not less than 60 minutes.

Water should be continuously delivered to the outside sprinkler system at the required application rate for the duration of exposure expected from a fire in the adjacent building or structure or other exposure source, such as an outside flammable liquid storage tank or combustible exterior storage. The duration may be in excess of 60 minutes if fire department operations cannot quickly bring streams into play on the exposed surface and cannot promptly arrest the exposing fire.

6-1.1.2 When automatic systems of sprinklers are installed, water supplies shall be from an automatic source.

6-1.1.3 When the water supply feeds other fire protection appliances, it shall be capable of furnishing total demand for such appliances as well as the outside sprinkler demand.

A sprinkler system designed to protect against exposure fires will reduce the radiated heat through a window opening by 50 to 90 percent, depending on whether the glass remains in place. The radiated heat transmitted through the opening will normally be sufficient to fuse sprinklers in the exposed building. Thus, simultaneous water supply requirements expected for the fire department, for the automatic sprinkler system, and for other systems and appliances in the exposed building should be added to the expected demand of the outside sprinklers in order to determine that the water supply will be of adequate rate and duration.

6-1.1.4 When fire department connections are used for water supply, they shall be so located that they will not be affected by the exposing fire.

6-1.2 Control.

6-1.2.1 Each system of outside sprinklers shall have an independent control valve. When more than one system is required, the division between systems shall be vertical and not horizontal.

Exception: When more than six lines are installed, the systems shall be divided horizontally with independent risers.

Division of sprinkler piping and sprinklers into systems should take into account the extent of exposure from each adjacent building or structure if it should become involved in fire. Further division of sprinkler piping may be necessary if the surface to be protected is quite large. The water supply may limit the area that can be protected. The exposing fire may be assumed to be limited to a smaller area of heat radiation and convection than the whole of an adjacent building side if the exposing building is compartmented by construction features such as fire walls, partitions, or enclosed stair shafts.

Flames from an adjacent building or structure are assumed to be convected upward in a rectangular plume area. Each outside sprinkler system should be arranged to simultaneously discharge water over the corresponding rectangular plume area, assuming that the exposure will occur from ground level upward.

6-1.2.2 Manually controlled open sprinklers shall be used only where constant supervision is present.

Where automatic actuation of outside sprinkler systems cannot be provided, the rapidity of operation should be ensured by manual operation of strategically located and identified valves. Responsible personnel who have been instructed in the operation and importance of these systems should be constantly in attendance in the building or on the premises.

6-1.2.3 Sprinklers may be of the open or closed type. Closed sprinklers in areas subject to freezing shall be on dry-pipe systems conforming to Section 5-2 or antifreeze systems conforming to Section 5-5.

6-1.2.4* Automatic systems of open sprinklers shall be controlled by the operation of fire detection devices designed for the specific application.

A-6-1.2.4 Spacing between approved fire detection devices should not exceed 30 ft (9.1 m) on buildings of less than three stories in height, and not exceed 40 ft (12.1 m) on buildings three or more stories in height. On buildings in excess of eight stories in height, there shall be at least one line of fire detectors for each eight stories with fire detectors staggered. One line of fire detectors should be located close up under the cornice, eave, or outside parapet.

Fire detection devices with weatherproof fixtures should be located on the exterior of the exposed building in such a manner as to rapidly detect radiation or convected heat from the burning building or structure. Spacing and location of detectors should be closer than that specified by the manufacturer if there is any obstruction to the line-of-sight view of surfaces of the exposing building or structure protected by the outside sprinkler system.

6-2 System Components.

6-2.1* Valves.

A-6-2.1 Valves should be so located as to be easily accessible.

6-2.1.1 Control valves shall be of the listed indicating type and shall be distinctively marked by letters not less than ½ in. (13 mm) high to clearly explain their use.

The control valves of automatically actuated outside sprinkler systems should be kept open to ensure that water will be immediately discharged from all operating sprinklers when radiation from fire impinges on the protected building's exterior surface. These valves should be supervised in accordance with 3-9.2.3 and identified in accordance with 3-9.3, for example:

"THIS VALVE SHOULD BE KEPT OPEN AT ALL TIMES. THIS VALVE CONTROLS THE WATER SUPPLY TO THE OUTSIDE SPRINKLER SYSTEM LOCATED ON THE EXTERIOR OF THE _____ WALL OF THIS BUILDING."

When a manually controlled outside sprinkler system is provided, the location of the system controlled by the valve and the importance of rapid opening of the valve should be included in the wording of the sign, for example:

"THIS VALVE MUST BE OPENED FULLY TO SUPPLY WATER TO SPRINKLERS LOCATED ON THE OUTSIDE OF THE _____ WALL OF THIS BUILDING. IMMEDIATE OPERATION WILL RETARD SPREAD OF FIRE TO THIS BUILDING FROM THE ADJACENT BUILDING/STRUCTURE."

6-2.1.2 Drain Valve. Each system of outside sprinklers shall have a separate drain valve installed on the system side of each control valve. Drain valves shall be in accordance with 3-6.2, except that in no case shall valves be smaller than 1 in. (25.4 mm).

6-2.1.3 Check Valves. When sprinklers run on two adjacent sides of a building, protecting against two separate and distinct exposures, with separate control valves for each side, the end lines shall be connected with check valves located so that one sprinkler around the corner will operate. The intermediate pipe between the two check valves shall be arranged to drain. As an alternate solution, an additional sprinkler shall be installed on each system located around the corner from the system involved.

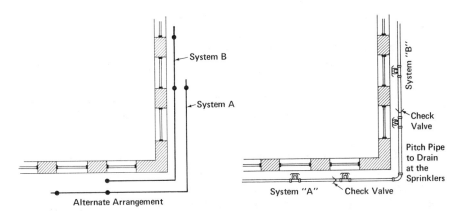

Figure 6-2.1.3   Arrangement of Check Valves.

The "trapped" section of pipe and sprinklers located between two check valves must be drained automatically to prevent freezing of water in the piping after system operation. This may be accomplished by providing a downturned tee outlet and ⅜-in. (9.5-mm) orifice sprinkler located in the piping at a low point. If the ⅜-in. orifice sprinkler has not fused during sprinkler system operation, it must be removed to drain the piping and then replaced.

6-2.1.4   When one exposure affects two sides of the protected structure, the system shall not be subdivided between the two sides, but rather shall be arranged to operate as a single system.

Heat transfer from one building or structure to another is typically accomplished by convection or radiation. Radiation is direct along a line-of-sight, while convection is via fire gases, which may be flaming and deflected by drafts or wind.

Therefore, outside sprinkler systems may be required to extend either beyond the wall area directly opposite the exposing building or structure, or to surfaces other than those parallel to the exposing building wall or structure surface, or both.

6-2.2 Pipe and Fittings.   Approved corrosion-resistant pipe and fittings shall be used for the equipment as far back as the control valve on the water supply.

Since outside sprinkler systems are exposed to the exterior environment, the exterior of the piping will be subject to corrosion. The interior of the piping will also be subject to corrosion if the sprinklers are open, as in a manually or automatically controlled deluge system. Pipe scale formed in the interior of the system could plug the sprinklers when the system is activated, so galvanized pipe and fittings are normally used for this application.

6-2.3 Strainers.   An approved strainer shall be provided in the riser or feed main, which supplies sprinklers having orifices smaller than ⅜ in. (9.5 mm).

Scale, mud, or other debris may become dislodged from piping by the rapidly moving water of a deluge or open sprinkler system. Sprinklers with orifices smaller than ⅜-in. (9.5-mm) diameter may become plugged. Strainers should be of an approved type, located adjacent to the control valve, and capable of being flushed without stopping water flow while the outside sprinkler system is operating.

6-2.4 Gage Connections.   An approved pressure gage conforming to 2-9.2.2 shall be installed immediately below the control valve of each system.

6-3 Sprinklers.[1]   Only sprinklers of such type as are approved for window, cornice, sidewall, or ridge pole service shall be installed for such use except where adequate coverage by use of other types of approved sprinklers and/or nozzles has been demonstrated.   Sprinklers may be of small orifice [¼ in., ⁵⁄₁₆ in., and ⅜ in. (6.4 mm, 7.9 mm, and 9.5 mm)] or large orifice [½ in., ⅝ in., and ¾ in. (12.7 mm, 15.9 mm, and 19.1 mm)].

Many sprinkler designs approved for use in outside sprinkler systems are equipped with heat-sensitive fusible elements. This will permit installation of wet-pipe systems in nonfreezing climates, and dry-pipe or antifreeze systems in areas where freezing conditions may occur during winter months.

Other sprinkler designs approved for use in outside sprinkler systems are not equipped with heat-sensitive fusible elements. Outside sprinkler systems that utilize these types of open sprinklers must be of the deluge type.

The principle of fire protection afforded by outside sprinkler systems is based on the ability of water to carry away heat that is transmitted from a fire in an adjacent building or structure. Water sprays from sprinklers have been shown to reduce radiat-

---

[1]For additional information on outside sprinklers see Appendix B-6-1.

ed heat passing through them by 50 to 70 percent, depending on waterflow rates. In addition, convected fire gases are cooled while passing through the sprinkler spray. The convected and radiated heat that passes through the water spray will expose the surface being protected. This heat must be absorbed by the water running down the protected surface. It is necessary, therefore, that the water issuing from an outside sprinkler system be forcefully applied to the surface; that the water remain in contact with the wall or window surface while wetting and running down the surface being protected; and that the water not be blown away or stream away without wetting all surfaces being protected.

B-6 Outside Sprinklers.

B-6-1 Type.

B-6-1.1 Small orifice sprinklers will normally be used where exposure is light or moderate, the area of coverage is small, or where one horizontal line of window sprinklers is installed at each floor level.

B-6-1.2 Large orifice sprinklers should be used where exposure is severe, or where one horizontal line of sprinklers is used to protect windows at more than one floor level.

B-6-2 Window Sprinklers.

B-6-2.1 When the exposure hazard is light or moderate, and only one horizontal line of sprinklers is installed, the sprinklers should have ⅜-in. (9.5-mm) orifices. Where conditions require more than one line of sprinklers, the sprinklers should have orifices as shown in the following table:

Table B-6-2.1

	2 Lines	3 Lines	4 Lines	5 Lines	6 Lines
Top line	⅜ in.	⅜ in.	⅜ in.	⅜ in.	⅜ in.
Next below	⁵⁄₁₆ in.	⁵⁄₁₆ in.	⅜ in.	⅜ in.	⅜ in.
Next below		¼ in.	⁵⁄₁₆ in.	⁵⁄₁₆ in.	⁵⁄₁₆ in.
Next below			¼ in.	⁵⁄₁₆ in.	⁵⁄₁₆ in.
Next below				¼ in.	¼ in.
Next below					¼ in.

For SI Units: 1 in. = 25.4 mm.

The definition of exposure severity is discussed in Appendix A of NFPA 80A, *Recommended Practice for Protection of Buildings from Exterior Fire Exposures.* Table 2-2.4(a) of NFPA 80A provides a guide for assessing the exposure severity from the burning building. Other factors, such as separation distance between buildings or structures, also affect the heat radiation and the rate of water application required for the exposed surface.

B-6-2.2 When there are more than six horizontal rows of windows, sprinklers over the first story may be omitted. Sprinklers may also be omitted over the second story windows if a field test indicates wetting of all surfaces.

B-6-2.3 Large orifice sprinklers may be used for protecting windows in two or three stories from one line of sprinklers. This will be determined by window and wall construction, such that all parts of the windows and frames will be thoroughly wetted by a single line of sprinklers.

B-6-2.4 For buildings not over three stories in height, one line of sprinklers will often be sufficient, located at the top story windows. For buildings more than three stories in height, a line of sprinklers may be used in every other story beginning at the top. With an odd number of stories, the lowest line can protect the first three stories. When several lines are used, the orifice should be decreased one size for each successive line below the top. In no case should an orifice less than ½ in. (13 mm) be used.

For window sprinklers to be effective, water must be uniformly distributed over the entire glass surface. Dry spots on the heated glass substantially increase the potential for glass breakage due to expansion of the glass.

Run-down of water applied to windows at upper levels is expected to cumulatively protect the lower level(s) of windows. The designer of the outside sprinkler system should be able to assure that water will flow over all obstacles in the path to lower level(s) of windows and that water will "sheet" across frames and glazing of lower windows.

B-6-2.5 For windows not exceeding 5 ft (1.5 m) wide protected by small orifice sprinklers, one sprinkler should be placed at the center near the top, so located that water discharged therefrom will wet the upper part of the window and, by running down over the sash and glass, wet the entire window. This may ordinarily be accomplished by placing one sprinkler in the center with the deflector about on a line with the top of the upper sash and 7, 8, and 9 in. (178, 203, and 229 mm) in front of the glass, with windows 3, 4, and 5 ft (0.9, 1.2, and 1.5 m) wide, respectively. When windows are over 5 ft (1.5 m) wide, or where mullions interfere, two or more sprinklers should be used.

B-6-2.6 When windows are 3 ft (0.9 m) or less in width, an orifice a size smaller than that required by B-6-2.1 may be used, but in no case may the orifice be smaller than ¼ in. (6.4 mm).

B-6-2.7 For windows up to 5 ft (1.5 m) wide protected by large orifice sprinklers, use one ½-in. (13-mm) sprinkler at the center of each window. For windows from 5 to 7 ft (1.5 to 2.1 m) wide, use a ⅝-in. (16-mm) sprinkler at the center of each window. For windows from 7 to 9½ ft (2.1 to 2.9 m) wide, use one ¾-in. (19-mm) sprinkler at the center of each window. For windows from 9½ to 12 ft (2.9 to 3.7 m) wide, use two ½-in (13-mm) sprinklers at each window.

B-6-2.8 Large orifice wide-deflector sprinklers should be placed with deflectors 2 in. (51 mm) below the top of the sash and 12 to 15 in. (305 to 380 mm) out from the glass. When the face of the glass is close to the exterior wall, cantilever brackets or similar type hangers may be used to maintain the window sprinklers 12 to 15 in. (305 to 380 mm) out from the glass.

The designer may utilize water spray patterns furnished by the sprinkler manufacturer to be assured that the location of the sprinkler and the orifice sizes selected will provide adequate density and complete coverage of the surface to be protected. The addition of a wetting agent to the water will assist in obtaining complete coverage intended in design.

B-6-3 Cornice Sprinklers.

B-6-3.1 The discharge orifice should be at least ⅜ in. (9.5 mm) in diameter except, when the exposure is severe, ½-in. or ⅝-in. (13-mm or 16-mm) cornice sprinklers should be installed.

B-6-3.2 Sprinklers should not be more than 8 ft (2.4 m) apart, except as noted in B-6-3.6. Projecting beams or other obstructions may make additional sprinklers necessary.

B-6-3.3 For cornices with bays up to 8 ft (2.4 m) wide, sprinklers should be placed in the center of each bay. For cornices with bays from 8 to 10 ft (2.4 to 3.0 m) wide, sprinkler orifices should be increased one size.

B-6-3.4 Cornice sprinklers should be located with deflectors approximately 8 in. (203 mm) below the roof plank.

B-6-3.5 When wood cornices are 30 in. (762 mm) or less above the windows, cornice sprinklers may be supplied by the same pipe used for window sprinklers.

See Figure B-6-3.5 on page 320. Cornice sprinklers may be of the open type, thus necessitating a piping system that relies on deluge operation—either automatically or manually controlled. The water spray patterns developed by the sprinkler

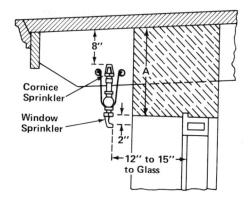

For SI Units: 1 in. = 25.4 mm.

Figure B-6-3.5   Location of Window and Cornice Sprinklers.

manufacturers should be used to establish adequate placement and orifice size.

B-6-3.6   Where the overhang of the cornice is not over 1 ft (0.3 m), window sprinklers should be used and be spaced as follows:

⅜-in. and ½-in. sprinklers . . . . Not more than 5 ft apart
⅝-in. sprinklers. . . . . Not more than 7 ft apart
¾-in. sprinklers. . . . . Not more than 9 ft apart

For SI Units: 1 in. = 25.4 mm; 1 ft = 0.3048 m.

B-6-3.7   The window sprinklers should be placed above the pipe near the outer edge of the cornice with deflectors not more than 3 in. (76 mm) down from the cornice and at such an angle as to throw the water upward and inward.

B-6-3.8   With an overhang of more than 1 ft (0.3 m), cornice sprinklers should be used.

6-4 Piping System.

6-4.1*   Pipe sizes of lines, risers, feed mains, and water supply shall be hydraulically calculated in accordance with Chapter 7 to furnish a minimum of 7 psi (0.5 bars) at any sprinkler with all sprinklers facing the exposure operating, or pipe sizes shall be in accordance with 6-4.2 and 6-4.3.

A-6-4.1   Hydraulic calculations should include all other fire protection systems or devices, such as inside sprinklers and hydrants, to determine that there is no danger of impairing their operation.

6-4.2 Branch line sizes on pipe schedule systems shall be as follows:

Table 6-4.2 Maximum Number of Sprinklers Supplied on Line

Size of Pipe Inches	Orifice Size—In. (mm)						
	¼ (6.4)	³⁄₁₆ (7.9)	⅜ (9.5)	⁷⁄₁₆ (11.1)	½ (12.7)	⅝ (15.9)	¾ (19.1)
1	4	3	2	2	1	1	1
1¼	8	6	4	3	2	2	1
1½		9	6	4	3	3	2
2				5	4	4	3

For SI Units: 1 in. = 25.4 mm.

6-4.3 Risers and feed main sizes on pipe schedule systems shall be as follows for central feed risers:

Table 6-4.3 Maximum Number of Sprinklers—Riser or Feed Mains

Pipe Size	Number of Sprinklers		
	⅜ in. (9.5 mm) or smaller orifice	½ in. (12.7 mm) orifice	¾ in. (19.1 mm) orifice
1½	6	3	2
2	10	5	4
2½	18	9	7
3	32	16	12
3½	48	24	17
4	65	33	24
5	120	60	43
6		100	70

For SI Units: 1 in. = 25.4 mm.

The piping should be sized and the water supply characteristics should be established for wet-pipe, dry-pipe, and deluge-type outside sprinkler systems. Such calculations may indicate the need for larger water supplies, the division of systems, or increased pipe sizes in order to simultaneously deliver water from all operating sprinklers to adequately absorb both radiated and convected heat from an adjacent burning structure.

6-5 Testing and Flushing.

6-5.1 Tests.

6-5.1.1 All piping shall be tested hydrostatically as specified in 1-11.2.

6-5.1.2   Operating tests shall be made of the system when completed, except where such tests may risk water damage.

Acceptance tests should reveal that all surfaces to be protected from radiant and convective heat are uniformly wetted and that the most remote sprinkler has an adequate discharge pressure. Water should be so applied that the effects of wind and drafts will not seriously detract from the intended water discharge and spray patterns of the sprinklers and from the wetting of all surfaces being protected.

6-5.2 Flushing.   Flushing shall be conducted in accordance with 1-11.1.

REFERENCE CITED IN COMMENTARY

The following publication is available from the National Fire Protection Association, Batterymarch Park, Quincy, MA 02269.

NFPA 80A-1987, *Recommended Practice for Protection of Buildings from Exterior Fire Exposures.*

7

Hydraulically Designed Sprinkler Systems

7-1 General.

7-1.1 Definition.

7-1.1.1 A hydraulically designed sprinkler system is one in which pipe sizes are selected on a pressure loss basis to provide a prescribed density, in gallons per minute per square foot [(L/min)/m²], distributed with a reasonable degree of uniformity over a specified area. This permits the selection of pipe sizes in accordance with the characteristics of the water supply available. The stipulated design density and area of application will vary with occupancy hazard.

Hydraulically designed sprinkler systems provide for an accepted method of sizing the pipe network. The pipe schedule technique, discussed in Chapter 8, has been available in one form or another since the first edition of this standard was published in 1896. Hydraulic designs allow for a more precise analysis of the piping system and allow one to optimize the selection of pipe sizes while also showing if the water supply is adequate for a given sprinkler system demand.

To hydraulically design a sprinkler system, the following must be known:

(a) The design density and the assumed area of sprinkler operation. This information is contained in Table 2-2.1.1(b) for Light, Ordinary (Groups 1, 2, and 3), and Extra (Groups 1 and 2) Hazard Occupancies. Other NFPA standards that prescribe density and area of operation are NFPA 30, *Flammable and Combustible Liquids Code;* NFPA 231, *Standard for General Storage;* NFPA 231C, *Standard for Rack Storage of Materials;* NFPA 231D, *Standard for Storage of Rubber Tires;* and NFPA 409, *Standard on Aircraft Hangars.* Some of the major insurance companies also have publications that specify density and area of application for properties insured by them.

(b) The characteristics of the water supply available. The types of acceptable water supplies as stated in Chapter 2 are:

1. Connections to water works systems. Information on the characteristics of the water supply may be obtained from

the water department, fire department, insurance companies, or by conducting a waterflow test as described in B-2-1.

2. Gravity tanks. The available pressure is determined by the height of the tank and the friction loss, based on the required flow, is easily calculated. The determination of the water supply characteristics is a fairly simple matter with gravity tanks.

For more information refer to NFPA 22, *Standard for Water Tanks for Private Fire Protection.*

3. Pumps. The fire pump curve determines the characteristics of the water supply. When pumps take suction from tanks, the elevation of the bottom of the tank with relationship to the pump must be considered. Where suction is from water mains, the fire pump curve is added to the water supply curve for the water main. This addition may be done mathematically or graphically.

For more information refer to NFPA 20, *Standard for the Installation of Centrifugal Fire Pumps.*

4. Pressure tanks. In this case the designer establishes the characteristics required, chooses a tank of proper volume, and establishes the air pressure from the procedure given in A-2-6.3.

For more information refer to NFPA 22, *Standard for Water Tanks for Private Fire Protection.*

This standard does not require that a factor of safety be applied to the water supply characteristics curve. A prudent designer may choose to leave some margin below the curve, particularly when the water works system comprises the sole supply. Typical system designs are completed "to the supply curve." The use of a buffer is considered good practice but is not required by this standard. The development of adjacent properties in the future may tend to decrease the supply curve.

7-1.1.2* The design basis for such a system or addition to an existing system supersedes the rules in the sprinkler standard governing pipe schedules, except that all systems continue to be limited by area, and pipe sizes shall be no less than 1 in. (25.4 mm) nominal for ferrous piping and ¾ in. (19 mm) nominal for copper tubing. The size of pipe, number of sprinklers per branch line, and number of branch lines per cross main are otherwise limited only by the available water supply. However, sprinkler spacing and all other rules covered in this and other applicable standards shall be observed.

Under no circumstances may a pipe schedule system (*see Chapter 8*) be connected to or supplied from a hydraulically designed system. Extending an existing pipe schedule system to a hydraulically designed system may be accomplished in some cases where a careful analysis of the resultant system has been accomplished. This would include shifting the remote area design to include a part of the existing pipe schedule design. It should be noted that ¾-in. (19-mm) nonmetallic piping is currently listed.

A-7-1.1.2   When additional sprinkler piping is added to an existing system, the existing piping does not have to be increased in size to compensate for the additional sprinklers, provided the new work is calculated and the calculations include that portion of the existing system as may be required to carry water to the new work.

In 1940 the use of ¾-in. (19-mm) steel pipe was eliminated from this standard in order to improve water discharge at end sprinklers and to reduce the danger of clogging. This restriction still prevails in all systems, including hydraulically designed systems.

The number of sprinklers per branch line is limited (*see 8-2.1, 8-3.1, and 8-4*) on pipe schedule systems, and prior to 1966 the number of branch lines on a cross main was limited to 14. These rules do not apply to hydraulically designed systems.

7-1.2* Nameplate Data.   The installer shall properly identify a hydraulically designed automatic sprinkler system by a permanently attached nameplate indicating the location(s) and the basis of design(s) [discharge density(ies) over designed area(s) of discharge, including gallons per minute and residual pressure demand at base of riser] and hose stream demand supplied by the sprinkler piping. Such nameplates shall be placed at the controlling alarm, dry-pipe, or preaction valve, for the system containing the hydraulically designed layout(s).

In more cases than not, "as-built" drawings become lost or misplaced over time. By keeping a permanent record of the design parameters attached to the system riser, any future modifications or work on the system will be much easier to perform. The information contained on the nameplate is also of vital importance in accessing the capability of the system to control fires as the occupancy of the building changes or the water supply weakens over the years.

A-7-1.2   Embossed plastic tape, pencil, ink, crayon, etc. should not be considered permanent markings. The pressure values should be rounded to the nearest psi (0.1 bar) and discharge flow to the nearest 5 gpm (20 L/min) increment. The nameplate should be secured to the riser with durable wire, chain, or equivalent.

See Figure A-7-1.2 on page 326.

This system as shown on ....company

print no. ....dated...........

for ....

at .... contract no.........

is designed to discharge at a rate of ....gpm

(L/min) per sq ft of floor area over a maximum

area of .... sq ft (m²) when supplied

with water at a rate of .... gpm (L/min)

at .... psi (bars) at the base of the riser.

Hose stream allowance of.......................

gpm (L/min) is included in the above.

Figure A-7-1.2   Sample Nameplate.

7-2 Information Required.

7-2.1 Basic Design Information. Basic design criteria for hydraulically designed sprinkler systems shall be obtained from this or other applicable standards. Where no standards exist, the authority having jurisdiction shall be consulted.

7-2.2 Sprinkler System Requirements. The following information shall be included when applicable:

(a) Area of water application, sq ft

(b) Minimum rate of water application (density), gpm/sq ft

(c) Area per sprinkler, sq ft

(d) Allowance for inside hose and outside hydrants, gpm

(e) Allowance for in-rack sprinklers, gpm.

If this information is incorrectly supplied for a given occupancy, the other details relating to the system design will be insufficient for the system layout.

7-2.3* **Water Supply Information.** The following information shall be included: waterflow data with existing or proposed water supply, dead-end or circulating:

(a) Location and elevation of static and residual test gage with relation to the riser reference point

(b) Flow location

(c) Static pressure, psi

(d) Residual pressure, psi

(e) Flow, gpm

(f) Date

(g) Time

(h) Test conducted by or information supplied by. . . .

Waterflow test data must be current, analyzed during peak flow conditions, and conducted in such a manner that it represents true orientation (direction) of the supply to the proposed system.

A-7-2.3 Designers should consult with the authority having jurisdiction on the water supply to be used in system calculations prior to system design and calculation.

7-2.4 Information Required on the Drawings.

7-2.4.1 In addition to the requirements of Section 1-9, the drawings shall also contain the information mentioned in the remainder of 7-2.4.

7-2.4.2 Hydraulic Reference Points. Reference points may be shown by a number and/or letter designation and shall correspond with comparable reference points shown on the hydraulic calculation sheets.

Hydraulic reference points are important to the designer, as they help ensure that parts of the system are not overlooked. They are essential to the plan checker, particularly in the case of gridded systems. Good coordination between calculations and plans is essential should it be necessary to reevaluate the system several years after it has been designed or should additions to the system become necessary.

7-2.4.3 Sprinklers. Description of sprinklers used.

7-2.4.4 System Design Criteria. The minimum rate of water application (density), the design area of water application, in-rack sprinkler demand, and the water required for hose streams both inside and outside shall be included.

The system design criteria must be on the drawings for the benefit of the plan checker and for future reference. This includes:
(a) Density;
(b) Area of water application;
(c) In-rack sprinkler demand, if applicable;
(d) Inside hose demand, if applicable; and
(e) Outside hose demand (This would be zero when applying 2-2.3.8).

7-2.4.5 Actual Calculated Requirements. The total quantity of water and the pressure required shall be noted at a common reference point for each system.

The actual calculated demand is normally referenced at the base of the riser in buildings containing several systems. In single system buildings, the reference point may be at the base of the riser or the point of connection to the city main.

7-2.4.6 Elevation Data. Relative elevations of sprinklers, junction points, and supply or reference points shall be noted.

7-3 Data Sheets and Abbreviations.

7-3.1 General. Hydraulic calculations shall be prepared on form sheets that include a summary sheet, detailed work sheets, and a graph sheet. (*See copy of typical forms, Figures A-7-3.3 and A-7-3.4.*)

7-3.2 Summary Sheet. The summary sheet shall contain the following information, when applicable:

(a) Date

(b) Location

(c) Name of owner and occupant

(d) Building number or other identification

(e) Description of hazard

(f) Name and address of contractor or designer

(g) Name of approving agency

(h) System design requirements

    1. Design area of water application, sq ft

    2. Minimum rate of water application (density), gpm per sq ft

    3. Area per sprinkler, sq ft

(i) Total water requirements as calculated including allowance for inside hose and outside hydrants

(j) Water supply information.

7-3.3* Detailed Work Sheets. Detailed work sheets (*for sample work sheet, refer to Figure A-7-3.3*) or computer printout sheets shall contain the following information:

(a) Sheet number

(b) Sprinkler description and discharge constant (K)

(c) Hydraulic reference points

(d) Flow in gpm

(e) Pipe size

(f) Pipe lengths, center to center of fittings

(g) Equivalent pipe lengths for fitting and devices

(h) Friction loss in psi per ft of pipe

(i) Total friction loss between reference points

(j) In-rack sprinkler demand

(k) Elevation head in psi between reference points

(l) Required pressure in psi at each reference point

(m) Velocity pressure and normal pressure if included in calculations

(n) Notes to indicate starting points, reference to other sheets, or to clarify data shown

(o)* Diagram to accompany gridded system calculations to indicate flow quantities and directions for lines with sprinklers operating in the remote area. [*See Figure A-7-3.3(o).*]

Contract No. _____   Sheet No. _____ of _____

Name & Location _____

Nozzle Type & Location	Flow in GPM (L/min)	Pipe Size in.	Fitting & Devices	Pipe Eqiv. Length	Friction Loss psi/ft. (bars/m)	Req. Psi. (bars)	Normal Pressure	Notes
q ——— Q				lgth. ——— ftg. ——— tot.		Pt Pf Pe	Pt Pv Pn	
q ——— Q				lgth. ——— ftg. ——— tot.		Pt Pf Pe	Pt Pv Pn	
q ——— Q				lgth. ——— ftg. ——— tot.		Pt Pf Pe	Pt Pv Pn	
q ——— Q				lgth. ——— ftg. ——— tot.		Pt Pf Pe	Pt Pv Pn	
q ——— Q				lgth. ——— ftg. ——— tot.		Pt Pf Pe	Pt Pv Pn	
q ——— Q				lgth. ——— ftg. ——— tot.		Pt Pf Pe	Pt Pv Pn	
q ——— Q				lgth. ——— ftg. ——— tot.		Pt Pf Pe	Pt Pv Pn	
q ——— Q				lgth. ——— ftg. ——— tot.		Pt Pf Pe	Pt Pv Pn	
q ——— Q				lgth. ——— ftg. ——— tot.		Pt Pf Pe	Pt Pv Pn	
q ——— Q				lgth. ——— ftg. ——— tot.		Pt Pf Pe	Pt Pv Pn	
q ——— Q				lgth. ——— ftg. ——— tot.		Pt Pf Pe	Pt Pv Pn	
q ——— Q				lgth. ——— ftg. ——— tot.		Pt Pf Pe	Pt Pv Pn	
q ——— Q				lgth. ——— ftg. ——— tot.		Pt Pf Pe	Pt Pv Pn	
q ——— Q				lgth. ——— ftg. ——— tot.		Pt Pf Pe	Pt Pv Pn	
q ——— Q				lgth. ——— ftg. ——— tot.		Pt Pf Pe	Pt Pv Pn	
q ——— Q				lgth. ——— ftg. ——— tot.		Pt Pf Pe	Pt Pv Pn	
q ——— Q				lgth. ——— ftg. ——— tot.		Pt Pf Pe	Pt Pv Pn	
q ——— Q				lgth. ——— ftg. ——— tot.		Pt Pf Pe	Pt Pv Pn	
q ——— Q				lgth. ——— ftg. ——— tot.		Pt Pf Pe	Pt Pv Pn	

Figure A-7-3.3   Sample Work Sheet.

Figure A-7-3.3 is just one of several work sheets used in the sprinkler industry today. Many of the computer printouts are in this same form.

Figure A-7-3.3(o) depicts a gridded system that has been calculated. The dashed lines outline the calculated area and the numbers in circles are reference points. The arrows on the pipe indicate the direction of flow and the numbers alongside indicate quantity in gallons per minute.

A diagram of this type must accompany hydraulic calculations for gridded systems to facilitate checking.

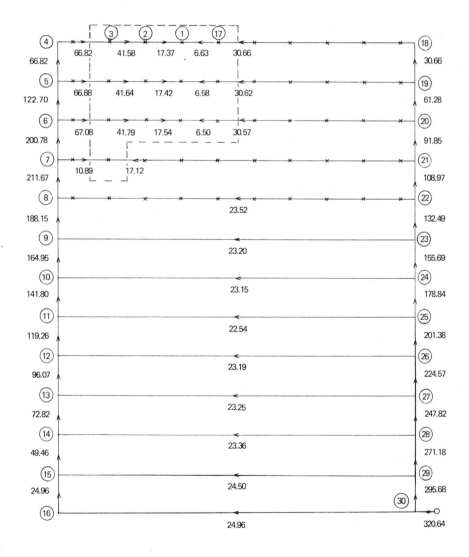

Figure A-7-3.3(o)   Flow Quantities and Direction.

7-3.4* **Graph Sheet.** Water supply curves and system requirements, plus hose and in-rack sprinkler demand when applicable, shall be plotted on semi-logarithmic graph paper ($Q^{1.85}$) so as to present a graphic summary of the complete hydraulic calculation.

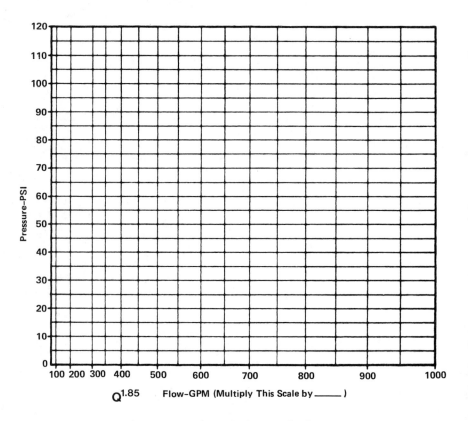

Figure A-7-3.4   Sample Graph Sheet.

This graph sheet is used to plot the characteristics of the water supply and the sprinkler system demand. (*See Figure A-7-3.4.*) It is referred to as hydraulic graph paper. The scale along the abscissa (X-axis) must be to the 1.85 power since, in the Hazen-Williams formula, pressure (P) in pounds per square inch is proportional to the flow (Q) in gallons per minute to the 1.85 power. Without this exponent adjustment, the water supply graph would be in the shape of a curve (similar to a fire pump curve) vice a straight line function.

7-3.5 Abbreviations and Symbols. The following standard abbreviations and symbols shall be used on the calculation form:

Symbol or Abbreviation	Item
p	Pressure in psi
gpm	U.S. Gallons per minute
q	Flow increment in gpm to be added at a specific location
Q	Summation of flow in gpm at a specific location
P_t	Total pressure in psi at a point in a pipe
P_f	Pressure loss due to friction between points indicated in location column
P_e	Pressure due to elevation difference between indicated points. This can be a plus value or a minus value. Where minus, the ($-$) shall be used; where plus, no sign need be indicated
P_v	Velocity pressure in psi at a point in a pipe
P_n	Normal pressure in psi at a point in a pipe
E	90° Ell
EE	45° Ell
Lt.E	Long Turn Elbow
Cr	Cross
T	Tee — flow turned 90 degrees
GV	Gate Valve
BV	Butterfly Valve
Del V	Deluge Valve
ALV	Alarm Valve
CV	Swing Check Valve
WCV	Butterfly (Wafer) Check Valve
St	Strainer
psi	pounds per square inch
v	Velocity of water in pipe in feet per second

7-4 Calculation.

7-4.1 Formulas.

7-4.1.1 Friction Loss Formula. Pipe friction losses shall be determined on the basis of the Hazen-Williams formula.

$$p = \frac{4.52 \, Q^{1.85}}{C^{1.85} \, d^{4.87}}$$

where p is the frictional resistance in pounds pressure per square inch per foot of pipe, Q is the gallons per minute flowing, and d is the actual internal diameter of pipe in inches with C as the friction loss coefficient.

$$\text{For SI Units: } P_m = 6.05 \times \frac{Q_m^{1.85}}{C^{1.85} \, d_m^{4.87}} \times 10.^5$$

P_m is the frictional resistance in bars per meter of pipe, Q_m is the flow in L/min, and d_m is the actual internal diameter in mm with C as the friction loss coefficient.

The Hazen-Williams formula is the most common of the empirical formulas used to determine relationships between flow, friction loss, and available pressure. The Darcy-Weisbach equation may provide a more precise method for establishing these relationships, but is not the selected method in fire protection calculations.

Hazen-Williams is dependent upon the relationship between the pipe type (the C factor), the pipe diameter, and any given flow. The C factor describes the relative roughness of the pipe interior and is similar to the ϵ factor used in the Reynolds Number calculation in the Darcy-Weisbach equation. C factors for various pipe types are shown in Table 7-4.3.1.4.

Formal Interpretation

Question: Is it acceptable to calculate sprinkler systems without any limit on flow velocities?

Answer: Yes, NFPA 13 does not limit the velocity of water in pipe.

While this standard does not stipulate a maximum limit on velocities, some insurance companies modify this standard and incorporate such limits. Twenty feet per second (6.1 m/sec) maximum velocities may result in superior designs. At velocities beyond this, the Hazen-Williams formula is not as conservative as other methods, such as Darcy-Weisbach.

7-4.1.2 Velocity Pressure Formula. Velocity pressure shall be determined on the basis of the formula

$$P_v = 0.001\ 123\ Q^2/D^4$$

where:

P_v = velocity pressure psi.

Q = flow in gpm.

D = the inside diameter in inches.

For SI units: 1 in. = 25.4 mm; 1 gal = 3.785 L; 1 psi = 0.0689 bar.

NFPA 15, *Standard for Water Spray Fixed Systems for Fire Protection*, defines velocity pressure as "a measure of energy required to keep the water in a pipe in motion." The basic formula for velocity pressure is:

$$P_v = 0.433 \frac{V^2}{2g}$$

Where: P_v = Velocity pressure in psi
V = Velocity in ft per second
g = 32.2 ft per second per second

By substituting $\frac{Q}{A}$ for V in this formula, and using proper units, the more usable formula that appears in the text is found.

7-4.1.3 Normal pressure P_n shall be determined on the basis of the formula

$$P_n = P_t - P_v$$

where:
P_t = total pressure in psi (bars)
P_v = velocity pressure in psi (bars)

Calculations that use total pressure, P_t, are the most common. NFPA 15 provides specific information on the relations between velocity pressure, normal pressure, and total pressure. These pressures relate to the direction (horizontal or vertical) of force in which a closed conduit pressure applies itself.

7-4.1.4 Hydraulic Junction Points. Pressures at hydraulic junction points shall balance within 0.5 psi (0.03 bar). The highest pressure at the junction point, and the total flows as adjusted, shall be carried into the calculations.

The value of 0.5 psi (0.03 bar) was selected as a reasonable degree of accuracy for balancing at hydraulic junction points. This is easily obtained by most computer programs without an excessive number of iterations. By requiring a balance at the hydraulic junction points, the designer will be assured that an excess system demand is not concealed in the calculations.

7-4.2 Equivalent Pipe Lengths of Valves and Fittings.

7-4.2.1 Table 7-4.2 shall be used to determine the equivalent length of pipe for fittings and devices unless manufacturer's test data indicate that other factors are appropriate. For saddle-type fittings having friction loss greater than that shown in Table 7-4.2, the increased friction loss shall be included in hydraulic calculations.

Table 7-4.2   Equivalent Pipe Length Chart

Fittings and Valves													
	¾ in.	1 in.	1¼ in.	1½ in.	2 in.	2½ in.	3 in.	3½ in.	4 in.	5 in.	6 in.	8 in.	10 in.	12 in.
45° Elbow	1	1	1	2	2	3	3	3	4	5	7	9	11	13
90° Standard Elbow	2	2	3	4	5	6	7	8	10	12	14	18	22	27
90° Long Turn Elbow	1	2	2	2	3	4	5	5	6	8	9	13	16	18
Tee or Cross (Flow Turned 90°)	3	5	6	8	10	12	15	17	20	25	30	35	50	60
Butterfly Valve	—	—	—	—	6	7	10	—	12	9	10	12	19	21
Gate Valve	—	—	—	—	1	1	1	1	2	2	3	4	5	6
Swing Check*	—	5	7	9	11	14	16	19	22	27	32	45	55	65

For SI Units: 1 ft = 0.3048 m.

*Due to the variations in design of swing check valves, the pipe equivalents indicated in the above chart are considered average.

NOTE: This table applies to all types of pipe listed in Table 7-4.3.1.4.

The fitting and valve losses shown in Table 7-4.2 are calculated values that have been rounded to a whole number for convenience. The exception is the swing check valve, in which several makes of valves were averaged to produce the numbers in the tables. Where possible, the designer should use the friction loss value for the specific check valve that is to be installed.

Some 1-in. (25.4-mm) saddle-type fittings have a friction loss equivalent of 21 ft (6.4 m) of pipe as opposed to the 5 ft (1.5 m) shown in the table. This is another case where the listing information for a given product is of utmost importance. An appreciable error would be introduced into the calculations if this increased loss is ignored.

7-4.2.2 Use Table 7-4.2 with Hazen-Williams C = 120 only. For other values of C, the values in Table 7-4.2 shall be multiplied by the factors indicated below:

Value of C	100	130	140	150
Multiplying Factor	0.713	1.16	1.33	1.51

(This is based upon the friction loss through the fitting being independent of the C factor available to the piping.)

Friction loss caused by a fitting occurs at some point downstream from the fitting and not in the fitting itself. Therefore, a C factor is not assigned to the fitting but a correction for "C" is made to account for the presence of a fitting.

The example on page 338 illustrates the use of the correction for "C" in a 2-in. (51-mm) diameter tee fitting:

$$\text{The base formula is } P_f = L \times \frac{4.52Q^{1.85}}{C^{1.85} \, d^{4.87}}$$

in which P_f is the friction loss in psi, L is the length in feet, and the other factors are as described in 7-4.1.1. The example looks at the friction loss for C = 100 and C = 120, and a flow of 100 gpm.

C = 100:

$$P_{f(100)} = 10 \times \frac{4.52 \, (100^{1.85})}{100^{1.85} \, 2^{4.87}} = 1.5 \text{ psi}$$

From 7-4.2.2, the correction for C = 100 (0.713) is now made so that

$$P_{f(100)} = 1.5 \times 0.713 = 1.1 \text{ psi}$$

C = 120:

$$P_{f(120)} = 10 \times \frac{4.52(100^{1.85})}{120^{1.85} \, 2^{4.87}} = 1.1 \text{ psi}$$

The correction factor for C = 120 is technically 1.0.

7-4.2.3 Specific friction loss values or equivalent pipe lengths for alarm valves, dry-pipe valves, deluge valves, strainers, and other devices shall be made available to the authority having jurisdiction.

7-4.3* Calculation Procedure.

Figure A-7-4.3(a) shows the floor plan and elevation of the sprinkler system in a small building (130 ft × 200 ft) that has been hydraulically designed. The density of 0.16 gpm/sq ft [6.3 (L/min)/m²] and the calculated area of 1500 sq ft (139 m²) are both obtained from Table 2-2.1.1(b) for an Ordinary Hazard (Group 1) Occupancy.

To determine the number of sprinklers to be calculated, it is necessary to divide the calculated area by the coverage per sprinkler which, in this case, is 1500 ÷ 130 = 11.54 sprinklers, which is rounded off to 12. The number of sprinklers to calculate per branch line is determined by dividing 1.2 $\sqrt{A}$ (*see* 7-4.3.1*) by the distance between sprinklers on the line which, in this case, is 1.2 $\sqrt{1500}$ ÷ 13 = 3.58. Therefore, use four sprinklers. Any fractional number in this calculation technique is always rounded up to the next whole number.

The calculated area is the hydraulically most remote rectangle encompassing four sprinklers on the branch line and three branch lines, to equal the required total of twelve. This area is indicated by crosshatching, and since this system is symmetrical about the cross main, this area may be located as shown or it could be in the lower left corner of the building.

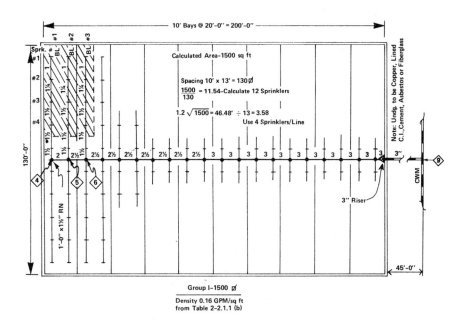

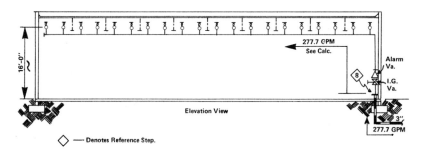

◇ — Denotes Reference Step.

For SI Units: 1 in. = 25.4 mm; 1 ft = 0.3048 m.

Figure A-7-4.3(a)   Hydraulic Calculation Example.

HYDRAULIC CALCULATIONS

FOR

ABC COMPANY

CONTRACT NO. *4001*

DATE *1-7-75*

DESIGN DATA—

      OCCUPANCY CLASSIFICATION *ORD. GR. 1*

      DENSITY *0.16* GPM/SQ. FT.

      AREA OF APPLICATION *1500* SQ. FT.

      COVERAGE PER SPRINKLER *130* SQ. FT.

      NO. OF SPRINKLERS CALCULATED *12*

      TOTAL WATER REQUIRED *650* GPM.
      INCLUDING HOSE STREAMS.

NAME OF CONTRACTOR _____

NAME OF DESIGNER _____

AUTHORITY HAVING JURISDICTION _____

Figure A-7-4.3(b)   Calculation Coversheet.

    This sheet is normally used as the coversheet for the calculations and contains the information required by 7-3.2.

STEP NO.	NOZZLE IDENT. AND LOCATION	FLOW IN G.P.M.	PIPE SIZE	PIPE FITTINGS AND DEVICES	EQUIV. PIPE LENGTH	FRICTION LOSS P.S.I./ FOOT	PRESSURE SUMMARY	NORMAL PRESSURE	NOTES	REF. STEP
									CONTRACT NAME _GROUP I 1500 #_ **SHEET** _2_ **OF** _3_	
									D = 0.16 GPM/#	
									K = 5.65	
1	1 BL-1	q	1		L 13.0	C = 120	Pt 13.6	Pt	q = 130 X .16 = 20.8	
					F		Pe	Pv		
		Q 20.8			T 13.0	.140	Pf 1.8	Pn		
2	2	q 22.2	1¼		L 13.0		Pt 15.4	Pt	q = 5.65 √15.4	
					F		Pe	Pv		
		Q 43.0			T 13.0	.141	Pf 1.8	Pn		
3	3	q 23.4	1½		L 13.0		Pt 17.2	Pt	q = 5.65 √17.2	
					F		Pe	Pv		
		Q 66.4			T 13.0	.149	Pf 1.9	Pn		
4	4 DN RN	q 24.7	1½	2T-16	L 20.5		Pt 19.1	Pt	q = 5.65 √19.1	4
					F 16.0		Pe	Pv		
		Q 91.1			T 36.5	.267	Pf 9.7	Pn		
5	CM TO BL-2	q	2		L 10.0		Pt 28.8	Pt	K = $\frac{91.1}{\sqrt{28.8}}$	5
					F		Pe	Pv	K = 16.98	
		Q 91.1			T 10.0	.079	Pf .8	Pn		
6	BL-2 CM TO BL-3	q 92.4	2½		L 10.0		Pt 29.6	Pt	q = 16.98 √29.6	6
					F		Pe	Pv		
		Q 183.5			T 10.0	.122	Pf 1.2	Pn		
7	BL-3 CM	q 94.2	2½		L 70.0		Pt 30.8	Pt	q = 16.98 √30.8	7
					F		Pe	Pv		
		Q 277.7			T 70.0	.262	Pf 18.3	Pn		
8	CM TO F/S	q	3	E 5 AV 15 GV 1	L 119.0 F T 140.0	.091	Pt 49.1 Pe 6.5 Pf 12.7	Pt Pv Pn	Pe = 15 x .433	8
		Q 277.7								
9	THRU UNDER-GROUND TO CITY MAIN	q	3	E 5 GV 1 T 15	L 50.0 F 32.0 T 82.2	C = 150 TYPE "M" .069	Pt 68.3 Pe Pf 5.7	Pt Pv Pn	COPPER 21 X 1.51 = 32	9
		Q 277.7								
		q			L		Pt 74.0	Pt		
					F		Pe	Pv		
		Q			T		Pf	Pn		
		q			L		Pt	Pt		
					F		Pe	Pv		
		Q			T		Pf	Pn		
							Pt			

Figure A-7-4.3(c)   Hydraulic Calculations.

In Step 1, sprinkler one and branch line (BL) one are identified. The flow for the first sprinkler is determined by multiplying the density by the coverage per sprinkler, which in this

case is 0.16 gpm/sq ft × 130 sq ft = 20.8 gpm. The pipe size is determined by trial and error. The fitting directly connected to a sprinkler is not included in the calculations [see 7-4.3.1.4(d)]; therefore, nothing is shown under the "Pipe Fittings and Devices" column. The equivalent pipe length is the center-to-center distance between sprinklers, which in this case is 13 ft. The friction loss (psi/ft) is determined by using the Hazen-Williams formula. When using Schedule 40 pipe and C = 120 the formula becomes:

P = 5.10 × 10⁻⁴ × Q^{1.85} = 0.140 psi/ft for a Q of 20.8 gpm.

By multiplying 0.14 psi/ft × 13 ft a friction loss of 1.8 psi from sprinkler one to sprinkler two is determined.

The total pressure (P_t) is determined by using the formula

$$P_t = \left[\frac{Q}{K}\right]^2$$

which in this case is $\left[\dfrac{20.8}{5.65}\right]^2 = 13.6.$

The pressure at sprinkler two is 13.6 psi (0.94 bar) plus 1.8 psi (0.12 bar) friction loss, or 15.4 psi (1.06 bar). The flow (Q) from sprinkler two is equal to K × $\sqrt{p}$, 5.65 × $\sqrt{15.4}$, or 22.2 gpm (84 L/min). This flow is added to sprinkler one and the same procedures are followed as above. In Step 4, the four sprinklers have been calculated, and it is necessary to carry the analysis of the pipe segment back to the cross main. The pipe length (L) included is 13.0 ft (4.0 m) between sprinklers plus 6.5 ft (2.0 m) for a starter piece plus 1.0 ft (0.3 m) for a riser nipple for a total of 20.5 ft (6.3 m). The tees at the top and bottom of the riser nipple are included and are equal to 8 ft (2.4 m) of pipe each. (See Table 7-4.2.) A K Factor is established for branch line one by dividing the flow (Q) by the square root of the pressure (P), which in this case is 91.1 divided by the square root of 28.8, or 16.98. This K Factor is used in predicting the flow in subsequent branch lines that are identical to branch line one. In Step 8, the pressure due to elevation (P_e) of 6.5 psi (= .433 × 15 ft) is added at one point. This can be done only if the branch lines are basically level as they are assumed to be in this case. In buildings with pitched roofs or branch lines at different elevations, the pressure due to elevation (P_e) would have to be applied where it occurs. In Step 9, note that copper pipe is used, so the C Factor is 150 (see Table 7-4.3.1.4) and the fitting and device equivalent lengths are totalled and multiplied by 1.51 (see 7-4.2.2).

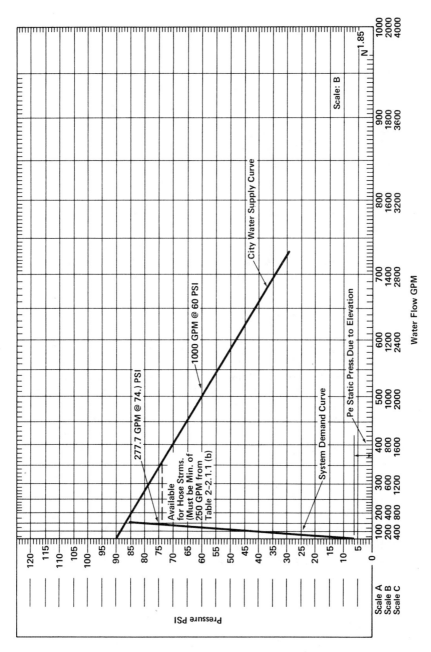

Figure A-7-4.3(d)   Hydraulic Graph.

In Figure A-7-4.3(d), the city water supply curve is based upon 90 psi (6.2 bar) static pressure and a 60 psi (4.1 bar) residual pressure with 1000 gpm (3785 L/min) flowing.

The system demand curve is started at the 6.5 psi (0.5 bar) point on the ordinate to consider the pressure due to elevation, P_e, and continued to the demand point of 277.7 gpm (1051 L/min) at 74.0 psi (5.1 bar). The system demand curve then represents the friction loss in the system. The horizontal dashed line drawn from the above demand point to the city water supply curve represents the amount of water available for hose streams. In this case, about 450 gpm (1703 L/min) are available and only 250 gpm (946 L/min) are required. This represents a factor of safety of approximately 7 psi (0.5 bar). If this factor of safety is warranted (*see commentary to 7-1.1.1*), the design is complete. If not, some pipe sizes may be reduced to absorb the 7 psi. If the system demand point falls above the city water supply curve, some of the piping must be enlarged so the friction loss will decrease. Other factors that could be used to reduce the pressure to a point at or below the supply curve are sprinkler orifice size and reduced sprinkler spacing. These would normally be used only in higher hazard occupancies.

7-4.3.1*   For all systems the design area shall be the hydraulically most demanding based on the criteria of 2-2.1.3. When the design is based on area/density method, the calculation shall be a rectangular area having a dimension parallel to the branch lines at least 1.2 times the square root of the area of sprinkler operation used. This may include sprinklers on both sides of the cross main. Any fractional sprinkler shall be carried to the next higher whole sprinkler. When the design is based on the room design method, the calculation shall be based on the room that is the hydraulically most demanding.

The 1966 edition of this standard was the first to include a Chapter 7. The following is from that edition: "The design area shall be the hydraulically most remote area and shall include all sprinklers on both sides of the cross main."

This meant that, if the system shown in Figure A-7-4.3(a) was to be calculated, the ten sprinklers on the end of the system (five on each side of the cross main) plus two on the next line would be calculated.

In cases of sprinkler spacing of 13 ft (3.96 m) on the line and 10 ft (3.05 m) between branch lines and where the design area included twelve sprinklers, some designers were calculating three sprinklers on the line (3 × 13 = 39) over four lines (4 × 10 = 40). This was almost a perfect square, but the design would probably be inadequate if four sprinklers operated on the branch line; hence, the change.

The first mention of gridded systems appeared in the 1975 edition as follows: "7-4.3.1 Exception No. 2: For gridded systems, the design area shall be the hydraulically most remote area which approaches a square." This definition was adopted since it did not seem logical to calculate all of the sprinklers on one line of a grid as had been done on "tree" systems since 1966.

In later editions of the standard, the characteristic square design area was changed to represent a rectangular area. The rectangle is required to have one side which is at least 1.2 times greater in length than its adjacent side. This approach results in the requirement for more sprinklers to be calculated on a given branch line, which tends to be a more conservative calculation than the original square area.

This design method establishes a sense of the number of sprinklers that need to be calculated. The actual geometric area for system design, which is determined by Table 2-2.1.1(b), must also be verified. In some cases, the methods discussed in this chapter may result in a given number of sprinklers to calculate but this must also be verified against the geometry of the design area. In all cases which use the area/density method, the actual geometric design area must be proven. This is frequently misinterpreted and misapplied. Part of the rationale behind remote design areas is an attempt to predict where the heat may travel for a given fire scenario.

Exception No. 1:   Where the design area under consideration consists of a corridor protected by one row of sprinklers, the maximum number of sprinklers that need be calculated is 5, unless openings from the corridor are unprotected. (See 2-2.3.)

Exception No. 1 has been in the standard for several years; however, in 1978 the maximum number of sprinklers was reduced from seven to five. It seemed highly unlikely that seven sprinklers in a corridor in a fully sprinklered building would be opened by one fire; hence, the change. This Exception would only be used when the design is based on the area of the largest room with automatic or self-closing doors as described in 2-2.3. Then the designer would have to calculate the largest room as well as the five sprinklers in the corridor to determine which requires the higher water supply.

However, if openings from the corridor are unprotected as outlined in 2-2.3, then the calculations must include sprinklers within the adjoining spaces, up to the area of application selected from Table 2-2.1.1(b).

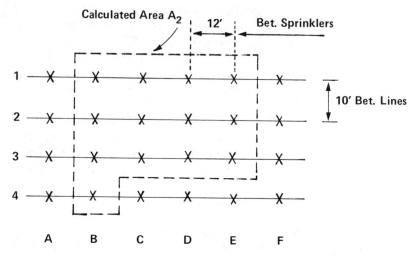

Figure A-7-4.3.1(a)   Example of Determining the Number of Sprinklers to Be Calculated.

NOTE 1:   For gridded systems, the extra sprinkler (or sprinklers) on branch line 4 may be placed in any adjacent location from B to E at the designer's option.

NOTE 2:   For tree and looped systems, the extra sprinkler on line 4 should be placed closest to the cross main.

Figure A-7-4.3.1(a) Example:

Assume a remote area of 1,500 sq ft with sprinkler coverage of 120 sq ft

Total sprinklers to calculate $= \dfrac{\text{Design Area}}{\text{Area per Sprinkler}}$

$= \dfrac{1500}{120} = 12.5$, calculate 13

Number of sprinklers on branch line $= \dfrac{1.2\sqrt{A}}{S}$

Where A = Design Area
      S = Distance Between Sprinklers on Branch Line

Number of sprinklers on branch line $= \dfrac{1.2\sqrt{1500}}{12} = 3.87$, calculate 4

For SI Units: 1 ft = 0.3048 m; 1 sq ft = 0.0929 m².

Figure A-7-4.3.1(a) is an example of the number of sprinklers to calculate on a gridded system. In order for this figure to apply to a tree or loop system, the branch line would terminate at the last sprinkler on the right in the calculated area and the cross main would be to the left. Also, the note would only apply to grids. The extra sprinkler on line four would have to be "B" for tree and loop systems as it is closest to the supply and would be hydraulically most demanding.

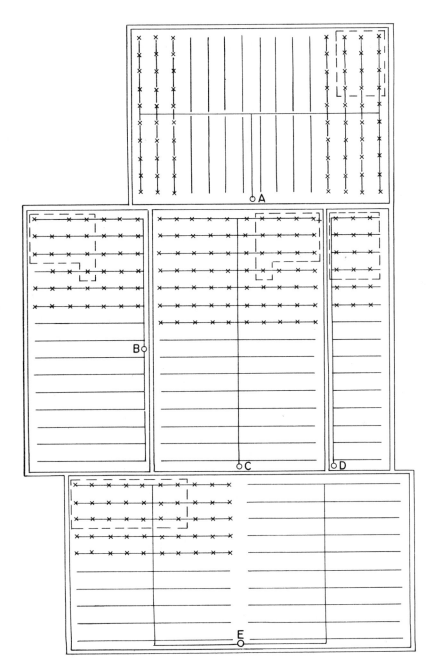

Figure A-7-4.3.1(b)   Example of Hydraulically Most Demanding Area.

Shown in Figure A-7-4.3.1(b) on page 347, System A shows design area for twelve sprinklers with three branch lines and four sprinklers per branch line. Systems B and C indicate the location of the extra sprinkler on the fourth branch line.

System D illustrates 7-4.3.1 Exception No. 2.

System E illustrates that sprinklers on both sides of the cross main may be required to fulfill the required design area and the 1.2 $\sqrt{A}$ restriction on the sprinklers on the branch line.

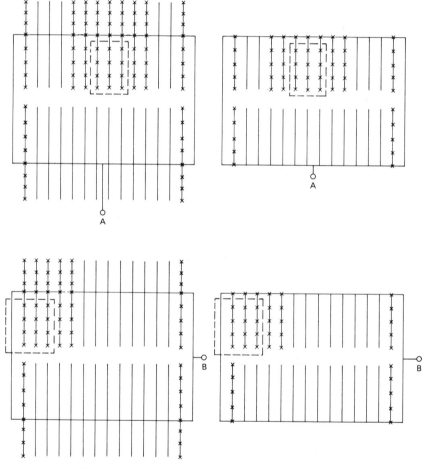

Figure A-7-4.3.1(c)   Example of Hydraulically Most Demanding Area.

This figure illustrates the design area for looped systems with various riser locations and branch line configurations.

Exception No. 2: In systems having branch lines with an insufficient number of sprinklers to fulfill the 1.2 $\sqrt{A}$ requirement, the design area shall be extended to include sprinklers on adjacent branch lines supplied by the same cross main.

Exception No. 2 was added in 1980 for clarification. Figure A-7-4.3.1(b), System D shows an example.

Exception No. 3: Where the design is based on the criteria of 2-2.3, including the Exceptions to that section, the above dimensional requirements do not apply.

Exception No. 4: Where the design area under consideration consists of a building service chute supplied by a separate riser, the maximum number of sprinklers that need be calculated is 3.

Exception No. 5: Residential sprinkler designs shall be in accordance with 7-4.4.

Exception No. 6: Designs for ESFR sprinklers shall be in accordance with Chapter 9.

7-4.3.1.1*   For gridded systems, the designer shall verify that the hydraulically most demanding area is being used. A minimum of two additional sets of calculations shall be submitted to demonstrate peaking of demand area friction loss when compared to areas immediately adjacent on either side along the same branch lines.

Exception: Computer programs that show the peaking of the demand area friction loss shall be acceptable based on a single set of calculations.

In the 1975 standard, the design area was shown as a square area located along the far cross main (not the supply cross main). It was quickly learned that this was not necessarily the hydraulically most remote area. Because of bi-directional flows, which are often found on the same branch line in a gridded system, hydraulically remote areas are not always obvious. This section requires the area to be moved one sprinkler to either side of the selected hydraulically most remote area.

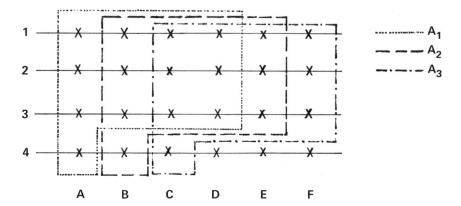

Figure A-7-4.3.1.1   Example of Determining the Most Remote Area for a Gridded System.

Figure A-7-4.3.1.1 Example:

If Area A_2 is selected as the most remote area, submit calculations to show that Areas A_1 and A_3 are subject to less friction loss.

7-4.3.1.2 System piping shall be hydraulically designed using design densities and areas of operation in accordance with Table 2-2.1.1(b) as required for the occupancies involved.

(a)* The density shall be calculated on the basis of floor area. The area covered by any sprinkler for use in hydraulic design and calculations shall be determined as follows:

1. Along Branch Lines. Determine distance to next sprinkler (or to wall in case of end sprinkler on branch line) upstream and downstream. Choose larger of either twice (1) the distance to the wall, or (2) distance to the next sprinkler. Call this "S."

2. Between Branch Lines. Determine the perpendicular distance to sprinklers on branch lines (or to wall in case of the last branch line) on each side of the branch line on which the subject sprinkler is positioned. Choose the larger of (1) the larger distance to the sprinklers on the next branch line, or (2) in the case of the last branch line, twice the distance to the wall. Call this "L."

Exception: For sidewall sprinklers, L will be the distance to the wall opposite the sprinklers, or in the case where sprinklers are provided on two sides, half the distance between the two sides.

3. Design Area for Sprinkler = S × L.

Exception: This does not apply to small rooms. (See 4-4.19.)

The first sentence in (a) has been in the standard for several years and applies to cases where the sprinklers are installed

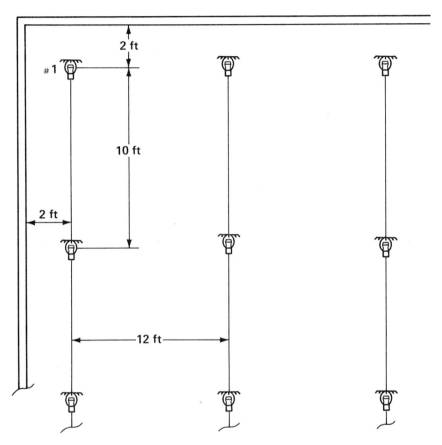

For SI Units: 1 ft = 0.3048 m.

Figure A-7-4.3.1.2(a)   Sprinkler Design Area.

A-7-4.3.1.2(a)   Example: Design Area for Sprinkler 1.
[See Figure A-7-4.3.1.2(a).]

1. Along Branch Lines—	2 ft, 10 ft	(0.6 m, 3.0 m)
	S = 10	(S = 3.0)
2. Between Branch Lines—	2 ft, 12 ft	(0.6 m, 3.7 m)
	L = 12	(L = 3.7)
3. Design Area—	10 × 12 =	(3.0 × 3.7) =
	120 sq ft	11.1 m²

For SI Units: 1 ft = 0.3048 m; 1 sq ft = 0.0929 m².

under a sloping roof. The area of coverage per sprinkler is based on the projected area on the floor. Even though the one line in Figure A-7-4.3.1.2(a) is only 2 ft (0.61 m) from the wall, the area of coverage is assumed to be 10 ft × 12 ft, or 120 sq ft (11.1 m²).

As referenced in the Exception, a small room is defined in 4-4.19. Also stated in the Exception, this method of calculation does not apply to small rooms. In small rooms the area of coverage per sprinkler is based on the area of the room divided by the number of sprinklers in the room.

(b)* When sprinklers are installed above and below a ceiling or in a case where more than two areas are supplied from a common set of branch lines, the branch lines and supplies shall be calculated to supply the largest water demand.

A-7-4.3.1.2(b) This subsection contemplates a ceiling constructed so as to reasonably assure that a fire on one side of the ceiling will operate sprinklers on one side only. When a ceiling is sufficiently open or of such construction that operation of sprinklers above and below the ceiling may be anticipated, the operation of such additional sprinklers should be considered in the calculations.

Formal Interpretation

Question: When hydraulically calculating sprinkler number one in Figure 1 below, should the area covered by this sprinkler be based on 12 ft, 0 in. × 10 ft, 0 in. spacing, or should it be calculated on the actual square foot coverage of 7 ft, 0 in. × 8 ft, 0 in.? This is a difference of 64 square feet.

Answer: 120 sq ft.

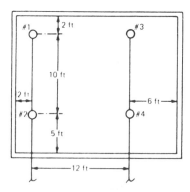

Figure 1.

(c) When sprinklers are installed above and below temporary obstructions such as overhead doors, the branch lines and the supply shall be calculated to supply the sprinklers both above and below the temporary obstruction.

> Since it is difficult to predict whether an overhead door will be up or down at the time of fire and, regardless of the door position, whether sprinklers above or below will operate, it was considered more prudent to include the sprinklers below the door in the calculated area with the roof sprinklers.

7-4.3.1.3* Each sprinkler in the design area and the remainder of the hydraulically designed system shall discharge at a flow rate at least equal to the stipulated minimum water application rate (density). Begin calculations at the hydraulically most remote sprinkler. Discharge at each sprinkler shall be based on the calculated pressure at that sprinkler.

> When applying velocity pressure, sometimes the discharge from the second sprinkler from the end of the line may be lower than that from the end sprinkler. If this is a significant amount (over 3 percent of design flow), then either the pipe sizing should be changed or the end sprinkler discharge increased slightly to compensate.
>
> The last sentence in the section is to prevent discharge averaging more than design area in a gridded system. For example, if it is assumed that twelve sprinklers in the design area discharge 20.8 gpm [(847 L/min)/m²] each, then the grid piping may be simply sized to accommodate this flow. In actuality, however, every sprinkler will be discharging at a slightly different pressure, so the rate would be 20.8 gpm plus the hydraulic increase caused by a higher available operating pressure. This increases the sophistication of the calculations.

A-7-4.3.1.3 When it is not obvious by comparison that the design area selected is the hydraulically most remote, additional calculations should be submitted. The most remote area, distance-wise, is not necessarily the hydraulically most remote area.

> All systems are not as uniform as the one in Figure A-7-4.3(a), so it may be necessary to calculate several portions of a system in order to properly size the pipe and determine the required water supply.

7-4.3.1.4 Calculate pipe friction loss in accordance with the Hazen-Williams formula with C values from Table 7-4.3.1.4.

(a) Include pipe, fittings, and devices such as valves, meters, and strainers, and calculate elevation changes that affect the sprinkler discharge.

Every component or elevation change that results in a change in pressure must be included in the calculations.

(b) Calculate the loss for a tee or a cross where flow direction change occurs based on the equivalent pipe length of the piping segment in which the fitting is included. The tee at the top of a riser nipple shall be included in the branch line; the tee at the base of a riser nipple shall be included in the riser nipple; and the tee or cross at a cross main–feed main junction shall be included in the cross main. Do not include fitting loss for straight through flow in a tee or cross.

Figure 7.1 helps in explaining this section.
Tee A is included with the 1½-in. starting piece.
Tee B is included with the 2-in. riser nipple.
Tees C and D are not included since the flow from the supply is straight through.
Tee F is included with the 3-in. piece of cross main.

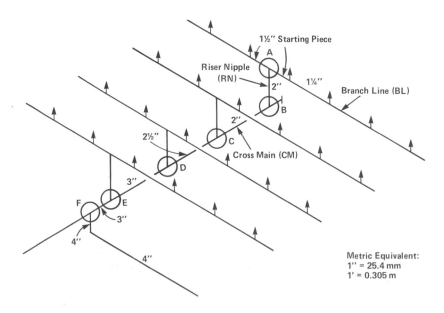

Figure 7.1.   Diagram of a typical sprinkler system used in making sample calculations.

Formal Interpretation

Question: Regarding 7-4.3.1.4(b) — in calculating the friction loss for a tee where the flow is turned 90°, is there any additional equivalent feet of pipe required for friction loss calculation because a bushing is used instead of a tee without a bushing? Referring to the sketch, would the equivalent feet of pipe be greater using a 1½ × 1½ × 2½ tee with 1½ × 1¼ bushing, as shown, than if a 1½ × 1¼ × 2½ tee were used?

Answer: The tee at the top of a riser nipple, where flow is turned 90°, should be considered part of the branch line segment. For example, if a 2½-in. riser supplies 1½-in. branch line piping on one side of the riser and 1¼-in. branch line piping on the other side, the loss through the tee should be represented by equivalent feet of 1½-in. pipe in the one direction and by 1¼-in. pipe in the other.

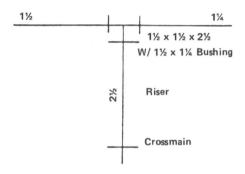

Figure 2.

The use of bushings is discouraged by the standard, as indicated by 3-8.3. However, when used in crosses or tees, the method should be no different from that described in 7-4.3.1.4(b). For a change in direction, the fitting friction loss equivalent length is sized according to the pipe supplied, not from any of the other outlet sizes.

(c) Calculate the loss of reducing elbows based on the equivalent feet value of the smallest outlet. Use the equivalent feet value for the *standard elbow* on any abrupt 90-degree turn, such as the screw-type pattern. Use the equivalent feet value for the *long-turn elbow* on any sweeping 90-degree turn, such as a flanged, welded, or mechanical joint-elbow type. (*See Table 7-4.2.*)

(d) Friction loss shall be excluded for the fitting directly connected to a sprinkler.

On a branch line with upright sprinklers screwed into the line tee, the fitting losses are excluded from the calculations because it is assumed that one fitting is included in the approval tests of a sprinkler.

In cases where sprinklers are installed on sprigs or drops, the reducing coupling at the sprinkler is ignored but the tee on the branch line must be included.

Table 7-4.3.1.4   Hazen-Williams C Values

Pipe or Tube	C Value*
Unlined Cast or Ductile Iron	100
Black Steel (Dry Systems including Preaction)	100
Black Steel (Wet Systems including Deluge)	120
Galvanized (all)	120
Plastic (listed)—All	150
Cement Lined Cast or Ductile Iron	140
Copper Tube or Stainless Steel	150

*The authority having jurisdiction may recommend other C values.

C values for steel and iron pipe materials tend to decrease with time. The values shown represent C values for new pipe.

Stainless steel pipe is often used in special sprinkler and water spray systems.

(e) Losses through a pressure-reducing valve shall be included based on the normal inlet pressure condition. Pressure loss data from the manufacturer's literature shall be used.

7-4.3.1.5*   Orifice plates or sprinklers of different orifice sizes shall not be used for balancing the system, except for special use such as exposure protection, small rooms or enclosures, or directional discharge. (*See 4-4.19 for definition of small rooms.*)

Orifice plates are prohibited because they may be removed while modifying a system and not be reinstalled, and this would nullify the hydraulic design. If installed in a horizontal run of pipe they can collect debris on the supply side and prevent adequate drainage on the system side.

Using sprinklers of different orifice sizes along the lines or on lines closer to the water supply is prohibited. It is feared that, either during original installation or on replacement of sprinklers, the selection of the wrong sprinkler orifice size could adversely affect the hydraulics of the system.

A-7-4.3.1.5 The use of sprinklers with differing orifice sizes in situations where different protection areas are needed is not considered balancing. An example would be a room that could be protected with sprinklers having differing orifice size in closet, foyer, and room areas. However, this procedure introduces difficulties when restoring a system to service after operation since it is not always clear which sprinklers go where.

7-4.3.1.6 Sprinkler discharge in closets, washrooms, and similar small compartments requiring only one sprinkler may be omitted from hydraulic calculations within the area of application. [Sprinklers in these small compartments shall, however, be capable of discharging minimum densities in accordance with Table 2-2.1.1(b).]

Exception: This shall not apply when areas of application are less than 1500 sq ft (140 m²).

This special rule is only applicable to the area/density method and cannot be utilized when using the room design method. In compartmented areas it is highly unlikely that all of the sprinklers in the design area will operate simultaneously. For example, if a hotel room is the largest room being considered, the sprinklers in the foyer, closet, and bathroom could operate from a fire in the bedroom. If these sprinklers have not been included in the calculations, they will rob water from the sprinkler(s) in the bedroom, possibly with adverse results.

The sprinkler discharge may be omitted from the calculations, but the design area does include the square footage of the small compartments.

It is not the intent of this section to eliminate from the calculations rooms that constitute the primary occupancy.

Small rooms such as bathrooms, closets, or foyers in hotels may be protected with small orifice sprinklers even though the sprinklers in the bedroom and hallway are ½-in. (13-mm) orifice or larger.

7-4.3.1.7 Where sprinklers are provided above and below permanent obstructions such as wide ducts, tables, etc., the water supply for one of the levels of sprinklers may be omitted from the hydraulic ceiling design calculations within the area of application. In any case, the most hydraulically demanding arrangement shall be calculated.

The presence of a permanent obstruction is treated in similar manner to a ceiling configuration that requires sprinklers above and below. It is of limited potential to design both sets of sprinklers to operate simultaneously in a fire condition.

7-4.3.1.8* Velocity pressure (P_v) may or may not be included in the calculations at the discretion of the designer. If velocity pressures are used, they shall be used on both branch lines and cross mains where applicable.

A-7-4.3.1.8 When velocity pressure is included in the calculations, the following assumptions are to be used:

(a) At any flowing outlet along a pipe, except the end outlet, only the normal pressure (P_n) can act on the outlet. At the end outlet the total pressure (P_t) can act. The following are to be considered end outlets:

1. The last flowing sprinkler on a dead-end branch line.

2. The last flowing branch line on a dead-end cross main.

3. Any sprinkler where a flow split occurs on a gridded branch line.

4. Any branch line where a flow split occurs on a loop system.

(b) At any flowing outlet along a pipe except the end outlet, the pressure acting to cause flow from the outlet is equal to the total pressure (P_t) minus the velocity presssure (P_v) on the upstream side.

(c) To find the normal pressure (P_n) at any flowing outlet except the end outlet, assume a flow from the outlet in question and determine the velocity pressure (P_v) for the total flow on the upstream side. Because normal pressure (P_n) equals total pressure (P_t) minus velocity pressure, the value of the normal pressure (P_n) so found should result in an outlet flow approximately equal to the assumed flow. If not, a new value should be assumed and the calculations repeated.

The total pressure will act on the remote outlet at the point where the flow enters the outlet from opposing directions.

Velocity pressure is small compared to the total pressure, and normally will not have a major effect on the end result of the hydraulic calculations.

7-4.3.2 Minimum operating pressure of any sprinkler shall be 7 psi (0.5 bar).

Exception: When higher minimum operating pressure for the desired application is specified in the listing of the sprinkler, it shall govern.

When sprinklers are submitted to the testing laboratories for examination and listing, they are subjected to water distribution and fire tests. These tests are conducted at a minimum flow of 15 gpm (57 L/min) per sprinkler. The pressure required to produce this flow through the ½-in. (13-mm) nominal orifice sprinkler is approximately 7 psi (0.5 bar). The Exception recognizes that specially listed sprinklers will have operating pressures beyond 7 psi, in some cases as high as 28 psi (1.93 bars).

$$P = \frac{(Q)^2}{(K)^2} = \frac{(15)^2}{(5.6)^2} = 7.17 \text{ psi}$$

7-4.4 Dwelling Units. When residential sprinklers are used, design shall comply with this section.

Residential sprinklers are tested with a different set of pass-fail criteria than other sprinklers. Therefore, conventional rules for densities, areas, and the like are not applicable to residential sprinklers.

Formal Interpretation

Question: When residential sprinklers are installed within dwelling units having combustible concealed spaces, shall the discharge and number of sprinklers be designed in accordance with 7-4.4?

Answer: Yes.

This indicates the intent that for dwelling unit designs with residential sprinklers the 3000 sq ft (232 m²) minimum rule [*see* 2-2.3.2(b)] does not apply.

7-4.4.1 Design Discharge. The system shall provide a discharge of not less than 18 gal/min (68 L/min) to any operating sprinkler and not less than 13 gal/min (49 L/min) per sprinkler to all operating sprinklers in the design area. Other discharge rates may be used in accordance with flow rates indicated in individual residential sprinkler listings.

These values of system discharge are based primarily on the Los Angeles and North Carolina fire tests. The flows of 18 gpm (68 L/min) and 13 gpm (49 L/min), respectively, were found to be successful in controlling the types of fires evaluated during these test programs.

7-4.4.2* Number of Design Sprinklers. The number of design sprinklers shall include all sprinklers within a compartment to a maximum of 4 sprinklers. When a compartment contains less than 4 sprinklers, the number of design sprinklers shall include all sprinklers in that compartment plus sprinklers in adjoining compartments to a total of 4 sprinklers. Adjoining corridors, if used in the calculations, shall be considered adjoin-

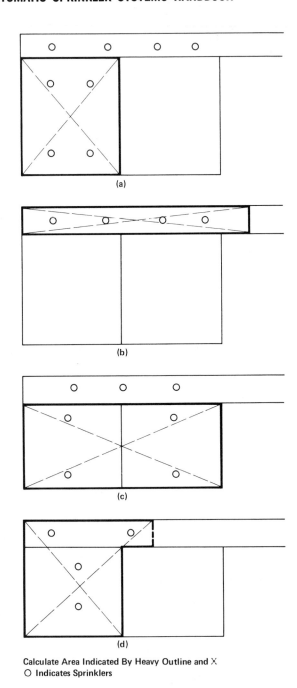

Calculate Area Indicated By Heavy Outline and X
O Indicates Sprinklers

Figure A-7-4.4.2   Examples of Design Area for Dwelling Units.

ing compartments solely for the purposes of these calculations. In all cases the design area shall include the 4 most hydraulically demanding sprinklers. (*See Figure A-7-4.4.2.*)

Note that in each case all sprinklers within a compartment are considered prior to going beyond the compartment.

NFPA 13D, *Standard for the Installation of Sprinkler Systems in One- and Two-Family Dwellings and Mobile Homes*, allows for a design to be based on two sprinklers operating. This is satisfactory for a one- or two-family dwelling, but the four sprinklers required for an NFPA 13 system allow for a sufficient factor of safety to allow for variables that might occur in a larger dwelling occupancy such as an apartment or hotel. It is noted that 3-11.2.9 refers to NFPA 13D for spacing, location, and position of residential sprinklers.

7-4.4.3 The definition of compartment for use in 7-4.4.2 to determine the number of design sprinklers is a space completely enclosed by walls and a ceiling. The compartment enclosure may have openings to an adjoining space if the openings have a minimum lintel depth of 8 in. (203 mm) from the ceiling.

The 8-in. (203-mm) minimum depth is to ensure that heat will be "banked," the result being that it will stay confined to the room of origin, enhancing operation over the fire.

7-4.4.4 Water Demand. The water demand for the dwelling unit shall be determined by multiplying the design discharge of 7-4.4.1 by the number of design sprinklers specified in 7-4.4.2.

7-4.4.5 Other Areas. When areas such as attics, basements, or other types of occupancies are outside of dwelling units but within the same structure, these areas shall be protected in accordance with all sections of this standard including appropriate water supply requirements of Table 2-2.1.1(b).

B-7 Sprinkler System Performance Criteria.

The following material is included in the standard in order to give guidance regarding the factors involved in using a conservative approach to the design of a sprinkler system.

B-7-1 Sprinkler system performance criteria have been based on test data. The factors of safety are generally small and are not definitive, and can depend on expected (but not guaranteed) inherent characteristics of the sprinkler systems involved. These inherent factors of safety consist of the following:

(a) The flow-declining pressure characteristic of sprinkler systems whereby the initial operating sprinklers discharge at a higher flow than with all sprinklers operating within the designated area.

(b) The flow-declining pressure characteristic of water supplies. This is particularly steep where fire pumps are the water source. This characteristic similarly produces higher than design discharge at the initially operating sprinklers.

The user of these standards may elect an additional factor of safety if the inherent factors are not considered adequate.

B-7-1.1 Performance specified sprinkler systems as opposed to scheduled systems can be designed to take advantage of multiple loops or gridded configurations. This results in minimum line losses at expanded sprinkler spacing, in contrast to the older *tree-type* configurations, where advantage cannot be taken of two-way type flows.

Where the water supply characteristics are relatively flat with pressures being only slightly above the required sprinkler pressure at the spacing selected, gridded systems with piping designed for minimal economic line losses can all but eliminate the inherent flow-declining pressure characteristic generally assumed to exist in sprinkler systems. In contrast, the economic design of a *tree-type* system would likely favor a system design with closer sprinkler spacing and greater line losses, demonstrating the inherent flow-declining pressure characteristic of the piping system.

Elements that enter into the design of sprinkler systems include:

1. Selection of density and area of application.

2. Geometry of the area of application (remote area).

3. Permitted pressure range at sprinklers.

4. Determination of the water supply available.

5. Ability to predict expected performance from calculated performance.

6. Future upgrading of system performance.

7. Size of sprinkler systems.

In developing sprinkler specifications, each of these elements needs to be considered individually. The most conservative design will be based on the application of the most stringent conditions for each of the elements.

B-7-1.2 Selection of Density and Area of Application. Specifications for density and area of application are developed from NFPA and other standards. It is desirable to specify densities rounded upward to the nearest 0.005 gpm/sq ft.

Prudent design should consider reasonable-to-expect variations in occupancy. This would include not only variations in types of occupancy, but, in the case of warehousing, the anticipated future range of materials to be stored, clearances, types of arrays, packaging, pile height, and pile stability, as well as other factors.

Design also considers some degree of adversity at the time of a fire. To take this into account, the density and/or area of application may be increased. Another way is to use a dual performance specification where, in addition to the normal primary specifications, a secondary density and area of application is specified. The objective of such a selection is to control the declining pressure-flow characteristic of the sprinkler system beyond the primary design flow.

A case can be made for designing feed and cross mains to lower velocities than branch lines to achieve the same result as specifying a second density and area of application.

B-7-1.3 Geometry of the Area of Application (Remote Area). It is expected that over any portion of the sprinkler system equivalent in size to the area of application, the system will achieve the minimum specified density for each sprinkler within that area.

Where a system is computer-designed, ideally the program should verify the entire system by shifting the area of application the equivalent of one sprinkler at a time so as to cover all portions of the system. Such a complete computer verification of performance of the system is most desirable, but unfortunately not all available computer verification programs currently do this.

The selection of the proper Hazen-Williams coefficient is important. New unlined steel pipe has a Hazen-Williams coefficient close to 140. However, it quickly deteriorates to 130, and after a few years of use, to 120. Hence, the basis for normal design is a Hazen-Williams coefficient of 120 for steel-piped wet systems. A Hazen-Williams coefficient of 100 is generally used for dry-pipe systems because of the increased tendency for deposits and corrosion in these systems. However, it should be realized that a new system will have fewer line losses than calculated, and the distribution pattern will be affected accordingly.

Conservatism can also be built into systems by intentionally designing to a lower Hazen-Williams coefficient than those indicated.

B-7-1.4 Ability to Predict Expected Performance from Calculated Performance. Ability to accurately predict the performance of a complex array of sprinklers on piping is basically a function of the pipe line velocity. The greater the velocity the greater is the impact on difficult-to-assess pressure losses. These pressure losses are presently determined by empirical means that lose validity as velocities increase. This is especially true for fittings with unequal and more than two flowing ports.

The inclusion of velocity pressures in hydraulic calculations improves the predictability of the actual sprinkler system performance. Calculations should come as close as practicable to predicting actual performance. Conservatism in design should be arrived at intentionally by known and deliberate means. It should not be left to chance.

B-7-1.5 Future Upgrading of System Performance. It may be desirable in some cases to build into the system the capability to achieve a higher level of sprinkler performance than needed at present. If this is to be a consideration in conservatism, consideration needs to be given to maintaining sprinkler operating pressures on the lower side of the optimum operating range, and/or by designing for low pipe line velocities, particularly on feed and cross mains, to facilitate future reinforcement.

REFERENCES CITED IN COMMENTARY

The following publications are available from the National Fire Protection Association, Batterymarch Park, Quincy, MA 02269.

NFPA 13D-1989, *Standard for the Installation of Sprinkler Systems in One- and Two-Family Dwellings and Mobile Homes*
NFPA 15-1985, *Standard for Water Spray Fixed Systems for Fire Protection*
NFPA 20-1987, *Standard for the Installation of Centrifugal Fire Pumps*
NFPA 22-1987, *Standard for Water Tanks for Private Fire Protection*
NFPA 30-1987, *Flammable and Combustible Liquids Code*
NFPA 231-1987, *Standard for General Storage*
NFPA 231C-1986, *Standard for Rack Storage of Materials*
NFPA 231D-1986, *Standard for Storage of Rubber Tires*
NFPA 409-1985, *Standard on Aircraft Hangars.*

8*

Pipe Schedule Systems

A-8 Long Runs of Pipe. When the construction or conditions introduce unusually long runs of pipe or many angles in risers or feed mains, an increase in pipe size over that called for in the schedules may be required to compensate for increased friction losses.

8-1* General. The pipe schedule sizing provisions shall not apply to hydraulically designed systems. Sprinkler systems having sprinklers with orifices other than ½ in. (13 mm) nominal, or piping material other than that covered in Table 3-1.1.1, shall be hydraulically designed. (*See Chapter 7.*)

Exception: Exposure protection systems designed in accordance with Chapter 6.

Sections 8-1, 8-2, 8-3, and 8-4 are only applicable to pipe schedule systems designed for use in conjunction with water supplies outlined in Table 2-2.1.1(a). Systems designed hydraulically in accordance with Chapter 7 permit the selection of pipe sizes appropriate to the characteristics of the water supply indicated by Table 2-2.1.1(b) and supersede the requirements of the pipe schedule tables. The pipe schedule sizing is based on ½-in. (13-mm) orifice sprinklers and piping material indicated in Table 3-1.1.1. It must not be assumed that these schedules will prove adequate when sprinklers of other than ½-in. orifice or piping materials other than those indicated in Table 3-1.1.1 are used.

A-8-1 The demonstrated effectiveness of pipe schedule systems is limited to their use with ½-in. (13-mm) orifice sprinklers. The use of other size orifices requires hydraulic calculations to prove their ability to deliver the required amount of water within the available water supply.

8-1.1 The number of automatic sprinklers on a given pipe size on one floor shall not exceed the number given in Sections 8-2, 8-3, or 8-4 for a given occupancy.

It should be noted that the numbers of sprinklers shown in the tables are maximum numbers for the minimum size pipes shown. The "graduated" type of system (larger pipe servicing increasingly smaller pipe) is not mandatory, as long as the minimum pipe sizes established in the pipe schedule tables are met.

Formal Interpretation

Question 1: Referring to the following figure (Figure 1), does Section 8-1.1 permit the piping supplying multiple floors to be sized for the maximum number of sprinklers on one floor or must it be sized for all sprinklers connected to it?

Answer: It may be sized to supply the maximum number of sprinklers on one floor. Referring to the figure, the pipe in question may be 1¼ in. for Light Hazard or Ordinary Hazard Occupancies.

Question 2: Is the sizing of piping to supply sprinklers on any one floor restricted to system risers?

Answer: No.

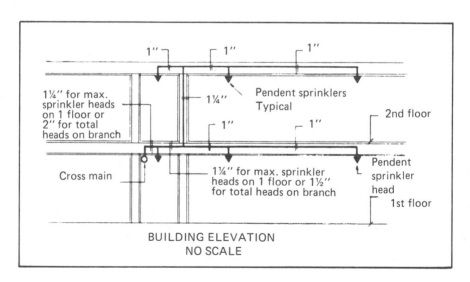

Figure 1.

8-1.2 Size of Risers. Each system riser shall be sized to supply all sprinklers on the riser on any one floor as determined by the standard schedules of pipe sizes in Sections 8-2, 8-3, or 8-4.

8-1.3 Slatted Floors, Large Floor Openings, Mezzanines, and Large Platforms.   Buildings having slatted floors, or large unprotected floor openings without approved stops, shall be treated as one area with reference to the pipe sizes, and the feed mains or risers shall be of the size required for the total number of sprinklers.

Fire will spread readily through grated or slatted floors. Therefore, an area with such divisions must be treated as a single fire area when sizing pipe.

8-2 Schedule for Light Hazard Occupancies.

8-2.1   Branch lines shall not exceed 8 sprinklers on either side of a cross main.

Exception: When more than 8 sprinklers on a branch line are necessary, lines may be increased to 9 sprinklers by making the two end lengths 1 in. (25.4 mm) and 1¼ in. (33 mm) respectively, and the sizes thereafter standard. Ten sprinklers may be placed on a branch line making the two end lengths 1 in. (25.4 mm) and 1¼ in. (33 mm), respectively, and feeding the tenth sprinkler by a 2½ in. (64 mm) pipe.

The amount of pressure lost to friction in long runs of small diameter pipe is significant. To ensure that pressure is not excessively reduced, the number of sprinklers on a branch line is limited to 8 sprinklers in Light Hazard Occupancies. The Exception allows up to 10 sprinklers on branch lines in Light Hazard Occupancies providing that pipe sizing is increased to reduce the friction loss.

8-2.2   Pipe sizes shall be in accordance with Table 8-2.2.

Exception: Each area requiring more sprinklers than the number specified for 3½ in. (89 mm) pipe in Table 8-2.2 and without subdividing partitions (not necessarily fire walls) shall be supplied by mains or risers sized for Ordinary Hazard Occupancies.

Table 8-2.2   Light Hazard Pipe Schedules

Steel		Copper	
1 in.	2 sprinklers	1 in.	2 sprinklers
1¼ in.	3 sprinklers	1¼ in.	3 sprinklers
1½ in.	5 sprinklers	1½ in.	5 sprinklers
2 in.	10 sprinklers	2 in.	12 sprinklers
2½ in.	30 sprinklers	2½ in.	40 sprinklers
3 in.	60 sprinklers	3 in.	65 sprinklers
3½ in.	100 sprinklers	3½ in.	115 sprinklers
4 in.	See 4-2.2.1	4 in.	See 4-2.2.1

A system in a Light Hazard Occupancy is limited to protection of 52,000 sq ft (4831 m²). Piping is sized in accordance with Table 8-2.2, unless there are more than 100 sprinklers in an area without subdividing partitions. Piping supplying such sprinklers must be sized in accordance with Section 8-3 to reduce the friction loss.

8-2.3 When sprinklers are installed above and below ceilings [*see Figures 8-2.3(a), (b), and (c)*] and such sprinklers are supplied from a common set of branch lines or separate branch lines from a common cross main, such branch lines shall not exceed 8 sprinklers above and 8 sprinklers below any ceiling on either side of the cross main. Pipe sizing up to and including 2½ in. (64 mm), shall be as shown in Table 8-2.3 utilizing the greatest number of sprinklers to be found on any two adjacent levels.

Formal Interpretation

Question: Is the ceiling, as referenced in Section 8-2.3, required to be a rated assembly (i.e., one-hour, two-hours, etc.)?

Answer: No.

A fire either above or below a ceiling in a Light Hazard Occupancy will normally be contained in that area by operating sprinklers. The standard permits, in Light Hazard pipe schedule systems, more sprinklers on a branch line in the above/below configuration than are permitted on a branch line protecting a single area, and also permits more total sprinklers to be supplied by a given pipe size. Paragraph 4-4.4 should be consulted when determining if sprinkler protection is needed above the ceiling.

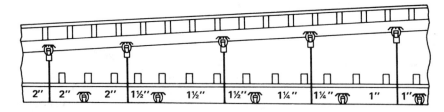

For SI Units: 1 in. = 25.4 mm.

Figure 8-2.3(a)   Arrangement of Branch Lines Supplying Sprinklers Above and Below a Ceiling.

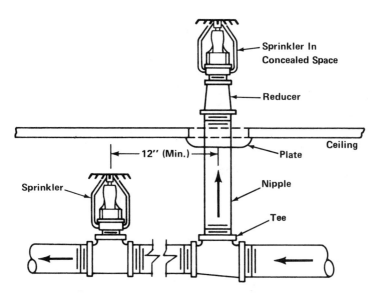

For SI Units: 1 in. = 25.4 mm.

Figure 8-2.3(b)   Sprinkler on Riser Nipple from Branch Line in Lower Fire Area.

Table 8-2.3   Number of Sprinklers Above and Below a Ceiling

Steel		Copper	
1 in.	2 sprinklers	1 in.	2 sprinklers
1¼ in.	4 sprinklers	1¼ in.	4 sprinklers
1½ in.	7 sprinklers	1½ in.	7 sprinklers
2 in.	15 sprinklers	2 in.	18 sprinklers
2½ in.	50 sprinklers	2½ in.	65 sprinklers

8-2.3.1* When the total number of sprinklers above and below a ceiling exceeds the number specified in Table 8-2.3 for 2½-in. (64-mm) pipe, the pipe supplying such sprinklers shall be increased to 3 in. (76 mm) and sized thereafter according to the schedule shown in Table 8-2.2 for the number of sprinklers above or below a ceiling, whichever is larger.

A-8-2.3.1   For example, a 2½-in. (64-mm) steel pipe, which is permitted to supply 30 sprinklers, may supply a total of 50 sprinklers where not more than 30 sprinklers are above or below a ceiling.

The probability of a fire involving the areas both above and below a ceiling in a Light Hazard Occupancy and opening in excess of 50 sprinklers is extremely remote. When sizing

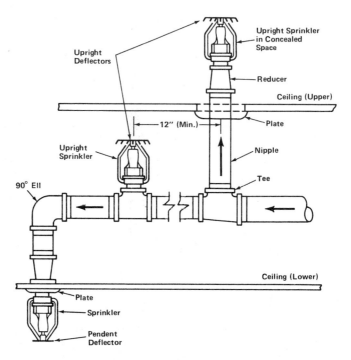

For SI Units: 1 in. = 25.4 mm.

Figure 8-2.3(c) **Arrangement of Branch Lines Supplying Sprinklers Above and Below Ceilings.**

piping for more than 50 sprinklers the pipe is increased to 3 in. (76 mm) to reduce friction loss.

8-3 Schedule for Ordinary Hazard Occupancies.

8-3.1 Branch lines shall not exceed 8 sprinklers on either side of a cross main.

Exception: When more than 8 sprinklers on a branch line are necessary, lines may be increased to 9 sprinklers by making the two end lengths 1 in. (25.4 mm) and 1¼ in. (33 mm), respectively, and the sizes thereafter standard. Ten sprinklers may be placed on a branch line making the two end lengths 1 in. (25.4 mm) and 1¼ in. (33 mm) respectively, and feeding the tenth sprinkler by a 2½ in. (64 mm) pipe.

The amount of pressure lost to friction in long runs of small diameter pipe is significant. To ensure that pressure is not excessively reduced, the number of sprinklers on a branch line is limited to 8 sprinklers in Ordinary Hazard Occupancies. The Exception allows up to 10 sprinklers on branch lines in Ordinary Hazard Occupancies providing that pipe sizing is increased to reduce friction loss.

8-3.2 Pipe sizes shall be in accordance with Table 8-3.2(a).

The pipe sizes specified in Table 8-3.2(a) were originally developed taking into consideration "normal" water supplies and friction loss. The pipe sizes tend to be somewhat smaller in branch lines and larger in cross mains, feed mains, and system risers than with systems that are designed hydraulically, as outlined in Chapter 7.

Table 8-3.2(a) Ordinary Hazard Pipe Schedule

Steel	Copper
1 in. . . . . 2 sprinklers	1 in. . . . . 2 sprinklers
1¼ in. . . . . 3 sprinklers	1¼ in. . . . . 3 sprinklers
1½ in. . . . . 5 sprinklers	1½ in. . . . . 5 sprinklers
2 in. . . . . 10 sprinklers	2 in. . . . . 12 sprinklers
2½ in. . . . . 20 sprinklers	2½ in. . . . . 25 sprinklers
3 in. . . . . 40 sprinklers	3 in. . . . . 45 sprinklers
3½ in. . . . . 65 sprinklers	3½ in. . . . . 75 sprinklers
4 in. . . . . .100 sprinklers	4 in. . . . . .115 sprinklers
5 in. . . . . .160 sprinklers	5 in. . . . . .180 sprinklers
6 in. . . . . .275 sprinklers	6 in. . . . . .300 sprinklers
8 in. . . . . .See 4-2.2.1 and 4-2.2.1 Exception	8 in. . . . . .See 4-2.2.1 and 4-2.2.1 Exception

Exception: When the distance between sprinklers on the branch line exceeds 12 ft (3.7 m), or the distance between the branch lines exceeds 12 ft (3.7 m), the number of sprinklers for a given pipe size shall be in accordance with Table 8-3.2(b).

This Exception recognizes the increased friction loss when either the distance between sprinklers on branch lines or the distance between branch lines exceeds 12 ft (3.7 m). It also recognizes the need for adjacent sprinklers to reinforce each

Table 8-3.2(b) Number of Sprinklers—Greater than 12 ft Separations

Steel	Copper
2½ in. . . . . .15 sprinklers	2½ in. . . . . .20 sprinklers
3 in. . . . . .30 sprinklers	3 in. . . . . .35 sprinklers
3½ in. . . . . .60 sprinklers	3½ in. . . . . .65 sprinklers
For other pipe and tube sizes, see Table 8-3.2(a).	

other toward the edges of the patterns in order to provide a more uniform overall density. Increasing the pipe sizing as required therefore reduces the pressure loss due to friction, making available greater flowing pressure with a resultant increased discharge and improved distribution.

8-3.3 When sprinklers are installed above and below ceilings and such sprinklers are supplied from a common set of branch lines or separate branch lines supplied by a common cross main, such branch lines shall not exceed 8 sprinklers above and 8 sprinklers below any ceiling on either side of the cross main. Pipe sizing up to and including 3 in. (76 mm) shall be as shown in Table 8-3.3 [*see Figures 8-2.3(a), (b), and (c)*] utilizing the greatest number of sprinklers to be found on any two adjacent levels.

Table 8-3.3   Number of Sprinklers Above and Below a Ceiling

Steel		Copper	
1 in.	2 sprinklers	1 in.	2 sprinklers
1¼ in.	4 sprinklers	1¼ in.	4 sprinklers
1½ in.	7 sprinklers	1½ in.	7 sprinklers
2 in.	15 sprinklers	2 in.	18 sprinklers
2½ in.	30 sprinklers	2½ in.	40 sprinklers
3 in.	60 sprinklers	3 in.	65 sprinklers

A fire either above or below a ceiling in an Ordinary Hazard Occupancy will normally be contained in that area by operating sprinklers. The standard permits, in Ordinary Hazard pipe schedule systems, more sprinklers on a branch line in the above/below configuration than are permitted on a branch line protecting a single area, and also permits more total sprinklers to be supplied by a given pipe size. Paragraph 4-4.4 should be consulted when determining if sprinkler protection is needed above the ceiling.

8-3.3.1* When the total number of sprinklers above and below a ceiling exceeds the number specified in Table 8-3.3 for 3-in. (76-mm) pipe, the pipe supplying such sprinklers shall be increased to 3½ in. (89 mm) and sized thereafter according to the schedule shown in Table 8-2.2 or Table 8-3.2(a) for the number of sprinklers above or below a ceiling, whichever is larger.

Exception: When the distance between the sprinklers protecting the occupied area exceeds 12 ft (3.7 m) or the distance between the branch lines exceeds 12 ft (3.7 m), the branch lines shall be sized in accordance with either Table 8-3.2(b), taking into consideration the sprinklers protecting the occupied area only, or paragraph 8-3.3, whichever requires the greater size of pipe.

A-8-3.3.1   For example, a 3-in. (76-mm) steel pipe, which is permitted to supply 40 sprinklers in an Ordinary Hazard area, may supply a total of 60 sprinklers when not more than 40 sprinklers protect the occupied area.

The probability of a fire involving the area both above and below a ceiling in an Ordinary Hazard Occupancy and opening in excess of 60 sprinklers is extremely remote. When sizing piping or tubing for systems with more than 60 sprinklers with steel pipe, and 65 sprinklers with copper tubing, the pipe or tubing size is increased to reduce friction loss.

8-4 Schedule for Extra Hazard Occupancies.   Branch lines shall not exceed 6 sprinklers on either side of a cross main. The number of sprinklers for a given pipe size shall be in accordance with Table 8-4.

Table 8-4   Extra Hazard Pipe Schedule

Steel		Copper	
1 in.	1 sprinkler	1 in.	1 sprinkler
1¼ in.	2 sprinklers	1¼ in.	2 sprinklers
1½ in.	5 sprinklers	1½ in.	5 sprinklers
2 in.	8 sprinklers	2 in.	8 sprinklers
2½ in.	15 sprinklers	2½ in.	20 sprinklers
3 in.	27 sprinklers	3 in.	30 sprinklers
3½ in.	40 sprinklers	3½ in.	45 sprinklers
4 in.	55 sprinklers	4 in.	65 sprinklers
5 in.	90 sprinklers	5 in.	100 sprinklers
6 in.	150 sprinklers	6 in.	170 sprinklers
8 in.	See 4-2.2.1	8 in.	See 4-2.2.1

The number of sprinklers on a branch line is limited to 6 sprinklers in Extra Hazard Occupancies. Fewer sprinklers are allowed on each size pipe than are allowed for Ordinary Hazard Occupancies. Of the three hazard categories, it is expected that this group will exhibit the highest flow requirements, thereby making control of the friction loss characteristics imperative.

8-5 Deluge Systems.   Open sprinkler and deluge systems shall be hydraulically calculated according to applicable standards.

Exception: Open sprinklers for exposure protection installed in conformance with Chapter 6.

Pipe schedule systems are designed with a margin of safety which presumes that only some of the sprinklers will operate in a fire. In open and deluge systems, water discharges from all sprinklers including those outside the fire area. Therefore, the system must be specifically designed to provide the density of coverage required to extinguish or control the fire over the entire protected area.

9

Large-Drop and Early Suppression Fast Response (ESFR) Sprinklers

9-1 General.

9-1.1 Applications. This chapter provides requirements for the installation of large-drop and early suppression fast response (ESFR) sprinklers. Listed sprinklers other than large-drop and ESFR sprinklers are not covered by this chapter.

Chapter 9 deals with two very specific types of sprinklers which are commonly used in buildings that contain storage in concentrated arrangements. NFPA 231, *General Storage*, NFPA 231C, *Rack Storage*, and A-9-2.4 and A-9-3.4.1 of NFPA 13 discuss the appropriateness of large-drop and ESFR sprinklers in warehouse facilities.

The development, use, and design parameters of each of these sprinkler types is somewhat unique and, therefore, it is of paramount importance to closely follow the requirements for the design of each type.

9-2 Large-Drop Sprinklers.

Large-drop sprinklers were developed and engineered with careful attention directed at their mechanical and hydraulic distribution properties under no-fire conditions. Testing in actual fire conditions established analysis of the effectiveness of droplet size and its impact to penetrate a given thermal plume, as well as the ability of this type of sprinkler to avoid skipping conditions (the phenomenon in which sprinklers beyond the fire area operate while some sprinklers within or adjacent to the fire area do not).

Large-drop sprinklers typically have a K factor of 11.0 to 11.5. Listing of these sprinklers also takes into consideration the ability of the sprinkler discharge pattern to penetrate a high-velocity fire plume. Listed extra large orifice sprinklers will also be in the K factor range of 11.0 to 11.5, but they are not evaluated as to their ability to penetrate the high-velocity fire plume. Because of this, only listed large-drop sprinklers are to be used in accordance with the design parameters of Section 9-2.

9-2.1* Definition. *Large-Drop Sprinkler.* A listed large-drop sprinkler is characterized by a K factor between 11.0 and 11.5, and proven ability to meet prescribed penetration, cooling, and distribution criteria prescribed in the large-drop sprinkler examination requirements. The deflector/discharge characteristics of the large-drop sprinkler generate large drops of such size and velocity as to enable effective penetration of the high-velocity fire plume.

The terms large-drop, large orifice, and extra large orifice connote three different types of sprinkers and must never be thought of as being interchangeable. Systems utilizing large orifice and extra large orifice sprinklers may be designed in accordance with Chapters 1 through 7 of this standard.

A-9-2.1 Large-drop sprinkler development began with the idea that a sprinkler that would produce a high proportion of large drops would be more effective against high-challenge fires than standard nominal ½-in. and $^{17}\!/_{32}$-in. sprinklers. Large drops are better able to fight their way through the strong updraft generated by a severe fire. The general philosophy behind this idea was to create an offensive weapon for direct attack against the burning fuel, but with as little diminution of defensive capability as possible. As originally conceived, it was supposed that the large drops were the predominant reason for increased effectiveness. It later became evident, however, that other factors such as distribution and pressure play an equal (and possibly more important) role. A consideration of all characteristics indicated that the desired sprinkler would provide high penetration, adequate cooling, reasonable distribution, low skipping potential, and increased discharge capacity.

9-2.2* Applicability.

A-9-2.2 Large-drop sprinklers were developed for use against severe fire challenges requiring sprinkler discharges of 55 gal/min (208 L/min) per sprinkler or more. The sprinklers were designed to deliver a heavy discharge through the strong updrafts generated by high-challenge fires. They depend primarily on direct attack of the burning fuel to gain rapid control of the fire, and to a lesser extent on prewetting and cooling.

Large-drop sprinklers will provide excellent protection against high-challenge fires, but careful design and close adherence to the rules given in Chapter 9 are required. This is especially important because experimental findings indicate that the characteristics of large-drop sprinklers are not the same as those attributed to standard sprinklers.

9-2.2.1*   Large-drop sprinklers are suitable for use with the hazards listed in Table A-9-2.4 and may be used in other specific hazard classifications and configurations only when proven by large-scale fire testing.

A-9-2.2.1   Fire tests have not been conducted with large-drop sprinklers over Ordinary and Extra Hazard Occupancies. Therefore, the protection requirements remain unknown.

At a given pressure, large-drop sprinklers will discharge approximately 100 percent and 40 percent more water than standard (½-in.) and large orifice (¹⁷⁄₃₂-in.) sprinklers, respectively.

Large-drop sprinklers were designed to cope with high-challenge fires where high levels of penetration and cooling are required. Since both characteristics fall off sharply at low pressures, large-drop sprinklers cannot be used effectively to compensate for weak water supplies by taking advantage of the increased discharge capacity.

9-2.2.2   Requirements of this standard shall apply, except those portions dealing with subjects specifically addressed in this chapter.

9-2.3 Installation.

Large-drop sprinklers are effective against high-challenge fires primarily because of the quality of the sprinkler discharge and the high degree of water penetration through the fire plume. Both of these characteristics are dependent upon maintaining a minimum sprinkler discharge pressure at all times. Large-drop sprinkler protection designs specified in this chapter are determined on the basis of a minimum pressure and a minimum number of sprinklers. Traditional area/density design concepts do not apply to large-drop sprinklers.

Large-drop sprinkler development tests investigated the area/density concept. Tests conducted with varying discharge pressures and sprinkler spacings, but with an identical sprinkler discharge density, showed that when pressure was decreased below the minimum required in this chapter, both the number of sprinklers that opened and the measured roof steel temperatures were unacceptable, and both were materially higher than in the tests with higher sprinkler discharge pressures.

9-2.3.1 Operating Pressure.

9-2.3.1.1   Large-drop sprinkler systems shall be designed such that the minimum operating pressure is not less than 25 psi (1.7 bar), unless large-scale testing proves a lesser pressure to be adequate for a particular hazard.

9-2.3.1.2 For design purposes, 95 psi (6.5 bar) shall be the maximum discharge pressure used at the starting point of the hydraulic calculations.

9-2.3.2 Type of System.

9-2.3.2.1 Large-drop sprinkler systems shall be limited to wet-pipe or preaction systems.

The fires that have been used in testing large-drop sprinklers grow in an exponential manner (power law fires or T² fires); therefore, a substantial escalation in heat release rates and thermal velocities would be expected in relatively short time periods. The delay that is inherent in a dry-pipe system would be so detrimental that most fires would have grown to such proportions that, for all practical purposes, the design abilities of the system (quantity of water discharge and optimum droplet size to penetrate the upward thermal plumes) would be overrun to the point of being ineffective. Excessive numbers of operating sprinklers would deplete the water supply, and fire growth would continue unabated to a point of severe damage to contents or to the point of introducing thermal damage to structural steel members.

9-2.3.2.2 Galvanized steel or copper pipe and fittings shall be used in preaction systems to avoid scale accumulation.

9-2.3.3 System Design.

9-2.3.3.1 Pipe shall be sized by hydraulic calculation.

9-2.3.3.2* The nominal diameter of branch line pipes (including riser nipples) shall be not less than 1¼ in. (33 mm) or greater than 2 in. (51 mm), except starter pieces, which may be 2½ in. (64 mm).

Exception: When branch lines are larger than 2 in. (51 mm), the sprinkler shall be supplied by a riser nilpple to elevate the sprinkler 13 in. (330 mm) for 2½-in. (64-mm) pipe and 15 in. (380 mm) for 3-in. (76-mm) pipe. These dimensions are measured from the center line of the pipe to the deflector. In lieu of this, sprinklers may be offset horizontally a minimum of 12 in. (305 mm).

The intent of limiting the maximum size of branch lines or providing riser nipples is to minimize the effect of pipe shadow below the sprinkler, which can result in significant areas with materially reduced water application and loss of sprinkler water penetration.

A-9-2.3.3.2 This test data was developed with 2-in. (51-mm) diameter pipe maximum. Riser nipples are required when branch lines exceed 2 in. (51 mm) in diameter.

9-2.3.4 Temperature Rating. Sprinkler temperature ratings shall be the same as those used in large-scale fire testing to determine the protection requirements for the hazard involved.

Exception: Sprinklers of intermediate and high temperature ratings shall be installed in specific locations as required by 3-11.6.3.

9-2.3.5* Spacing.

A-9-2.3.5 It is important that sprinklers in the immediate vicinity of the fire center not skip, and this requirement imposes certain restrictions on the spacing.

Spacing requirements were determined by testing at various sprinkler spacings and discharge pressures, on the basis of the following parameters:

(a) Maintain a minimum distance between sprinklers to ensure sufficient overlapping of sprinkler discharge patterns for adequate penetration of the fire plume.

(b) Maintain sufficient separation of sprinklers to prevent sprinkler skipping.

(c) Maintain sufficient cooling at ceiling level.

The importance of satisfying these three criteria cannot be overstated. They are so closely intertwined that it is difficult to evaluate one without considering the effect on the others, and vice versa. Certain combinations of spacing dimensions, when used with the discharge pressures and fire hazards contemplated here, will result in skipping, loss of sprinkler water penetration, and excessively high ceiling temperatures. If any of these adverse conditions occur, the adequacy of the sprinkler protection may be seriously compromised.

9-2.3.5.1* The area of coverage shall be limited to a minimum of 80 sq ft (7.4 m²) and a maximum of 130 sq ft (12.18 m²).

A-9-2.3.5.1 Tests involving areas of coverage over 100 sq ft (9.3 m²) are limited in number, and the use of areas of coverage over 100 sq ft (9.3 m²) should be carefully considered.

9-2.3.5.2 The distance between branch lines and between sprinklers on the branch lines shall be limited to not more than 12 ft (3.7 m) nor less than 8 ft (2.4 m).

Exception: Under open wood joist construction, the maximum distance shall be limited to 10 ft (3.0 m).

9-2.3.6 Clear Space Below Sprinklers. At least 36 in. (914 mm) shall be maintained between sprinkler deflectors and the top of storage.

At clearances of less than 36 in. (914 mm), large-drop sprinklers installed within the allowable spacing guidelines will not provide sufficient overlapping of sprinkler discharge patterns between adjacent sprinklers.

9-2.3.7* Distance Below Ceiling. Sprinklers shall be positioned so that the tops of deflectors are in conformance with Table 9-2.3.7.

Table 9-2.3.7   Minimum and Maximum Distance of Deflectors Below Ceiling for Various Construction Types

Construction Type[1]	Minimum Distance, In. (mm)	Maximum Distance, In. (mm)
Smooth ceiling and bar joist	6 (152)	8 (203)
Beam and girder	6 (152)	12 (305)
Panel up to 300 sq ft (27.9 m²)	6 (152)	14 (358)
Open wood joist	1 (25) below bottom of joists	6 (152) below bottom of joists

[1]See Chapter 4 for definitions of construction types.

A-9-2.3.7 If all other factors are held constant, the operating time of the first sprinkler will vary exponentially with the distance between the ceiling and deflector. At distances greater than 7 in. (178 mm), for other than open wood joist construction, the delayed operating time will permit the fire to gain headway, with the result that substantially more sprinklers operate. At distances less than 7 in. (178 mm), other effects come into play. Changes in distribution, penetration, and cooling nullify the advantage gained by faster operation. The net result is again increased fire damage accompanied by an increase in the number of sprinklers operated. The optimum clearance between deflectors and ceiling is, therefore, 7 in. (178 mm). For open wood joist construction the optimum clearance between deflectors and the bottom of joists is 3½ in. (89 mm).

9-2.3.8 Location of Sprinklers in Beam and Girder and Panel Construction.

9-2.3.8.1 Under beam and girder construction and under panel construction, the branch lines may run across the beams, but sprinklers shall be located in the bays and not under the beams.

Unlike standard spray upright and standard spray pendent sprinklers, large-drop sprinklers cannot be located under beams. The ceiling effect that a beam surface provides for other sprinkler types is not effective for large-drop sprinklers. The net effect is that the initial delay permits a larger fire, causing more sprinklers to operate. A discussion on the operating time of large-drop sprinklers is included in A-9-2.3.7.

Since the sprinklers must be located within the bays, the beam table and associated rules of Chapter 4 must still be adhered to. *(See 9-2.3.8.2.)*

9-2.3.8.2 The maximum distance of deflector above the bottom of beams shall be limited to the values specified in Chapter 4.

9-2.3.9* Obstructions in Piping.

Any debris that might find its way into the system can partially obstruct and defeat the specially designed orifice/deflector combination.

A-9-2.3.9 The plugging of a single sprinkler in the vicinity of the fire origin can cause a significant increase in the number of operating sprinklers as well as an increase in fire damage. Therefore, it is essential that the sprinkler piping be kept free of obstructing material.

Fouling or plugging of piping or heads may be prevented by several methods: sound installation practices, including the provision of screens or strainers when necessary; avoidance of dry-pipe systems with their history of plugging; and a high level of maintenance.

9-2.3.9.1 Screens located in the inlet piping directly connected to rivers, lakes, ponds, reservoirs, uncovered tanks, and similar sources (*see NFPA 24, Standard for the Installation of Private Fire Service Mains and Their Appurtenances*) shall be cleaned and serviced at least annually and immediately after any work has been performed on nearby underground mains.

9-2.3.9.2* Visual and/or flushing investigations shall be conducted of all systems for foreign material, at intervals not exceeding five years.

A-9-2.3.9.2 Investigations should be conducted more frequently when justified by local conditions.

9-2.3.10* Obstructions to Distribution.

A-9-2.3.10 To a great extent, large-drop sprinklers rely on direct attack to gain rapid control of both the burning fuel and ceiling temperatures. Therefore, interference with the discharge pattern and obstructions to the distribution should be avoided.

9-2.3.10.1 Obstructions Located at the Ceiling. When sprinkler deflectors are located above the bottom of beams, girders, ducts, fluorescent lighting fixtures, or other obstructions located at the ceiling, the sprinklers shall be positioned so that the maximum distance from the bottom of the obstruction to the deflectors does not exceed the value specified in Chapter 4.

9-2.3.10.2 Obstructions Located Below the Sprinklers.

9-2.3.10.2.1 Sprinklers shall be positioned with respect to fluorescent lighting fixtures, ducts, and obstructions more than 24 in. (610 mm) wide and located entirely below the sprinklers so that the minimum horizontal distance from the near side of the obstruction to the center of the sprinkler is not less than the value specified in Table 9-2.3.10.2.1. (*See Figure 9-2.3.10.2.1.*)

This section applies to obstructions no more than 24 in. (610 mm) below the sprinklers. The intent is to minimize obstruction to the lateral distribution from individual sprinklers.

Table 9-2.3.10.2.1   Position of Sprinklers in Relation to Obstructions Located Entirely Below the Sprinklers

Distance of Deflector Above Bottom of Obstruction	Minimum Distance to Side of Obstruction, ft (m)
Less than 6 in. (152 mm)	1½ (0.5)
6 in. (152 mm) to less than 12 in. (305 mm)	3 (0.9)
12 in. (305 mm) to less than 18 in. (457 mm)	4 (1.2)
18 in. (457 mm) to less than 24 in. (610 mm)	5 (1.5)

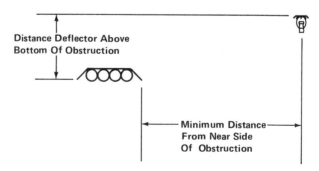

Figure 9-2.3.10.2.1   Position of Sprinklers in Relation to Obstructions Located Entirely below the Sprinklers. (To be used with Table 9-2.3.10.2.1)

9-2.3.10.2.2 When the bottom of the obstruction is located 24 in. (610 mm) or more below the sprinkler deflectors:

(a) Sprinklers shall be positioned so that the obstruction is centered between adjacent sprinklers. (*See Figure 9-2.3.10.2.2.*)

(b) The obstruction shall be limited to a maximum width of 24 in. (610 mm). (*See Figure 9-2.3.10.2.2.*)

Exception: When obstruction is greater than 24 in. (610 mm) wide, one or more lines of sprinklers shall be installed below the obstruction.

(c) The obstruction shall not extend more than 12 in. (305 mm) to either side of the midpoint between sprinklers. (*See Figure 9-2.3.10.2.2.*)

Exception: When extensions exceed 12 in. (305 mm), one or more lines of sprinklers shall be installed below the obstruction.

(d) At least 18 in. (457 mm) clearance shall be maintained between the top of storage and the bottom of the obstruction. (*See Figure 9-2.3.10.2.2.*)

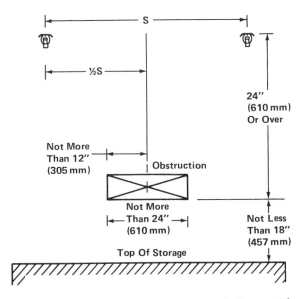

Figure 9-2.3.10.2.2   Position of Sprinklers in Relation to Obstructions Located 24 in. (610 mm) or more below Deflectors.

At distances more than 24 in. (610 mm) below the sprinklers, obstructions will interfere with the distribution from more than one sprinkler, due to characteristics of the sprinkler's discharge pattern. The provisions here will ensure that sufficient overlap of discharge patterns from adjacent sprinklers is achieved, and thus ensure that adequate water distribution is available to attain fire plume penetration.

9-2.3.10.3 Obstructions Parallel to and Directly Below Branch Lines.   In the special case of an obstruction running parallel to and directly below a branch line:

(a) The sprinkler shall be located at least 36 in. (914 mm) above the top of the obstruction. (*See Figure 9-2.3.10.3.*)

(b) The obstruction shall be limited to a maximum width of 12 in. (305 mm). (*See Figure 9-2.3.10.3.*)

(c) The obstruction shall be limited to a maximum extension of 6 in. (152 mm) to either side of the center line of the branch line. (*See Figure 9-2.3.10.3.*)

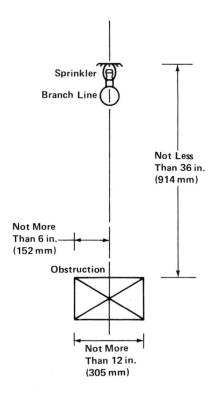

Figure 9-2.3.10.3   Position of Sprinklers in Relation to Obstructions Running Parallel to and Directly below Branch Lines.

The obstruction to distribution as expected under the conditions specified in 9-2.3.10.3 would not be significantly greater than that which would occur due to pipe shadow.

9-2.4* Protection Requirements.

A-9-2.4   Large scale fire testing has shown that large-drop sprinkler systems designed per the rules of Chapter 9 to provide the minimum operating pressures for the number of design sprinklers for the specific hazards of Table A-9-2.4 will adequately control the fire.

Protection requirements for the various hazards listed in Table A-9-2.4 *(see pages 386 and 387)* were, for the most part, developed from results of large-scale tests. In some cases, where large-scale tests were conducted on only one specific commodity for a particular storage arrangement, judgments were applied in determining protection requirements for lesser hazard commodities stored in the same manner.

9-2.4.1*   Protection shall be provided as specified in Table A-9-2.4 or appropriate NFPA standards in terms of minimum operating pressure and the number of sprinklers to be included in the design area.

A-9-2.4.1   In testing where discharge density and all other conditions (except pressure and spacing) remained constant, sprinkler demand area, steel temperatures, and fire damage all exhibited a marked increase as the spacing was decreased. The importance of these findings cannot be overemphasized. They state clearly that the term density has no meaning when applied to large-drop sprinklers and it (density) cannot be used as a design parameter.

In testing the effect of discharge pressure, series of tests were run with two different fuel types. Although the relationship of decreased number of operated sprinklers with increased pressure held, the associated curves did not parallel each other. Therefore, it is not possible to extrapolate the protection requirements from one hazard to a higher hazard, nor is it possible to develop a curve from one test point. However, the protection requirements developed for a particular hazard (through large-scale fire tests) may be applied safely to hazards known to be less severe, provided that the less severe hazards are arranged in the same configuration.

9-3 Early Suppression Fast Response (ESFR) Sprinklers.

9-3.1*   **Definition.**   *Early Suppression Fast Response (ESFR) Sprinkler.* A listed ESFR sprinkler is a thermosensitive device designed to react at a predetermined temperature by automatically releasing a stream of water and distributing it in a specified pattern and quantity over a designated area so as

Table A-9-2.4 Pressure and Number of Design Sprinklers for Various Hazards (Note 6) for Large-Drop Sprinklers

Minimum Operating Pressure (Note 1), psi (bar)	25 (1.7)	50 (3.4)	75 (5.2)
Hazard (Note 2)	Number Design Sprinklers		
Palletized Storage			
Class I, II, and III commodities up to 25 ft (7.6 m) with maximum 10 ft (3.0 m) clearance to ceiling	15	Note 3	Note 3
Class IV commodities up to 20 ft (6.1 m) with maximum 10 ft (3.0 m) clearance to ceiling	20	15	Note 3
Unexpanded plastics up to 20 ft (6.1 m) with maximum 10 ft (3.0 m) clearance to ceiling	25	15	Note 3
Idle wood pallets up to 20 ft (6.1 m) with maximum 10 ft (3.0 m) clearance to ceiling	15	Note 3	Note 3
Solid-Piled Storage			
Class I, II, and III commodities up to 20 ft (6.1 m) with maximum 10 ft (3.0 m) clearance to ceiling	15	Note 3	Note 3
Class IV commodities and unexpanded plastics up to 20 ft (6.1 m) with maximum 10 ft (3.0 m) clearance to ceiling	Does Not Apply	15	Note 3
Double-Row Rack Storage with Minimum 5½ ft (1.7 m) Aisle Width (Note 4)			
Class I and II commodities up to 25 ft (7.6 m) with maximum 5 ft (1.5 m) clearance to ceiling	20	Note 3	Note 3
Class I, II, and III commodities up to 20 ft (6.1 m) with maximum 10 ft (3.0 m) clearance to ceiling	15	Note 3	Note 3
Class IV commodities up to 20 ft (6.1 m) with maximum 10 ft (3.0 m) clearance to ceiling	Does Not Apply	20	15
Unexpanded plastics up to 20 ft (6.1 m) with maximum 10 ft (3.0 m) clearance to ceiling	Does Not Apply	30	20

Commodity			
Unexpanded plastics up to 20 ft (6.1 m) with maximum 10 ft (3.0 m) clearance to ceiling (Note 7)	Does Not Apply	20	Note 3
Class IV commodities and unexpanded plastics up to 20 ft (6.1 m) with maximum 5 ft (1.5 m) clearance to ceiling	Does Not Apply	15	Note 3
On-End Storage of Roll Paper (Note 5)			
Heavyweight paper, in closed array, banded in open array, or banded or unbanded in a standard array, up to 26 ft (7.9 m) with a maximum 34 ft (10.4 m) clearance to ceiling	Does Not Apply	15	Note 3
Any grade of paper, EXCEPT LIGHTWEIGHT paper, with stacks in closed array, or banded or unbanded standard array up to 20 ft (6.1 m) with maximum 10 ft (3.0 m) clearance to ceiling	Does Not Apply	15	Note 3
Medium weight paper completely wrapped (sides and ends) in one or more layers of heavyweight paper, or lightweight paper in two or more layers of heavyweight paper with stacks in closed array, banded in open array, or banded or unbanded in a standard array, up to 26 ft (7.9 m) with maximum 34 ft (10.4 m) clearance to ceiling	Does Not Apply	15	Note 3
Record Storage			
Paper records and/or computer tapes in multitier steel shelving up to 5 ft (1.5 m) in width and with aisles 30 in. (76 cm) or wider, without catwalks in the aisles, up to 15 ft (4.6 m) with maximum 5 ft (1.5 m) clearance to ceiling	15	Note 3	Note 3
Same as above, but with catwalks of expanded metal or metal grid with minimum 50 percent open area, in the aisles	Does Not Apply	15	Note 3

Notes:
1. Open Wood Joist Construction. Testing with open wood joist construction showed that each joist channel should be fully firestopped to its full depth at intervals not exceeding 20 ft (6.1 m). In unfirestopped open wood joist construction, or if firestops are installed at intervals exceeding 20 ft (6.1 m), the minimum operating pressures should be increased by 40 percent.
2. Building steel required no special protection for the occupancies listed.
3. The higher pressure will successfully control the fire, but the required number of design sprinklers should not be reduced from that required for the lower pressure.
4. In addition to the transverse flue spaces required by NFPA 231C, minimum 6 in. (152 mm) longitudinal flue spaces were maintained.
5. See NFPA 231F for definitions.
6. Unless otherwise specified the sprinklers used in the tests were high temperature rating.
7. Based on tests using sprinklers of ordinary temperature rating.

to provide early suppression of a fire when installed on the appropriate sprinkler piping.

ESFR sprinklers evolved by combining the concepts of fast response and sufficient fire plume penetration by sprinkler discharge to achieve early fire suppression. The theory is that sufficient sprinkler water discharge in the early phases of a fire can suppress the fire before a severe fire plume is generated. A severe fire plume adversely affects sprinkler protection in two ways: 1) the strong updraft created by the fire plume diminishes the amount of water actually reaching the burning surfaces of the combustibles by either vaporizing the water spray or carrying it away, making fire suppression difficult to achieve; and 2) the high heat release associated with a severe fire plume causes sprinklers distant from the actual fire area to open, resulting in an increased water demand and excessive water damage.

Standard sprinklers, because they operate more slowly and provide less effective sprinkler discharge than ESFR sprinklers, operate on the concept of fire control. With standard sprinklers, a severe fire plume will develop during the course of the fire and persist for some time until fire control is achieved. The concept of fire control is to provide sufficient sprinkler discharge to retard fire spread through the operation of approximately 20 to 30 sprinklers. According to this concept, sprinklers are expected to prevent fire from spreading by slowly reducing its intensity and by sufficiently prewetting the surrounding combustibles so that they do not ignite. At the same time, sprinkler water is expected to absorb heat and to cool the burning structure, including steelwork, thereby preventing structural failure and collapse.

ESFR sprinklers, on the other hand, operate earlier in the fire than standard sprinklers, and provide discharge adequate to suppress the fire before a severe fire plume develops. In principle, early suppression is determined by three factors, which can be independently measured: sprinkler response time (RTI), actual delivered density (ADD), and required delivered density (RDD). An ESFR sprinkler must possess specific properties related to these three factors. Systems that have not been designed to address all three of these factors cannot be relied upon to achieve early fire suppression. A discussion of each of these three factors follows.

Response Time Index. Response time index (RTI) is a measurable expression of the sensitivity, or responsiveness to temperature change, of the thermal sensing element of a sprinkler. The more responsive the element is to temperature change, the lower the RTI value. The RTI of a sprinkler model generally varies little with the sprinkler temperature rating. See the discussion on RTI contained on page 175 of this handbook.

Sprinkler response time for a given fire situation is determined by the sprinkler RTI value, the sprinkler temperature rating, and the location of the sprinkler relative to the fire. However, because of thermal lag of the thermal sensing element's mass, the gas temperature near the sprinkler may substantially exceed the rated temperature before the sprinkler operates. This may negate the ability of a sprinkler with a high RTI to achieve early fire suppression, because water application from the first sprinkler that operates will occur too late, while the heat release rate and fire plume velocity are increasing rapidly.

Small fires are generally easier to suppress than large ones. It follows, then, that the sooner the sprinkler system effectively discharges water, the less water will be needed to suppress the fire. For a given temperature rating, sprinkers with a lower RTI will operate sooner in a rapidly growing fire, providing sprinkler water discharge onto a smaller fire than would be the case with sprinklers having a higher RTI.

Required Delivered Density. Required delivered density (RDD) is the measure of a particular hazard's suppressibility. RDD can be defined as the delivered density required to achieve fire suppression. The value of RDD depends on the fire size at the time of the application of water.

Actual Delivered Density. Actual delivered density (ADD) is the rate at which water is actually deposited from the operating sprinklers onto the top horizontal surface of a burning combustible array. ADD is a measure of the sprinkler distribution pattern and penetration capability in the presence of a fire plume. As the delay in sprinkler actuation lengthens, the fire size increases and the ADD decreases. ADD is a function of fire plume velocity, the drop momentum and size, and the distance the drops must travel.

Three measurements — RTI, ADD, and RDD — are the controlling factors that define the time-dependent nature of early fire suppression. In theory, the earlier the water is applied to a growing fire, the lower the required delivered density will be and the greater the actual delivered density will be. In other words, the faster the sprinkler response (low RTI), the lower the RDD and the higher the ADD. Conversely, the later the water is applied (high RTI), the higher the RDD and the lower the ADD. When ADD is less than RDD, the sprinkler discharge will no longer be effective enough to achieve early fire suppression. Thus, it is clear that early fire suppression depends on correct installation of ESFR sprinklers which have been shown to meet the performance criteria related to the RTI, ADD, and RDD factors for the hazard being protected. Figure 9.1 shows the generalized concept that early suppression is achieved when the ADD exceeds the RDD.

Caution must be exercised to avoid confusing ESFR sprinklers with other types of sprinklers which are equipped with fast responding thermal sensing elements, but which have not been evaluated to ensure proper relationships between RTI, ADD, and RDD.

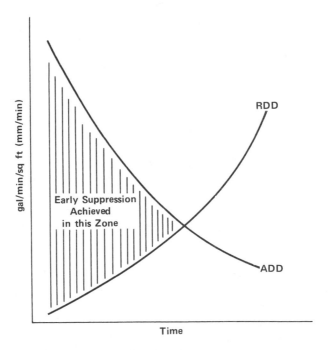

Figure 9.1.   Generalized RDD-ADD concept.

A-9-3.1   It is important to realize that the effectiveness of these highly tested and engineered sprinklers depends on the combination of fast response and the quality and uniformity of the sprinkler discharge. It should also be realized that ESFR sprinklers cannot be relied upon to provide fire "control," let alone suppression, if they are used outside the guidelines specified in Section 9-3.

9-3.2 Applicability.

It is important that ESFR sprinklers be used where it has been shown by appropriate testing that they will perform satisfactorily.

Certain types of storage and storage arrangements exhibit fire characteristics very different from most common storage areas. Specific ESFR protection guidelines for these configurations may be significantly different than the requirements in this chapter; in some cases, ESFR sprinklers may not be appropriate since full-scale testing has not been conducted at this point. ESFR sprinklers are not appropriate when: 1) fires cannot be readily extinguished by water, such as flammable liquids; 2) storage arrangements create shielded conditions which prevent water penetration; and 3) fast-spreading flame fronts would be expected to be generated ahead of operating sprinklers, such as with exposed foam plastic insulation on the underside of a roof or where oil or dust deposits are present on structural roof members.

Also, ESFR sprinklers should not be used in the same area as other types of sprinklers. A fire occurring in the non-ESFR area may cause operation of ESFR sprinklers remote from the actual fire, unnecessarily depleting water supplies. A physical cutoff between ESFR and non-ESFR sprinklered areas will prevent unnecessary ESFR sprinkler operations.

9-3.2.1 ESFR sprinklers are suitable for use with the hazards listed in Table A-9-3.4.1 and may be used in other specific hazard classifications and configurations only when proven by large-scale or other suitable fire testing.

9-3.2.2 Requirements of this standard shall apply, except those portions dealing with subjects specifically addressed in this chapter.

9-3.3 Installation.

9-3.3.1 Construction.

9-3.3.1.1 Roof Construction. ESFR sprinklers are suitable for use in buildings with the following types of roof construction:

(a) Smooth ceiling,

(b) Bar joist,

(c) Beam and girder,

(d) Panel.

9-3.3.1.2 Roof Slope. Roof slope cannot exceed ¼ in. per ft (21 mm/m).

Roof slope requirements take into consideration the effect of slope on sprinkler operating patterns due to skewed distribution of heat from a fire beneath the roof. With greater than acceptable roof slopes, heat may travel up the slope away from sprinklers, which need to operate early in order to suppress a fire, delaying their operation and preventing achievement of early fire suppression.

Full-scale testing of the ESFR sprinkler did not evaluate the impact of vents or draft curtains in an occupancy protected with ESFR sprinklers. Introduction of such elements produces an unknown quantity, namely the synergistic effect on RTI, RDD, and ADD. The position taken in NFPA 204M, *Guide for Smoke and Heat Venting*, is indeterminate in establishing if venting is a positive or negative factor in buildings protected with automatic sprinklers. This position even further enhances the importance of limiting ESFR sprinkler use to those ceiling configurations presented in 9-3.3.1.

9-3.3.2 Operating Pressure.   ESFR sprinkler systems shall be designed such that the minimum operating pressure is not less than 50 psi (3.4 bars).

9-3.3.3 System Type.   ESFR sprinkler systems shall be limited to wet-pipe systems only.

Since fires in the occupancies contemplated by this chapter grow extremely fast with large increases in heat release rates over a short period of time, and ESFR sprinklers are very quick to respond to growing fires, it is paramount that water be applied immediately after the first sprinklers operate. Otherwise, the fire will continue to grow, operating many sprinklers and decreasing the water supply to a point of noneffectiveness.

Calculations have shown that the time delays inherent in delivering water to sprinklers in both dry-pipe and preaction systems will allow many more ESFR sprinklers to open than is contemplated here.

9-3.3.4 System Design.   Pipe shall be sized by hydraulic calculation to supply the most hydraulically remote 12 sprinklers, flowing 4 sprinklers on 3 branch lines.

9-3.3.5 Temperature Rating.   Sprinkler temperature ratings shall be nominal 165°F (74°C).

Exception: Sprinklers of intermediate and high temperature ratings shall be installed in specific locations as required by 3-11.6.3.

9-3.3.6* Spacing.

Spacing requirements are determined on the basis of the following parameters:

(a)   Maintain sufficient actual delivered density (ADD) for various spacings, clearances between sprinklers and the top of storage, and possible ignition locations, whether centered between four sprinklers, centered between two sprinklers, or directly under one sprinkler, to ensure that early fire suppression will be achieved.

(b)   Maintain sufficient separation of sprinklers to prevent skipping.

(c)   Maintain sufficient cooling at ceiling level.

A-9-3.3.6 Limitations on spacing and distances between sprinklers are based on testing that shows that ESFR sprinklers cannot distribute water sufficiently to achieve suppression when these limitations are exceeded.

9-3.3.6.1   The protection area per sprinkler shall be limited to a minimum of 80 sq ft (7.4 m²) and a maximum of 100 sq ft (9.3 m²).

9-3.3.6.2   The distance between branch lines and between sprinklers on the branch lines shall be limited to not more than 12 ft (3.7 m) nor less than 8 ft (2.4 m).

9-3.3.7 Distances.

9-3.3.7.1 Clear Space Below Sprinklers.   At least 36 in. (914 mm) shall be maintained between sprinkler deflectors and the top of storage.

At clearances of less than 36 in. (914 mm), ESFR sprinklers installed within the allowable spacing guidelines will not provide sufficient overlapping of sprinkler discharge patterns between adjacent sprinklers.

9-3.3.7.2 Distances Below Ceiling.   Sprinklers shall be positioned so that deflectors are a maximum 14 in. (356 mm) and a minimum 6 in. (152 mm) below the ceiling.

At distances outside the allowable guidelines sprinkler response will be delayed, allowing the fire to grow too large before sprinklers operate. This will increase the required delivered density (RDD) while decreasing the actual delivered density (ADD), and early fire suppression may not be achieved.

9-3.3.8 Location of Sprinklers in Beam and Girder and Panel Construction.
Under beam and girder construction and under panel construction, the branch lines may run across the beams, but sprinklers shall be located in the bays and not under the beams.

9-3.3.9 Obstruction in Piping.

9-3.3.9.1 Screens located in the inlet piping directly connected to rivers, lakes, ponds, reservoirs, uncovered tanks, and similar sources (*see NFPA 24, Standard for Installation of Private Fire Service Mains and Their Appurtenances*) shall be cleaned and serviced at least annually and after any work has been performed on fire protection water supply.

9-3.3.9.2 Visual and/or flushing investigations shall be conducted of all systems for foreign material at intervals not exceeding five years.

9-3.3.10 Obstruction to Distribution.

9-3.3.10.1 Obstructions Located at or Near the Ceiling. When sprinkler deflectors are located above the bottom of beams, girders, ducts, fluorescent lighting fixtures, or other obstructions located at the ceiling, the sprinklers shall be positioned so that the maximum distance from the bottom of the obstruction to the deflector does not exceed the value specified in Chapter 4.

> The effect of pattern interruption via an obstruction must be minimized and reconciled using the applicable values from Chapter 4. "Shadow areas" in a high-piled storage occupancy provide enough of a fuel supply to a growing fire that the effect is not considered minimal; thus, great care in positioning ESFR sprinklers with respect to lights, structural members, and the like must be exercised. For obstructions that are located entirely below the sprinkler, the values discussed in 9-3.3.10.2.1 must be used.

9-3.3.10.2 Obstructions Located Entirely Below the Sprinklers.

9-3.3.10.2.1 Sprinklers shall be positioned with respect to any fluorescent lighting fixtures, ducts, or any other obstruction more than 12 in. (305 mm) wide and located entirely below the sprinklers so that the minimum horizontal distance from the near side of the obstruction to the center of the sprinkler is not less than the value specified in Table 9-3.3.10.2.1. (*See Figure 9-2.3.10.2.1.*)

Table 9-3.3.10.2.1   Position of Sprinklers in Relation to Obstructions
Located Entirely Below the Sprinklers

Distance of Deflector Above Bottom of Obstruction	Minimum Distance to Side of Obstruction, ft (m)
Less than 6 in. (152 mm)	1½ (0.5)
6 in. (152 mm) to less than 12 in. (305 mm)	3 (0.9)
12 in. (305 mm) to less than 18 in. (457 mm)	4 (1.2)
18 in. (457 mm) to less than 24 in. (610 mm)	5 (1.5)
24 in. (610 mm) to less than 30 in. (660 mm)	6 (1.8)

9-3.3.10.2.2 When the bottom of obstructions more than 12 in. (305 mm) wide is located below the sprinkler deflectors such that the deflector distances above obstructions specified in Table 9-3.3.10.2.1 are exceeded, install additional sprinklers beneath obstructions, and include such sprinklers in the water demand.

Exception No. 1:   Obstructions up to 12 in. (305 mm) wide and located below a single sprinkler, but not below two or more adjacent sprinklers (including diagonally).

Exception No. 2:   Continuous obstructions, such as utility piping or ductwork, up to 12 in. (305 mm) wide, offset by at least 24 in. (610 mm) from both the center line of sprinklers and the center line of any flue spaces for rack storage.

9-3.4 Protection Requirements.

9-3.4.1* Occupancies.   ESFR sprinklers may be used to protect the occupancies indicated in Table A-9-3.4.1. ESFR sprinklers cannot be used to protect rack storage with solid shelves or with solid, five-sided, open-top containers.

Since NFPA 13 is an installation standard, the "when to use" ESFR sprinklers is delegated to the appendix. The appropriateness of allowing ESFR sprinklers is left to an occupancy standard such as NFPA 231, *General Storage*, or NFPA 231C, *Rack Storage*. See Table A-9-3.4.1 on page 396.

9-3.4.2 Hose Streams.   Small hose stations shall be provided. Hose stream water demand is not required to be added to total water demand.

9-3.4.3 Water Supply Duration.   Water supply duration shall be at least 60 minutes.

REFERENCES CITED IN COMMENTARY

The following publications are available from the National Fire Protection Association, Batterymarch Park, Quincy, MA 02269.

NFPA 204M-1985, *Guide for Smoke and Heat Venting*
NFPA 231-1987, *Standard for General Storage*
NFPA 231C-1986, *Standard for Rack Storage of Materials*.

Table A-9-3.4.1   Protectible Occupancies for ESFR Sprinklers

Type of Storage	Commodity	Maximum Height of Storage	Maximum Height of the Building
Single-, double-, and multiple-row and portable rack storage (no open-top containers or solid shelves), and solid-piled or palletized storage.	Cartoned unexpanded plastics Class I through IV commodities. Cartoned polyurethane (*not* polystyrene) foamed-in-place packaging (encapsulated or nonencapsulated)	25	30
Roll paper on end, open/standard/closed array, banded or unbanded	Heavyweight paper	20	30
Roll paper on end, open/standard/closed array, banded or unbanded	Mediumweight paper	20	30
Aerosol storage	To be determined		

10

Referenced Publications

10-1 The following documents or portions thereof are referenced within this standard and shall be considered part of the requirements of this document. The edition indicated for each reference shall be the current edition as of the date of the NFPA issuance of this document.

10-1.1 NFPA Publications. National Fire Protection Association, Batterymarch Park, Quincy, MA 02269.

NFPA 13A-1987, *Recommended Practice for the Inspection, Testing and Maintenance of Sprinkler Systems*

NFPA 13D-1989, *Standard for the Installation of Sprinkler Systems for One- and Two-Family Dwellings and Mobile Homes*

NFPA 14-1986, *Standard for the Installation of Standpipe and Hose Systems*

NFPA 20-1987, *Standard for the Installation of Centrifugal Fire Pumps*

NFPA 22-1987, *Standard for Water Tanks for Private Fire Protection*

NFPA 24-1987, *Standard for the Installation of Private Fire Service Mains and Their Appurtenances*

NFPA 51B-1984, *Standard for Fire Prevention in Use of Cutting and Welding Processes*

NFPA 71-1987, *Standard for the Installation, Maintenance, and Use of Signaling Systems for Central Station Service*

NFPA 72A-1987, *Standard for the Installation, Maintenance, and Use of Local Protective Signaling Systems for Guard's Tour, Fire Alarm, and Supervisory Service*

NFPA 72B-1986, *Standard for the Installation, Maintenance, and Use of Auxiliary Protective Signaling Systems for Fire Alarm Service*

NFPA 72C-1986, *Standard for the Installation, Maintenance, and Use of Remote Station Protective Signaling Systems*

NFPA 72D-1986, *Standard for the Installation, Maintenance, and Use of Proprietary Protective Signaling Systems*

NFPA 72E-1987, *Standard on Automatic Fire Detectors*

NFPA 81-1986, *Standard for Fur Storage, Fumigation and Cleaning*

NFPA 96-1987, *Standard for the Installation of Equipment for the Removal of Smoke and Grease-Laden Vapors from Commercial Cooking Equipment*

NFPA 220-1985, *Standard on Types of Building Construction*

NFPA 231-1987, *Standard for General Storage*

NFPA 231C-1986, *Standard for Rack Storage of Materials*

NFPA 1963-1985, *Standard for Screw Threads and Gaskets for Fire Hose Connections.*

10-1.2   The following additional NFPA codes and standards contain specific sprinkler design criteria. (*See 2-2.1.1 Exception No. 4 and A-2-1.*)

NFPA 15-1985, *Standard for Water Spray Fixed Systems for Fire Protection*

NFPA 16-1986, *Standard on Deluge Foam-Water Sprinkler and Foam-Water Spray Systems*

NFPA 40-1988, *Standard for the Storage and Handling of Cellulose Nitrate Motion Picture Film*

NFPA 43A-1980, *Code for the Storage of Liquid and Solid Oxidizing Materials*

NFPA 45-1986, *Standard on Fire Protection for Laboratories Using Chemicals*

NFPA *101®*-1988, *Life Safety Code®*

NFPA 214-1988, *Standard on Water-Cooling Towers*

NFPA 231-1987, *Standard for General Storage*

NFPA 231C-1986, *Standard for Rack Storage of Materials*

NFPA 231D-1986, *Standard for Storage of Rubber Tires*

NFPA 231F-1987, *Standard for Storage of Roll Paper*

NFPA 307-1985, *Standard for Construction and Fire Protection of Marine Terminals, Piers, and Wharves*

NFPA 409-1985, *Standard on Aircraft Hangars*

NFPA 423-1983, *Standard for Construction and Protection of Aircraft Engine Test Facilities.*

10-1.3 Other Codes and Standards.

10-1.3.1 ANSI Publications.   American National Standards Institute, Inc., 1450 Broadway, New York, New York 10018.

ANSI B1.20.1-1983, *Pipe Threads, General Purpose*

ANSI B16.1-1975, *Cast Iron Pipe Flanges and Flanged Fittings, Class 25, 125, 250 and 800*

ANSI B16.3-1985, *Malleable Iron Threaded Fittings, Class 150 and 300*

ANSI B16.4-1985, *Cast Iron Threaded Fittings, Classes 125 and 250*

ANSI B16.5-1981, *Pipe Flanges and Flanged Fittings*

ANSI B16.9-1986, *Factory-Made Wrought Steel Buttwelding Fittings*

ANSI B16.11-1980, *Forged Steel Fittings, Socket-Welding and Threaded*

ANSI B16.18-1984, *Cast Copper Alloy Solder Joint Pressure Fittings*

ANSI B16.22-1980, *Wrought Copper and Copper Alloy Solder Joint Pressure Fittings*

ANSI B16.25-1986, *Buttwelding Ends*

ANSI B36.10M-1985, *Welded and Seamless Wrought Steel Pipe.*

10-1.3.2 ASTM Publications.   American Society for Testing and Materials, 1916 Race Street, Philadelphia, PA 19105.

ASTM A53-1987, *Standard Specification for Welded Pipe, Steel, Black and Hot-Dipped, Zinc-Coated and Seamless Steel Pipe*

ASTM A234-1987, *Standard Specification for Piping Fittings of Wrought-Carbon Steel and Alloy Steel for Moderate and Elevated Temperatures*

ASTM A795-1985, *Specification for Black and Hot-Dipped Zinc-Coated (Galvanized) Welded and Seamless Steel Pipe for Fire Protection Use*

ASTM B32-1987, *Standard Specification for Solder Metal, 95-5 (Tin-Antimony-Grade 95TA)*

ASTM B75-1986, *Standard Specification for Seamless Copper Tube*

ASTM B88-1986, *Standard Specification for Seamless Copper Water Tube*

ASTM B251-1987, *Standard Specification for General Requirements for Wrought Seamless Copper and Copper-Alloy Tube*

ASTM E380-1986, *Standard for Metric Practice.*

10-1.3.3 AWS Publications. American Welding Society, 2501 N.W. 7th Street, Miami, FL 33125.

AWS A5.8-1981, *Specification for Brazing Filler Metal*

AWS D10.9-1980, *Specification for Qualification of Welding Procedures and Welders for Piping and Tubing.*

Appendix A

This Appendix is not a part of the requirements of this NFPA document, but is included for information purposes only.

The material contained in Appendix A of the 1989 edition of the *Standard for the Installation of Sprinkler Systems* is included in the text within this Handbook and therefore is not repeated here.

Appendix B

This Appendix is not a part of the requirements of this NFPA document, but is included for information purposes only.

The material contained in Appendix B of the 1989 edition of the *Standard for the Installation of Sprinkler Systems* is included in the text within this Handbook and therefore is not repeated here.

Appendix C

Referenced Publications

C-1 The following documents or portions thereof are referenced within this standard for informational purposes only and thus are not considered part of the requirements of this document. The edition indicated for each reference should be the current edition as of the date of the NFPA issuance of this document.

C-1.1 NFPA Publications.   National Fire Protection Association, Batterymarch Park, Quincy, MA 02269.

NFPA 13A-1987, *Recommended Practice for the Inspection, Testing and Maintenance of Sprinkler Systems*

NFPA 231C-1986, *Standard for Rack Storage of Materials*

NFPA 231F-1987, *Standard for the Storage of Roll Paper.*

C-2 The following NFPA Recommended Practices contain specific srprinkler design criteria on various subjects. *(See 2-2.1.1 Exception No. 4 and A-2-1.)*

NFPA 16A-1988, *Recommended Practice for the Installation of Closed-Head Foam-Water Sprinkler Systems*

NFPA 231E-1989, *Recommended Practice for the Storage of Baled Cotton.*

NFPA 13A

Recommended Practice for the

Inspection, Testing and Maintenance

of Sprinkler Systems

1987 Edition

NOTICE: Information on referenced publications can be found in the Appendix.

Foreword

An automatic sprinkler system provides for the extinguishment of fire in a building by the prompt and continuous discharge of water directly upon burning material. This is accomplished by means of an arrangement of pipes to which are attached outlet devices known as automatic sprinklers. These sprinklers are so constructed as to open automatically whenever the surrounding temperature reaches a predetermined point.

In general, there are two types of automatic sprinkler equipment, dry-pipe systems and wet-pipe systems. In locations that are not subject to freezing temperatures, wet-pipe systems, in which the pipe lines contain water under pressure, may be installed, but in buildings or portions thereof that are subject to freezing temperatures, the dry-pipe system is ordinarily used. In the latter type of system, water is admitted to the pipes automatically after elevated ceiling temperature has caused the automatic sprinkler to operate.

If not properly maintained, a sprinkler system may become inoperative. The following offers advice and suggestions relative to the inspection, testing, and maintenance of sprinkler equipment upon which the safety of life and property may depend.

A sprinkler system serves in a passive capacity until a fire emergency occurs. Often this silent vigil can be quite lengthy, and indeed most systems are never called upon to extinguish a fire. It is this long period of inactivity that makes inspections, tests, and maintenance so crucial to the proper operation of a system when a fire does occur.

1

General Information

1-1 Scope. This recommended practice provides for the inspection, testing, and maintenance of sprinkler systems.

Because this publication is a recommended practice, it is advisory only. Unlike a standard, which specifies regulations, this document provides guidance only and the word *shall*, which indicates a requirement, is not used. Regulations governing the installation of the components of sprinkler systems are found in the appropriate standards. It is the feeling of many that requirements for adequate maintenance of automatic sprinkler systems should be mandatory rather than advisory. Unfortunately, the multiplicity of conditions which actually exist in the field would make the application of a mandatory standard in all instances difficult, if not impossible, and thereby render it meaningless. It is, therefore, felt that the guidance provided by the recommended practice is more appropriate. Further this guidance is considered so important to the proper operation of the system that a copy of this recommended practice is required to be furnished to the owner of the sprinkler system by 1-5.2 of NFPA 13.

1-2 Purpose. The purpose of this recommended practice is to provide guidance for inspection, testing, and maintenance of sprinkler systems.

The responsibility for properly maintaining a sprinkler system is the obligation of the owner(s) of the property. Suggestions as to details for proper maintenance by the owner are expanded in Section 1-5.

1-3 Definitions.

Antifreeze System. A system employing automatic sprinklers attached to a piping system containing an antifreeze solution and connected to a water supply. The antifreeze solution, followed by water, discharges immediately from sprinklers opened by a fire.

See Section 5-5 of NFPA 13, *Installation of Sprinkler Systems.*

Approved.   Means "acceptable to the authority having jurisdiction."

NOTE: The National Fire Protection Association does not approve, inspect or certify any installations, procedures, equipment or materials nor does it approve or evaluate testing laboratories. In determining the acceptability of installations or procedures, equipment or materials, the authority having jurisdiction may base acceptance on compliance with NFPA or other appropriate standards. In the absence of such standards, said authority may require evidence of proper installation, procedure or use. The authority having jurisdiction may also refer to the listings or labeling practices of an organization concerned with product evaluations which is in a position to determine compliance with appropriate standards for the current production of listed items.

Authority Having Jurisdiction.   The "authority having jurisdiction" is the organization, office, or individual responsible for "approving" equipment, an installation, or a procedure.

NOTE: The phrase "authority having jurisdiction" is used in NFPA documents in a broad manner since jurisdictions and "approval" agencies vary as do their responsibilities. Where public safety is primary, the "authority having jurisdiction" may be a federal, state, local, or other regional department or individual such as a fire chief, fire marshal, chief of a fire prevention bureau, labor department, health department, building official, electrical inspector, or others having statutory authority. For insurance purposes, an insurance inspection department, rating bureau, or other insurance company representative may be the "authority having jurisdiction." In many circumstances the property owner or his designated agent assumes the role of the "authority having jurisdiction"; at government installations, the commanding officer or departmental official may be the "authority having jurisdiction."

Butterfly Valve.   An indicating-type control valve incorporating wafer-type body with gear-operated, quarter-turn disc in the waterway.

Cold Weather Valve.   An indicating-type valve for the control of 10 sprinklers or less in a wet system protecting an area subject to freezing. The valve is normally closed and the system drained during freezing weather.

Control Valve.   A valve that may be opened or closed to regulate the flow of water to all or part of a sprinkler system.

See 3-9.2 of NFPA 13, *Standard for the Installation of Sprinkler Systems.*

Deluge System.   A system employing open sprinklers installed in a water supply through a valve that is opened by the operation of a fire detection system installed in the same areas as the sprinklers. When this valve opens, water flows into the piping system and discharges from all sprinklers attached thereto.

See Section 5-3 of NFPA 13, *Standard for the Installation of Sprinkler Systems.*

Dry-Pipe System.   A system employing automatic sprinklers installed in a piping system containing air or nitrogen under pressure, the release of which, as from the opening of a sprinkler, permits the water pressure to open a valve known as a dry-pipe valve. The water then flows into the piping system and out the opened sprinklers.

See Section 5-2 of NFPA 13, *Standard for the Installation of Sprinkler Systems.*

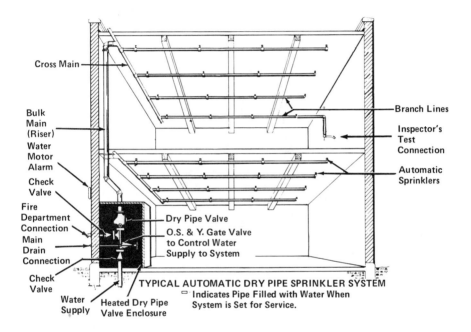

Cross Main

Bulk
Main
(Riser)

Water
Motor
Alarm

Check
Valve

Fire
Department
Connection

Main
Drain
Connection

Check
Valve

Water
Supply

Heated Dry Pipe
Valve Enclosure

Dry Pipe Valve

O.S. & Y. Gate Valve
to Control Water
Supply to System

Branch Lines

Inspector's
Test
Connection

Automatic
Sprinklers

TYPICAL AUTOMATIC DRY PIPE SPRINKLER SYSTEM
⌐ Indicates Pipe Filled with Water When
System is Set for Service.

Figure 1-3(a)   Dry-pipe System.

Emergency Impairment.   A condition wherein a sprinkler system or a portion thereof is out of order due to an unexpected occurrence such as a ruptured pipe, operated sprinkler, interruption of water supply to the system, etc.

Indicator Post.   A control extending above ground or through a wall for operating sprinkler control valves. A target or indicator visible through an opening in the post shows whether the valve is open or shut.

Inspection.   A visual examination of a sprinkler system or portion thereof to verify that it appears to be in operating condition and is free from physical damage.

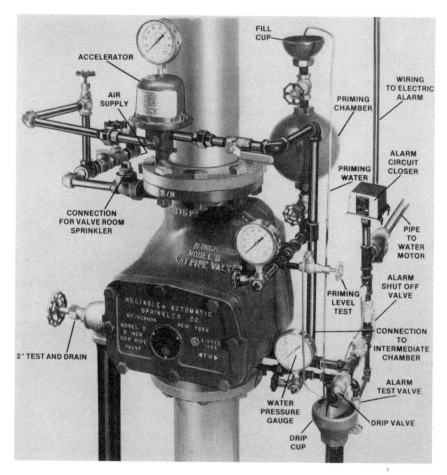

Figure 1-3(b)   Typical Dry-pipe Valve with Trimmings.

Listed. Equipment or materials included in a list published by an organization acceptable to the "authority having jurisdiction" and concerned with product evaluation, that maintains periodic inspection of production of listed equipment or materials and whose listing states either that the equipment or material meets appropriate standards or has been tested and found suitable for use in a specified manner.

NOTE: The means for identifying listed equipment may vary for each organization concerned with product evaluation, some of which do not recognize equipment as listed unless it is also labeled. The "authority having jurisdiction" should utilize the system employed by the listing organization to identify a listed product.

Locks and Chains. Heavy-duty locks and chains sufficiently durable to resist any cutting device less than heavy-duty bolt cutters.

Maintenance. Work performed to keep equipment operable, or to make repairs.

Outside Screw and Yoke (O. S. & Y.) Valve. A gate valve with a rising stem that indicates if the valve is open or closed.

Preaction System. A closed sprinkler system containing air that may or may not be under pressure, with a supplemental fire detection system installed in the same areas as the sprinklers. Actuation of the fire detection system opens a valve that permits water to flow into the sprinkler piping and to be discharged from any sprinklers that may be open.

> See 5-3.6 of NFPA 13, *Standard for the Installation of Sprinkler Systems.*

Preplanned Impairment. A condition where a sprinkler system or a portion thereof is out of service due to work that has been planned in advance such as revisions to the water supply or sprinkler system piping.

Qualified Inspection Service. A service program provided by a fire protection contractor and/or owner's representative in which all of the Chapter 7 provisions are included.

Quick-Opening Device. A listed device such as an accelerator or an exhauster used to cause a dry-pipe valve to operate more rapidly.

> See 5-2.4 of NFPA 13, *Standard for the Installation of Sprinkler Systems.*

Roadway Box. A sleeve providing access to an underground control valve.

Seals. The use of small-diameter wire that is threaded and knotted within a lead seal. Alternately, plastic devices are available for the same purpose.

The closing of the seal makes it impossible to turn a valve unless the seal is broken. The broken seal indicates that the valve may have been moved from its normal position.

Should. Indicates a recommendation or that which is advised but not required.

Sprinkler System. A sprinkler system, for fire protection purposes, is an integrated system of underground and overhead piping designed in accordance with fire protection engineering standards. The installation includes a water supply such as a gravity tank, fire pump, reservoir or pressure tank, and/or connection by underground piping to a city main. The portion of the sprinkler system aboveground is a network of specially sized or hydraulically designated piping installed in a building, structure or area, generally overhead, and to which sprinklers are connected in a systematic pattern. The system includes a controlling valve and a device for actuating an alarm when the system is in operation. The system is usually activated by heat from a fire and discharges water over the fire area.

Tamper Switch.   An electrical device for control-valve supervision which initiates an alarm when the control valve is moved from the normal position.

Testing.   Conducting periodic physical checks on the sprinkler system such as water flow tests, alarm tests, or dry-pipe valve trip tests.

Waterflow Alarm.   A listed device so constructed and installed that any flow of water from a sprinkler system equal to or greater than that from a single automatic sprinkler will result in an alarm signal.

See Section 3-12 of NFPA 13, *Standard for the Installation of Sprinkler Systems.*

Wet-Pipe System.   A system employing automatic sprinklers installed in a piping system containing water and connected to a water supply. Water discharges immediately from sprinklers opened by a fire.

See Section 5-1 of NFPA 13, *Standard for the Installation of Sprinkler Systems.*

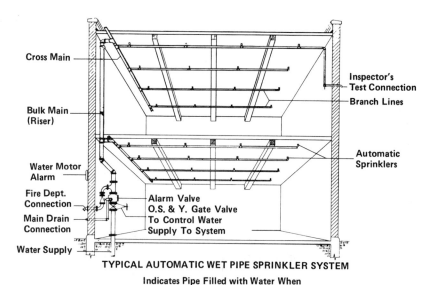

TYPICAL AUTOMATIC WET PIPE SPRINKLER SYSTEM
Indicates Pipe Filled with Water When
System is Set for Service.

Figure 1-3(c)   Wet-pipe System.

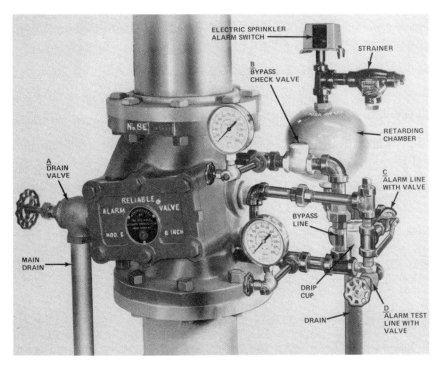

Figure 1-3(d)   Typical Alarm Valve and Trimmings.

1-4 Units.   Metric units of measurement in this recommended practice are in accordance with the modernized metric system known as the International System of Units (SI). Two units (liter and bar), outside of but recognized by SI, are commonly used in international fire protection. These units are listed in Table 1-4 with conversion factors.

Table 1-4

Name of Unit	Unit Symbol	Conversion Factor
liter	L	1 gal = 3.785 L
cubic decimeter	dm³	1 gal = 3.785 dm³
pascal	Pa	1 psi = 6894.757 Pa
bar	bar	1 psi = 0.0689 bar
bar	bar	1 bar = 10⁵ Pa

For additional conversions and information see ASTM E380, *Standard for Metric Practice.*

1-4.1   If a value for measurement as given in this recommended practice is followed by an equivalent value in other units, the first stated is to be regarded as the requirement. A given equivalent value may be approximate.

1-4.2 The conversion procedure for the SI units has been to multiply the quantity by the conversion factor and then round the result to the appropriate number of significant digits.

1-5 Responsibility of the Owner or Occupant.

1-5.1 The responsibility for properly maintaining a sprinkler system is the obligation of the owners of the property.

By means of periodic tests, the equipment is shown to be in good operating condition or any defects or impairments are revealed. Such tests are made, however, at the owner's responsibility and risk. Intelligent cooperation in the performance of these tests shows evidence of the owner's interest in property conservation.

1-5.2 Automatic sprinkler systems installed in accordance with NFPA standards require a minimum of inspection, testing, and maintenance; however, deterioration or impairment may result from neglect. Definite provision for periodic competent attention is a prime requirement if the system is to serve its purpose effectively.

1-5.3 Arrangements should be made to keep all stock piles, racks, and other possible obstructions the proper distance below sprinklers. [The minimum recommended distance below sprinkler deflectors at the ceiling is 18 in. (457 mm). The minimum recommended distance below sprinkler deflectors in racks can be found in NFPA 231, *Standard for General Storage*; NFPA 231C, *Standard for Rack Storage*; and NFPA 231D, *Standard on Storage of Rubber Tires*.]

1-5.4 A competent and reliable employee should be given the responsibility of regularly inspecting, testing, and maintaining the system and reporting any troubles or defects to his employer. This employee should have proper instruction and training and a general understanding of the mechanical requirements of operation.

1-5.5 Support personnel should be trained in inspection, testing, and maintenance and be fully capable of taking over the functions at any time when the authorized individual is unavailable.

1-5.6 Public Fire Department.

1-5.6.1 It is advisable to notify the fire department of the installation of automatic sprinkler equipment so that it may become familiar with the system. The fire department should know the extent of the protection and the location and arrangement of the control valves and the connections for fire department use.

The fire department should also be notified if the system or a major portion of it is temporarily taken out of service. This notification allows the fire department to preplan in the event

of any emergency and also provides it an opportunity for making suggestions for provision of emergency or temporary water supplies during the impairment period.

1-5.7 Security Personnel.

1-5.7.1 Instruct security personnel in the following:

(a) Location and use of control valves, drain valves, and alarm devices.

(b) Prompt transmittal of a fire alarm to a fire department or brigade, before attempting to extinguish the fire.

(c) Proper notification in case of fire or impairment of sprinkler equipment.

(d) Daily visual inspection of all sprinkler control valves on the guard's first round to ascertain that they are open.

(e) Proper notification immediately of any valve found closed.

(f) Proper notification when sprinkler alarms operate, to determine the cause of water flow.

(g) Do not close sprinkler control valves until it has definitely been established that there is no fire.

(h) During cold weather, verify that windows or other openings are closed and that proper temperature is being maintained to prevent freezing.

The importance of inspection to ascertain that all sprinkler control valves are open cannot be overemphasized. Closed valves are by far the greatest cause of sprinkler system failure. It is also important that security personnel not close valves until it has been established that a fire has been completely extinguished. There have been a significant number of large losses where the sprinkler valve was prematurely closed, either because of concern regarding water damage or because the party closing the valve incorrectly thought the fire had been extinguished. When closed valves are found with no detectable reason for their being closed, security personnel should be especially alert for possible arson attempts.

1-6 Sprinkler Inspection Service.

1-6.1 The level of reliability of the protection offered by an automatic sprinkler system is promoted when there is a qualified inspection service. Qualified inspection service should include:

(a) Four visits per year, at regular intervals.

(b) All services indicated in summary Table 7-3.

(c) The completion of a report form with copies furnished to the property owner. (*See Chapter 7, Report of Inspection, Exhibit I.*)

1-6.2 The outside inspection services are an adjunct to, and are not intended to replace, the owners' obligations.

2

Water Supplies

2-1 General. The source and quantity of water is of fundamental importance. To ensure the continued existence of proper flow, it is necessary that periodic inspections and tests be conducted by qualified personnel.

2-2 Gravity Tanks and Suction Tanks. *(See NFPA 22, Standard for Water Tanks for Private Fire Protection.)*

2-2.1 Monthly inspections should be made to check the maintenance of water at the proper level in the tank.

Constant maintenance of a full supply of water in gravity tanks is necessary not only to ensure proper performance of the sprinkler system in the event of a fire, but to prevent shrinkage of wooden tanks and minimize corrosion of steel tanks.

2-2.2 Heating devices should be kept in order and the water temperature in the tank should be checked daily during freezing weather to maintain a minimum temperature of 40°F (4°C).

2-2.3 The tank roof should be kept tight and in good repair, with the hatches fastened closed and the frostproof casing of the tank riser in good repair.

2-2.4 Ice should not be allowed to form on any part of the tank structure. The prevention of freezing in the riser or the formation of ice in the tank itself is extremely important. Freezing in the riser of an elevated tank may obstruct the flow of water from the tank. The formation of a layer of ice on the water of elevated or suction tanks also may impede or prevent the flow from the tank. The formation of heavy icicles through leaking of the tank is dangerous as tank collapse may ensue or people may be endangered by falling icicles.

2-2.5 The bases of the tower columns should be kept free from dirt and rubbish that would permit the accumulation of moisture with consequent corrosion. The tops of foundation piers should always be at least 6 in. (152 mm) above the ground level.

Coal or ashes or combustible material of any kind should not be piled near the columns as this may cause failure of the steelwork due to fire,

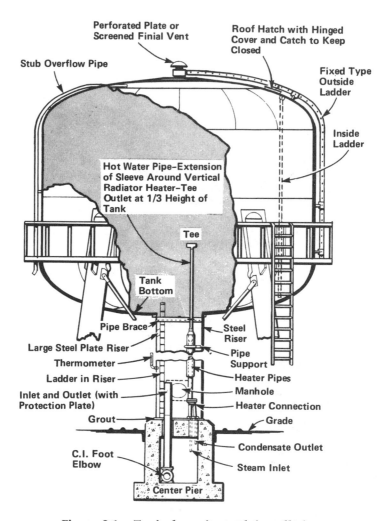

Figure 2.1.   Typical gravity tank installation.

heating, or corrosion. The tank site should be kept cleared of weeds, brush, and grass.

2-2.6 Before repainting, the surface should be thoroughly dried and all loose paint, rust, scale, and other surface contamination should be removed. After proper surface preparation, the original paint system should be restored. It may be necessary or economical to repaint the entire inside surface. On the exterior, normal maintenance will involve local patching and periodic application of one complete finish coat when the preceding has weathered thin or for improved appearance after patching.

The painters should not allow any scrapings or other foreign material to fall down the riser or outlet. If the opening is covered for protection, only a few sheets of paper tied over the end of the settling-basin-stub should be used. The paper should be removed upon the completion of the job.

For detailed information refer to NFPA 22, *Standard for Water Tanks for Private Fire Protection,* Care and Maintenance Section.

2-2.7 Necessary periodic emptying of steel tanks for repainting can be minimized by use of a cathodic corrosion prevention system that counteracts the natural electrolytic action that is the basis for most corrosion. Such a system needs periodic attention to the condition of suspended electrodes. If chemical water additives are used to inhibit corrosion, semi-annual chemical analysis of the water should be made. (*See also NFPA 22, Standard for Water Tanks for Private Fire Protection, Section A-2-7.13.*)

If cathodic protection is maintained in a steel tank, the tank should be cleaned out sufficiently often to prevent sediment and scale entering the discharge pipe.

2-2.8 The authority having jurisdiction should always be notified in advance when and for how long the tank is to be out of service.

2-3 Pressure Tanks. (*See NFPA 22, Standard for Water Tanks for Private Fire Protection.*)

2-3.1 Pressure tanks should be inspected regularly, checking the water level and air pressure monthly.

2-3.2 The interior of pressure tanks should be inspected carefully at three-year intervals to determine if corrosion is taking place and if repainting or repairing is needed. When necessary, they should be thoroughly scraped and wire brushed and repainted with an approved metal-protective paint.

2-3.3 Applicable safety codes should be consulted with respect to the maintenance and testing of pressure tanks.

2-3.4 The tank should be pressure tested at intervals as required by the ASME, *Non-Fired Pressure Vessel Code.*

2-3.5 Sight gage valves should be kept closed except when a test for water level is being made.

By keeping these valves closed, the water and air in the tank are isolated from the sight glass, so that breaking of the glass will not affect the volume of water nor the pressure available for fire protection.

2-3.6 The tank and its supports should be examined and painted as recommended for gravity tanks.

2-3.7 The heat within the tank enclosure should be checked daily during cold weather to maintain a 40°F (4°C) room temperature.

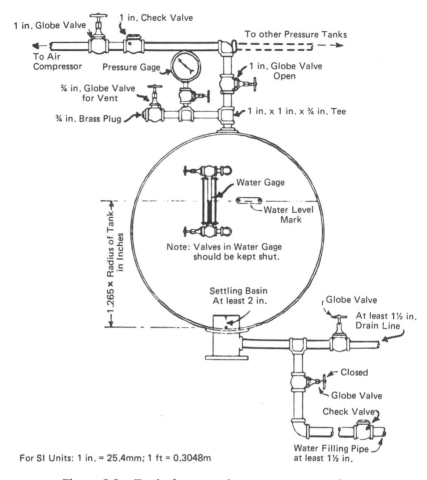

For SI Units: 1 in. = 25.4mm; 1 ft = 0.3048m

Figure 2.2. Typical connection to pressure tanks.

2-4 Fire Pumps. (*See NFPA 20, Standard for Installation of Centrifugal Fire Pumps and NFPA 21, Standard for the Operation and Maintenance of National Standard Steam Fire Pumps.*)

2-4.1 General.

2-4.1.1 The pump room should be kept clean and accessible at all times. The fire pump, driver, and controller should be protected against possible interruption of service through damage caused by explosion, fire, flood, earthquake, rodents, insects, windstorm, freezing, vandalism, and other adverse conditions.

2-4.1.2 The suction pipes, intakes, foot valves, and screens of fire pumps should be examined frequently to make sure that they are free from any obstruction. Mud, gravel, leaves, and other foreign material entering the suction pipe may cause damage to the pump or obstruction of the piping of the sprinkler system. The formation of ice may also impair the operation of the pump.

NOTE: Horizontal pumps should be provided with water under a positive head.

2-4.1.3 Suitable means should be provided for maintaining the temperature of a pump room or pump house, where required, above 40°F (4°C). Where pumps are driven by internal combustion engines the temperature of the pump room, pump house, or area where engines are installed should never be less than the minimum recommended by the engine manufacturer.

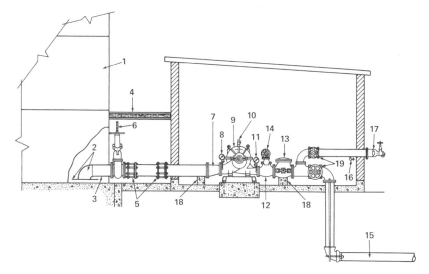

Figure 2.3. Horizontal split case fire pump installation with the water supply under a positive head: 1) aboveground suction tank; 2) entrance elbow; 3) suction pipe; 4) frostproof casing; 5) flexible couplings for strain relief; 6) O.S. & Y. gate valve; 7) eccentric reducer; 8) suction gage; 9) horizontal split case fire pump; 10) automatic air release; 11) discharge gage; 12) reducing discharge tee; 13) discharge check valve; 14) relief valve if required; 15) discharge pipe; 16) drain valve or ball drip; 17) hose valve manifold with hose valves; 18) pipe supports; 19) indicating gate or indicating butterfly valve.

2-4.1.4 Pump rooms and pump houses should be dry and free of condensate. Accumulation of water in the steam pump supply line or drainage equipment may be dangerous and should be avoided. Where condensate is a problem some heat should be provided.

2-4.1.5 Fire pumps should be operated only in connection with fire protection service and not for plant use.

2-4.1.6 Oil in internal combustion engine pumps should be changed in accordance with manufacturer's instructions, but not less than annually.

2-4.1.7 Storage batteries should be tested frequently to determine the condition of battery cells and the amount of charge in the battery. Only distilled water should be used in battery cells. The plates should be kept submerged at all times.

2-4.1.8 Fuel storage tanks should be kept full at all times.

2-4.2 Periodic Operation and Testing.

2-4.2.1 The pump should be operated every week at rated speed. Inspect the condition of the pump, bearings, stuffing boxes, suction pipe strainers, and the various other details pertaining to the driver and control equipment. The examination should be extended to include the condition and reliability of the electric power supply and, if the pump is engine driven, the storage batteries, lubrication system, and oil and fuel supplies.

Exception: Electric motor driven fire pumps should be tested monthly.

The packing glands of horizontal shaft centrifugal pumps are part of the pump's lubrication system and should drip slowly when the pump is in operation.

2-4.2.2 When automatically controlled pumping units are to be tested weekly by manual means, at least one start should be accomplished by reducing the water pressure either with the test drain on the pressure sensing line or with a larger flow from the system.

The pressure switch in the control panel that activates the pump automatically is connected to the system through small diameter, noncorrosive piping. The test drain valve will be found in that piping.

2-4.2.3 If the driver has an internal combustion engine, it should be run for at least 30 minutes to bring it up to normal running temperature and to make sure it is running smoothly at rated speed. Automatically controlled equipment should be arranged to automatically start the engine with the initiating means being a solenoid valve drain on the pressure control line.

2-4.2.4 Steam pumps should be operated until water is discharged freely from the relief valve. Regular inspections should be made: checking the maintenance of ample pressure; proper supply of lubricating oil; operative condition of relief valve and level of water in the priming tank.

2-4.2.5 A yearly flow test should be made to ensure that neither pump nor suction pipe is obstructed and the pump is operating properly. When the water supply is from a public service main, pump operation should not reduce the suction head at the pump below the pressure allowed by the local authority. At this time both the static and pumping water level of vertical shaft pumps should be determined.

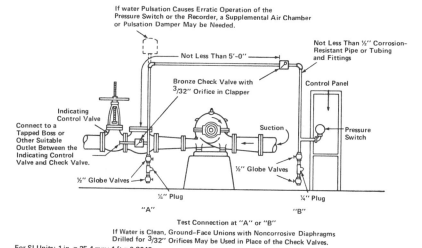

If water Pulsation Causes Erratic Operation of the Pressure Switch or the Recorder, a Supplemental Air Chamber or Pulsation Damper May be Needed.

Not Less Than 5'-0"

Not Less Than ½" Corrosion-Resistant Pipe or Tubing and Fittings

Bronze Check Valve with ³/32" Orifice in Clapper

Control Panel

Indicating Control Valve

Connect to a Tapped Boss or Other Suitable Outlet Between the Indicating Control Valve and Check Valve.

Suction

Pressure Switch

½" Globe Valves

½" Globe Valves

¼" Plug

¼" Plug

"A"

"B"

Test Connection at "A" or "B"

If Water is Clean, Ground–Face Unions with Noncorrosive Diaphragms Drilled for ³/32" Orifices May be Used in Place of the Check Valves.

For SI Units: 1 in. = 25.4 mm; 1 ft = 0.3048 m

Figure 2.4. Piping connection for each automatic pressure switch for fire pumps or jockey pumps. Solenoid drain valve used for engine-driven pumps may be at "A," "B," or inside of controller enclosure.

Arrangements must be made for the safe discharge and disposal of the large volume of water.

2-5 Hydrants. (*See NFPA 24, Standard for Private Fire Service Mains and Their Appurtenances.*)

2-5.1 Inspection of Hydrants.

(a) Public hydrants near the building should be observed for any signs of damage or vandalism.

(b) Private hydrants should be inspected monthly to verify that they are visible and readily accessible with caps in place.

2-5.2 Maintenance.

(a) Lubricate private hydrants twice yearly.

(b) Private hydrants should be serviced as recommended by manufacturers.

2-5.3 Testing. At least annually, private hydrants should be opened and closed to ensure proper water flow and drainage.

2-6 Riser Flow Tests.

2-6.1   Water flow tests should be made quarterly from water supply test pipes (main drain valves).

Test at the main drain valves includes noting of pressure gage readings with unrestricted flow of water with the drain valve wide open, as compared with the reading with the drain valve closed. If the readings vary materially from those previously established or from normal readings, the condition should be investigated. These tests are intended to show whether or not the normal water supply is available on the system and to indicate the possible presence of closed valves or other obstructions in the supply pipe.

NOTE: Water flow test of a system having a direct connnection to central station or fire department should be made only after proper notice is given to the signal receiving station.

See **Figure 2-9.1 of NFPA 13**, *Standard for the Installation of Sprinkler Systems.*

2-7 Control Valves.   (*See NFPA 26, Recommended Practices for the Supervision of Valves Controlling Water Supplies for Fire Protection.*)

2-7.1 General.

2-7.1.1   Valves should be numbered and each should have a sign indicating the portion of the system it controls.

2-7.1.2   A valve seal and tag system should be used in connection with the supervision and maintenance of a sprinkler system.

2-7.1.3   Each control valve in the sprinkler system should be secured in its normal or open position by means of a seal, lock, or tamper switch.

2-7.1.4   All control valves of the sprinkler system should be inspected at regular intervals.

   (a) Sealed valves — weekly

   (b) Locked valves and valves with tamper switches — monthly.

Paragraphs 2-7.1.3 and 2-7.1.4 are the most important in the entire recommended practice. Closed valves are the greatest cause of sprinkler system failure. A conscientiously applied program of securing control valves in the open position in combination with a regular inspection procedure will minimize the likelihood of such an occurrence.

2-7.1.5 If a normally open sprinkler valve is closed, thus shutting off any part of the system, the owner or manager of the property should be notified immediately so that the owner may follow his normal valve supervision procedure, including notifying the authority having jurisdiction. (*See Chapter 6, Impairments.*)

2-7.1.6 Valves should be kept in normal position and the sprinkler system in service to the greatest extent possible during alterations and repairs.

When alterations or repairs would otherwise take a system out of service for more than a few hours, the portion of the system being modified should be blanked off and the remainder of the system retained in service. Whenever possible, the portion of the system being modified should be returned to service at the completion of each day's work.

When the alteration or repair isolates the system from its supply, a temporary supply, such as connecting the opened main drain valve to an opened hydrant with fire hose, should be provided when possible.

2-7.1.7 After any alterations or repairs, an inspection should be made to ensure that the valves are in the fully open position, properly sealed, locked, or equipped with a tamper switch, and the system is in commission.

2-7.1.8 Valve stems should be oiled or greased at least once a year. At this time, completely close and reopen the valve to test its operation and distribute the lubricant.

2-7.2 Valve Inspection Report.

2-7.2.1 A valve inspection report should show that the valves are:

(a) In normal open or closed position

(b) Properly sealed, locked, or equipped with a tamper switch

(c) In good operating condition

(d) Readily accessible

(e) Provided with wrenches where required.

2-7.3 Indicator Post.

2-7.3.1 Quarterly, each post indicator valve should be opened until spring or torsion is felt in the rod, indicating that the rod has not become detached from the valve. Valves should be backed one-quarter turn from the wide open position to prevent jamming.

2-7.4 Underground Gate Valves with Roadway Boxes.

2-7.4.1   Quarterly, each valve should be operated with a T-handle wrench to verify that it is in the open position.

2-7.4.2   The location of each such valve should be clearly indicated by a sign on a nearby wall or by a marker. The sign should also indicate direction of valve opening, clockwise or counterclockwise.

2-7.4.3   The roadway box for the valve should always be readily accessible, and the cover should be kept in place.

2-8 Fire Department Connections.

See Section 2-7 of NFPA 13, *Standard for the Installation of Sprinkler Systems.*

2-8.1   Fire department connections should be visible and accessible at all times. They should be inspected monthly.

2-8.2   Caps or plugs should be in place, threads in good condition, ball drip or drain in working order, and check valve not leaking. Prior to replacing caps or plugs, ensure that waterway is clear of foreign material.

2-9 Hose and Hose Stations.   (*See NFPA 1962, Standard for the Care, Use and Maintenance of Fire Hose Including Connections and Nozzles.*)

See NFPA 14, *Standard for the Installation of Standpipe and Hose Systems.*

2-9.1   Hose stations should be inspected monthly to ensure that all equipment is in place and in good condition. Hose racks or reels and nozzles should be checked for obvious signs of mechanical damage. Hose station control valves should be checked for signs of leakage.

2-9.2   Hose including gaskets should be removed and re-racked at least annually.

3

Automatic Sprinklers

See Section 3-11 of NFPA 13, *Standard for the Installation of Sprinkler Systems.*

3-1 General.

3-1.1 Sprinklers should be visually checked regularly. Sprinklers should be free from corrosion, foreign material, and paint, and not bent or damaged.

3-1.2 The standard sprinkler is the type manufactured since 1953, incorporating a uniform, hemispherical discharge pattern. Water is discharged in all directions below the plane of the deflector. Little or no water is discharged upward to wet the ceiling. Sprinkler deflectors are stamped as follows:

Upright Sprinkler Marked SSU

Pendent Sprinkler Marked SSP

3-1.3 The old-style sprinkler is the type manufactured before 1953. It discharges approximately 40 percent of the water upward to the ceiling. It can be installed in either the upright or pendent position.

3-1.4 Only listed sprinklers may be used. Sprinklers may not be altered in any respect nor have any type of ornamentation, paint, or coatings applied after shipment from the place of manufacture.

3-1.5 Corrosion-resistant or specially coated sprinklers are installed in locations where chemicals, moisture, or other corrosive vapors exist.

3-1.6 Temperature Ratings.

3-1.6.1 The standard temperature rating of automatic sprinklers is shown in Table 3-1.6.1. Automatic sprinklers are manufactured with their frame arms colored in accordance with color code designated in Table 3-1.6.1.

Table 3-1.6.1 is shown on page 428.

3-1.6.2 When higher temperature sprinklers are necessary to meet extraordinary conditions, special sprinklers rated as high as 650°F (343°C) are available and may be used.

Table 3-1.6.1

Temperature Ratings, Classifications, and Color Codings

Maximum Ceiling Temperature		Temperature Rating		Temperature Classification	Color Code
°F	°C	°F	°C		
100	38	135 to 170	57 to 77	Ordinary	Uncolored
150	66	175 to 225	79 to 107	Intermediate	White
225	107	250 to 300	121 to 149	High	Blue
300	149	325 to 375	163 to 191	Extra-High	Red
375	191	400 to 475	204 to 246	Very Extra-High	Green
475	246	500 to 575	260 to 302	Ultra-High	Orange

3-1.6.3 Information regarding the highest temperature that may be encountered in any location in a particular installation should be obtained by use of a thermometer, which should be hung for several days in the questionable location under the normal ambient temperature condition.

3-2 Replacement Sprinklers.

3-2.1 Care should be taken to ensure that replacement sprinklers have the proper characteristics for the location:

(a) Style

(b) Orifice size

(c) Temperature rating

(d) Coating, if any

(e) Deflector type (upright, pendent, sidewall, etc.).

Advances in technology have produced fast-acting sprinkler operating elements. The sensitivity of such elements is important to the operation of the appropriate number of sprinklers in a compartment *(as defined in 7-4.4.3 of NFPA 13)*. Therefore, residential sprinklers in a compartment must be of the same manufacturer and have the same heat-response elements (including temperature rating). This must be considered when providing replacement sprinklers.

Standard upright or pendent sprinklers having the characteristics indicated in 3-2.1 in common may be used interchangeably regardless of model or manufacturer. The listing specifications of other types of sprinklers should be checked if the use

of a replacement sprinkler of a different model or manufacturer is contemplated. The five characteristics listed materially
affect the operation of the sprinkler and, therefore, its ability
to extinguish a particular fire. Orifice size is of particular
importance, especially when replacing other than ½-in.
(13-mm) orifices (when replacing either large or small orifice
sprinklers), due to the considerably different hydraulic and
discharge characteristics. Replacement of sprinklers with improper temperature rating could result in premature opening
where the higher temperature sprinkler was originally specified or installed, due to exposure to some particular heat source
such as a unit heater, boiler, etc. This is also important when
replacing sprinklers in storage occupancies where higher
temperature sprinklers may have been specified to reduce the
number of sprinklers opening in the event of fire in this type of
high-heat-release exposure.

In recent years, special sprinklers having protection areas or
distances between sprinklers different from those specified for
standard sprinklers in NFPA 13 have been listed. Extreme care
must be taken to ensure that such sprinklers are replaced with
sprinklers having comparable characteristics.

3-2.2 Standard sprinklers manufactured after 1952 may be used to replace
old-style sprinklers manufactured prior to 1953.

*Exception: Piers and wharves. See 3-11.2.8.1 of NFPA 13, Installation of
Sprinkler Systems.*

Old-style sprinklers are used for the protection of fur vaults.
(*See 4-4.16 of NFPA 13, Standard for the Installation of Sprinkler
Systems.*)

3-2.3 Old-style sprinklers may be used to replace existing old-style
sprinklers.

3-2.4 Old-style sprinklers should not be used to replace standard sprinklers without a complete engineering review of the system.

Because of their improved discharge characteristics,
standard sprinklers are installed to protect a greater coverage
area per sprinkler than was permitted for the old-style sprinklers. Therefore, old-style sprinklers installed in systems designed for standard sprinklers will not provide adequate coverage.

3-2.5 Secondhand sprinklers should not be used.

3-3 Automatic Sprinkler Replacement and Testing Program.

3-3.1 Representative samples of solder-type sprinklers with temperature classification of Extra High (325°F)(163°C) or greater that are exposed on a semicontinuous to continuous maximum allowable ambient temperature condition should be tested at 5-year intervals for operation by a testing laboratory acceptable to the authority having jurisdiction.

The fusing element in 360°F (182°C) solder-type sprinklers will, if exposed to temperatures approaching the fusing element's rating, gradually change its melting point.

3-3.1.1 A representative sample of sprinklers should normally consist of a minimum of two per floor or individual riser, and in any case not less than four, or 1 percent of the number of sprinklers per individual sprinkler system, whichever is greater.

3-3.2 All automatic sprinklers should be replaced when painted, corroded, damaged, or loaded with foreign materials, or when representative samples fail to meet test requirements.

3-3.3 When sprinklers have been in service for 50 years, representative samples should be submitted to a testing laboratory acceptable to the authority having jurisdiction for operational testing. Test procedure should be repeated at 10-year intervals.

3-3.3.1 Sprinklers made previous to 1920 should be replaced.

3-3.4 When residential or quick response sprinklers have been in service for 20 years, representative samples should be submitted to a testing laboratory acceptable to the authority having jurisdiction for operational testing and checks on sensitivity. Test procedures should be repeated at 10-year intervals.

The more conservative recommendations for residential and quick-response sprinklers is because of their newness and their lighter components when compared to standard sprinklers.

3-4 Sprinkler Guards.   Sprinklers so located as to be subject to mechanical injury should be protected with approved sprinkler guards.

Generally, sprinklers that are located closer than 7 ft (2.1 m) from the floor are considered to be subject to mechanical injury, and guards should be considered in these cases. Also, sprinklers located under the rakes of stairwells at lower levels should be provided with guards. Guards are also important in storage locations and for sprinklers installed in storage racks.

3-5 Stock of Spare Sprinklers.

3-5.1 A supply of spare sprinklers (never less than six) should be stored in a cabinet on the premises for replacement purposes. The cabinet should be so located that it will not be exposed to moisture, dust, corrosion, or a temperature exceeding 100°F (38°C).

3-5.1.1 The stock of spare sprinklers should be as follows:

(a) For buildings having not over 300 sprinklers - not less than 6 sprinklers

(b) For buildings having 300 to 1,000 sprinklers - not less than 12 sprinklers

(c) For buildings having over 1,000 sprinklers - not less than 24 sprinklers

(d) Stock of spare sprinklers should include all types and ratings installed.

The stock of spare sprinklers required is a minimum. Spare sprinklers of all types and ratings installed should be available. For an occupancy with a variety of types and ratings of sprinklers installed, the stock of spare sprinklers should be increased above the minimum.

3-5.1.2 A special sprinkler wrench should be provided and kept in the cabinet, to be used in the removal and installation of sprinklers. Other types of wrenches may damage the sprinklers.

3-5.1.3 Automatic sprinklers and fusible links protecting commercial-type cooking equipment and their associated ventilation systems should be inspected twice yearly and replaced annually.

Bulb-type sprinklers and bulb-type spray nozzles showing no build-up of grease or other material need not be replaced.

3-5.1.4 Sprinklers protecting spraying areas should be clean and protected against overspray residue so that they will operate quickly in the event of fire. If covered, polyethylene or cellophane bags having a thickness of 0.003 in. (0.076 mm) or less, or thin paper bags, should be used. Coverings should be replaced or heads cleaned frequently so that heavy deposits or residue do not accumulate. If not covered, the sprinklers should be replaced annually.

4

Sprinkler System Components

4-1 General.

4-1.1 The sprinkler contractor provides instructional literature describing operation and proper maintenance of fire protection devices. This instructional literature should be posted near the system riser.

4-2 Piping.

See Section 3-1 of NFPA 13, *Standard for the Installation of Sprinkler Systems.*

4-2.1 General Provisions. Piping should be kept in good condition and free from mechanical injury. Sprinkler piping should not be used for support of ladders, stock, or other material.

4-2.2 When the piping is subject to corrosive atmosphere, a protective coating that resists corrosion should be provided and maintained in proper condition.

4-2.3 When the age or service conditions of the sprinkler equipment warrant, an internal examination of the piping should be made. When it is necessary to flush a part or all of the piping system, this work should be done by sprinkler contractors or other qualified workers.

4-3 Hangers.

See Section 3-10 of NFPA 13, *Standard for the Installation of Sprinkler Systems.*

4-3.1 Hangers should be kept in good repair. Broken or loose hangers should be replaced or refastened.

4-3.2 Broken or loose hangers may put undue strain on piping and fittings, cause breaks, and interfere with proper drainage.

4-4 Gages.

See 2-9.2 of NFPA 13, *Standard for the Installation of Sprinkler Systems.*

4-4.1   Gages on wet-pipe sprinkler systems should be checked monthly to ensure that normal water supply pressure is being maintained. Gages on dry, preaction, and deluge systems should be inspected weekly to ensure that normal air and water pressures are being maintained.

A pressure reading on the gage on the system side of an alarm valve in excess of the pressure recorded on the gage on the supply side of the valve is normal, as the highest pressure from the supply will get trapped in the system. Equal gage readings could indicate a leak in the system. If there are no visible leaks, it is a good possibility that the alarm valve itself is leaking or that pressure has been recently drained from the system side of the alarm check valve, as would occur during alarm tests. [*See Figure 1-3(d).*]

4-4.2   Gages should be checked with an inspector's gage every five years.

4-5 Water Flow Alarm Devices.

See 3-12.3 of NFPA 13, *Standard for the Installation of Sprinkler Systems.*

4-5.1   Water-flow alarm devices include mechanical water motor gongs, vane-type water flow devices, and pressure switches that provide audible and/or visual signals.

4-5.2   Valves controlling water supply to alarm devices should be sealed or locked in the normally open position.

4-5.3   Water-flow alarm devices should be tested at least quarterly, weather permitting.

4-6 Notification to Supervisory Service.

4-6.1   To avoid false alarms where supervisory service is provided, including proprietary, remote alarm receiving facility, or fire department, the central station should always be notified before operating any valve or otherwise disturbing the sprinkler system.

4-7 Wet Systems—Alarm Valves.

4-7.1   Test alarms quarterly by opening the inspector's test connection.

Exception: Where weather conditions or other circumstances prohibit using the inspector's test connection, the by-pass test connection may be used.

4-7.2   Cold weather valves should be closed at the approach of freezing weather. Drain the piping in the area subject to freezing. The drain valves on the exposed piping should be left slightly open. (Automatic protection should be restored when danger of freezing is past.)

NOTE: To provide year-round protection, it is recommended that cold weather valves be replaced with dry-pipe valves or antifreeze systems.

One manufacturer refers to its 2-in. (51-mm) dry-pipe valve as being a cold weather valve. This section refers only to manually operated control valves.

4-7.3   The freezing point of solutions in antifreeze systems should be checked annually by measuring the specific gravity with a hydrometer, and adjusting the solutions if necessary. The use of antifreeze solutions should be in conformity with any state or local health regulations.

See Figures 5-5.3.3(a), 5-5.3.3(b), 5-5.4, and A-5-5.3.3 of NFPA 13, *Standard for the Installation of Sprinkler Systems.*

4-7.4   Buildings should be inspected to verify that windows, skylights, doors, ventilators, and other openings and closures will not unduly expose sprinkler piping to freezing. Blind spaces, unused attics, stair towers, low spaces under buildings and roof houses are often subject to freezing.

4-8 Dry Systems—Dry Valves, Accelerators, Exhausters.

See Section 5-2 of NFPA 13, *Standard for the Installation of Sprinkler Systems.*

4-8.1   Dry-pipe systems should not be converted to wet-pipe during warm weather. This will cause corrosion and accumulation of foreign matter in the pipe system and loss of alarm service.

4-8.2 Inspection and Maintenance.

4-8.2.1   The priming water should be inspected quarterly and maintained at the proper level as recommended by the dry valve manufacturer.

4-8.2.2   Grease or other sealing material must not be used on seats of dry-pipe valves. Force should not be used in attempting to make dry valves tight.

4-8.2.3   Test water flow and low air pressure alarms and perform a water-flow test through the main drain connection quarterly.

A valved bypass is provided in the dry-pipe valve trim to facilitate waterflow alarm tests. It should be utilized when the alarms are to be tested without a trip test of the dry-pipe valve. [*See Figure 1-3(b).*]

4-8.2.4 The air or nitrogen pressure on each dry-pipe system should be checked at least once a week and maintained as per manufacturer's instructions. All leakage resulting in pressure loss greater than 10 psi (0.7 bar) per week should be repaired.

4-8.2.5 The dry-pipe valve enclosure should be maintained at a minimum temperature of 40°F (4°C).

4-8.2.6 Before and during freezing weather, all low-point drains on dry-pipe systems should be drained as frequently as required to remove all moisture. This process should be repeated daily until all condensate has been removed. The freezing of a small amount of water in the system piping may cause rupture of the sprinkler system resulting in extensive damage to the sprinkler system and water damage to the building and contents. Drum drip assemblies should be in a warm area or in a heated enclosure, when practical.

> See Figure 3-6.3.3 of NFPA 13, *Standard for the Installation of Sprinkler Systems.*
>
> The dry-pipe valve need not be tripped for moisture to enter a dry-pipe system. It will condense out of the air pressurizing the piping. When draining drum drips, the normally open top valve is closed to isolate the drum drip from the system.

4-8.3 Trip Tests.

4-8.3.1 Trip tests of each dry-pipe valve, including quick-opening devices, if any, should be done in the spring to allow all condensate to drain from the system piping. At this time, thoroughly clean the dry-pipe valve, renew parts as required, and reset the valve.

4-8.3.2 Each dry-pipe valve should be trip tested with the control valve partially open, and cleaned and reset at least once each year during warm weather. The shutoff valve should be kept open at least far enough to permit full flow of water at good pressure through the main drain when it is fully opened.

4-8.3.3 Before any dry-pipe valve is tripped or tested, the water supply line to it should be thoroughly flushed. The main drain below the valve should be opened wide, and water at full pressure should be discharged long enough to clear the pipe of any accumulation of scale or foreign material. If there is a hydrant on the supply line, this hydrant should be flushed before the main drain is opened.

4-8.3.4 Caution. The tripping of dry-pipe valves with throttled water supplies will not completely operate some models that require a high rate of flow to complete movement of the clapper assemblies.

4-8.3.5 All dry-pipe valves should have a tag or card attached showing the date on which the valve was last tripped and showing the name of the person and the organization making the test. Separate records of initial air and water pressures, tripping time and tripping air pressure, and dry-pipe valve operating condition should be kept for comparison with previous test records.

4-8.4 Trip Test Full Flow. Each dry-pipe valve should be trip tested with control valve wide open at least once every three years or when the system is altered. This test should be conducted by opening the inspector's test pipe. The test should be terminated when the dry-pipe valve has tripped and clean water is flowing at the inspector's test connection.

A full flow trip test is recommended only once every three years with a restricted flow trip test the other two years because, while full flow tests must be periodically conducted, they have some undesirable side effects. The high-velocity flow will tend to draw foreign material into the system, and the wetting of the pipe wall will result in the development of scale. There is also the problem of draining all the water from the system.

4-9 Air Compressor.

4-9.1 An air compressor should be lubricated only if recommended by the manufacturer and in accordance with his instructions. The motor unit should be kept dirt free. Filters and strainers should be cleaned as required. Crystals in air dryers should be replaced when color changes indicate they have absorbed moisture.

4-10 Air Maintenance Device.

4-10.1 Strainers, filters, and restriction orifices should be cleaned as required. If regulator is provided with a drain cock, periodically remove condensation.

4-11 Quick-Opening Devices (Accelerator or Exhauster).

See 5-2.4 of NFPA 13, *Standard for the Installation of Sprinkler Systems.*

4-11.1 The quick-opening device should be tested at least twice a year.

The manufacturer's instructions for testing and resetting the device should be carefully followed. If the device does not operate properly when tested, the dry-pipe system should be kept in service and the device repaired or replaced immediately. Repair parts or a replacement device should be obtained from the original manufacturer.

4-12 Deluge, Preaction, and Automatic On-Off Preaction Systems.

See Section 5-3 of NFPA 13, *Standard for the Installation of Sprinkler Systems.*

4-12.1   Complete charts are furnished by the installing company, showing the proper method of operating and testing these systems. Only competent mechanics fully instructed with respect to the details and operation of such systems should be employed in their repair and adjustment. It is highly advisable for the owner to arrange with the installing company for at least annual inspection and testing of the equipment.

4-12.2   In preaction systems when it is necessary to repair the actuating system, as distinguished from the piping system itself, the water may be turned into the sprinkler piping, and automatic sprinkler protection thus maintained without alarm service, provided there is no danger of freezing.

4-12.3   Test detection systems semiannually and alarms quarterly according to the procedures suggested by the manufacturer.

5

Flushing

5-1 Flushing.

5-1.1 For effective control and extinguishment of fire, automatic sprinklers should receive an unobstructed flow of water. Although the overall performance record of automatic sprinklers has been very satisfactory, there have been numerous instances of impaired efficiency because sprinkler piping or sprinklers were plugged with pipe scale, mud, stones, or other foreign material. If the first sprinklers to open in a fire are plugged, the fire in that area will not be extinguished, an excessive number of sprinklers will operate causing increased water damage, and possibly the fire will spread out of control.

5-2 Types of Obstruction Material.

5-2.1 Obstructions may consist of compacted fine materials, such as rust, mud, or sand. Pipe scale is found more frequently in dry-pipe than in wet systems. Dry-pipe systems that have been maintained wet or dry alternately over a period of years are particularly susceptible to the accumulation of scale. Also, in systems continuously dry, condensation of moisture in the air supply may result in the formation of a hard scale along the bottom of the piping. When sprinklers open, the scale is broken loose and carried along the pipe, plugging some of the sprinklers or forming obstructions at the fittings.

5-2.2 Stones of various sizes, cinders, cast-iron pipe tubercles, chips of wood, or other coarse materials may be found. Sprinkler piping is sometimes partially obstructed by such objects as pieces of wood, paint brushes, broken pump valves or springs, or excess materials from improperly poured pipe joints. Materials may be sucked from the bottom of streams or reservoirs by fire pumps with poorly arranged or inadequately screened intakes and forced into the system. Sometimes floods damage intakes. Other materials may be permitted to enter by careless workers during installation or extensions of mains.

5-3 Preventing Entrance of Obstructive Material.
The following measures should be taken to assure, as far as possible, that sprinkler systems are clear of obstructive foreign matter and will remain unobstructed.

5-3.1 Take care when installing underground mains, both public and private, to prevent entrance of stones, soil, or other foreign material. As assurance that such material has not entered newly installed sprinkler systems from underground mains, installers are required as a condition of acceptance to flush all newly installed mains before connecting the inside piping. Private fire service mains should also be flushed after repairs or when breaks have occurred in public mains.

5-3.2 Screen pump suction supplies and maintain screens in good condition. Equip connections from penstocks with strainers or grids, unless the penstock inlets themselves are so equipped.

5-3.3 Keep dry-pipe systems on air the year round, instead of alternately on air or water, to inhibit formation of rust and scale.

5-3.4 Use extreme care when cleaning tanks and open reservoirs to prevent material from entering piping. Materials removed from the interior of gravity tanks during cleaning should not be permitted to enter the discharge pipe.

5-4 Conditions Showing Need for Investigation. Although precautions for preventing entrance of obstructive materials are generally followed at well-maintained premises, evidence based on fire experience and hundreds of flushing investigations shows that some sprinkler systems are obstructed to an extent that would seriously impair their effectiveness during a fire.

5-4.1 Conditions that may indicate the need of investigation include the following:

(a) Defective intake screens for fire pumps taking suction from streams and reservoirs.

(b) Discharge of obstructive material during routine water tests.

(c) Foreign material in fire pumps, in dry-pipe valves, or in check valves.

(d) Heavy discoloration of water during drain tests or plugging of inspector's test connections.

(e) Plugging of sprinklers.

(f) Plugged piping in sprinkler systems dismantled during building alterations.

(g) Failure to flush underground mains following installations or repairs.

(h) A record of broken public mains in the vicinity.

(i) Abnormally frequent tripping of dry-pipe valve.

5-4.2 Sprinkler systems should be examined internally at periodic intervals for obstructions. Where unfavorable conditions such as those itemized above are found, the system should be examined at five-year intervals after installation or possibly sooner. Where conditions are favorable, dry-pipe systems should be examined at ten-year intervals after installation.

5-4.3 Dry-pipe systems found obstructed should be flushed and reexamined at intervals of not more than five years.

5-5 Precautions. When sprinkler systems are to be shut off for investigation or for flushing, take all the precautions outlined earlier. To prevent accidental water damage, control valves should be shut tight and the system completely drained before sprinkler fittings are removed or pipes disconnected. Cover stock and machinery susceptible to water damage, and provide equipment for mopping up any accidental discharge of water.

5-5.1 Large quantities of water are required for effective flushing by the hydraulic method, and it is important to plan in advance the most convenient methods of disposal.

5-6 Investigation Procedure.

5-6.1 From the plan of the fire protection system, determine the sources of water supply, age of mains and sprinkler systems, types of systems, and general piping arrangement.

5-6.2 Examine the fire pump suction supply and screening arrangements. If needed, have the suction cleaned before using the pump in tests and flushing operations. Gravity tanks should be inspected internally, except steel tanks recently cleaned and painted. If possible, have the tank drained and determine whether there is loose scale on the shell or sludge or other obstruction on the tank bottom. Cleaning and repainting may be required, particularly if not done within five years.

5-7 Test—Flushing Mains.

5-7.1 Use hydrants near the ends of mains for flow tests to determine whether mains contain obstructive material. If such material is found, mains should be thoroughly flushed before investigating sprinkler systems. Connect two lengths of 2½-in. hose to the hydrant. Attach burlap bags to free ends of the hose from which the nozzles have been removed to collect any material flushed out, and flow water long enough to determine condition of the main being investigated. If there are several sources of water supply, investigate each independently, avoiding any unnecessary interruptions to sprinkler protection. On extensive layouts repeat the tests at several hydrants to determine general conditions.

5-8 Testing Sprinkler Systems.

5-8.1 Investigate the dry systems first. Tests on several carefully selected, representative systems usually are sufficient to indicate general conditions throughout the premises. When preliminary investigations indicate considerable obstructive material, this would justify investigating all systems (both wet and dry) before outlining needed flushing operations.

5-8.2   In selecting specific systems or pipes for investigation, consider:

(a) Pipes found obstructed during a fire or during maintenance work

(b) Systems adjacent to points of recent repair to yard mains, particularly if hydrant flow shows material in the main

(c) Pipes involving long horizontal runs of feed and cross mains. Obstructions are most likely to be found in the most remote branch lines at the end of the longest cross main from the longest feed main, particularly if the branch lines are lower than part of the feed main, as under a deck or platform.

5-8.3   Tests should include flows through 2½-in. fire hose directly from cross mains [*see Figures 5-8.3(a) and 5-8.3(b)*] and flows through 1½-in. hose from representative branch lines.

5-8.4   The fire pump should be operated for the large volume flows, as maximum pressure is desirable. Burlap bags should be used to collect dislodged material as is done in the investigation of yard mains, and each flow should be continued long enough to show the condition of the piping interior. After a test, leave all valves open and locked or sealed.

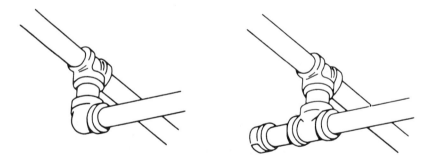

Figure 5-8.3(a)   **Replacement of Elbow at End of Cross Main with a Flushing Connection Consisting of a 2-in. Nipple and Cap.**

5-9 Dry-Pipe Systems.

5-9.1   Having selected the test points of a dry-pipe system, close the main control valve and release air from the system. Check the piping visually with a flashlight while it is being dismantled. Attach hose valves and 1½-in. hose to ends of branch lines to be tested, shut these valves, and have air pressure restored on the system and the control valve reopened. Open the hose valve on the end branch line allowing the system to trip in simulation of normal

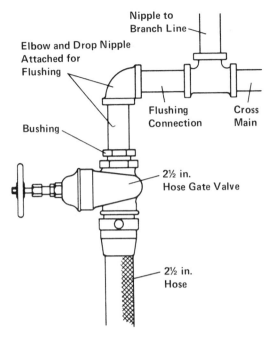

Figure 5-8.3(b)   Connection of 2½-in. Hose Gate Valve with 2-in. Bushing, and Nipple and Elbow to 2-in. Cross Main.

action. If this test plugs the hose or piping, the extent of plugging should be noted and cleared from the branch line before proceeding with further tests.

5-9.2   After flowing the small end line, shut its hose valve and test the feed or cross main by discharging water through a 2½-in. fire hose, collecting any foreign material in a burlap bag.

5-9.3   After the test, the dry-pipe valve should be internally cleaned and reset in the normal manner. Its control valve should be opened, sealed, and a drain test made.

5-10 Wet Systems.

5-10.1   Testing of wet systems is similar to that of dry systems except that the system must be drained after closing the control valve to permit installation of hose valves for the test. Slowly reopen the control valve and make a small hose flow as prescribed for the branch line, followed by the 2½-in. hose flow for the cross main.

5-10.2   In any case, if lines become plugged during the tests, piping must be dismantled and cleaned, the extent of plugging noted, and a clear flow obtained from the branch line before proceeding further.

5-10.3 Make similar tests on representative systems to indicate the general condition of the wet systems throughout the installation, keeping a detailed record of what is done.

5-11 Outside Sprinklers for Protection Against Exposure Fires.

5-11.1 Outside or open sprinkler equipment should be flow tested once each year during warm weather. Before making flow tests, proper precautions should be taken to prevent damage from water discharge. Flow tests will determine that the sprinklers and the system piping are in good condition and free of obstructions. Obstructed sprinklers or piping should be cleared immediately.

5-12 Flushing Procedure.

5-12.1 If investigation indicates the presence of sufficient material to obstruct sprinklers, a complete flushing program should be carried out. The work should be done by qualified competent personnel. The source of the obstructing material should be determined and steps taken to prevent further entrance of such material. This entails such work as inspection and cleaning of pump suction screening facilities or cleaning of private reservoirs. If recently laid public mains appear to be the source of the obstructing material, waterwork authorities should be requested to flush their system. For recommendations and procedures for cleaning pump suctions see NFPA 20, *Standard for the Installation of Centrifugal Fire Pumps.*

5-13 Private Fire Service Mains.

5-13.1 Mains should be thoroughly flushed before flushing any interior piping. Flush piping through hydrants at dead ends of the system or through blowoff valves, allowing the water to run until clear. If the water is supplied from more than one direction or from a looped system, close divisional valves to produce a high velocity flow through each single line. A velocity of at least 6 ft per second (1.8 m/s) is necessary for cleaning the pipe and for lifting foreign material to an aboveground flushing outlet. Use the flow specified in Table 5-13.1 for the size of the main under investigation.

Table 5-13.1   Waterflow Recommended for Flushing Piping

Size of Pipe	Flow	
In.	gpm	(L/min)
4	400	(1514)
6	750	(2839)
8	1000	(3785)
10	1500	(5678)
12	2000	(7570)

5-13.2 Connections from main to sprinkler riser should be flushed. Although flow through a short open-ended 2-in. drain may create sufficient velocity in a 6-in. main to move small obstructing material, the restricted waterway of the globe valve usually found on a sprinkler drain may not allow stones and other large objects to pass. If presence of large size material is suspected, a larger outlet will be needed to pass such material and to create the 750-gpm (2839-L/min) flow necessary to move it. Fire department connections on sprinkler risers can be used as flushing outlets by removing the clappers. Mains can also be flushed through a temporary siamese fitting attached to the riser connection before the sprinkler system is installed (*see Figure 5-13.2*).

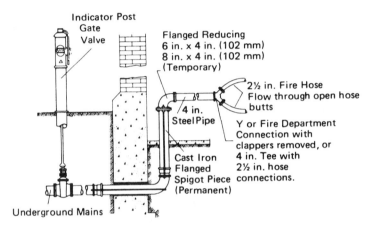

Figure 5-13.2   Arrangement for Flushing Branches from Underground Mains to Sprinkler Risers.

5-13.3 Sprinkler Piping.

5-13.3.1 Two methods are commonly used for flushing sprinkler piping: (a) the hydraulic method and (b) the hydropneumatic method.

(a) The hydraulic method consists of flowing water progressively from the mains, sprinkler risers, feed mains, cross mains, and finally the branch lines in the same direction in which it would flow during a fire.

(b) The hydropneumatic method utilizes special equipment and compressed air to blow a charge of approximately 30 gal (114 L) of water from the ends of branch lines back into feed mains and down the riser, washing the foreign material out of an opening at the base of the riser.

The hydraulic method of flushing gridded systems is explained in 5-13.4.8. This recommended practice does not address hydropneumatic flushing of gridded systems, as there has not yet been enough experience to establish a preferred procedure.

5-13.3.2   The choice of method depends on conditions at the individual premises. If examination indicates the presence of loose sand, mud, or moderate amounts of pipe scale, the piping can generally be satisfactorily flushed by the hydraulic method. Where the material is more difficult to remove and available water pressures are too low for effective scouring action, the hydropneumatic method is generally more satisfactory.

5-13.3.3   In some cases, where obstructive material is solidly packed or adheres tightly to the walls of the piping, the pipe will have to be dismantled and cleaned by rodding or other positive means.

5-13.3.4   Successful flushing by either the hydraulic or hydropneumatic method is dependent on establishing sufficient velocity of flow in the respective pipes to remove silt, scale, and other obstructive material. With the hydropneumatic method, this is accomplished by the air pressure behind the charge of water.

5-13.3.5   When flushing a branch line through the end pipe, sufficient water should be discharged to scour the largest pipe in the branch line. Lower rates of flow may reduce the efficiency of the flushing operation. To establish the recommended flow, remove small end piping and connect hose to larger section, if necessary.

5-13.3.6   To determine that the piping is clear after it has been flushed, representative branch lines and cross mains should be investigated, using both visual examination and sample flushings.

5-13.3.7   Whenever any section of piping is found severely or completely obstructed with packed material, such as hard scale, cinders, or gravel, the piping will usually have to be disassembled to remove the material.

5-13.3.8   Where pipe scale indicates internal or external corrosion, a section of the pipe affected should be thoroughly cleaned to determine if the walls of the pipe have seriously weakened.

5-13.4 Hydraulic Method.   After the mains have been thoroughly cleared, flush risers, feed mains, cross mains, and finally the branch lines. Following this sequence will prevent drawing obstructing material into the interior piping.

5-13.4.1   Water should be turned into dry-pipe systems for one to two days before flushing, if possible, to soften pipe scale and deposits. When alarm is turned off due to this procedure, consideration should be given to providing watch service during the unattended hours. To flush risers, feed mains, and cross mains, attach 2½-in. hose gate valves to the extreme ends of these lines [see Figure 5-8.3(b).] Such valves usually can be procured from the manifold of fire pumps or hose standpipes. As an alternative, an adapter with 2½-in. hose thread and standard pipe thread can be used with a regular gate valve. A length of fire hose without a nozzle should be attached to the flushing connection. To prevent kinking of the hose and to obtain maximum flow, an elbow should usually be installed between the end of the sprinkler pipe and the hose gate valve. Attach the valve and hose so that no excessive strain will be placed on the threaded pipe and fittings. Support hose lines properly.

5-13.4.2 Where feed and cross mains and risers contain pipe 4, 5, and 6 in. in size, it may be necessary to use a fire department connection with two hose connections to obtain sufficient flow to scour this larger pipe.

5-13.4.3 In multistory buildings, systems should be flushed by starting at the lowest story and working up. Branch line flushing in any story may follow immediately the flushing of feed and cross mains in that story, allowing one story to be completed at a time.

5-13.4.4 Where a repetition of the trouble is probable, leave a 2-in. capped nipple at the ends of the cross mains for flushing piping. Sprinkler installation rules require that a flushing connection be provided at the end of each cross main terminating in 1¼-in. or larger pipe.

5-13.4.5 Flush branch lines after feed and cross mains have been thoroughly cleared. This will avoid drawing obstructing material from these pipes into the branches. Equip the ends of several branch lines with gate valves, and flush individual lines of the group consecutively. This will eliminate the need for shutting off and draining the sprinkler system to change a single hose line. The hose should be 1½ in. and as short as practicable. Branch lines may be flushed in any order that will expedite the work.

5-13.4.6 Branch lines may also be flushed through pipe 1½ in. or larger extending through a convenient window. If pipe is used, 45° elbows should be provided at the ends of branch lines. When flushing branch lines, hammering the pipes is an effective method of moving obstructions.

5-13.4.7 All pendent sprinklers should be removed and cleaned of obstructions.

5-13.4.8 Flushing Gridded Sprinkler Systems. All new gridded sprinkler systems should be arranged so that they can be thoroughly flushed. Figure 5-13.4.8 should be used as a guide. Other arrangements accomplishing the same results are acceptable.

In the case of a system such as in Figure 5-13.4.8, the flushing procedure is as follows:

(a) Disconnect all branch lines close to the secondary cross main, and cap or valve all open ends supplied by the primary cross main. Visually examine the interior of each branch line connected to the secondary cross main and plug or cap.

(b) Flush the primary cross main, first one end, then the other.

(c) Flush each branch line independently.

(d) Flush the secondary cross main from an auxiliary water source.

(e) Reconnect branch lines.

(f) Flush the secondary cross main, first one end, then the other.

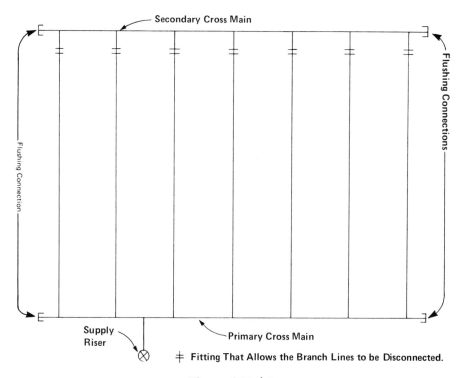

Figure 5-13.4.8

5-13.5 Hydropneumatic Method. The apparatus used for hydropneumatic flushing consists of a hydropneumatic machine, a source of compressed air, 1-in. (25-mm) air supply hose, 1½-in. hose for connecting to the sprinkler system, and 2½-in. hose.

5-13.5.1 The hydropneumatic machine (*see Figure 5-13.5.1*) consists of a 30-gal (114-L) water tank mounted over a 25-cu ft [(185-gal) (700-L)] compressed air tank. The compressed air tank is connected to the top of the water tank through a 2-in. lubricated plug cock. The bottom of the water tank is connected through hose to a suitable water supply. The compressed air tank is connected through suitable air hose to either the plant air system or a separate air compressor.

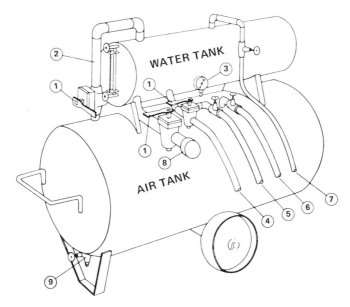

Figure 5-13.5.1   Hydropneumatic Machine.

1. Lubricated plug cocks.

2. Pipe connection between air and water tanks. This connection is open when flushing sprinkler system.

3. Air pressure gage.

4. 1-in. (25-mm) rubber hose (air type). Used to flush sprinkler branch lines.

5. Hose connected to source of water. Used to fill water tank.

6. Hose connected to ample source of compressed air. Used to supply air tank.

7. Water tank overflow hose.

8. 2½-in. pipe connection. When flushing large interior piping, connect woven jacket fire hose here and close 1-in. (25-mm) plug cock hose connection (4) used for flushing sprinkler branch lines.

9. Air tank drain valve.

5-13.5.2   To flush the sprinkler piping, the water tank is filled with water, the pressure raised to 100 psi (6.90 bars) in the compressed air tank, and the plug cock between tanks opened to put air pressure on the water. The water tank is connected by hose to the sprinkler pipe to be flushed. Then the lubricated plug cock on the discharge outlet at the bottom of the water tank is snapped open, permitting the water to be "blown" through the hose and sprinkler pipe by the compressed air. The water tank and air tank should be recharged after each blow.

5-13.5.3   Outlets for discharging water and obstructing material from the sprinkler system must be arranged. With the clappers of dry-pipe valves and alarm check valves on their seats and cover plates removed, sheet metal fittings can be used for connection to 2½-in. hose lines or for discharge into a drum. [Maximum capacity per blow is about 30 gal (114 L).] If the main riser drain is to be used, the drain valve should be removed and a direct hose connection made. For wet-pipe systems with no alarm check valves, the riser should be taken apart just below the drain opening and a plate inserted to prevent foreign material from dropping to the base of the riser. Where dismantling of a section of the riser for this purpose is impractical, the hydropneumatic method should not be used.

5-13.5.4 Before starting a flushing job, each sprinkler system to be cleaned should be studied and a schematic plan prepared showing the order of the blows.

5-13.5.5 Hydropneumatic Method of Flushing Branch Lines. With the mains already flushed or known to be clear, the branch sprinkler lines should next be flushed. The order of cleaning individual branch lines must be carefully laid out if an effective job is to be done. In general, flush the branch lines starting with the line closest to the riser and work toward the dead end of the cross main (*see Figure 5-13.5.5*).

The order of flushing the branch lines is shown by the circled numerals. In this example the southeast quadrant is flushed first, then the southwest, next the northeast, and last the northwest.

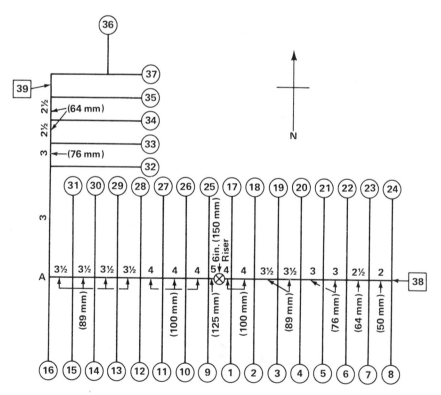

Figure 5-13.5.5   Schematic Diagram of Sprinkler System Showing Sequence to be Followed when Hydropneumatic Method is to be Used.

5-13.5.5.1 Air hose 1 in. (25 mm) in diameter is used to connect the machine with the end of the branch line being flushed. This hose should be as short as practicable. When the blow is made, the air pressure should be allowed to drop to 85 psi (5.9 bars) before the valve is closed. The resulting short slug of water will have less friction loss and a high velocity and hence do a more effective cleaning job than if the full 30 gal (114 L) of water is used. One blow is made for each branch line.

5-13.5.6 Hydropneumatic Method of Flushing Large Interior Piping. When flushing cross mains, completely fill the water tank and raise the pressure in the air receiver to 100 psi (6.9 bars). Connect the machine to the end of the cross main to be flushed with not over 50 ft (15.2 m) of 2½-in. hose. After opening the valve, allow air pressure in the machine to drop to zero. Two to six blows are necessary at each location, depending on the size and length of the main.

5-13.5.6.1 In Figure 5-13.5.5, the numerals in squares indicate the location and order of the cross main blows. Since the last branch line blows were west of the riser, clean the cross main east of the riser first. Where large cross mains are to be cleaned, it is suggested, if practical, to make one blow at 38, one at 39, the next at 38, then at 39, alternating in this manner until the required number of blows has been made at each location.

5-13.5.6.2 Cross mains are best flushed by introducing the blow at a point where water moving through the piping will make the least number of right-angle bends. In Figure 5-13.5.5, blows at 39 should be adequate to flush the cross mains back to the riser. Do not attempt to clean the cross mains from A to the riser by backing out branch line 16 and connecting the hose to the open side of the tee. If this were done, a considerable portion of the blow would pass northward up the 3-in. pipe list supplying branches 32 to 37, and the portion passing eastward to the riser could be ineffective. When the size, length, and condition of cross mains require blowing from a location corresponding to A, the connection should be made directly to the cross main corresponding to the 3½-in. pipe so that the entire flow would travel to the riser.

5-13.5.6.3 When flushing through a tee, always flush the run of tee after flushing the branch. Note the location of blows 35, 36, and 37 in Figure 5-13.5.5.

5-13.5.6.4 When flushing feed mains, arrange the work so that the water will pass through a minimum of right-angle bends.

5-13.5.6.5 The importance of doing a thorough flushing job should be strongly emphasized to those in charge of the work. In a number of instances, sprinklers in systems that had supposedly been flushed clear became plugged during a subsequent fire, permitting it to get out of control and cause serious damage.

6

Impairments

6-1 General. Valves controlling the water supply to all or part of a sprinkler system should be electrically supervised, or sealed, locked, or equipped with tamper switches and should be inspected frequently because of their importance to fire protection. The closing of control valves without proper authorization or preparation can seriously jeopardize operations. If sprinkler systems, fire hydrants, ground storage tanks, gravity tanks, fire pumps, etc. are impaired, the consequences may result in loss of life and damage to property. It is essential that adequate measures are taken during a fire protection impairment to ensure that the increased risks are minimized.

6-2 Impairment Coordinator. A representative of the building owner, manager, or tenant should be assigned to coordinate all impairments and restoration of protection.

6-3 Preplanned Impairment Programs.

6-3.1 All preplanned impairments should be authorized by the Coordinator. Before authorization is given, he should be responsible for verifying that the following is accomplished:

(a) Determine the exact extent of the intended impairment.

(b) Inspect the area or buildings to be involved and determine the increased risks.

(c) Submit recommendations to management. Consideration should be given to the need for temporary protection, termination of all hazardous operations, and frequent inspections of the area involved with 24-hour per day watchman service.

(d) Notify the fire department.

(e) Notify the insurance carrier, the alarm company, and other appropriate authority and implement Tag Impairment System (if such system is in use).

(f) Notify the supervisors in the areas to be affected.

6-3.2   When all impaired equipment is restored to normal working order, the following should be accomplished:

(a) Verify that all control valves are fully opened and locked, sealed, or equipped with a tamper switch.

(b) Conduct a main drain and alarm test on each sprinkler riser affected.

(c) Maintain as large a portion of the system in service as possible.

(d) Advise supervisors that protection has been restored.

(e) Advise the fire department that protection has been restored.

(f) Advise the insurance carrier, the alarm company, and other appropriate authorities that protection has been restored.

6-4 Emergency Impairments.   Emergency impairments include sprinkler leakage, frozen or ruptured piping, equipment failure, etc. When this occurs, appropriate emergency action should be taken. The Coordinator should be contacted, and he should proceed to the extent possible to implement the preplanned impairment program including the restoration of sprinkler protection.

6-5 Restoring Systems to Service after Disuse.

6-5.1   Occasionally, automatic sprinkler systems in idle or vacant properties are shut off and drained. When the equipment in such properties is restored to service, it is recommended that such work be performed by a responsible and experienced sprinkler contractor. In such cases, the following procedures are recommended:

6-5.1.1   All lines of sprinkler piping should be traced from the extremities of the system to the main connections with a careful check for blank gaskets in flanges, closed valves, corroded or damaged sprinklers or piping, insecure or missing hangers, and insufficient support. Proper repairs or adjustments should be made and needed extensions or alterations of the equipment should be completed.

6-5.1.2   Air should be used to test the system for leaks before turning on the water. Water should be admitted slowly to the system, with proper precautions against damage by escape of water from previously undiscovered defects. When the system has been filled under normal service pressure, drain valve tests should be made to detect any closed valve that possibly could have been overlooked. All available test pipes then should be flushed, and where such pipes are not provided in accordance with the present standards, the proper equipment should be installed.

6-5.1.3   Where the sprinkler system has been long out of service, damaged by freezing, or subject to extensive repairs or alterations, the entire system should be hydrostatically tested in accordance with NFPA 13, *Standard for the Installation of Sprinkler Systems.* Special care should be taken to detect any sprinklers showing leaks and to make replacements where necessary.

6-5.1.4 Dry-pipe valves, quick-opening devices, alarm valves and all alarm connections should be examined, put in proper condition and tested. Fire pumps, pressure and gravity tanks, reservoirs, and other water supply equipment should receive proper attention before being placed in service. Each supply should be tested separately.

6-5.1.5 An investigation for obstruction or stoppage in the sprinkler system piping should be made. (*See Chapter 5.*)

6-5.1.6 All control valves should be operated from closed to fully open position and should be left sealed, locked, or equipped with a tamper switch.

6-6 Sprinkler System Alterations.

6-6.1 Alterations will usually involve an impairment to all or part of the sprinkler system. Any alteration to a sprinkler system should be done in accordance with NFPA 13, *Standard for the Installation of Sprinkler Systems*, or other applicable NFPA standards.

7

Fire Records

7-1 Protection Records.

7-1.1 In all businesses, it is desirable to keep records of inspection, testing, and maintenance of protection equipment. The exact program for any building or set of buildings should be tailored to a specific plan for the particular building, considering occupancy, types of protection, alarms provided, etc. In the development of a fire protection record plan, it is advisable to consider advice from various sources including the:

Authority Having Jurisdiction (Rating Bureaus, Fire Prevention Bureaus, Fire Marshals, etc.)

Manufacturers of Various Devices

Fire Insurance Companies

Independent Fire Protection Consultants

Sprinkler Contractors

Other Applicable NFPA Codes as outlined in the Appendix.

7-1.2 An individual within the organization should be designated (Protection Record Administrator) to implement inspection, testing, and maintenance programs. Some firms may elect to do the basic functions themselves and contract the more technically involved operations.

7-1.3 The person in charge should consider the use of master control records, including a copy of the manufacturer's instructions covering all devices.

7-2 Valve Tag Systems.

7-2.1 Closed valves have caused over 30 percent of all sprinkler system failures. Adoption of the valve tag systems should visually highlight and minimize this significant cause of unsatisfactory sprinkler system performance. (*See Figure 7-2.1.*)

Figure 7-2.1 is shown on page 458.

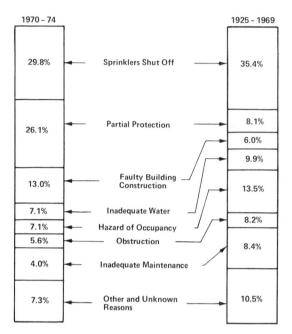

Figure 7-2.1   Reasons for Unsatisfactory Sprinkler System Performance.

7-2.2 Tag Impairment System.   This system usually includes a three-part tag that is easily identifiable. One part is tied to a valve to be closed temporarily. The second part is sent to the authority having jurisdiction. The third part is displayed in the protection record administrator's office and sent to the authority having jurisdiction when protection is restored.

7-2.3 Valve Record Tag System.   A tag on each valve indicating date sealed or locked, and date and results of maintenance procedures, should be provided. This provides a chronological record of valve maintenance.

7-3 Inspection Records.   Inspection and maintenance records of the following activities should be kept by the protection record administrator:

As pointed out in the commentary to 3-3.4, because of the new technology that has resulted in smaller, faster responding operating elements used in residential and other quick-response sprinklers, there is a need to monitor the field performance of these sprinklers. Since sensitivity is crucial to the performance of these sprinklers, it will be important to know if there has been a deterioration in such sensitivity over the years. This is another reason for the recommendation to test a sample of these types of sprinklers after a 20-year interval.

Table 7-3

Summary: Minimum Inspection, Testing and Maintenance

Records–Inspection = Visual Observation
      Testing = Handling Equipment, etc.
      Maintenance = Periodic Servicing and Repair

For Guidance on Specific Valves, Pumps, Hydrants, etc. Refer to the Manufacturer's Instructions

Parts	Activity	Frequency	Section Number
Flushing Piping	Test	5 years	5.4.2
Fire Department Connections	Inspection	Monthly	2.8
Control Valves	Inspection	Weekly-Sealed	2-7.1.4
	Inspection	Monthly-Locked	2-7.1.4
	Inspection	Monthly-Tamper Switch	2-7.1.4
	Maintenance	Yearly	2-7.1.8
Indicator Post Valve	Test	Quarterly	2-7.3.1
Valves in Roadway Boxes	Test	Quarterly	2-7.4.1
Main Drain	Flow Test	Quarterly	2-6.1
Open Sprinklers	Test	Annual	5-11.1
Pressure Gage	Calibration Test	5 years	4-4.2
Sprinklers	Test	50 years	3-3.3
Sprinklers—High Temp.	Test	5 years	3-3.1
Sprinklers—Residential	Test	20 years	3-3.4
Water Flow Alarms	Test	Quarterly	4-5.3
			4-7.1
			4-12.3
Preaction/Deluge Detection Systems	Test	Semiannually	4-12.3
Preaction/Deluge Systems	Test	Annually	4-12.1

(continued)

Table 7-3 (Continued)

Parts	Activity	Frequency	Section Number
Hydrants	Inspection	Monthly	2-5.1
	Test (Open and Close)	Annually	2-5.3
	Maintenance	Semi-annually	2-5.2
Antifreeze Solution	Test	Annually	4-7.3
Cold Weather Valves	Open and Close Valves	Fall, Close; Spring, Open	4-7.2
Dry/Preaction/Deluge systems Air Pressure and			
Water Pressure	Inspection	Weekly	4-8.2.4
Enclosure	Inspection	Daily-Cold Weather	4-8.2.5
Priming Water Level	Inspection	Quarterly	4-8.2.1
Low-Point Drains	Test	Fall	4-8.2.6
Dry-Pipe Valves	Trip Test	Annual-Sprigng	1-6.1
			4-8.3.1
Dry-Pipe Valves	Full Flow Trip	3 years-Spring	4-8.4
Quick Opening Devices	Test	Semi-annually	4-11.1
Gravity Tank			
Water Level	Inspection	Monthly	2-2.1
Heat	Inspection	Daily-Cold Weather	2-2.2
Condition	Inspection	Biannual	NFPA 22
Pressure Tank			
Water Level & Pressure	Inspection	Monthly	2-3.1
Heat Enclosures	Inspection	Daily-Cold Weather	2-3.7
Condition	Inspection	3 years	2-3.2
Pump	Test Flow	Annually	2-4.2.5
Engine Drive	Test Operate	Weekly	2-4.2.1
Motor Drive	Test Operate	Monthly	2-4.2.1
Steam Drive	Test Operate	Weekly	NFPA 21

Exhibit I Report of Inspection

Owner's Section (To be answered by Owner or Occupant)

A. Explain any occupancy hazard changes since the previous inspection.

B. Describe fire protection modifications made since last inspection.

C. Describe any fires since last inspection.

D. When was the system piping last checked for stoppage, corrosion, or foreign material?

E. When was the dry-piping system last checked for proper pitch?

F. Are dry valves adequately protected from freezing?

Inspector's Section (All responses reference current inspection)

1. General

    a. Is the building occupied?

    b. Are all systems in service?

    c. Is there a minimum of 18 in. (457 mm) clearance between the top of the storage and the sprinkler deflector?

    d. In areas protected by a wet system, does the building appear to be properly heated in all areas, including blind attics and perimeter areas, where accessible? Do all exterior openings appear to be protected against freezing?

    e. Does the hand hose on the sprinker system appear to be satisfactory?

2. Control Valves (*See Item 14.*)

    a. Are all sprinkler system control valves and all other valves in the appropriate open or closed position?

    b. Are all control valves in the open position and locked, sealed, or equipped with a tamper switch?

3. Water Supplies (*See Item 15.*)

    a. Was a water flow test of the main drain made at the sprinkler riser?

4. Tanks, Pumps, Fire Department Connections

a. Are fire pumps, gravity tanks, reservoirs, and pressure tanks in good condition and properly maintained?

b. Are fire department connections in satisfactory condition, couplings free, caps in place, and check valves tight? Are they accessible and visible?

5. Wet Systems (*See Item 13.*)

a. Are cold weather valves (O.S.&Y.) in the appropriate open or closed position?

b. Have antifreeze system solutions been tested?

c. Were the antifreeze test results satisfactory?

6. Dry Systems (*See Items 10 to 14.*)

a. Is the dry valve in service?

b. Are the air pressure and priming water level in accordance with the manufacturer's instructions?

c. Has the operation of the air or nitrogen supply been tested? Is it in service?

d. Were low points drained during this inspection?

e. Did quick-opening devices operate satisfactorily?

f. Did the dry valve trip properly during the trip pressure test?

g. Did the heating equipment in the dry-pipe valve room operate at the time of inspection?

7. Special Systems — as defined in Section 1-3 (*See Item 16.*)

a. Did the deluge or preaction valves operate properly during testing?

b. Did the heat-responsive devices operate properly during testing?

c. Did the supervisory devices operate during testing?

8.   Alarms

   a.   Did water motor and gong test satisfactorily?

   b.   Did electric alarm test satisfactorily?

   c.   Did supervisory alarm service test satisfactorily?

9.   Sprinklers

   a.   Are all sprinklers free from corrosion, loading, or obstruction to spray discharge?

   b.   Are sprinklers over 50 years old, thus requiring sample testing?

   c.   Is stock of spare sprinklers available?

   d.   Does the exterior condition of the sprinkler system appear to be satisfactory?

   e.   Are sprinklers of proper temperature ratings for their locations?

10.   Date dry-pipe valve trip tested (control valve partially open)

_____

(See Trip Test Table that follows.)

11.   Date dry-pipe valve trip tested (control valve fully open)

_____

(See Trip Test Table that follows.)

12.   Date quick-opening device tested

_____

(See Trip Test Table that follows.)

13.   Date deluge or preaction valve tested

_____

(See Trip Test Table that follows.)

14.   Review Control Valve Maintenance

_____

(See Control Valve Maintenance Table that follows.)

		DRY VALVE			Q.O.D.		
		MAKE	MODEL	SERIAL NO.	MAKE	MODEL	SERIAL NO.

		TIME TO TRIP THRU TEST CONNECTION*		WATER PRESSURE	AIR PRESSURE	TRIP POINT AIR PRESSURE	TIME WATER REACHED TEST OUTLET*		ALARM OPERATED PROPERLY	
DRY PIPE OPERATING TEST		MIN.	SEC.	PSI	PSI	PSI	MIN.	SEC.	YES	NO
	Without Q.O.D.									
	With Q.O.D.									
	IF NO, EXPLAIN									

	OPERATION							
	☐ PNEUMATIC ☐ ELECTRIC ☐ HYDRAULIC							
	PIPING SUPERVISED ☐ YES ☐ NO DETECTING MEDIA SUPERVISED					☐ YES	☐ NO	
DELUGE & PREACTION VALVES	DOES VALVE OPERATE FROM THE MANUAL TRIP AND/OR REMOTE CONTROL STATIONS					☐ YES	☐ NO	
	IS THERE AN ACCESSIBLE FACILITY IN EACH CIRCUIT FOR TESTING			IF NO, EXPLAIN				
	☐ YES ☐ NO							
	MAKE	MODEL	DOES EACH CIRCUIT OPERATE SUPERVISION LOSS ALARM		DOES EACH CIRCUIT OPERATE VALVE RELEASE		MAXIMUM TIME TO OPERATE RELEASE	
			YES	NO	YES	NO	MIN.	SEC.

Control Valve Maintenance Table

Control Valves	Number	Type	Open	Secured	Closed	Signs	Explain Abnormal Condition
City Connection Control Valve							
Tank Control Valves							
Pump Control Valves							
Sectional Control Valves							
System Control Valves							
Other Control Valves							

15. Water Flow Test at Sprinkler Riser

Water Supply Source		City	Tank	Pump
Date	Test Pipe Location	Size Test Pipe	Static Pressure	Residual (Flow) Pressure
Last Water Flow Test				
This Water Flow Test				

16. Heat Responsive Devices

Test Method _____

_____

Type of Equipment _____

Manufacturer _____

Test Results:

Valve No. __A B C D E F__   Valve No. __A B C D E F__

Valve No. __A B C D E F__   Valve No. __A B C D E F__

Valve No. __A B C D E F__   Valve No. __A B C D E F__

Valve No. __A B C D E F__   Valve No. __A B C D E F__

Auxiliary Equipment: No.? ___Type? ___Location? ___Test Result? ___

17. Explain any "No" answers and comments. _____

_____

_____

_____

_____

18. Adjustments or corrections made during this inspection: _____

_____

_____

_____

_____

19. Although these comments are not the result of an engineering review, the following desirable improvements are recommended:

_____

_____

_____

_____

Signature: _____

Date: _____

Appendix

This Appendix is not part of the requirements of this NFPA document, but is included for information purposes only.

A-1 Referenced Publications. The following documents or portions thereof are referenced within this document for informational purposes only and thus are not considered part of the recommendations of this document. The edition indicated for each reference should be the current edition as of the date of the NFPA issuance of this document. These references should be listed separately to facilitate updating to the latest edition by the user.

A-1-1 NFPA Publications. National Fire Protection Association, Batterymarch Park, Quincy, MA 02269.

NFPA 13-1987, *Standard for the Installation of Sprinkler Systems*

NFPA 13E-1984, *Recommendations for Fire Department Operations in Properties Protected by Sprinkler and Standpipe Systems*

NFPA 20-1987, *Standard for the Installation of Centrifugal Fire Pumps*

NFPA 21-1982, *Standard for the Operation and Maintenance of National Standard Steam Fire Pumps*

NFPA 22-1984, *Standard for Water Tanks for Private Fire Protection*

NFPA 24-1984, *Standard for Private Fire Service Mains and Their Appurtenances*

NFPA 26-1983, *Recommended Practices for the Supervision of Valves Controlling Water Supplies for Fire Protection*

NFPA 33-1985, *Standard for Spray Application Using Flammable and Combustible Materials*

NFPA 231D-1986, *Standard for Storage of Rubber Tires*

NFPA 1962-1979, *Standard for the Care, Maintenance and Use of Fire Hose Including Connections and Nozzles.*

A-2 Other Publications.

A-2-1 A selected list of other NFPA publications related to the inspection, testing, and maintenance of sprinkler systems is as follows:

NFPA 14-1986, *Standard for the Installation of Standpipe and Hose Systems*

NFPA 71-1985, *Standard for the Installation, Maintenance and Use of Central Station Signaling Systems*

NFPA 72A-1985, *Standard for the Installation, Maintenance, and Use of Local Protective Signaling Systems*

NFPA 72B-1986, *Standard for the Installation, Maintenance and Use of Auxiliary Protective Signaling Systems*

NFPA 307-1985, *Construction and Fire Protection of Marine Terminals, Piers and Wharves*

NFPA 231-1987, *Standard for Indoor General Storage*

NFPA 231C-1980, *Standard for Rack Storage of Materials*

NFPA 231D-1986, *Standard for the Storage of Rubber Tires.*

A-2-2 Other Codes and Standards.   This publication makes reference to the following codes and standards.

FP-27, *Dry-Pipe Valves, Trip Tests and Quick Opening Devices*, available from American Insurance Services Group, Inc., Engineering and Safety Service Division, 85 John Street, New York, NY 10038.

ASME, *Non-Fired Pressure Vessel Code*, available from the American Society of Mechanical Engineers, East 47th Street, New York, NY 10017.

ASTM E380-1976, *Standard for Metric Practice*, available from the American Society for Testing and Materials, 1916 Race Street, Philadelphia, PA 19103.

REFERENCES CITED IN COMMENTARY

The following publications are available from the National Fire Protection Association, Batterymarch Park, Quincy, MA 02269.

NFPA 13-1989, *Standard for the Installation of Sprinkler Systems*

NFPA 14-1986, *Standard for the Installation of Standpipe and Hose Systems.*

NFPA 13D

Standard for the

Installation of Sprinkler Systems in One- and Two-Family Dwellings and Mobile Homes

1989 Edition

NOTICE: An asterisk (*) following the number or letter designating a paragraph indicates explanatory material on that paragraph in Appendix A. Material from Appendix A is integrated with the text and is identified by the letter A preceding the subdivision number to which it relates.

Information on referenced publications can be found in Chapter 5 and Appendix B.

Preface

It is intended that this standard provide a method for those individuals wishing to install a sprinkler system for additional life safety and property protection. It is not the purpose of this standard to require the installation of an automatic sprinkler system. This standard assumes that one or more smoke detectors will be installed in accordance with NFPA 74, *Standard for the Installation, Maintenance, and Use of Household Fire Warning Equipment*.

National attention focused on the residential fire problem — primarily the result of a report published by the Presidential Commission on Fire Prevention & Control titled "America Burning" (1973) — caused the NFPA Committee on Automatic Sprinklers to direct its attention to the residential fire problem. Thus, in the summer of 1973, the NFPA Sprinkler Committee established a Subcommittee on Residential and Light Hazard Occupancies. The subcommittee was charged with developing a standard that would produce a reliable but inexpensive sprinkler system for these occupancies, where the majority of fire deaths were and are occurring. In its first meeting, basic philosophies were established for residential systems that carry through to the current edition of NFPA 13D.

- Cost was of major importance. A system having slightly less reliability and fewer operational features than that described in NFPA 13 but which could be installed at a substantially lower cost was necessary if acceptance of a residential system was to be achieved.
- Life safety would be the primary goal of NFPA 13D, with property protection as a secondary goal.
- System design should be such that a fire could be controlled for sufficient time to enable people to escape; i.e., a 10-minute stored water supply with an adequate local audible alarm.
- Piping arrangements, components, and hangers must be compatible with residential construction techniques; combined sprinkler/plumbing systems were acceptable from a fire protection standpoint.
- The fire record could reasonably serve as a baseline to permit omission of sprinklers in areas of low incidents of fire deaths — thus saving cost.

The first draft document produced by the subcommittee actually encompassed residential systems for one- and two-family dwellings, mobile homes, and multifamily housing up to four stories in height. However, when finally adopted in 1975, the multifamily housing portion had been eliminated because of strong feelings that such systems needed to be designed in accordance with NFPA 13. This first standard was put together largely without formalized test data and adopted by NFPA at its May 1975 meeting. Some 18 years later, however, the need for an installation standard covering these low-rise residential occupancies was evident and NFPA 13R was developed. *(See page 537 for information on this new standard.)*

Beginning in 1976, the National Fire Prevention & Control Administration (later renamed the U.S. Fire Administration) funded a number of research programs to evaluate the residential fire problem in depth and determine the "what" and "how" of accomplishing a satisfactory solution. The NFPCA/USFA programs included studies to assess the probable impact of using sprinklers to reduce the incidence of deaths and injuries in residential fires.[1] Other studies evaluated the design, installation, practical usage, and user acceptance factors that would impact on accomplishing the installation of reliable and acceptable systems;[2] studies to evaluate minimum water discharge rates, automatic sprinkler flow, and response sensitivity design criteria;[3, 4] and full-scale tests to test prototype systems.[6, 7, 8, 9]

Collectively these research efforts provided an extensive data base from which a complete revision of the standard was developed and published in 1980. This portion of this handbook deals with the 1989 edition of the standard and is intended to give the reader some insight as to the rationale of the Committee in arriving at the standard's requirements as well as the data base that supports this rationale. Since 1980, the Committee has begun to broaden the standard to deal with special situations such as protection of piping in areas subject to freezing and placement of sprinklers beneath other than smooth, flat ceilings.

The serious student is encouraged to study the various publications listed in the bibliography that appears at the end of this section. The student is also cautioned that extensive research continues in this rapidly changing area of residential fire protection and that current data may necessitate the updating of the information contained herein.

1

General Information

1-1* Scope. This standard deals with the design and installation of automatic sprinkler systems for protection against the fire hazards in one- and two-family dwellings and mobile homes.

This standard applies only to one- and two-family dwellings and mobile homes as defined under Section 1-3. By allowing selective omission of sprinklers from certain areas and permitting the use of 10-minute water supplies, the prescribed level of protection afforded by this standard is inappropriate for multi-family (three or more) occupancies. Such systems should be designed in accordance with NFPA 13 or NFPA 13R.

Formal Interpretation

Question: Is NFPA 13D appropriate for use in multiple (three or more) attached dwellings under any condition?

Answer: No. NFPA 13D is appropriate for use only in one- and two-family dwellings and mobile homes. Buildings which contain more than two dwelling units shall be protected in accordance with NFPA 13. 3-11.2.9 of NFPA 13 permits residential sprinklers to be used in residential portions of other buildings provided all other requirements of NFPA 13, including water supplies, are satisfied.

Note: Building codes may contain requirements such as 2-hour fire separations which would permit adjacent dwellings to be considered unattached.

A-1-1 NFPA 13D is appropriate for the protection against fire hazards only in one- and two-family dwellings and mobile homes. Residential portions of any other building may be protected with residential sprinklers in accordance with 3-11.2.9 of NFPA 13-1989, *Standard for the Installation of Sprinkler Systems*. Other portions of such buildings should be protected in accordance with NFPA 13.

The criteria in this standard are based on full-scale fire tests of rooms containing typical furnishings found in residential living rooms, kitchens, and bedrooms. The furnishings were arranged as typically found in dwelling units in a manner similar to that shown in Figures A-1-1(a), (b), and (c). Sixty full-scale fire tests were conducted in a two-story dwelling in Los Angeles, California and 16 tests were conducted in a 14-ft (4.3-m) wide mobile home in Charlotte, North Carolina. Sprinkler systems designed and installed according to this standard are expected to prevent flashover within the compartment of origin if sprinklers are installed in the compartment. A sprinkler system designed and installed according to this standard cannot, however, be completely expected to control a fire involving unusually higher average fuel loads than typical for dwelling units [10 lb/sq ft (49 kg/m²)] and where the interior finish has an unusually high flame spread rating (greater than 225).

For protection of multifamily dwellings, refer to NFPA 13, *Standard for the Installation of Sprinkler Systems*, or NFPA 13R, *Standard for the Installation of Sprinkler Systems in Residential Occupancies up to Four Stories in Height.*

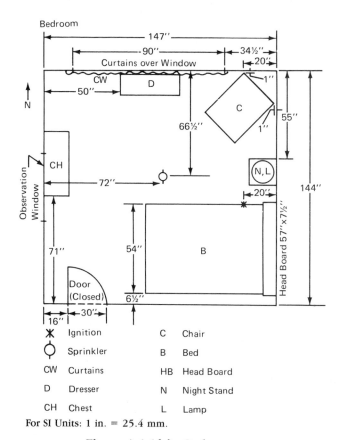

For SI Units: 1 in. = 25.4 mm.

Figure A-1-1(a)   Bedroom.

Mobile Home Bedroom

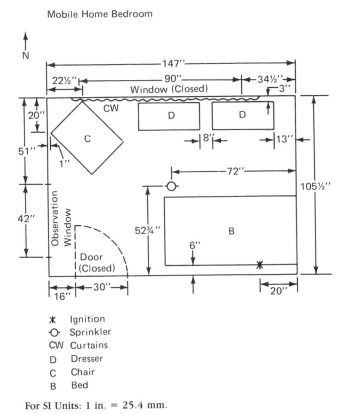

Figure A-1-1(b)   Mobile Home Bedroom.

While the requirements of this standard are based on the Los Angeles and North Carolina fire tests,[7, 8, 9, 10] earlier exploratory tests by Factory Mutual Research Corporation and Battelle Columbus Laboratories[3, 4] indicated that there is a potential for systems to be developed that will control residential fires at application rates as low as 0.025 gpm per sq ft [(1.02 L/min)/ m²]. These tests usually involved smoldering fires in typical residential room configurations and flaming fires using a proto-type (i.e., a gas burner) rather than a real-world fire array. Subsequent tests at Factory Mutual Research Corporation[6] and in test facilities involving fires in room configurations at Los Angeles and in North Carolina[7, 8, 9, 10] established that the design had to be based on the density and area of application found in 4-1.1 and 4-1.2 of this standard. The probability of successfully

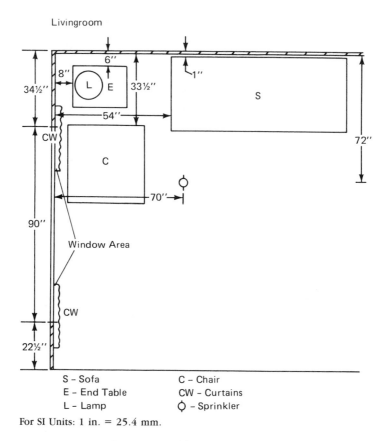

S – Sofa   C – Chair
E – End Table   CW – Curtains
L – Lamp   ○ – Sprinkler

For SI Units: 1 in. = 25.4 mm.

Figure A-1-1(c)   Living Room.

reducing deaths, injuries, and property damage is lowered
when the very low densities from the early testing referred to
above are provided.

1-2* Purpose. The purpose of this standard is to provide a sprinkler
system that will aid in the detection and control of residential fires and thus
provide improved protection against injury, life loss, and property damage.
A sprinkler system installed in accordance with this standard is expected to
prevent flashover (total involvement) in the room of fire origin, when
sprinklered, and to improve the chance for occupants to escape or be
evacuated.

Guidelines are established for the design and installation of sprinkler
systems for one- and two-family dwellings and mobile homes. Nothing in
this standard is intended to restrict new technologies or alternate arrangements providing the level of safety prescribed by the standard is not
lowered.

A-1-2 Levels of Protection. Various levels of firesafety are available to dwelling occupants to provide life safety and property protection.

This standard recommends, but does not require, sprinklering of all areas in a dwelling; it permits sprinklers to be omitted in certain areas. These areas are the ones shown by NFPA statistics [*see Table A-1-2(a)*] to be the ones where the incidence of life loss from fires in dwellings is low. Such an approach produces a reasonable degree of firesafety. Greater protection to both life and property will be achieved by sprinklering all areas.

Guidance for installation of smoke detectors and fire detection systems may be found in NFPA 74, *Standard for the Installation, Maintenance, and Use of Household Fire Warning Equipment.*

Both Underwriters Laboratories Inc. and Factory Mutual Research Corporation have developed test standards for the evaluation of residential sprinklers.[11, 12] These standards have significantly different fire test requirements from those found in the UL and FM standards for commercial sprinklers. The residential sprinkler criteria are:
- Maximum ceiling air temperature—approximately 600°F (315°C).
- Maximum temperature at 5 ft 3 in. (1.6 m) above floor—200°F (93°C).
- Maximum of two sprinklers in test area to operate during test.

The results of these criteria center on:

(a) Capability of the sprinkler to prevent flashover;

(b) Maintaining a 200°F temperature at "eye level," which will result in a lower, survivable temperature at or near the floor; and

(c) Achieving control of the fire with one or two operating sprinklers, thus allowing for a more "reasonable" water supply.

The laboratory test configuration is a 12-ft × 24-ft (3.7-m × 7.3-m) room with a combustible array simulating residential furnishings.

The Committee adopted the concept of "levels of protection" in order to achieve a low-cost system and to avoid having to install dry-pipe sprinkler systems in cold climates. The areas where omission of sprinklers is permitted per Section 4-6 of this standard are those where the record justifies such omission. [*See Table A-1-2(a).*] The Committee felt that dry-pipe systems were a less desirable form of protection because of the apparent delay in application of water. (This was noted in early Factory Mutual tests and substantiated in the Los Angeles and

Table A-1-2(a)
Causal Factors in
One- and Two-Family Dwelling
Fires Which Caused One or More Deaths

Area of Origin

Living Room	41%
Bedroom	27%
Kitchen	15%
Storage Area	4%
Heating Equipment Room	3%
Structural Area	2%
Other Areas	8%

Based on 6066 incidents where area of origin was reported

Form of Heat Ignition

Smoking Materials	36%
Heat from Fuel - Fire or Powered Object	25%
Heat from Miscellaneous Open Flame (Including Match)	15%
Heat from Electrical Equipment Arcing or Overload	14%
Hot Objects Including Properly Operating Electrical Equipment	7%
Other	3%

Based on 5016 incidents where form of heat of ignition was reported

Form of Material Ignited

Furniture	27%
Bedding	18%
Combustible Liquid or Gas	13%
Interior Finish	9%
Structural Member	9%
Waste, Rubbish	4%
Clothing, on a Person	3%
Cooking Materials	3%
Electrical Insulation	2%
Curtains, Drapery	2%
Other	10%

Based on 5080 incidents where form of material ignited was reported

Total number of incidents reported   10,194

Source: FIDO Data Base 1973 to 1982, NFPA Fire Analysis Department.

Table A-1-2(b)
1980-84 One- and Two-Family Dwellings and Mobile Homes
Annual Averages

Fires-544,000   Civilian Deaths-3,900   Civilian Injuries-14,100

Area of Origin (901 Code)	Civilian Deaths (Used for Ranking)	Percentages by Area of Origin		Civilian Injuries
		Fires		
Living room, den, lounge (14)	40.2	11.6		21.9
Bedroom (21-22)	24.1	11.6		20.9
Kitchen (24)	14.0	20.6		27.5
Structural Area (70-79)	5.8	15.5		7.4
[Crawl space (71)]	(1.5)	(3.2)		(2.9)
[Unspecified (79)]	(1.0)	(1.0)		(0.7)
[Balcony, porch (72)]	(0.9)	(1.1)		(0.9)
[Ceiling/Floor assembly (73)]	(0.7)	(0.8)		(0.5)
[Ceiling/Roof assembly (74)]	(0.6)	(2.3)		(0.7)
[Wall assembly (75)]	(0.6)	(2.0)		(0.8)
Dining room (23)	2.3	1.1		1.6
Heating equipment room (62)	1.9	3.7		3.6
Bathroom (25)	1.2	1.7		1.9
Hallway, corridor (01)	1.2	0.9		1.1
Garage* (47)	1.1	3.4		3.7
Interior stairway (03)	1.0	0.4		0.4
Closet (42)	0.9	1.2		1.3
Other known single area	4.2	26.6		7.5
[Chimney (51)]	(0.4)	(18.9)		(0.7)
Multiple areas (97)	0.8	0.7		0.6
Unclassified, not applicable (98-99)	1.3	1.0		0.6
Total:	100.0	100.0		100.0

* Does not include dwelling garages coded as property type, which is a large number.

North Carolina fire tests.[7, 8, 9, 10]) The key to achieving control of the residential fire is the rapid application of water. Further, residential systems must be simple and easy to maintain— factors not associated with dry-pipe systems.

1-3* Definitions.

Approved.   Acceptable to the "authority having jurisdiction."

NOTE: The National Fire Protection Association does not approve, inspect or certify any installations, procedures, equipment, or materials nor does it approve or evaluate testing laboratories. In determining the acceptability of installations or procedures, equipment or materials, the authority having jurisdiction may base acceptance on compliance with NFPA or other appropriate standards. In the absence of such standards, said authority may require evidence of proper installation, procedure or use. The authority having jurisdiction may also refer to the listings or labeling practices of an organization concerned with product evaluations which is in a position to determine compliance with appropriate standards for the current production of listed items.

Authority Having Jurisdiction.   The "authority having jurisdiction" is the organization, office or individual responsible for "approving" equipment, an installation or a procedure.

NOTE: The phrase "authority having jurisdiction" is used in NFPA documents in a broad manner since jurisdictions and "approval" agencies vary as do their responsibilities. Where public safety is primary, the "authority having jurisdiction" may be a federal, state, local or other regional department or individual such as a fire chief, fire marshal, chief of a fire prevention bureau, labor department, health department, building official, electrical inspector, or others having statutory authority. For insurance purposes, an insurance inspection department, rating bureau, or other insurance company representative may be the "authority having jurisdiction." In many circumstances the property owner or his designated agent assumes the role of the "authority having jurisdiction"; at government installations, the commanding officer or departmental official may be the "authority having jurisdiction."

Check Valve.   A valve which allows flow in one direction only.

Control Valve.*   A valve employed to control (shut) a supply of water to a sprinkler system.

A-1-3   System control valves should be of the indicating type, such as plug valves, ball valves, butterfly valves, or O. S. & Y. gate valves.

The type of control valve used should be an indicating type, i.e., one that has some external means that will indicate to the occupant of a home that the valve is in the open position.

Design Discharge.   Rate of water discharged by an automatic sprinkler expressed in gallons per minute.

Dry System. A system employing automatic sprinklers attached to a piping system containing air under atmospheric or higher pressures. Loss of pressure from the opening of a sprinkler or detection of a fire condition causes the release of water into the piping system and out the opened sprinkler.

It should be pointed out that because of the concern for delayed water delivery and the fact that there are no residential sprinklers listed for dry system application, residential dry systems are not allowed, per the Exception to 3-5.1. Further, currently available data from residential fire tests does not support the four-sprinkler design that was included in the 1980 edition of this standard. Rather than deleting all references to dry systems in the standard, selective wording has been left in place to allow a window of opportunity for development and listing of a dry-pipe residential sprinkler.

Dwelling. Any building which contains not more than one or two "dwelling units" intended to be used, rented, leased, let or hired out to be occupied, or which are occupied for habitation purposes.

Dwelling Unit. One or more rooms arranged for the use of one or more individuals living together as in a single housekeeping unit, normally having cooking, living, sanitary, and sleeping facilities.

Labeled. Equipment or materials to which has been attached a label, symbol or other identifying mark of an organization acceptable to the "authority having jurisdiction" and concerned with product evaluation, that maintains periodic inspection of production of labeled equipment or materials and by whose labeling the manufacturer indicates compliance with appropriate standards or performance in a specified manner.

Listed. Equipment or materials included in a list published by an organization acceptable to the "authority having jurisdiction" and concerned with product evaluation, that maintains periodic inspection of production of listed equipment or materials and whose listing states either that the equipment or material meets appropriate standards or has been tested and found suitable for use in a specified manner.

NOTE: The means for identifying listed equipment may vary for each organization concerned with product evaluation, some of which do not recognize equipment as listed unless it is also labeled. The "authority having jurisdiction" should utilize the system employed by the listing organization to identify a listed product.

Mobile Home. A factory-assembled structure equipped with service connections and made so as to be readily movable as a unit on its running gear and designed to be used as a dwelling unit with or without a foundation.

Multipurpose Piping Systems.   Piping systems within dwellings and mobile homes intended to serve both domestic and fire protection needs.

Preengineered System.   A packaged sprinkler system including all components connected to the water supply designed to be installed according to pretested limitations.

Pump.   A mechanical device that transfers and/or raises the pressure of a fluid (water).

Residential Sprinkler.   An automatic sprinkler which has been specifically listed for use in residential occupancies.

Residential sprinklers are specifically listed by UL and FM for residential service and have been tested for compliance with the residential sprinkler standard.[11, 12] Typically, they are fast response sprinklers, and they may be of the upright, pendent, or sidewall configuration. (*See the discussion and chart on fast response sprinklers in the commentary to 3-11.2 of NFPA 13.*)

Shall.   Indicates a mandatory requirement.

Should.   Indicates a recommendation or that which is advised but not required.

Sprinkler—Automatic.   A fire suppression device which operates automatically when its heat-actuated element is heated to or above its thermal rating allowing water to discharge over a specific area.

Sprinkler System.   An integrated system of piping connected to a water supply, with listed sprinklers that will automatically initiate water discharge over a fire area. When required, the sprinkler system also includes a control valve and a device for actuating an alarm when the system operates.

Standard.   A document containing only mandatory provisions using the word "shall" to indicate requirements. Explanatory material may be included only in the form of "fine print" notes, in footnotes, or in an Appendix.

Supply Pressure.   Pressure within the supply (i.e., city or private supply water source).

Supply pressure is the pressure that one can expect to attain from the city or other water supply, as explained in A-4-4.3.

System Pressure.   A pressure within the system (i.e., above the control valve).

System pressure is determined by subtracting friction losses between the street connection and the water supply valve in the residence from the water pressure in the street after making any necessary adjustments for changes in elevation, as explained in A-4-4.3 of this standard.

Waterflow Alarm. A sounding device activated by a waterflow detector or alarm check valve and arranged to sound an alarm that will be audible in all living areas over background noise levels with all intervening doors closed.

Waterflow Detector. An electric signaling indicator or alarm check valve actuated by water flow in one direction only.

Wet System. A system employing automatic sprinklers attached to a piping system containing water and connected to a water supply so that water discharges immediately from sprinklers opened by a fire.

1-4* Maintenance. The owner is responsible for the condition of a sprinkler system and shall keep the system in normal operating condition.

A-1-4 The responsibility for properly maintaining a sprinkler system is the obligation of the owner or manager who should understand the sprinkler system operation. A minimum monthly maintenance program should include the following:

(a) Visually inspect all sprinklers to ensure against obstruction of spray.

(b) Inspect all valves to assure that they are open.

(c) Test all waterflow devices.

(d) The alarm system, if installed, should be tested.

NOTE: When it appears likely that the test will result in a response of the fire department, notification to the fire department should be made prior to the test.

(e) Pumps, where employed, should be operated. (*See NFPA 20, Standard for the Installation of Centrifugal Fire Pumps.*)

(f) The pressure of air used with dry systems should be checked.

(g) Water level in tanks should be checked.

(h) Care should be taken to see that sprinklers are not painted either at the time of installation or during subsequent redecoration. When painting sprinkler piping or painting in areas next to sprinklers, the sprinklers may be protected by covering with a bag which should be removed immediately after painting has been finished.

(i) For further information see NFPA 13A, *Recommended Practice for the Inspection, Testing and Maintenance of Sprinkler Systems.*

A good guide for the maintenance of sprinkler systems is found in NFPA 13A. The property owner may obtain further guidance on maintenance of the sprinkler system from his local sprinkler contractor. Generally, because of its simplicity, maintenance of a residential sprinkler system does not require any greater care than that normally exercised in maintaining a residential plumbing system.

1-5 Design and Installation.

1-5.1 Devices and Materials.

1-5.1.1* Only new residential sprinklers shall be employed in the installation of sprinkler systems.

As with NFPA 13 and 13R systems, only new sprinklers are permitted to be used. Used sprinklers are not easily traceable as to the environment of their prior use. The presence of new, listed residential sprinklers provides a high degree of confidence in this important component.

A-1-5.1.1 At least three spare sprinklers of each type, temperature rating, and orifice size used in the system should be kept on the premises. When fused sprinklers are replaced by the owner, fire department, or others, care should be taken to assure that the replacement sprinkler has the same operating characteristics.

The sprinklers used in residential systems are special types that have been tested for residential use.[12] They are essentially one-time operating valves that must be treated carefully when they are being installed. When the system is originally installed, only new sprinklers should be used to ensure that they are of the proper type and in the best condition.

On those occasions when it is necessary to replace a sprinkler because it has operated (fused), or has been damaged, one should carefully check the orifice size, the temperature rating, and the deflector configuration to be certain that the exact same type of sprinkler replaces the one that has operated. Many sprinklers have unique operating characteristics (i.e., upright, pendent, or sidewall) and some have special limitations on their area of coverage.

1-5.1.2 Only listed and approved devices and approved materials shall be used in sprinkler systems.

Exception: Listing may be waived for tanks, pumps, hangers, waterflow detection devices, and waterflow valves.

At the time the original edition of the standard was prepared, tanks, pumps, hangers, waterflow devices, and water control valves in the size range suitable for residential use had not been tested or listed by testing laboratories. This holds true to this day. It was the feeling of the Committee that listing was desirable but not mandatory for these devices. The Committee anticipated that the type of equipment used would be similar to that found in common use in residential plumbing systems.

The economic impact of requiring listed equipment for tanks, pumps, and valves may increase the system cost 500 percent in some cases. Such action would eradicate the "low-cost" goal established by USFA when the residential sprinkler program was set up and agreed to by the Committee in preparing the standard. This cost increase in turn would discourage homeowners from installing such a system.

1-5.1.3 Preengineered systems shall be installed within the limitations which have been established by the testing laboratories where listed.

Approval laboratories have sometimes tested and listed complete systems that include all of the components necessary for an application. In this case, the system might include sprinklers, water supply tanks, valves, pumps, special piping limitations, etc. This section permits a manufacturer to develop such a system for application as a residential sprinkler system, but does not require the use of such a package system.

1-5.1.4* All systems shall be tested for leakage at normal system operating water pressure.

Exception: When a fire department pumper connection is provided, hydrostatic pressure tests shall be provided in accordance with NFPA 13, Standard for the Installation of Sprinkler Systems.

A-1-5.1.4 Testing of a system can be accomplished by filling the system with water and checking visually for leakage at each joint or coupling.

Fire department connections are not required for systems covered by this standard, but may be installed at the discretion of the owner. In these cases hydrostatic tests in accordance with NFPA 13, *Standard for the Installation of Sprinkler Systems*, are required.

Dry systems should also be tested by placing the system under air pressure. Any leak that results in a drop in system pressure greater than 2 psi (0.14 bar) in 24 hours should be corrected. Check for leaks using soapy water brushed on each joint or coupling. Leaks will be shown by the presence of bubbles. This test should be made prior to concealing of piping.

> Although this appendix material describes the test procedures for a dry system, it should be pointed out that at this time dry systems are not permitted. (*See 3-5.1 and commentary to 4-1.2.*)
>
> Unlike the requirements of NFPA 13, and in some cases NFPA 13R, residential systems do not require a hydrostatic test at 50 psi (3.4 bars) above the operating pressure. If such had been required, it would have added significantly to the installed cost of the system and would have made "do-it-yourself" installation difficult. Plumbing systems have been successfully installed in homes without special hydrostatic testing for years, and since it is intended that the residential sprinkler system be similar to, if not a part of, the plumbing system, the Committee concluded that a special hydrostatic test was not generally required.
>
> While a fire department connection to a residential system is not required, it can be installed. When this occurs, the system pressure can be increased using a fire department pumper, and in such instances it should be verified that the system has the integrity to withstand such pressures. Thus, a hydrostatic test should be done in accordance with NFPA 13 when a fire department pumper connection is included as a part of the system. Normally, one would expect this to be at a pressure of 200 psi (13.8 bars).

1-6 Units.   Metric units of measurement in this standard are in accordance with the modernized metric system known as the International System of Units (SI). Two units (liter and bar), outside of but recognized by SI, are commonly used in international fire protection. These units are listed in Table 1-6 with conversion factors.

Table 1-6

Name of Unit	Unit Symbol	Conversion Factor
liter	L	1 gal = 3.785 L
pascal	Pa	1 psi = 6894.757 Pa
bar	bar	1 psi = 0.0689 bar
bar	bar	1 bar = 10^5 Pa

For additional conversions and information see ASTM E380, *Standard for Metric Practice.*

1-6.1 If a value for measurement as given in this standard is followed by an equivalent value in other units, the first stated is to be regarded as the requirement. A given equivalent value may be approximate.

1-6.2 The conversion procedure for the SI units has been to multiply the quantity by the conversion factor and then round the result to the appropriate number of significant digits.

2

Water Supply

2-1 General Provisions. Every automatic sprinkler system shall have at least one automatic water supply. When stored water is used as the sole source of supply, the minimum quantity shall equal the water demand rate times 10 minutes. (*See 4-1.3.*)

Since any sprinkler system is only as good as its water supply, this supply must be automatic and reliable. The adequacy of a public water supply is determined in the manner described in 4-4.3 and A-4-4.3 of this standard. When a captured water supply is used, the amount should provide at least 10 minutes duration [at least 260 gal (984 L)]. A fuel oil tank of 275 gal (1041 L) capacity would make a satisfactory storage tank. This provides sufficient water to hold a fire in control while giving occupants sufficient time to evacuate the residence. Concurrently, some factor of safety is built in to the 10-minute supply which accounts for the delay of an occupant while recognizing the fire condition.

2-2* Water Supply Sources. The following water supply sources are acceptable:

(a) A connection to a reliable water works system.

(b) An elevated tank.

(c) A pressure tank installed in accordance with NFPA 13, *Standard for the Installation of Sprinkler Systems*, and NFPA 22, *Standard for Water Tanks for Private Fire Protection*.

(d) A stored water source with an automatically operated pump.

A-2-2 Connection for fire protection to city mains is often subject to local regulation concerning metering and backflow prevention requirements. Preferred and acceptable water supply arrangements are shown in Figures A-2-2(a), (b), and (c). When a meter must be used between the city water main and the sprinkler system supply, an acceptable arrangement is shown in Figure A-2-2(c). Under these circumstances, the flow characteristics of the meter must be included in the hydraulic calculation of the system. [*See*

Table 4-4.3(d).] When a tank is used for both domestic and fire protection purposes, a low water alarm actuated when the water level falls below 110 percent of the minimum quantity specified in Section 2-1 should be provided.

Figures A-2-2(a), (b), and (c) show preferred and acceptable arrangements of connections to city mains and include control valves, meters, domestic take-offs, waterflow detectors, pres-

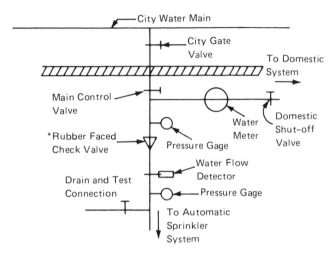

Figure A-2-2(a)   Preferable Arrangement.

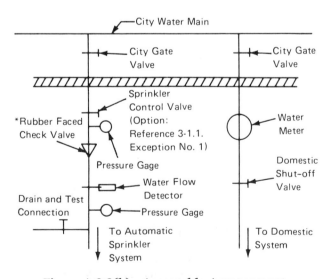

Figure A-2-2(b)   Acceptable Arrangement.

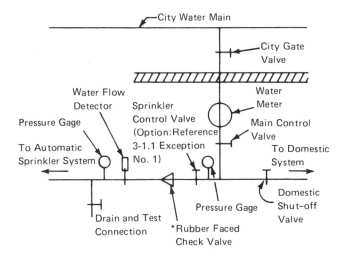

*Rubber Faced Check Valves Optional.

Figure A-2-2(c)   Acceptable Arrangement.

sure gages, and check valves appropriately located. When back-flow preventors are required by local water company require-ments, they will normally be located in the position where the rubber-faced valve is shown on the diagrams. The piping ar-rangement, including valves and fittings, must be taken into account when performing the calculations for system adequacy described in 4-4.3.

From a fire protection standpoint, meters are undesirable because of their high friction loss characteristic, which must be included in the hydraulic calculation of the system in accor-dance with the values shown in Table 4-4.3(d). Unfortunately, many water authorities require metering of all water connec-tions to a residence.

Occasionally, a system may be supplied from an elevated tank, and when this is done, the tank must have an elevation sufficient to provide adequate pressure to supply the system. This is determined by the methods described in 4-4.3 and 4-4.4. When a pressure tank is used for water supply, the amount of water and air in the tank is determined by following 2-6.3 and A-2-6.3 of NFPA 13 and 4-4.4 of this standard.

A stored water source with an automatically operated pump must have a pump with a capacity sufficient to supply the system's demand as described in 4-1.1, 4-1.2, and 4-1.3, and must have adequate pressure, determined in accordance with 4-4.4.

2-3* Multipurpose Piping System. A piping system serving both sprinkler and domestic needs shall be acceptable when:

(a) In common water supply connections serving more than one dwelling unit, 5 gpm (19 L/min) is added to the sprinkler system demand to determine the size of common piping and the size of the total water supply requirements.

> **This section clarifies when the domestic use must be added to the sprinkler system demand.**

(b) Smoke detectors are provided in accordance with NFPA 74, *Standard for the Installation, Maintenance, and Use of Household Fire Warning Equipment.*

(c) All piping in the system conforms to the piping specifications of this standard.

(d) Permitted by the local plumbing or health authority.

A-2-3(a) In dwellings where long-term use of lawn sprinklers is common, provision should be made for such usage.

> **In setting this requirement, the Committee recognized that typical domestic water usage does not span a great length of time (except for an automatic long-term demand for a domestic use, such as a lawn sprinkling system or a washing machine) and is not likely to occur simultaneously with a fire. When one can anticipate an automatic domestic demand that might take place concurrent with operation of sprinklers to combat a fire, such demand should be taken into account in the system design. Another approach to this condition involves an automatically controlled solenoid valve on the lawn sprinkler system. Upon actuation of the system waterflow switch, an electrical signal can be sent to the solenoid valve, which closes, and the supply for the lawn sprinklers is terminated. The result is that no water is robbed from the fire protection sprinkler system.**
> **In permitting combined piping systems, i.e., systems serving both domestic and sprinkler needs concurrently, the Committee recognized that there is some value in ensuring that the sprinkler system will be in operation when common piping serves both sprinklers and domestic needs. In a residence, if one does not have water for normal usage, it is likely that something will be done about it, and promptly. The Committee also recognized that restrictions may be placed on such combined systems by local plumbing or health regulations because**

there are typically trapped piping sections at the extremities of a sprinkler system that are not permitted in such codes. Further, some plumbing regulations permit piping methods of a lesser quality level than are prescribed by this standard.

Figures A-2-3(a) and A-2-3(b) show two arrangements for a multipurpose piping system. Figure A-2-3(a) shows a common point when domestic water and fire protection water are divided. Figure A-2-3(b) shows a common piping line which is tapped to supply a laundry fixture. Figure A-2-3(c) shows an isometric view of a multilevel dwelling and one possible arrangement of the multipurpose piping. System design would account for simultaneous flow of the sprinklers as well as one or more domestic fixture outlets.

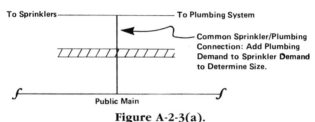

Figure A-2-3(a).

Figure A-2-3(b).

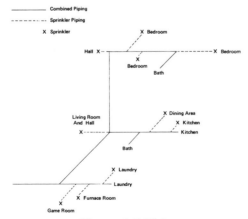

Figure A-2-3(c).

2-4 Mobile Home Water Supply. A water supply for a sprinklered dwelling manufactured off-site shall not be less than that specified on the manufacturer's nameplate. [*See 4-4.3(k) Exception.*]

The piping for a sprinkler system installed in a mobile home will typically be installed by the mobile home manufacturer. This standard anticipates that, when this is done, the mobile home manufacturer will specify the capacity and pressure needed to supply the system from a water supply to which the system must ultimately be connected.

3

System Components

3-1 Valves and Drains.

3-1.1   Each system shall have a single control valve arranged to shut off both the domestic and sprinkler systems, and a separate shutoff valve for the domestic system only.

A control valve permits the occupant to shut off the system in order to replace a sprinkler that has operated to extinguish a fire. The control valve must be in the open position at all other times.

With only one control valve, there is little possibility of inadvertently shutting off the sprinkler system because, with the domestic water shut off, something would be done rather quickly to get it back in operation. The separate control valve for the sprinkler system is acceptable when the valve is supervised as described in the following Exception.

Exception No. 1:   The sprinkler system piping may have a separate control valve where supervised by one of the following methods:

(a) Central station, proprietary, or remote station alarm service,

(b) Local alarm service that will cause the sounding of an audible signal at a constantly attended point, or

(c) Locking the valves open.

Exception No. 2:   A separate shutoff valve is not required for the domestic water supply in multipurpose piping systems.

On a multipurpose piping system, the availability of water is "self supervising." If water supply is not available to water closets, sinks, or similar equipment, it can also be concluded that the sprinkler system has no supply and action can be taken to put the entire system back into service.

3-1.2   Each sprinkler system shall have a ½ in. (13 mm) or larger drain and test connection with valve on the system side of the control valve.

This arrangement permits the occupant to open this valve to determine that water is actually being supplied to the system and to check operation of the alarm. It also permits the system piping to be drained when maintenance or repairs are being done.

3-1.3   Additional drains shall be installed for each trapped portion of a dry system which is subject to freezing temperatures.

Figure 3.1 illustrates a trapped section of piping and how a drain is arranged to comply with 3-1.3. It is important that such portions of the system be carefully drained before the onset of freezing weather each year. If water from the operation of the

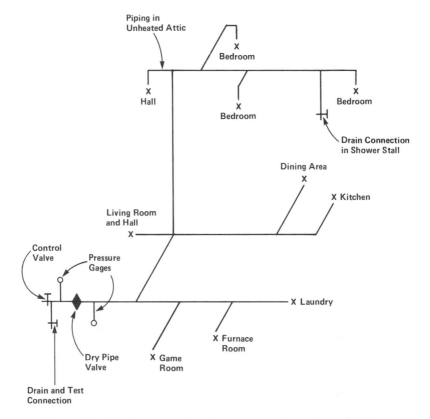

Figure 3.1. Drain connection for a trapped section of piping.

system or condensation collects in a trapped section of piping, it may freeze and break the piping, possibly causing extensive water damage.

3-2 Pressure Gages.

3-2.1 A pressure gage shall be installed to indicate air pressure on dry systems and on water supply pressure tanks.

Recalling that equipment does not exist yet for dry-pipe residential sprinkler systems, the focus of this requirement rests upon the gage for a pressure tank. This gage allows one to insure that adequate operating pressure is available to the system.

3-3 Piping.

3-3.1 Pipe or tube used in sprinkler systems shall be of the materials in Table 3-3.1 or in accordance with 3-3.2 through 3-3.5. The chemical properties, physical properties, and dimensions of the materials listed in Table 3-3.1 shall be at least equivalent to the standards cited in the table and designed to withstand a working pressure of not less than 175 psi (12.1 bars).

Table 3-3.1

Materials and Dimensions	Standard
Specifications for Welded and Seamless Steel Pipe	ASTM A53
Wrought-Steel Pipe	ANSI B36.10M
Specifications for Electric-Resistance Welded Steel Pipe	ASTM A135
Copper Tube (Drawn, Seamless)	
Specification for Seamless Copper Tube	ASTM B75
Specification for Seamless Copper Water Tube	ASTM B88
Specification for General Requirements for Wrought	
Seamless Copper and Copper-Alloy Tube	ASTM B251
Brazing Filler Metal (Classification BCuP-3 or BCuP-4)	AWS A5.8
Specification for solder Metal, 95-5	
(Tin-Antimony-Grade 95TA)	ASTM B32

3-3.2 Other types of pipe or tube may be used, but only those investigated and listed for this service by a testing and inspection agency laboratory.

This statement also appears in NFPA 13 and NFPA 13R to allow for the continued development and use of special pipe products, such as polybutylene, CPVC, and certain thin-walled steel piping.

3-3.3 Whenever the word pipe is used in this standard, it shall be understood to also mean tube.

3-3.4 Schedule 10 steel pipe may be joined with mechanical groove couplings approved for service with grooves rolled on the pipe by an approved groove rolling machine.

3-3.5 Fittings used in sprinkler systems shall be of the materials listed in Table 3-3.5 or in accordance with 3-3.7. The chemical properties, physical properties, and dimensions of the materials listed in Table 3-3.5 shall be at least equivalent to the standards cited in the table. Fittings used in sprinkler systems shall be designed to withstand the working pressures involved, but not less than 175 psi (12.1 bars) cold water pressure.

Table 3-3.5

Materials and Dimensions	Standard
Cast Iron	
Cast Iron Threaded Fittings, Class 125 and 250	ANSI B16.4
Cast Iron Pipe Flanges and Flanged Fittings	ANSI B16.1
Malleable Iron	
Malleable Iron Threaded Fittings, Class 150 and 300	ANSI B16.3
Steel	
Factory-made Wrought Steel Buttweld Fittings	ANSI B16.9
Buttwelding Ends for Pipe, Valves, Flanges, and Fittings	ANSI B16.25
Spec. for Piping Fittings of Wrought Carbon Steel and Alloy Steel for Moderate and Elevated Temperatures	ASTM A234
Pipe Flanges and Flanged Fittings, Steel Nickel Alloy and Other Special Alloys	ANSI B16.5
Forged Steel Fittings, Socket Welded and Threaded	ANSI B16.11
Copper	
Wrought Copper and Copper Alloy Solder-Joint Pressure Fittings	ANSI B16.22
Cast Copper Alloy Solder-Joint Pressure Fittings	ANSI B16.18

3-3.6 Joints for the connection of copper tube shall be brazed.

Exception: Soldered joints (95-5 solder metal) may be used for wet-pipe copper tube systems.

Previous editions allowed 50-50 solder because it was the type normally used in domestic plumbing systems. The 1986 amendments to the Federal Safe Drinking Water Act prohibit the use of 50-50 solder in plumbing systems. While this does not specifically apply to residential sprinkler systems, the Committee considered it poor practice to use this leaded solder in close proximity to potable water systems. The 95-5 solder has a higher melting point, no lead, and is allowed by the act.

3-3.7 Other types of fittings may be used, but only those investigated and listed for this service by a testing and inspection agency laboratory.

The types of pipe and fittings that are described in this section of the standard are those that have been found satisfactory either due to their service record or based on their having been tested and listed by a testing laboratory. Currently, the most commonly used materials are ordinary and lightweight steel pipe and Types K and L copper and copper-alloy tube. This section does not preclude the use of plastic pipe, if it has been tested and listed by a testing and inspection agency laboratory. (*See Sections 3-1 and 3-7 of NFPA 13 for additional information on pipe and fittings.*)

3-4 Piping Support.

3-4.1 Piping shall be supported from structural members using support methods comparable to those required by local plumbing codes.

Unlike the hanging requirements for sprinkler systems installed under NFPA 13, this standard simply requires adequate support by methods recognized in local plumbing codes for the piping materials listed in Table 3-3.1. Generally, residential systems deal with pipe sizes of ¾ in. (19 mm) to 1¼ in. (33 mm) and will never be expected to exceed 2 in. (51 mm). The dynamic loads and forces that may be generated by water flow in a residential sprinkler system are considerably lower than those experienced in an NFPA 13 or NFPA 13R system. Most plumbing code hangers that will be used will give adequate support for these piping materials.

Exception: Listed piping shall be supported in accordance with any listing limitations.

The Exception recognizes that listed piping may have specific hanger requirements as a condition of the listing.

3-4.2   Piping laid on open joists or rafters shall be secured to prevent lateral movement.

3-5 Sprinklers.

Since the listing of the first residential sprinkler, manufacturers have listed others, which currently include both pendent and sidewall types. The newer residential sprinklers are more aesthetically pleasing in design, which should alleviate the homeowner's concern for the appearance of these devices in the home. Examples of these newer designs are provided in the figures that follow.

Figure 3.2. The first two listed "residential" sprinklers. Shown are the Grinnell Model F954 (left) and Central "Omega" Model R-1 (right) pendent sprinklers.

3-5.1   Listed residential sprinklers shall be used. The basis of such a listing shall be tests to establish the ability of the sprinklers to control residential fires under standardized fire test conditions. The standardized room fires shall be based on a residential array of furnishings and finishes.

Both Underwriters Laboratories Inc. and Factory Mutual Research Corporation published standards for testing and evaluating residential sprinklers.[11, 12] These standards call for the sprinklers to pass room fire tests and have fast response characteristics. Standard commercial sprinklers of the type used in NFPA 13 sprinkler systems are tested to a different set of criteria. They are not evaluated for their ability to control the fire with only two sprinklers as the residential sprinkler is. The residential sprinkler test requirements were verified in testing at Factory Mutual Research Corporation,[6] and in the Los Angeles and North Carolina residential sprinkler fire tests referenced in the bibliography.[7, 8, 9, 10]

Figure 3.3. Viking Model M-1 horizontal sidewall residential sprinkler.

Exception No. 1:   Residential sprinklers shall not be used in dry systems unless specifically listed for that purpose.

Exception No. 2:   Listed dry sprinklers may be used in accordance with 4-3.2.

Exceptions No. 1 and 2 were added because of the concern for the expected delay in water delivery inherent in dry-pipe systems. There are no residential sprinklers available which are listed for use in a dry-pipe sprinkler system. Residential sprinklers utilized in NFPA 13, 13D, and 13R are all bound by this limitation.

3-5.2   Ordinary temperature rated residential sprinklers [135 to 170°F (57 to 77°C)] shall be installed where maximum ambient ceiling temperatures do not exceed 100°F (38°C).

The requirements of this paragraph are applicable to the environmentally controlled portions of the occupancy. The intent is to use ordinary temperature sprinklers in areas not subject to unusually high ambient temperatures.

3-5.3 Intermediate rated residential sprinklers [175 to 225°F (79 to 107°C)] shall be installed where maximum ambient ceiling temperatures are between 101 and 150°F (39 and 66°C).

Conditions in a residence which may result in high ambient ceiling temperatures are attributable to: track lighting units, fireplaces, wood burning stoves, cooking equipment, and attics with no air conditioning. Spacing and positioning of sprinklers in the immediate vicinity of these areas cannot be ignored. Steps must be taken to insure that sprinklers are correctly selected for such areas. Recommendations and advice from the suppliers of heat producing equipment and sprinklers should be solicited to allow for the placement of sprinklers in such areas. Paragraph 3-5.4 of this standard gives the necessary rules for certain applications of this concept.

3-5.4 The following practices shall be observed when installing residential sprinklers, unless maximum expected ambient temperatures are otherwise determined.

(a) Sprinklers under glass or plastic skylights exposed to direct rays of the sun shall be of intermediate temperature classification.

(b) Sprinklers in an unventilated concealed space under uninsulated roof, or in an unventilated attic, shall be of intermediate temperature classification.

3-5.5 Operated or damaged sprinklers shall be replaced with sprinklers having the same performance characteristics as original equipment.

Underwriters Laboratories Inc. has conducted a program in which field samples of automatic sprinklers have been tested in order to determine their operating characteristics. Experience gained from this program has shown that, when a sprinkler has been damaged or painted after leaving the factory, it is unlikely to operate properly at the time of a fire and thus should be replaced. Obviously, a sprinkler that has operated must also be replaced.

3-5.6 Painting and Ornamental Finishes.

It is extremely important that only manufacturers who have had their procedures listed apply paint or ornamental finishes to residential sprinklers. The installing contractor or home-owner does not have the facilities available to properly duplicate the manufacturer's procedures. "Do-it-yourself" applications could seriously impair the sprinkler's operation, render it inoperable, or impact on the characteristics of the discharge pattern.

3-5.6.1* Sprinkler frames may be factory painted or enameled as ornamental finish in accordance with 3-5.6.2; otherwise sprinklers shall not be painted and any sprinklers that have been painted, except for factory applied coatings, shall be replaced with new listed sprinklers.

A-3-5.6.1 Decorative painting of a residential sprinkler is not to be confused with the temperature identification colors as referenced in 3-11.6 of NFPA 13-1989, *Standard for the Installation of Sprinkler Systems.*

3-5.6.2 Ornamental finishes shall not be applied to sprinklers by anyone other than the sprinkler manufacturer, and only sprinklers listed with such finishes shall be used.

3-5.7 When nonmetallic sprinkler ceiling plates (escutcheons) or recessed escutcheons (metallic or nonmetallic) are used they shall be listed based on tests of the assembly as a residential sprinkler.

These devices are required to be tested and listed by the listing laboratory as an assembly with the sprinkler. The tests are intended to determine that the escutcheon will not interfere with the operation of the sprinkler.

3-6* **Alarms.**   Local waterflow alarms shall be provided on all sprinkler systems.

Exception: Dwellings or mobile homes having smoke detectors in accordance with NFPA 74, Standard for the Installation, Maintenance, and Use of Household Fire Warning Equipment.

A-3-6   Alarms should be of sufficient intensity to be clearly audible in all bedrooms over background noise levels with all intervening doors closed. The tests of audibility level should be conducted with all household equipment that may be in operation at night in full operation. Examples of such equipment are window air conditioners and room humidifiers. When off-premises alarms are provided, at least water flow and control valve position should be monitored.

Since the philosophy of this standard is to provide a system that will control a fire for 10 minutes while giving occupants an opportunity to evacuate, an alarm is needed to warn occupants and initiate that evacuation. This standard suggests two methods by which this can be done. If smoke detectors are installed in accordance with NFPA 74, this standard anticipates that they will give an adequate warning for purposes of evacuation.

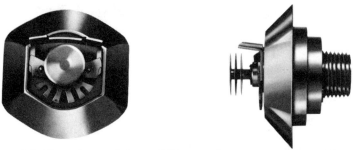

Figure 3.4. Two views of Central "Omega" HEC-12 residential horizontal sidewall sprinkler.

Figure 3.5. Central "Omega" EC-20A adjustable residential pendent sprinkler.

Figure 3.6. Viking Model M-1 pendent mounted residential sprinkler with glass bulb and orifice cover removed. (Sprinkler is shown in an inverted position.)

However, if smoke detectors are not installed, then the sprinkler system must be equipped with a waterflow device, which will sound an audible alarm throughout the residence to initiate evacuation.

To conduct audibility tests, have members of the family position themselves on each level of the house inside a room (such as a bedroom) with the door closed. Operate the alarm and have the family members report whether or not it can be heard. Bear in mind that the alarm must be of sufficient level to wake a person who is asleep.

4

System Design

4-1 Design Criteria.

Design criteria for residential sprinkler systems is a radical departure from the conventional area/density methods of NFPA 13. The goal for an NFPA 13D design is to gain fire control with a maximum of two sprinklers in operation. This chapter details how to accomplish that goal.

4-1.1 Design Discharge. The system shall provide a discharge of not less than 18 gpm (68 L/min) to any single operating sprinkler and not less than 13 gpm (49 L/min) per sprinkler to the number of design sprinklers.

This section prescribes a discharge of 18 gpm (68 L/min) for a single operating sprinkler and 13 gpm (49 L/min) per sprinkler for two sprinklers operating in the design area in a room. This may be extrapolated to a density of 0.125 gpm per sq ft [(5.1 L/min)/m²] for one sprinkler and 0.09 gpm per sq ft [(3.7 L/min)/m²] for two sprinklers at the maximum spacing of 144 sq ft (13.4 m²) per sprinkler. Test data indicate that the primary criterion should be flow rate, rather than density, for a residential system.[7, 8, 9, 10]

These values of system discharge are based primarily on the Los Angeles and North Carolina fire tests.[7, 8, 9, 10] A rather high percentage (95 percent) of the full-scale tests which were conducted were successful.

4-1.2* Number of Design Sprinklers. The number of design sprinklers shall include all sprinklers within a compartment to a maximum of two sprinklers.

Available data from fire tests of residential sprinklers do not support the four-sprinkler design criteria for dry systems that was originally (1980) included in this section as an Exception. Currently listed residential sprinklers are not listed for use in dry-pipe systems.

It should be clarified that even if more than two sprinklers are required for coverage in a compartment having a flat ceiling, the maximum number of sprinklers included as the number of design sprinklers for water supply calculations is two. It is frequently misinterpreted by many that room size must be limited to make sure that only two sprinklers will cover the area. Paragraph A-4-2.3 gives some guidance for sloped and beamed ceilings.

A-4-1.2 It is intended that the design area is to include the two adjacent sprinklers producing the greatest water demand within the compartment. It is also intended that the number of design sprinklers is limited to a maximum of two sprinklers even if there are more than two sprinklers in the compartment, except as noted in A-4-2.3.

In the Los Angeles and North Carolina fire tests,[7, 8, 9, 10] all of the test fires used typical residential room configurations. These were extinguished with one or two sprinklers when the waterflow rate, spacing, and discharge characteristics of the sprinklers were adequate. When one of these features was not adequate, the test failed because one environmental criterion was exceeded or because the water supply was overtaxed when an excessive number of sprinklers opened. Accordingly, design of the residential system should be based on two sprinklers operating (unless no compartment has more than one sprinkler). With two sprinkler designs, each sprinkler must produce a minimum flow of 13 gpm (49 L/min) when calculations are done in accordance with 4-4.3. With only one sprinkler in any room, design is based on an 18-gpm (68-L/min) flow rate.

A room compartment is determined by the enclosing walls or the enclosing geometrics of a soffet, lintel, or beam that forms a definite interruption in a smooth ceiling configuration. [*See Figures A-4-1.2(a) and A-4-1.2(b)*.] In open plan configurations, compartment boundaries may be determined by beams, soffets, or lintels over doorways as well as by walls or partitions. In any event, the designer selects no more than two sprinklers to define the design area. When only one sprinkler is within all compartment boundaries, it defines the design area.

4-1.2.1 The definition of compartment for use in 4-1.2 to determine the number of design sprinklers is a space that is completely enclosed by walls and a ceiling. The compartment enclosure may have openings to an adjoining space if the openings have a minimum lintel depth of 8 in. (203 mm) from the ceiling.

As previously described under 4-1.2, the design area may be determined either by walls and partitions or by beams, soffets,

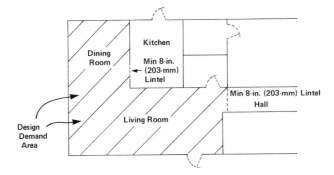

Figure A-4-1.2(a)   Sprinkler Design Areas for Typical Residential Occupancy.

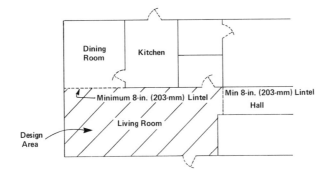

Figure A-4-1.2(b)   Sprinkler Design Areas for Typical Residential Occupancy.

or lintels provided they have a minimum depth of 8 in. (203 mm). The lintel depth described by this standard allows for rapid collection of heat in the area of fire origin and was confirmed in the Los Angeles test series.

4-1.3 Water Demand. The water demand for the system shall be determined by multiplying the design discharge of 4-1.1 by the number of design sprinklers of 4-1.2.

This calculation will produce a water demand of 18 gpm (68 L/min) for a single sprinkler and 26 gpm (98 L/min) for two sprinklers. These values are used in performing the calculations in 4-4.3 to determine pipe sizing or adequacy of public water supply pressure.

4-1.4 Sprinkler Coverage.

4-1.4.1 Residential sprinklers shall be spaced so that the maximum area protected by a single sprinkler does not exceed 144 sq ft (13.4 m²).

This section indicates that the design area for a single sprinkler should not exceed 144 sq ft (13.4 m²) and normally will be the actual area covered by the sprinklers included in the design and specified under 4-1.2. When test and listing indicate greater spacing, this is permitted by 4-1.6.

4-1.4.2 The maximum distance between sprinklers shall not exceed 12 ft (3.7 m) on or between pipelines and the maximum distance to a wall or partition shall not exceed 6 ft (1.8 m). The minimum distance between sprinklers within a compartment shall be 8 ft (2.4 m).

4-1.5 The minimum operating pressure of any sprinkler shall be in accordance with the listing information of the sprinkler and provide the minimum flow rates specified in 4-1.1.

4-1.6 Application rates, design areas, areas of coverage, and minimum design pressures other than those specified in 4-1.1, 4-1.2, 4-1.4, and 4-1.5 may be used with special sprinklers which have been listed for such specific residential installation conditions.

These limitations on spacing were determined from the Los Angeles and North Carolina fire tests.[7, 8, 9, 10] Since these tests were conducted with prototype fast response residential sprinklers, the Committee has recognized that other fast response residential sprinklers may be developed that will operate properly with different spacing limitations. Such sprinklers may be used provided that spacings which differ from those described in this standard are justified, based on testing conducted by a testing and inspection agency/laboratory. (*See definitions of "Labeled" and "Listed" in Section 1-3.*)

It should be noted that some residential sprinklers have been listed to cover areas as large as 324 sq ft (30 m²), approximately 225 percent more than the prototype sprinkler. While this is advantageous because fewer sprinklers would be necessary for a large compartment, the impact of larger water supplies [32 gpm (121 L/min) for one and 22.5 gpm (85.2 L/min) for two operating sprinklers] and pressures [36.3 psi (2.5 bars) for one and 18.0 psi (1.2 bars) for two operating sprinklers] must be weighed against the use of such sprinklers.

4-2 Position of Sprinklers.

4-2.1 Pendent and upright sprinklers shall be positioned so that the deflectors are within 1 to 4 in. (25.4 to 102 mm) from the ceiling.

Exception: Special residential sprinklers shall be installed in accordance with the listing limitations.

This limitation is based on the results of the Los Angeles and North Carolina fire tests.[7, 8, 9, 10]

4-2.2 Sidewall sprinklers shall be positioned so that the deflectors are within 4 to 6 in. (102 to 152 mm) from the ceiling.

Exception: Special residential sprinklers shall be installed in accordance with the listing limitations.

4-2.3* Sprinklers shall be positioned so that the response time and discharge are not unduly affected by obstructions such as ceiling slope, beams, or light fixtures.

A-4-2.3 Fire testing has indicated the need to wet walls in the area protected by residential sprinklers at a level closer to the ceiling than that accomplished by standard sprinkler distribution. Where beams, light fixtures, sloped ceilings, and other obstructions occur, additional residential sprinklers will be necessary to achieve proper response and distribution. In addition, for sloped ceilings, higher flow rates may be needed. Guidance may be obtained from the manufacturer.

A series of 33 full-scale tests were recently conducted in a test room of 12 ft × 24 ft (3.6 m × 7.2 m) floor area to determine the effect of cathedral (sloped) and/or beamed ceiling construction on fast response residential sprinkler performance. The testing was limited to one pendent-type residential sprinkler model, two ceiling slopes (0 and 14 degrees), and two beam configurations on the single enclosure size. In order to judge the effectiveness of sprinklers in controlling fires, two baseline tests, in which the ceiling was smooth and horizontal, were conducted with the pendent sprinklers installed and with a total water supply of 26 gpm (98 L/min) as required by NFPA 13D. The results of the baseline tests were compared with tests in which the ceiling was beamed and/or sloped and two pendent sprinklers were installed with the same water supply. Under the limited conditions tested, the comparison indicates that sloped and/or beamed ceilings represent a serious challenge to the fire protection afforded by fast response residential sprinklers. However, further tests with beamed ceilings indicated that fire control equivalent to that obtained in the baseline tests may be obtained if one sprinkler is centered in each bay formed by the beams and a total water supply of 36 gpm (136 L/min) is available. Fire control equivalent to that obtained in the baseline tests was obtained for the smooth sloped ceiling tests when three sprinklers were installed with a total water supply of 54 gpm (200 L/min). In a single smoldering-started fire test, the fire was suppressed.[1]

---

[1] Effects of Cathedral and Beamed Ceiling Construction on Residential Sprinkler Performance, FMRC J.I. M3N5.RA(3), by Bill Jr., R.G.; Kung, N-C; Brown, W.R.; and Hill, E., prepared for U.S. Fire Administration (Feb. 1988).

Table A-4-2.3   Maximum Distance from Spinkler Deflector
to Bottom of Ceiling Obstruction

Distance from Sprinkler to Side of Ceiling Obstruction	Maximum Distance from Sprinkler Deflector to Bottom of Ceiling Obstruction
Less than 6 in.	Not permitted
6 in. to less than 1 ft	0 in.
1 ft to less than 2 ft	1 in.
2 ft to less than 2 ft 6 in.	2 in.
2 ft 6 in. to less than 3 ft	3 in.
3 ft to less than 3 ft 6 in.	4 in.
3 ft 6 in. to less than 4 ft	6 in.
4 ft to less than 4 ft 6 in.	7 in.
4 ft 6 in. to less than 5 ft	9 in.
5 ft to less than 5 ft 6 in.	11 in.
5 ft 6 in. to less than 6 ft	14 in.

For SI Units: 1 in. = 25.4 mm; 1 ft = 0.3048 m.

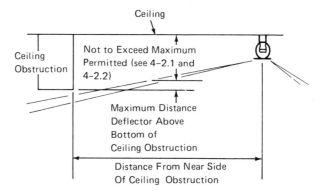

Figure A-4-2.3   Position of Deflector, Upright or Pendent, When Located Above Bottom of Ceiling Obstruction.

Location of sprinklers in violation of Table A-4-2.3 and Figure A-4-2.3 will cause obstruction of the discharge and may prevent the sprinkler from controlling a fire. Fire control characteristics of residential sprinklers depend to some extent on a high wall wetting characteristic. This feature is necessary to allow the pattern to reach the areas behind typical perimeter fuel load arrangements found in the home. Chairs, couches, desks, and shelves typically placed along a wall may harbor the source of ignition. The result is that a residential sprinkler has a superior horizontal discharge pattern in comparison to other sprinklers.

Sprinklers designed to have different discharge characteristics may be positioned closer to a beam than the distance permitted by Table A-4-2.3. However, test data must be supplied to support such different positioning.

4-3 System Types.

4-3.1 Wet-Pipe Systems.   A wet-pipe system shall be used when all piping is installed in areas not subject to freezing.

4-3.2 Dry-Pipe Systems.   Where system piping is located in unheated areas subject to freezing, a dry-pipe or antifreeze system shall be used.

Exception: Listed standard dry-pendent or dry sidewall sprinklers may be extended into unheated areas not intended for living purposes.

4-3.3 Antifreeze Systems.

This section has been added to the 1989 edition in order to allow the use of antifreeze systems. While long recognized in NFPA 13, antifreeze systems were not specifically discussed nor specifically prohibited in NFPA 13D. These system types have some additional costs associated with them and require more maintenance than a wet-pipe system. There is some concern that home owners may not check antifreeze levels on an annual basis and may not make attempts to locate and use the types of antifreeze specified by this section.

If these special precautions are noted and the commitment to maintenance is acknowledged, then an antifreeze system can be safely and effectively used in accordance with this standard.

See commentary on page 297 (Section 5-5 of NFPA 13) for more information on antifreeze systems.

4-3.3.1 Definition.   An antifreeze system is one employing automatic sprinklers attached to a piping system containing an antifreeze solution and connected to a water supply. The antifreeze solution, followed by water, discharges immediately from sprinklers opened by a fire.

4-3.3.2* Where Used.   The use of antifreeze solutions shall be in conformity with any state or local health regulations.

A-4-3.3.2   Antifreeze solutions may be used for maintaining automatic sprinkler protection in small unheated areas. Antifreeze solutions are recommended only for systems not exceeding 40 gallons (151 L).

Because of the cost of refilling the system or replenishing small leaks, it is advisable to use small dry valves where more than 40 gallons (151 L) are to be supplied.

Propylene glycol or other suitable material may be used as a substitute for priming water, to prevent evaporation of the priming fluid, and thus reduce ice formation within the system.

Again it must be recognized that with no residential sprinkler listed for dry-pipe use, dry-pipe systems may not be installed under this standard.

4-3.3.3 Antifreeze Solutions.

4-3.3.3.1 When sprinkler systems are supplied by public water connections the use of antifreeze solutions other than water solutions of pure glycerine (C.P. or U.S.P. 96.5 percent grade) or propylene glycol shall not be permitted. Suitable glycerine-water and propylene glycol-water mixtures are shown in Table 4-3.3.3.1.

**Table 4-3.3.3.1 Antifreeze Solutions
to be Used if Public Water is Connected to Sprinklers**

Material	Solution (by Volume)	Specific Gravity at 60°F (15.6°C)	Freezing Point °F	Freezing Point °C
Glycerine	50% Water	1.133	−15	−26.1
C.P. or U.S.P. Grade*	40% Water	1.151	−22	−30.0
	30% Water	1.165	−40	−40.0
	Hydrometer Scale 1.000 to 1.200			
Propylene Glycol	70% Water	1.027	+ 9	−12.8
	60% Water	1.034	− 6	−21.1
	50% Water	1.041	−26	−32.2
	40% Water	1.045	−60	−51.1
	Hydrometer Scale 1.000 to 1.200 (Subdivisions 0.002)			

*C.P. —Chemically Pure.
 U.S.P.—United States Pharmacopoeia 96.5%.

4-3.3.3.2 If public water is not connected to sprinklers, the commercially available materials indicated in Table 4-3.3.3.2 are suitable for use in antifreeze solutions.

4-3.3.3.3* An antifreeze solution shall be prepared with a freezing point below the expected minimum temperature for the locality. The specific gravity of the prepared solution shall be checked by a hydrometer with suitable scale.

A-4-3.3.3.3 Beyond certain limits, increased proportion of antifreeze does not lower the freezing point of solution. (*See Figure A-4-3.3.3.3.*)

Table 4-3.3.3.2   Antifreeze Solutions
to be Used if Public Water is not Connected to Sprinklers

Material	Solution (by Volume)	Specific Gravity at 60°F (15.6°C)	Freezing Point °F	°C
Glycerine	If glycerine is used, see Table 4-3.3.3.1			
Diethylene Glycol	50% Water	1.078	−13	−25.0
	45% Water	1.081	−27	−32.8
	40% Water	1.086	−42	−41.1
	Hydrometer Scale 1.000 to 1.120 (Subdivisions 0.002)			
Ethylene Glycol	61% Water	1.056	−10	−23.3
	56% Water	1.063	−20	−28.9
	51% Water	1.069	−30	−34.4
	47% Water	1.073	−40	−40.0
	Hydrometer Scale 1.000 to 1.120 (Subdivisions 0.002)			
Propylene Glycol	If propylene glycol is used, see Table 4-3.3.3.1			
Calcium Chloride 80% "Flake"	Lb CaCl$_2$ per gal of Water			
Fire Protection Grade*	2.83	1.183	0	−17.8
Add corrosion inhibitor	3.38	1.212	−10	−23.3
of sodium bichromate	3.89	1.237	−20	−28.9
¼ oz per gal water	4.37	1.258	−30	−34.4
	4.73	1.274	−40	−40.0
	4.93	1.283	−50	−45.6

*Free from magnesium chloride and other impurities.

Glycerine, diethylene glycol, ethylene glycol, and propylene glycol should never be used without mixing with water in proper proportions, because these materials tend to thicken near 32°F (0°C).

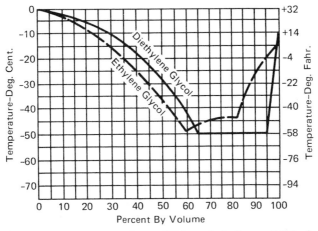

Figure A-4-3.3.3.3   Freezing Points of Water Solutions of Ethylene Glycol and Diethylene Glycol.

4-3.3.4* **Arrangement of Supply Piping and Valves.** All permitted anti-freeze solutions are heavier than water. At the point of contact (interface) the heavier liquid will be below the lighter liquid in order to prevent diffusion of water into the unheated areas. In most cases, this necessitates the use of a 5-ft (1.5-m) drop pipe or U-loop as illustrated in Figure 4-3.3.4. The preferred arrangement is to have the sprinklers below the interface between the water and the antifreeze solution.

If sprinklers are above the interface, a check valve with a ⅟32-in. (0.8-mm) hole in the clapper shall be provided in the U-loop. A water control valve and two small solution test valves shall be provided as illustrated in Figure 4-3.3.4. An acceptable arrangement of a filling cup is also shown.

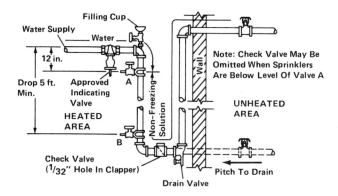

NOTE: The ⅟32-in. (0.8-mm) hole in the check valve clapper is needed to allow for expansion of the solution during a temperature rise and thus prevent damage to sprinkler heads.

For SI Units: 1 in. = 25.4 mm; 1 ft = 0.3048 m.

Figure 4-3.3.4 Arrangement of Supply Piping and Valves.

A-4-3.3.4 To avoid leakage, the materials and workmanship should be excellent, the threads clean and sharp, and the joints tight. Use only metal-faced valves.

4-3.3.5* **Testing.** Before freezing weather each year, the solution in the entire system shall be emptied into convenient containers and brought to the proper specific gravity by adding concentrated liquid as needed. The resulting solution may be used to refill the system.

A-4-3.3.5 Tests should be made by drawing a sample of the solution from valve B two or three times during the freezing season, especially if it has been necessary to drain the building sprinkler system for repairs, changes, etc. A small hydrometer should be used so that a small sample will be sufficient. When water appears at valve B or when the test sample indicates that the solution has become weakened, empty the entire system and recharge as previously described.

4-4 Pipe Sizing.

4-4.1 Piping shall be sized in accordance with 4-4.3 and 4-4.4. If more than one design discharge is required (*see 4-1.1*), the pipe sizing procedure shall be repeated for each design discharge.

Exception: When piping is sized hydraulically, calculations shall be made in accordance with the methods described in NFPA 13, Standard for the Installation of Sprinkler Systems.

The dual discharge requirements of 4-1.1 require two separate sets of calculations.

4-4.2 Minimum Pipe Size.

The requirements for the minimum pipe size are the same as those in NFPA 13 [i.e., ¾ in. (19 mm) for copper and 1 in. (25.4 mm) for steel]. This results in less friction loss than if smaller pipe and tubing sizes were permitted. Also, it lessens the possibility of pipe clogging from scale and rust.

The minimum size for types of pipe or tube other than for copper or steel, including nonmetallic tube, would be included as part of the listing information for that product. Because of the concerns with clogging and friction loss, it is doubtful that sizes less than ¾ in. would be listed.

4-4.2.1 Minimum pipe size, including copper piping, shall be ¾ in. (19 mm).

Exception: Minimum size of steel pipe shall be 1 in. (25.4 mm).

4-4.3* To size piping for systems connected to a city water supply, the following approximate method is acceptable. This procedure cannot be used for gridded or looped type systems. Hydraulic calculation procedures in accordance with Chapter 7 of NFPA 13 for grid-type or looped systems shall be used.

(a) Establish system flow rate in accordance with Section 4-1.

(b) Determine water pressure in the street.

(c) Select pipe sizes.

(d) Deduct meter pressure losses if any. [*See Table 4-4.3(d).*]

(e) Deduct pressure loss for elevation. (Building height above street in ft × 0.434 = psi.) (Building height above street in meters × 0.098 = bars.)

(f) Deduct pressure losses from the city main to the inside control valve by multiplying the factor from Table 4-4.3(a) or (b) by the total length(s) of pipe in ft (m). [Total length includes equivalent length of fittings as determined by applying Table 4-4.3(c).]

(g) Deduct pressure losses for piping within building by multiplying factor from Table 4-4.3(a) or (b) by the total length in ft (m) of each size of pipe between the control valve and the farthest sprinkler.

(h) Deduct valve and fitting pressure losses. Count the valves and fittings from the control valve to the farthest sprinkler. Determine the equivalent length for each valve and fitting as shown in Table 4-4.3(c) and add these values to obtain the total equivalent length for each pipe size. Multiply the equivalent length for each size by the factor from Table 4-4.3(a) or (b) and total these values.

(i) In multilevel buildings, steps (a) through (h) shall be repeated to size piping for each floor.

(j) If the remaining pressure is less than the operating pressure established by the testing laboratory for the sprinkler being used, a redesign is necessary. If this pressure is higher than required, smaller piping may be used when justified by calculations.

(k) The remaining piping shall be sized the same as the piping to the farthest sprinkler unless smaller sizes are justified by calculations.

Exception: For sprinklered dwellings manufactured off-site, the minimum pressure needed to satisfy the system design criteria on the system side of the meter shall be specified on a data plate by the manufacturer. (See Section 2-3.)

A-4-4.3 Determination of public water supply pressure should take into account probable minimum pressure condition prevailing during such periods as at night, or during summer months when heavy usage may occur; also, the possibility of interruption by floods, or ice conditions in winter.

This section of the standard describes a simplified method of sizing pipe in a residence to ensure that the flow rate from the sprinklers in the design area will meet the design discharge criteria of 4-1.1.

In performing the calculation for a specific system, it is necessary to go through the calculation procedure twice. The first calculation is made for a single sprinkler flowing at 18 gpm

Table 4-4.3(a)   Pressure Losses (psi/ft) Schedule 40 Steel Pipe.   C = 120

Flow Rate - GPM

Pipe Size in.	10	12	14	16	18	20	25	30	35	40	45	50
1	0.04	0.05	0.07	0.09	0.11	0.13	0.20	0.28	0.37	0.47	0.58	0.71
1¼	0.01	0.01	0.02	0.02	0.03	0.03	0.05	0.07	0.10	0.12	0.15	0.19
1½	0.01	0.01	0.01	0.01	0.01	0.02	0.02	0.03	0.05	0.06	0.07	0.09
2	–	–	–	–	–	0.01	0.01	0.01	0.01	0.02	0.02	0.03

For SI Units: 1 gal = 3.785 L; 1 psi = 0.0689 bar; 1 ft = 0.3048 m.

Table 4-4.3(b) Pressure Losses (psi/ft) Copper Tubing—Types K, L, & M. C = 150

Tubing Size in.	Type	Flow Rate - GPM											
		10	12	14	16	18	20	25	30	35	40	45	50
¾	M	0.08	0.12	0.16	0.20	0.25	0.30	0.46	0.64	0.85	—	—	—
	L	0.10	0.14	0.18	0.23	0.29	0.35	0.53	0.75	1.00	—	—	—
	K	0.13	0.18	0.24	0.30	0.38	0.46	0.69	0.97	1.28	—	—	—
1	M	0.02	0.03	0.04	0.06	0.07	0.08	0.13	0.18	0.24	0.30	0.38	0.46
	L	0.03	0.04	0.05	0.06	0.08	0.10	0.15	0.20	0.27	0.35	0.43	0.53
	K	0.03	0.04	0.06	0.07	0.09	0.11	0.17	0.24	0.31	0.40	0.50	0.61
1¼	M	0.01	0.01	0.02	0.02	0.03	0.03	0.05	0.07	0.09	0.11	0.15	0.17
	L	0.01	0.01	0.02	0.02	0.03	0.03	0.05	0.07	0.10	0.12	0.16	0.19
	K	0.01	0.01	0.02	0.02	0.03	0.04	0.06	0.08	0.11	0.13	0.17	0.20
1½	M	—	0.01	0.01	0.01	0.01	0.01	0.02	0.03	0.04	0.05	0.06	0.08
	L	—	0.01	0.01	0.01	0.01	0.01	0.02	0.03	0.04	0.05	0.07	0.08
	K	—	0.01	0.01	0.01	0.01	0.02	0.02	0.03	0.05	0.06	0.07	0.09
2	M	—	—	—	—	—	—	0.01	0.01	0.01	0.01	0.02	0.02
	L	—	—	—	—	—	—	0.01	0.01	0.01	0.01	0.02	0.02
	K	—	—	—	—	—	—	0.01	0.01	0.01	0.01	0.02	0.02

For SI Units: 1 gal. = 3.785 L; 1 psi = 0.0689 bar; 1 ft = 0.3048 m.

Table 4-4.3(c) Equivalent Length of Pipe in Feet for Steel and Copper Fittings and Valves

Fitting/Valve Diameter In.	Elbows			Tees		Valves					
	45 Degrees	90 Degrees	Long Radius	Flow Thru Branch	Flow Thru Run	Gate	Angle	Globe	Globe "Y" Pattern	Cock	Check
¾	1	2	1	4	1	1	10	21	11	3	3
1	1	3	2	5	2	1	12	28	15	4	4
1¼	2	3	2	6	2	2	15	35	18	5	5
1½	2	4	3	8	3	2	18	43	22	6	6
2	3	5	3	10	3	2	24	57	28	7	8

Based on Crane Technical Paper No. 410.
For SI Units: 1 ft = 0.3048 m.
(This is based upon the friction loss through the fitting being independent of the C Factor available to the piping.)

Table 4-4.3(d)
Pressure Losses in Water Meters

Meter (Inches)	Pressure Loss (psi) Flow (gpm)					
	18	23	26	31	39	52
⅝	9	14	18	26	*	*
¾	4	8	9	13	*	*
1	2	3	3	4	6	10
1½	**	1	2	2	4	7
2	**	**	**	1	2	3

NOTE: Higher pressure losses specified by the manufacturer should be used in place of those specified in the table. Lower pressure losses may be used when supporting data is provided by the meter manufacturer.
*Above maximum rated flow of commonly available meters.
**Less than 1 psi.
For SI Units: 1 gpm = 3.785 L/min; 1 in. = 25.4 mm.

(68 L/min), and the second calculation is made for two sprinklers flowing at 26 gpm (98 L/min) (13 gpm each). The pressure required to produce this flow at the sprinkler will depend on the coefficient of discharge (K Factor), which must be obtained from the manufacturer of the particular residential sprinkler that will be used. In multilevel dwellings, calculations will need to be done for all levels of the building to be certain that pipe is sized for the variations in flow that can occur due to elevation differences between floors and differences in piping geometry between floors.

The second step is to determine the water pressure in the street in front of the property. This is done by obtaining information on the public water supply pressure from the local water company or by arranging with a local plumber to place a pressure gage on the water supply inlet connection in the dwelling.

Next, pipe sizes are arbitrarily selected and valves and fittings laid out to meet the piping arrangements described in Figures A-2-2(a), (b), and (c), and A-2-3(a), (b), and (c). Losses in piping, fittings, meters, valves, and for elevation are then determined in accordance with Steps (d), (e), (f), (g), and (h), and these are deducted from the city pressure to determine the pressure at the sprinkler design area on each floor of the building. If this pressure is less than the operating pressure determined in the calculation to meet flow demands (*see Step 1 above*), a redesign is necessary and pipe must be upsized or the water supply improved. If the remaining pressure is higher

	Individual Loss	Net Total
a. Water Pressure in Street _____	_____	_____
b. Arbitrarily Select Pipe Size _____		
c. Deduct Meter Loss (Size)_____	_____	_____
d. Deduct Head Loss for Elevation System Control Valve* (_____ft × 0.434) _____		
e. Deduct Pressure Loss from City Main to Sprinkler _____	_____	_____

_____Pipe   – _____ft

_____Valves – _____ft

_____Elbows– _____ft

_____Tee   – _____ft

_____Total – _____ft ×   _____   _____   _____

f. Deduct Pressure Loss for Piping—Control
   Valve to Farthest Sprinkler*

Size	Quan.	Description	Total Equiv. Feet
	_____	90° Elbow	_____
	_____	45° Elbow	_____
	_____	Tee	_____
	_____	Check Valve	_____
	_____	Valve(_____)	_____
	_____	Total	

Ft × _____ = _____ _____

Size	Quan.	Description	Total Equiv. Feet
	_____	90° Elbow	_____
	_____	45° Elbow	_____
	_____	Tee	_____
	_____	Check Valve	_____
	_____	Valve(_____)	_____
	_____	Total	Ft × _____ = _____ _____

Remaining Pressure for Sprinkler Operation _____

*Factors from Tables 4-4.3(a), (b), (c), and (d).

For SI Units: 1 ft = 0.3048 m; 1 psi = 0.0689 bar.

Figure A-4-4.3(1)   Calculation Sheet.

	Individual Loss	Net Total
Water Pressure at Supply Outlet	_____	_____
a. Deduct Head Loss for Elevation (_____ft × 0.434)	_____	_____
b. Deduct Pressure Loss from Piping Within Building*		
Remaining Pressure for Sprinkler Operation	_____	_____

*Factors from Tables 4-4.3(a), (b), (c), and (d).

For SI Units: 1 ft = 0.3048 m; 1 psi = 0.0689 bar.

Figure A-4-4.3(2)   Calculation Sheet — Elevated Tank, Booster Pump, Pump Tank Supply.

than required to supply the demand rate at the sprinkler, the designer can either stop and accept the calculation or downsize the pipe to achieve optimum sizing.

A sample calculation for a typical residence follows:

This example is based on the dwelling illustrated by Figure 4.1, which is a plot plan also showing an existing city main. Figures 4.2 and 4.3 show the sprinkler and piping layouts for the basement and first floor of this dwelling.

Assume the following:
- That the sprinklers used are listed for application rates, design area, and area of coverage as indicated in 4-1.1, 4-1.2, and 4-1.4.
- That the flow from one sprinkler is 18 gpm (68 L/min) and from two sprinklers is 13 gpm (49 L/min) each (26 gpm total) per 4-1.1 of this standard.
- That the sprinkler used has a coefficient (K Factor) of 3.9. Thus, the pressure needed for the system at the flowing sprinkler is calculated as follows:

$$Q = K \sqrt{P} \qquad P = (Q/K)^2$$

For 18 gpm, $P = (18/3.9)^2 = (4.61)^2 = 21.3$ psi

For 13 gpm, $P = (13/3.9)^2 = (3.33)^2 = 11.1$ psi

4-4.4   To size piping for systems with an elevated tank, pump, or pump-tank combination, determine the pressure at the water supply outlet and proceed through steps (c), (e), (g), (h), (i), (j), and (k) of 4-4.3.

4-5 Piping Configurations.   Piping configurations may be looped, gridded, straight run, or combinations thereof.

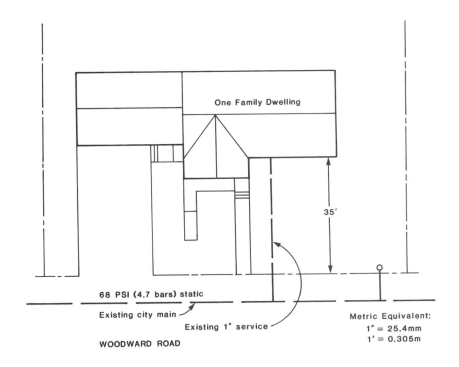

One Family Dwelling

35'

68 PSI (4.7 bars) static

Existing city main

Existing 1″ service

WOODWARD ROAD

Metric Equivalent:
1″ = 25.4mm
1′ = 0.305m

PLOT PLAN
Plan 1 of 3

Figure 4.1. Typical residence for sample calculation.

When piping configurations are looped or gridded, calculations must be conducted in accordance with 4-4.3 and the Exception to 4-4.1 of this standard. This means that hydraulic calculations to size the piping must be done in accordance with the method described in NFPA 13.

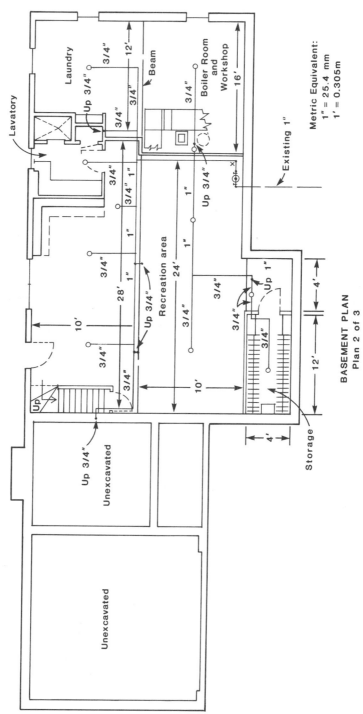

Figure 4.2. Basement floor plan for sample sprinkler calculation.

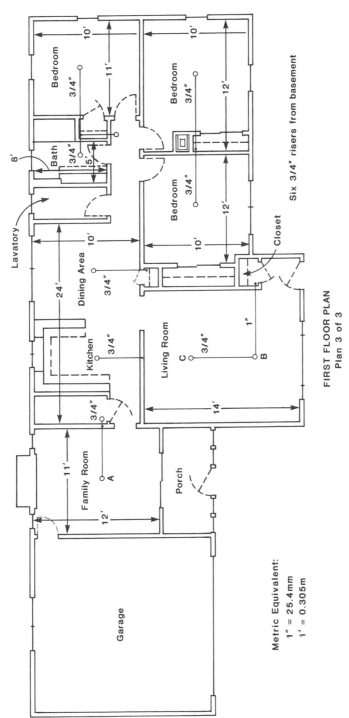

Figure 4.3. First floor plan for sample sprinkler calculation.

Calculation #1 at 18 gpm (21.3 psi) from Sprinkler "A"

1st Floor: Type K Copper				Individual loss	Net total
a. Water pressure in street				_____	_68_
b. Arbitrarily selected pipe size _1" + 3/4"_					
c. Deduct meter loss (_3/4_ size)				_4.0_	_64_
d. Deduct loss for elevation					
System control valve*					
(_15_ ft × 0.434)				_6.5_	_57.5_
e. Deduct loss from city main					

to control valve

_1"_	tube	_48_ ft	
_3_	valves	_3_ ft	
_3_	elbows	_9_ ft	
_1_	tee	_5_ ft	
	Total	_65_ ft × _0.09_	_5.9_ ___ _51.6_

f. Deduct loss for tubing: control
valve to farthest sprinkler*

3/4-in. tube: 19 ft	×	_0.38_	_7.2_
1-in. tube: 35 ft	×	_0.09_	_3.1_ ___ _41.3_

Size	Quan.	Description	Total equiv. ft		
_3/4"_	_4_	90° elbow	_8_		
	___	45° elbow	___		
	___	tee	___		
	___	check valve	___		
	___	valve (___)	___		
		Total	_8_ ft × _0.38_ = _3.0_		_38.3_

Size	Quan.	Description	Total equiv. ft		
_1"_	_1_	90° elbow	_3_		
	___	45° elbow	___		
	_1_	tee	_5_		
	_1_	check valve	_4_		
	___	valve (___)	___		
		Total	_12_ ft × _0.09_ = _1.1_		_37.2_

This exceeds 21.3 psi and design is acceptable.

Note: Repeat calculation for basement to verify adequacy of basement piping size.

Remaining pressure for sprinkler operations.

* Factors from Table 4.4.3(a), (c), and (d).

For SI Units: 1 ft = 0.3018 m; 1 psi = 0.0689 bar.

Figure 4.4. Calculation Example No. 1.

Calculation #2 at 26 gpm (11.1 psi) from Sprinklers "B and "C"

1st Floor: Type K Copper Tube	Individual loss	Net total
a. Water pressure in street		*68*
b. Arbitrarily selected pipe size *1" + 3/4"*		
c. Deduct meter loss (*3/4* size)	*9.0*	*59.0*
d. Deduct loss for elevation		
System control valve*		
(*15* ft X 0.434)	*6.5*	*52.5*

e. Deduct loss from city main

to control valve

_*1"*_ tube	*48* ft			
3 valves	*3* ft			
3 elbows	*9* ft			
1 tee	*5* ft			
Total	*65* ft	X	*0.20*	*13.0*

Net total: *39.5*

f. Deduct loss for tubing: control (13 gpm)

valve to farthest sprinkler*

3/4-in. tube: 12 ft	X	*0.19*	*2.3*	
1-in. tube: 14 ft	X	*0.20*	*2.8*	*34.4*

Size	Quan.	Description	Total equiv. ft
	1	90° elbow	*2*
		45° elbow	
3/4"		tee	
		check valve	
		valve (___)	
		Total	*2* ft X *.27* = *.54*

Net total: *33.8*

Size	Quan.	Description	Total equiv. ft
	1	90° elbow	*3*
		45° elbow	
1"	*1*	tee	*5*
	1	check valve	*4*
		valve (___)	
		Total	*12* ft X *0.20* = *2.4*

Net total: *31.4*

This exceeds 11.1 psi and design is acceptable.

Note: Repeat calculation for basement to verify adequacy of basement piping size.

Remaining pressure for sprinkler operations.

* Factors from Table 4.4.3(a), (c), and (d).

For SI Units: 1 ft = 0.3018 m; 1 psi = 0.0689 bar.

Figure 4.5 Calculation Example No. 2. The farthest sprinkler in this case is the farthest sprinkler in the only room having more than one sprinkler.

4-6 Location of Sprinklers. Sprinklers shall be installed in all areas.

Exception No. 1: Sprinklers may be omitted from bathrooms not exceeding 55 sq ft (5.1 m²) with noncombustible plumbing fixtures.

Exception No. 2: Sprinklers may be omitted from small closets where the least dimension does not exceed 3 ft (0.9 m) and the area does not exceed 24 sq ft (2.2 m²) and the walls and ceilings are surfaced with noncombustible or limited combustible materials as defined by NFPA 220, Standard on Types of Building Construction.

Exception No. 3: Sprinklers may be omitted from garages, open attached porches, carports, and similar structures.

Exception No. 4: Sprinklers may be omitted from attics, crawl spaces, and other concealed spaces that are not used or intended for living purposes or storage.

Exception No. 5: Sprinklers may be omitted from entrance foyers that are not the only means of egress.

Selective omission of sprinklers from certain areas raises concern that a reduced or insufficient level of protection is being considered. This standard does indeed recognize the presence and availability of the "levels of protection" concept which spans most fire protection codes and standards. Areas mentioned in these Exceptions are not selected at random but instead represent those areas in which fires do not result in a high percentage of fatalities. Table A-1-2(a) shows statistics for various fire deaths and their relation to the area of fire origin. In addition, the following is noted:

- Exception No. 1. Combustible fuel loading in most bathrooms is typically low.
- Exception No. 2. Small closets are usually impractical places to install sprinklers because of their relatively small size.
- Exceptions No. 3, 4, and 5. Mandatory sprinklering of these areas would necessitate the use of dry-pipe systems in areas where freezing weather is encountered. This would detract from the rapid response of the system within the occupied areas of the dwelling and thus detract from, rather than enhance, life safety. Further, most building codes require a 1-hour fire rated separation between garages and other portions of the dwelling.

5

Referenced Publications

5-1 The following documents or portions thereof are referenced within this standard and shall be considered part of the requirements of this document. The edition indicated for each reference is the current edition as of the date of the NFPA issuance of this document.

5-1.1 NFPA Publications. National Fire Protection Association, Battery-march Park, Quincy, MA 02269.

NFPA 13-1989, *Standard for the Installation of Sprinkler Systems*

NFPA 22-1987, *Standard for Water Tanks for Private Fire Protection*

NFPA 74-1989, *Standard for the Installation, Maintenance, and Use of Household Fire Warning Equipment*

NFPA 220-1985, *Standard on Types of Building Construction.*

5-1.2 Other Codes and Standards.

5-1.2.1 ANSI Publications. American National Standards Institute, Inc., 1450 Broadway, New York, NY 10018.

ANSI B16.1-1975, *Cast Iron Pipe Flanges and Flanged Fittings, Class 25, 125, 250 and 800*

ANSI B16.3-1985, *Malleable Iron Threaded Fittings, Class 150 and 300*

ANSI B16.4-1985, *Cast Iron Threaded Fittings, Classes 125 and 250*

ANSI B16.5-1981, *Pipe Flanges and Flanged Fittings*

ANSI B16.9-1986, *Factory-Made Wrought Steel Buttwelding Fittings*

ANSI B16.11-1980, *Forged Steel Fittings, Socket-Welding and Threaded*

ANSI B16.18-1984, *Cast Copper Alloy Solder Joint Pressure Fittings*

ANSI B16.22-1980, *Wrought Copper and Copper Alloy Solder Joint Pressure Fittings*

ANSI B16.25-1986, *Buttwelding Ends*

ANSI B36.10M-1985, *Welded and Seamless Wrought Steel Pipe.*

5-1.2.2 ASTM Publications.   American Society for Testing and Materials, 1916 Race Street, Philadelphia, PA 19105.

ASTM A53-1987, *Standard Specification for Pipe, Steel, Black, and Hot-Dipped, Zinc-Coated Welded and Seamless Steel Pipe*

ASTM A135-1986, *Standard Specification for Electric-Resistance-Welded Steel Pipe*

ASTM A234-1987, *Standard Specification for Piping Fittings of Wrought Carbon Steel and Alloy Steel for Moderate and Elevated Temperatures*

ASTM A795-1985, *Specification for Black and Hot-Dipped Zinc-Coated Welded and Seamless Steel Pipe for Fire Protection Use*

ASTM B32-1987, *Standard Specification for Solder Metal, 95-5 (Tin-Antimony-Grade 95TA)*

ASTM B75-1981, *Standard Specification for Seamless Copper Tube*

ASTM B88-1986, *Standard Specification for Seamless Copper Water Tube*

ASTM B251-1987, *Standard Specification for General Requirements for Wrought Seamless Copper and Copper-Alloy Tube*

ASTM E380-1986, *Standard for Metric Practice.*

5-1.2.3 AWS Publication.   American Welding Society, 550 NW LeJeune Road, Miami, FL 33135.

AWS A5.8-1981, *Specification for Brazing Filler Metal.*

Appendix A

The material contained in Appendix A of this standard is included within the text of this handbook, and therefore is not repeated here.

Appendix B
Referenced Publications

B-1 The following documents or portions thereof are referenced within this standard for informational purposes only and thus are not considered part of the requirements of this document. The edition indicated for each reference is the current edition as of the date of the NFPA issuance of this document.

B-1.1 NFPA Publications. National Fire Protection Association, Batterymarch Park, Quincy, MA 02269.

NFPA 13-1989, *Standard for the Installation of Sprinkler Systems*

NFPA 13A-1987, *Recommended Practice for the Inspection, Testing and Maintenance of Sprinkler Systems*

NFPA 13R-1989, *Standard for the Installation of Sprinkler Systems in Residential Occupancies up to Four Stories in Height*

NFPA 20-1987, *Standard for the Installation of Centrifugal Fire Pumps*

NFPA 74-1989, *Standard for the Installation, Maintenance, and Use of Household Fire Warning Equipment.*

REFERENCES CITED IN COMMENTARY

[1]Halpin, B.M., Dinan, J.J., and Deters, O.J., *Assessment of the Potential Impact of Fire Protection Systems on Actual Fire Incidents*, Johns Hopkins University-Applied Physics Laboratory (JHU/APL), Laurel, MD, October 1978.

This study describes an in-depth analysis of fires involving fatalities and includes an assessment of how use of detectors, sprinklers, or remote alarms would have changed the results.

[2]Yurkonis, Peter, *Study to Establish the Existing Automatic Fire Suppression Technology for Use in Residential Occupancies*, Rolf Jensen & Associates, Inc., Deerfield, IL, December 1978.

This study identified suppression systems and evaluated design and cost factors affecting practical usage and user acceptance.

[3]Kung, H., Haines, D., and Green, R. Jr., *Development of Low-Cost Residential Sprinkler Protection*, Factory Mutual Research Corporation (FMRC), Norwood, MA, February 1978.

This study is aimed primarily at development of low-cost residential sprinkler systems having minimal water discharge rates providing adequate life and property protection from smoldering fires.

[4]Henderson, N.C., Riegel, P.S., Patton, R.M., and Larcomb, D.B., *Investigation of Low-Cost Residential Sprinkler Systems*, Battelle Columbus Laboratories (BCL), Columbus, OH, June 1978.

Fire tests of commercial nozzles and evaluation of piping methods to achieve low cost systems based on propane burner, wood crib, and furniture fire tests.

[5]Clarke, Graham, *Performance Specifications for Low-Cost Residential Sprinkler System*, Factory Mutual Research, Norwood, MA, January 1978.

A review of data from the listed USFA research efforts as a basis to prepare a proposed standard to install residential sprinkler systems and test residential sprinklers.

[6]Kung, H. C., Spaulding, R. D., and Hill, E.E. Jr., *Sprinkler Performance in Residential Fire Tests*, Factory Mutual Research, Norwood, MA, December 1980.

Fire tests on representative living room and bedroom residential configuration with combinations of furnishings and with open and closed windows, and with variations in sprinkler application rates and response sensitivity of sprinklers. This work has been closely evaluated by the NFPA 13D Subcommittee during its conduct, and the Subcommittee was responsible for directing many of the test conditions that were evaluated.

[7]Cote, A., and Moore, D., *Field Test and Evaluation of Residential Sprinkler Systems, Los Angeles Test Series*, National Fire Protection Association, Boston, MA, April 1980. (A report for the NFPA 13D Subcommittee)

A series of fire tests in actual dwellings with sprinkler systems installed in accordance with the data resulting from all prior test work and the criteria included in the proposed 1980 revisions to NFPA 13D so as to evaluate the effectiveness of the system under conditions approaching actual use.

[8]Moore, D., *Data Summary of the North Carolina Test Series of USFA Grant 79027 Field Test and Evaluation of Residential Sprinkler Systems*. National Fire Protection Association, Boston, MA, September 1980. (A report for the NFPA 13D Subcommittee)

A series of fire tests in actual mobile homes with sprinkler systems installed in accordance with the data resulting from prior test work and following criteria described in NFPA 13D-1980 in order to evaluate the effectiveness of the system under conditions approaching actual use.

[9]Kung, H. C., Spaulding, R.D., Hill, E.E. Jr., and Symonds, A.P., *Technical Report, Field Evaluation of Residential Prototype Sprinkler, Los Angeles Fire Test Program*, Factory Mutual Research Corporation, Norwood, MA, February 1982.

[10]Cote, A. E., *Final Report on Field Test and Evaluation of Residential Sprinkler Systems*, National Fire Protection Association, Quincy, MA, July 1982.

[11]Factory Mutual Research Corporation/Underwriters Laboratories Inc., *Standard for Residential Automatic Sprinklers for Fire Protection Service*, Northbrook, IL, April 1980.

A study to develop a performance standard for testing residential sprinklers under room fire conditions that are in harmony with the requirements of NFPA 13D.

[12]Underwriters Laboratories Inc., *Standard for Residential Automatic Sprinklers for Fire Protection Service,* UL 1626, Northbrook, IL, October 1986.

This standard describes the requirements for the testing of residential sprinklers by a testing laboratory. Factory Mutual Engineering Association has a similar standard.

[13]Evans, D.D., and Madrzykowski, D., *Characterizing the Thermal Response of Fusible-Link Sprinklers,* U.S. Department of Commerce, National Bureau of Standards, Washington, DC, August 1981.

The following publications are available from the National Fire Protection Association, Batterymarch Park, Quincy, MA 02269.

NFPA 13-1989, *Standard for the Installation of Sprinkler Systems*
NFPA 13A-1987, *Recommended Practice for the Inspection, Testing and Maintenance of Sprinkler Systems*
NFPA 13R-1989, *Standard for the Installation of Sprinkler Systems in Residential Occupancies up to Four Stories in Height*
NFPA 74-1989, *Standard for the Installation, Maintenance, and Use of Household Fire Warning Equipment.*

NFPA 13R

Standard for the

Installation of Sprinkler Systems in Residential Occupancies up to Four Stories in Height

1989 Edition

NOTICE: An asterisk (*) following the number or letter designating a paragraph indicates explanatory material on that paragraph in Appendix A. Material from Appendix A is integrated with the text and is identified by the letter A preceding the subdivision number to which it relates.

Information on referenced publications can be found in Chapter 3 and Appendix B.

Preface

It is intended that this standard provide a method for those individuals wishing to install a sprinkler system for life safety and property protection. It is not the purpose of this standard to require the installation of an automatic sprinkler system. This standard assumes that one or more smoke detectors will be installed in accordance with NFPA 74, *Standard for the Installation, Maintenance, and Use of Household Fire Warning Equipment*.

Since the first edition of NFPA 13D, *Standard for the Installation of Sprinkler Systems in One- and Two-Family Dwellings and Mobile Homes*, there has been interest in establishing criteria for sprinkler installations in low-rise residential occupancies. Many of the basic philosophies upon which NFPA 13D was developed also apply to low-rise residential occupancies. Based on public proposals and evidence that jurisdictions were developing their own standards, it became readily apparent that the need existed for a national consensus standard to provide cost-effective sprinkler protection for low-rise residential occupancies with life safety as the primary goal and property protection as a secondary goal.

The NFPA Technical Committee on Automatic Sprinklers intends that NFPA 13R, *Standard for the Installation of Sprinkler Systems in Residential Occupancies up to Four Stories in Height*, contain requirements that provide an acceptable level of total fire protection with respect to life safety and property protection. NFPA 13R is intended to provide a high, but not absolute, level of life safety and a lesser level of property protection. Greater protection to both life and property could be achieved by providing sprinkler protection throughout all areas of the building in accordance with NFPA 13, provided residential sprinklers are installed in the residential areas as permitted by NFPA 13.

While NFPA 13R differs from NFPA 13 in several areas, such as extent of coverage and the mandatory use of residential sprinklers in most residential areas, the knowledgeable user of NFPA 13 will also notice many similarities. However, NFPA 13R does offer the advantage of highlighting the criteria for low-rise residential occupancies since it is a standard specifically written for such occupancies.

It should be noted that the standard does assume that one or more smoke detectors will be installed in accordance with NFPA 74, *Standard for the Installation, Maintenance, and Use of Household Fire Warning Equipment.* The detectors are intended to provide a life safety benefit from smoldering fires and from fires originating in areas not protected by automatic sprinklers.

1

General Information

1-1* Scope. This standard deals with the design and installation of automatic sprinkler systems for protection against fire hazards in residential occupancies up to four stories in height.

This standard deals with the design and installation of automatic sprinkler systems for protection against hazards in residential occupancies of up to four stories. It is based in part on tests referenced in Appendix A which have demonstrated the ability of such systems to suppress fires with a growth rate similar to that of residential furnishings provided the sprinkler response occurs in the earliest stages of fire growth. Typical curves illustrating the severity of energy release rates, including those for residential furnishings, are shown in Figure 1.1.[1]

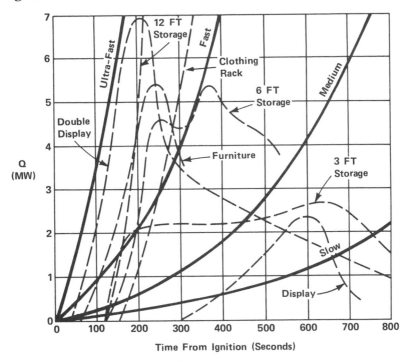

Figure 1.1. Energy release time curves.

The dark line curves on the graph are for reference and represent the fire growth curves frequently used to typify expected fire growth rates. (*See NFPA 72E, Standard on Automatic Fire Detectors.*)

The standard does not provide a definition of "story" for use in determining building height. It is anticipated that the height of the building will be determined in accordance with the applicable building code. With respect to large board and care occupancies, it should be noted that NFPA *101®*, *Life Safety Code®*, defines a method of determining height for purposes of applying that code.

As noted in Section 1-3 (Definitions), the use of the phrase "residential occupancies" for purposes of applying NFPA 13R differs from residential occupancies as defined in NFPA 101. The NFPA 13R definition of residential occupancies is consistent with NFPA 101 except that three categories of board and care occupancies and one- and two-family dwellings are excluded. The categories of board and care facilities are excluded due to a concern for overall building evacuation time and the fact that evacuation time in accordance with Chapter 21 of NFPA 101 is based on reaching a point of safety which may not necessarily be the exterior of the building. Due to the lack of complete sprinkler coverage, NFPA 13R may not provide adequate protection for occupants who require an extended period of time to evacuate or who are protected in place.

One- and two-family dwellings are exempted since they fall within the scope of NFPA 13D. However, NFPA 13R or NFPA 13 may be more appropriate than NFPA 13D for large houses which have areas not commonly found in family residences, such as large entertainment rooms, central cooking and laundry areas, or large storage rooms.

A-1-1   NFPA 13R is appropriate for use only in residential occupancies, as an option to NFPA 13, *Standard for the Installation of Sprinkler Systems*, as defined in this standard, up to four stories in height. Residential portions of any other building may be protected with residential sprinklers in accordance with 3-11.2.9 of NFPA 13, *Standard for the Installation of Sprinkler Systems*. Other portions of such sections should be protected in accordance with NFPA 13.

The criteria in this standard are based on full-scale fire tests of rooms containing typical furnishings found in residential living rooms, kitchens, and bedrooms. The furnishings were arranged as typically found in dwelling units in a manner similar to that shown in Figures A-1-1(a), (b), and (c). Sixty full-scale fire tests were conducted in a two-story dwelling in Los Angeles, California, and 16 tests were conducted in a 14-ft (4.3-m) wide mobile home in Charlotte, North Carolina. Sprinkler systems designed and installed according to this standard are expected to prevent flashover within the compartment of origin if sprinklers are installed in the compartment. A sprinkler system designed and installed according to this standard may not, however, be expected to control a fire involving unusually higher average fuel loads than typical for dwelling units [10 lb/sq ft (49 kg/m²)], configurations of fuels other than those with typical residential occupancies, or conditions where the interior finish has an unusually high flame spread rating (greater than 225).

To be effective, sprinkler systems installed in accordance with this standard must have the sprinklers closest to the fire open before the fire exceeds the ability of the sprinkler discharge to extinguish or control that fire. Conditions that allow the fire to grow beyond that point before sprinkler activation or that interfere with the quality of water distribution can produce conditions beyond the capabilities of the sprinkler system described in this standard. Unusually high ceilings or ceiling configurations that tend to divert the rising hot gases from sprinkler locations or change the sprinkler discharge pattern from its standard pattern can produce fire conditions that cannot be extinguished or controlled by the systems described in this standard.

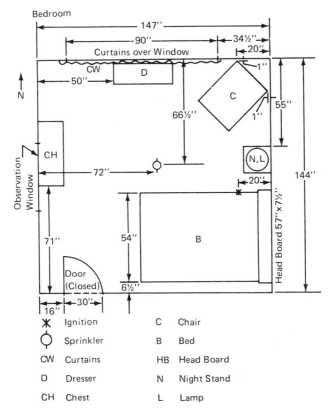

Figure A-1-1(a)   Bedroom.

Mobile Home Bedroom

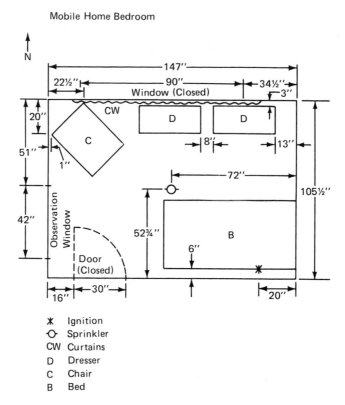

* Ignition
Ｏ Sprinkler
CW Curtains
D Dresser
C Chair
B Bed

Figure A-1-1(b)   Mobile Home Bedroom.

1-2* Purpose. The purpose of this standard is to provide design and installation requirements for a sprinkler system to aid in the detection and control of fires in residential occupancies and thus provide improved protection against injury, life loss, and property damage. A sprinkler system designed and installed in accordance with this standard is expected to prevent flashover (total involvement) in the room of fire origin, when sprinklered, and to improve the chance for occupants to escape or be evacuated.

Nothing in this standard is intended to restrict new technologies or alternate arrangements, providing that the level of safety prescribed by the standard is not lowered.

A-1-2 Levels of Protection. Various levels of sprinkler protection are available to provide life safety and property protection. The standard is designed to provide a high, but not absolute, level of life safety and a lesser level of property protection. Greater protection to both life and property could be achieved by sprinklering all areas in accordance with NFPA 13,

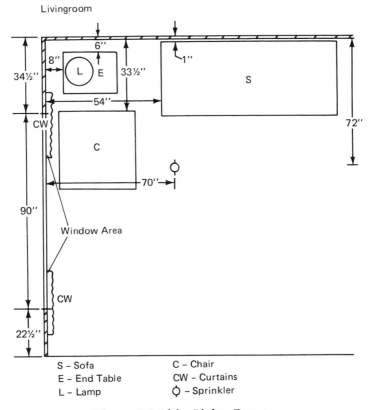

Figure A-1-1(c)   Living Room.

Standard for the Installation of Sprinkler Systems, which permits the use of residential sprinklers in residential areas.

This standard recommends, but does not require, sprinklering of all areas in the building; it permits sprinklers to be omitted in certain areas. These areas are the ones shown by NFPA statistics to be ones where the incidence of life loss from fires in residential occupancies is low. Such an approach provides a reasonable degree of fire safety to life. (*See Table A-1-2 for Deaths and Injuries in Multifamily Residential Buildings.*)

It should be recognized that the omission of sprinklers from certain areas could result in the development of untenable conditions in adjacent spaces. Where evacuation times may be delayed, additional sprinkler protection and other fire protection features, such as detection and compartmentation, may be necessary.

Both Underwriters Laboratories Inc. and Factory Mutual Research Corporation have developed test standards for the evalu-

Table A-1-2
Annual Averages of Deaths and Injuries in Apartments
1980–1984

Fires—123,000	Civilian Deaths—930	Percentages by Area of Origin	Civilian Injuries—5,470
Area of Origin (901 Code)	Civilian Deaths (Used for Ranking)	Fires	Civilian Injuries
Living room, den, lounge (14)	38.5	11.3	23.2
Bedroom (21-22)	28.7	17.4	27.1
Kitchen (24)	9.8	35.3	27.2
Hallway corridor (101)	4.3	3.2	3.4
Interior stairway (03)	3.2	1.0	1.1
Structural Area (70-79)	3.1	8.1	3.5
[Balcony, porch (72)]	(1.2)	(1.3)	(0.7)
[Unspecified (79)]	(1.0)	(0.5)	(0.2)
[Ceiling/Roof assembly (74)]	(0.3)	(0.7)	(0.3)
Lobby (05)	1.3	0.6	0.7
Dining room (23)	1.2	0.8	1.0
Closet (42)	1.2	1.9	1.9
Balcony, porch (72)	1.2	1.3	0.7
Other known single area	4.1	17.8	8.8
[Bathroom (25)]	(0.6)	(2.1)	(1.3)
Multiple areas (97)	1.6	0.7	0.9
Unclassified, not applicable (98-99)	1.8	0.6	0.5
Total:	100.0	100.0	100.0

ation of residential sprinklers.[2,3] These standards have significantly different fire test requirements than those found in the UL and FM standards for commercial sprinklers. The residential sprinkler criteria are:

- Maximum temperature 3 in. (94 mm) below the ceiling—600°F (316°C).
- Maximum temperature 5¼ ft (1.6 m) above the floor—200°F (93°C).
- Temperature at 5¼ ft above the floor shall not exceed 130°F (54°C) for more than any continuous two-minute period.
- Maximum temperature ¼ in. (6.4 mm) above the finished ceiling—500°F (260°C).
- Maximum of two sprinklers in test area to operate during test.
- Vertical walls within coverage area are to be wetted to at least within 28 in. (711 mm) of the ceiling.

The concept of "levels of protection" has been adopted to achieve a low-cost system and to avoid the necessity of installing dry-pipe sprinkler systems in cold climates. The areas where omission of sprinklers is permitted in accordance with Section 2-6 of this standard are those where the record justifies such omission. (*See Table A-1-2.*) The Committee felt that dry-pipe systems are a less desirable form of protection because of the delay in application of water. (This was noted in early Factory Mutual tests and substantiated in the Los Angeles and North Carolina fire tests.[4,5,6,7]) The key to achieving control of the residential fire is the rapid application of water. Further, residential systems must be simple and easy to maintain—factors not associated with dry-pipe systems.

It should be noted that Table A-1-2 contains data on life safety and not necessarily property protection. Code officials will need to evaluate reductions in building code criteria related to property protection when the system is installed in accordance with this standard. Of particular interest would be areas where sprinklers may be omitted. (*See Section 2-6.*)

1-3 Definitions.

Approved.   Acceptable to the "authority having jurisdiction."

NOTE: The National Fire Protection Association does not approve, inspect or certify any installations, procedures, equipment, or materials nor does it approve or evaluate testing laboratories. In determining the acceptability of installations or procedures, equipment or materials, the authority having jurisdiction may base acceptance on compliance with NFPA or other appropriate standards. In the absence of such standards, said authority may require evidence of proper installation, procedure or use. The authority having jurisdiction may also refer to the listings or labeling practices of an organization concerned with product evaluations which is in a position to determine compliance with appropriate standards for the current production of listed items.

A common error is the assumption that "approved" means listed or labeled or another similar designation. For purposes of this standard, "approved" means that which is acceptable to the authority having jurisdiction. Although the authority having jurisdiction may use a listing or label to assist in approving an item, it should not be assumed that all approvals are based on a listing procedure. It should also be noted that some testing agencies use the term "approval," which is really tantamount to a "listing."

Paragraph 2-2.1.2 requires that only listed or approved devices and materials as indicated by the standard shall be used in sprinkler systems. Various other sections require that specific components be listed (i.e., residential sprinkler, nonmetallic sprinkler ceiling plates, etc.).

Authority Having Jurisdiction.   The "authority having jurisdiction" is the organization, office or individual responsible for "approving" equipment, an installation or a procedure.

NOTE: The phrase "authority having jurisdiction" is used in NFPA documents in a broad manner since jurisdictions and "approval" agencies vary as do their responsibilities. Where public safety is primary, the "authority having jurisdiction" may be a federal, state, local or other regional department or individual such as a fire chief, fire marshal, chief of a fire prevention bureau, labor department, health department, building official, electrical inspector, or others having statutory authority. For insurance purposes, an insurance inspection department, rating bureau, or other insurance company representative may be the "authority having jurisdiction." In many circumstances the property owner or his designated agent assumes the role of the "authority having jurisdiction"; at government installations, the commanding officer or departmental official may be the "authority having jurisdiction."

In the simplest terms, the authority having jurisdiction (AHJ) is that person or office enforcing the standard. Commonly the AHJ is a fire marshal or building official where enforcement of the standard is mandatory. The AHJ can also be a contract officer, design engineer, insurance engineer, or other agency, especially when the standard is being applied or enforced on a nonregulatory basis.

Check Valve.   A valve that allows flow in one direction only.

Control Valve.   An indicating valve employed to control (shut) a supply of water to a sprinkler system.

The type of control valve used should be an indicating type, i.e., one that has some external means that will indicate that the valve is in the open position.

Design Discharge.   Rate of water discharged by an automatic sprinkler, expressed in gallons per minute.

Dry System.   A system employing automatic sprinklers that are attached to a piping system containing air under atmospheric or higher pressures. Loss of pressure from the opening of a sprinkler or detection of a fire condition causes the release of water into the piping system and out the opened sprinkler.

It should be pointed out that because of the concern for delayed water delivery and the fact that there are no residential sprinklers listed for dry system application, dry systems are not allowed for protection inside the dwelling units in accordance with Exception No. 1 to 2-4.5.1. However, dry systems may be installed in areas outside the dwelling unit in accordance with 2-5.2 and NFPA 13.

Dwelling Unit.   One or more rooms arranged for the use of one or more individuals living together as in a single housekeeping unit, normally having cooking, living, sanitary, and sleeping facilities.

For the purposes of this standard, a dwelling unit includes hotel rooms and suites, dormitory rooms, apartments, condominiums, sleeping rooms in board and care facilities, and similar living units.

Labeled.   Equipment or materials to which has been attached a label, symbol or other identifying mark of an organization acceptable to the "authority having jurisdiction" and concerned with product evaluation, that maintains periodic inspection of production of labeled equipment or materials and by whose labeling the manufacturer indicates compliance with appropriate standards or performance in a specified manner.

Listed.   Equipment or materials included in a list published by an organization acceptable to the "authority having jurisdiction" and concerned with product evaluation, that maintains periodic inspection of production of listed equipment or materials and whose listing states either that the equipment or material meets appropriate standards or has been tested and found suitable for use in a specified manner.

NOTE: The means for identifying listed equipment may vary for each organization concerned with product evaluation, some of which do not recognize equipment as listed unless it is also labeled. The "authority having jurisdiction" should utilize the system employed by the listing organization to identify a listed product.

One prominent testing inspection agency laboratory uses the designation "classified" for pipe. Material with this designation meets the intent of "listed." Also see the commentary to "Approved."

Multipurpose Piping Systems.   Piping systems within residential occupancies intended to serve both domestic and fire protection needs.

Residential Occupancies.   Residential occupancies as included in the scope of this standard include the following, as defined in NFPA *101, Life Safety Code*:

(1)   Apartment buildings.

(2)   Lodging and rooming houses.

(3)   Board and care facilities (slow evacuation type with 16 or less occupants and prompt evacuation type).

(4)   Hotels, motels, and dormitories.

For purposes of applying this standard, the definition of "residential occupancies" differs from the definition used in NFPA 101.
Specifically, the definition does not include the following occupancies which are considered residential occupancies by NFPA 101.
• One- and two-family dwellings
• Small, impractical board and care facilities
• Large, slow board and care facilities
• Large, impractical board and care facilities.
(See commentary to Section 1-1.)

Residential Sprinkler.   An automatic sprinkler that has been specifically listed for use in residential occupancies.

Residential sprinklers are specifically listed by UL and FM for residential service and have been tested for compliance with the residential sprinkler standard.[2,3] Typically, they are fast response sprinklers, and they may be of the upright, pendent, or sidewall configuration.

Shall.   Indicates a mandatory requirement.

Should.   Indicates a recommendation or that which is advised but not required.

Sprinkler—Automatic. A fire suppression device that operates automatically when its heat-actuated element is heated to or above its thermal rating, allowing water to discharge over a specific area.

Sprinkler System. An integrated system of piping connected to a water supply, with listed sprinklers that will automatically initiate water discharge over a fire area. When required, the sprinkler system also includes a control valve and a device for actuating an alarm when the system operates.

It is important to note that the definition of "sprinkler system" includes water supplies and underground piping. While tanks, pumps, and underground piping installation and design requirements are covered by other standards, when they are used in connection with overhead sprinkler piping they become an integral part of the sprinkler system. As critical components of the sprinkler system's performance, they must be treated as part of the system.

Standard. A document containing only mandatory provisions using the word "shall" to indicate requirements. Explanatory material may be included only in the form of "fine print" notes, in footnotes, or in an appendix.

Waterflow Alarm. A sounding device activated by a waterflow detector or alarm check valve.

Waterflow Detector. An electric signaling indicator or alarm check valve actuated by water flow in one direction only.

Wet System. A system employing automatic sprinklers that are attached to a piping system containing water and connected to a water supply, so that water discharges immediately from sprinklers opened by a fire.

1-4 Units. Metric units of measurement in this standard are in accordance with the modernized metric system known as the International System of Units (SI). Two units (liter and bar), outside of but recognized by SI, are

Table 1-4

Name of Unit	Unit Symbol	Conversion Factor
liter	L	1 gal = 3.785 L
pascal	Pa	1 psi = 6894.757 Pa
bar	bar	1 psi = 0.0689 bar
bar	bar	1 bar = 10^5 Pa

For additional conversions and information see ASTM E380, *Standard for Metric Practice.*

commonly used in international fire protection. These units are listed, with conversion factors, in Table 1-4.

1-4.1   If a value for measurement as given in this standard is followed by an equivalent value in other units, the first stated is to be regarded as the requirement. A given equivalent value may be approximate.

When and if the United States converts to SI, some of the values will probably be rounded off to the nearest whole number.

1-4.2   The conversion procedure for the SI units has been to multiply the quantity by the conversion factor and then round the result to the appropriate number of significant digits.

1-5 Piping.

1-5.1   Pipe or tube used in sprinkler systems shall be of the materials in Table 1-5.1 or in accordance with 1-5.2 through 1-5.5. The chemical properties, physical properties, and dimensions of the materials listed in Table 1-5.1 shall be at least equivalent to the standards cited in the table and designed to withstand a working pressure of not less than 175 psi (12.1 bars).

Table 1-5.1

Materials and Dimensions	Standard
Specification for Black and Hot-Dipped Zinc-Coated (Galvanized) Welded and Seamless Steel Pipe for Fire Protection Use	ASTM A795
Specification for Welded and Seamless Steel Pipe	ASTM A53
Wrought-Steel Pipe	ANSI B36.10M
Specification for Electric-Resistance Welded Steel Pipe	ASTM A135
Copper Tube (Drawn, Seamless) Specification for Seamless Copper Tube	ASTM B88
Specification for General Requirements for Wrought Seamless Copper and Copper-Alloy Tube	ASTM B251
Brazing Filler Metal (Classification BCuP-3 or BCuP-4)	AWS A5.8
Specification for Solder Metal, 95-5 (Tin-Antimony-Grade 95TA)	ASTM B32

1-5.2   Other types of pipe or tube may be used, but only those listed for this service.

> There is no prohibition against the use of any material, as long as it is listed for use in sprinkler systems. This section is intended to encourage development of either more efficient or cost-effective materials.
>
> Not all pipe made to a particular standard is listed. Listed piping is identified by the logo of the listing agency. Similar piping manufactured with less exacting quality control must not be used in sprinkler systems.
>
> At this time there are two synthetic piping materials that are listed for sprinkler system application, Chlorinated Polyvinyl Chloride (CPVC) and polybutylene pipe. It must be noted that the listings of these materials often include restrictions, such as: the pipe cannot be left "exposed." The listings are specific and do indicate for which standard (i.e., NFPA 13D) or hazard classification (i.e., Light Hazard in accordance with NFPA 13) the pipe is listed. Since this is the first edition of NFPA 13R, the laboratories will need to review the listing requirements for such special listed pipe to ensure suitability for use in accordance with the provisions of this standard.
>
> The user must refer to the extensive listing information with respect to these materials to accomplish correct installation.

1-5.3   Whenever the word pipe is used in this standard, it shall be understood to also mean tube.

> For the purposes of this standard, the word pipe refers to any conduit for transporting water.

1-5.4   Pipe joined with mechanical grooved fittings shall be joined by a listed combination of fittings, gaskets, and grooves. When grooves are cut or rolled on the pipe they shall be dimensionally compatible with the fittings.

Exception: Steel pipe with wall thicknesses less than Schedule 30 [in sizes 8 in. (203 mm) and larger] or Schedule 40 [in sizes less than 8 in. (203 mm)] shall not be joined by fittings used with pipe having cut grooves.

> When grooves are cut, material is lost and the use of thin-wall pipe could result in too little material remaining between the inside diameter and the root diameter of the groove. The Exception does permit an innovative threaded assembly for use with thinner-walled pipe based on a listing from a testing laboratory. It should be noted that 2-4.4 requires the piping support to be in accordance with NFPA 13, which contains

special criteria for hanger spacing when threaded thin-wall pipe is used. (*See 3-10.1.11.1 of NFPA 13.*)

1-5.5 Fittings used in sprinkler systems shall be of the materials listed in Table 1-5.5 or in accordance with 1-5.7. The chemical properties, physical properties, and dimensions of the materials listed in Table 1-5.5 shall be at least equivalent to the standards cited in the table. Fittings used in sprinkler systems shall be designed to withstand the working pressures involved, but not less than 175 psi (12.1 bars) cold water pressure.

Table 1-5.5

Materials and Dimensions	Standard
Cast Iron	
Cast Iron Threaded Fittings, Class 125 and 250	ANSI B16.4
Cast Iron Pipe Flanges and Flanged Fittings	ANSI B16.1
Malleable Iron	
Malleable Iron Threaded Fittings, Class 150 and 300	ANSI B16.3
Steel	
Factory-made Threaded Fittings, Class 150 and 300	ANSI B16.9
Buttwelding Ends for Pipe, Valves, Flanges, and Fittings	ANSI B16.25
Spec. for Piping Fittings of Wrought Carbon Steel and Alloy Steel for Moderate and Elevated Temperatures	ASTM A234
Pipe Flanges and Flanged Fittings, Steel Nickel Alloy and Other Special Alloys	ANSI B16.5
Forged Steel Fittings, Socket Welded and Threaded	ANSI B16.11
Copper	
Wrought Copper and Copper Alloy Solder-Joint Pressure Fittings	ANSI B16.22
Cast Copper Alloy Solder-Joint Pressure Fittings	ANSI B16.18

Fittings of the types and materials indicated in Table 1-5.5 must be manufactured to the standards indicated or to standards meeting or exceeding the indicated standards. Any fitting for use in a sprinkler system must be designed for a working pressure of at least 175 psi (12.1 bars).

1-5.6 Joints for the connection of copper shall be brazed.

Exception: Soldered joints (95-5 solder metal) may be used for wet-pipe copper tube systems.

The standard does not recognize 50-50 solder, which was prohibited for use in plumbing systems by the 1986 amendments to the Federal Safe Drinking Water Act. While the prohibition does not specifically extend to residential sprinkler systems, the Committee considered it a good practice to discontinue using this leaded solder in such close proximity to a potable water system.

1-5.7 Other types of fittings may be used, but only those listed for this service.

The types of pipe and fittings that are described in this section are those that have been found satisfactory either due to service record or based on their having been tested and listed by a testing laboratory. Sections 3-1 and 3-7 of NFPA 13 contain additional information on pipe and fittings. Paragraph 1-5.7 allows development of new fittings and their use after investigation and listing.

1-6 System Types.

1-6.1 Wet-Pipe Systems. A wet-pipe system shall be used when all piping is installed in areas not subject to freezing.

Wet-pipe systems are the most reliable and simplest of all sprinkler systems since no equipment other than the sprinklers themselves need operate. Only those sprinklers that have been operated by heat from the fire will discharge water. Wet-pipe systems are recommended wherever possible.

1-6.2 Provision shall be made to protect piping from freezing in unheated areas by use of one of the following acceptable methods:

(a) Antifreeze system.

State or local plumbing and health regulations may not allow the introduction of foreign materials into piping systems connected to public water. Where these regulations are in effect, the use of antifreeze and the type of solution permitted should be checked with local authorities. It may be permitted where piping arrangements make contamination of public water virtually impossible. In many instances, antifreeze systems may be the only practical solution to a fire protection problem in a small area of a building and deviations from codes may be allowed.

(b) Dry-pipe system.

Dry-pipe systems are installed in lieu of wet-pipe systems where piping is subject to freezing. Dry-pipe systems should not be used for the purpose of reducing water damage from pipe breakage or leakage since they operate too quickly to be of value for this purpose.

It should be pointed out that because of the concern for delayed water delivery and the fact that there are no residential sprinklers listed for dry system application, dry systems are not allowed for protection inside dwelling units in accordance with Exception No. 1 to 2-4.5.1. However, dry systems may be installed in areas outside the dwelling unit in accordance with 2-5.2 and NFPA 13.

Exception: Listed standard dry-pendent, dry upright, or dry sidewall sprinklers may be extended into unheated areas not intended for living purposes.

Dry-pendent, dry upright, and dry sidewall sprinklers are specially designed to prevent water from entering the pipe between the sprinkler supply pipe (branch line) and the operating mechanism of the sprinkler. These sprinklers may be used on wet-pipe systems where individual sprinklers are extended into spaces subject to freezing.

1-6.2.1   Antifreeze solutions shall be installed in accordance with 5-5.3 of NFPA 13, *Standard for the Installation of Sprinkler Systems.*

Tables 5-5.3.1 and 5-5.3.2 in NFPA 13 contain permitted antifreeze solutions depending on whether or not the sprinkler system is supplied by public water.

2

Working Plans, Design, Installation, Acceptance Tests, and Maintenance

2-1 Working Plans and Acceptance Tests.

2-1.1 Working Plans.

Working plans are prepared primarily for installers of the system. The plans also serve to protect the interest of the owner, who usually is not knowledgeable in sprinkler system installations and consequently will rely upon others to check the plans for conformance to this standard. The owner may decide to which other "authorities having jurisdiction," in addition to code enforcers, the plans are to be submitted, for example, consultants or insurance representatives.

When the installation is completed, it is checked against the plans to determine compliance with the approved plans and this standard. The owner should retain working plans and specifications for reference and to assist in the preventive maintenance program. In the event that alterations are undertaken sometime in the future, a good deal of expense and time can be saved, if accurate working plans and documentation are available, especially since these systems are hydraulically designed.

The symbols commonly used in working plans can be found in NFPA 172, *Standard Fire Protection Symbols for Architectural and Engineering Drawings.*

2-1.1.1 Working plans shall be submitted for approval to the authority having jurisdiction before any equipment is installed or remodeled. Deviations from approved plans will require permission of the authority having jurisdiction.

2-1.1.2 Working plans shall be drawn to an indicated scale, on sheets of uniform size, with a plan of each floor, made so that they can be easily duplicated, and shall show the following data:

(a) Name of owner and occupant.

(b) Location, including street address.

557

(c) Point of compass.

(d) Ceiling construction.

(e) Full height cross section.

(f) Location of fire walls.

(g) Location of partitions.

(h) Occupancy of each area or room.

(i) Location and size of concealed spaces, attics, closets, and bathrooms.

(j) Any small enclosures in which no sprinklers are to be installed.

(k) Size of city main in street, pressure and whether dead-end or circulating and, if dead-end, direction and distance to nearest circulating main, city main test results including elevation of test hydrant.

(l) Make, manufacturer, type, heat-response element, temperature rating, and nominal orifice size of sprinkler.

(m) Temperature rating and location of high-temperature sprinklers.

Consideration must be given to the location of sprinkler heads with respect to heat sources. Manufacturer's literature contains guidance which should be consulted when considering the proximity of sprinklers to heat sources in the design of systems.

(n) Number of sprinklers on each riser, per floor.

(o) Kind and location of alarm bells.

(p) Type of pipe and fittings.

(q) Type of protection for nonmetallic pipe.

(r) Nominal pipe size with lengths shown to scale.

NOTE: Where typical branch lines prevail, it will be necessary to size only one line.

(s) Location and size of riser nipples.

(t) Type of fittings and joints and location of all welds and bends.

(u) Types and locations of hangers, sleeves, braces, and methods of securing sprinklers, where applicable.

Nonmetallic pipe has necessitated the "securing" of sprinklers to the ceiling. Refer to the manufacturer's listing information.

(v) All control valves, check valves, drain pipes, and test connections.

(w) Underground pipe size, length, location, weight, material, point of connection to city main; the type of valves, meters, and valve pits; and the depth at which the top of the pipe is laid below grade.

(x) For hydraulically designed systems, the material to be included on the hydraulic data nameplate.

(y) Name and address of contractor.

2-1.2 Approval of Sprinkler Systems.

2-1.2.1 The installer shall perform all required acceptance tests (*see 2-1.3*), complete the Contractor's Material and Test Certificate(s) (*see Figure 2-1.2.1*), and forward the certificate(s) to the authority having jurisdiction, prior to asking for approval of the installation.

A material and test certificate (*see pages 560 and 561*) acknowledges that materials used and tests made are in accordance with the requirements of the approved plans. The certificate also provides a record of the test results, which can be used for comparison with tests that are conducted as part of a preventive maintenance program.

2-1.2.2 When the authority having jurisdiction desires to be present during the conducting of acceptance tests, the installer shall give advance notification of the time and date the testing will be performed.

Frequently the authority having jurisdiction wishes to be present when acceptance tests are conducted. The installer should be aware of the policies or procedures of the authority having jurisdiction with respect to witnessing acceptance tests. When the authority having jurisdiction desires to witness the tests, the installer should provide sufficient advance notice. Failure to do so may result in conducting the tests a second time.

2-1.3 Acceptance Tests.

2-1.3.1 Flushing of Underground Connections.

2-1.3.1.1 Underground mains and lead-in connections to system risers shall be flushed before connection is made to sprinkler piping, in order to

CONTRACTOR'S MATERIAL & TEST CERTIFICATE FOR **A**BOVEGROUND PIPING

PROCEDURE

Upon completion of work, inspection and tests shall be made by the contractor's representative and witnessed by an owner's representative. All defects shall be corrected and system left in service before contractor's personnel finally leave the job.

A certificate shall be filled out and signed by both representatives. Copies shall be prepared for approving authorities, owners, and contractor. It is understood the owner's representative's signature in no way prejudices any claim against contractor for faulty material, poor workmanship, or failure to comply with approving authority's requirements or local ordinances.

PROPERTY NAME		DATE

PROPERTY ADDRESS

PLANS	ACCEPTED BY APPROVING AUTHORITIES (NAMES)	
	ADDRESS	
	INSTALLATION CONFORMS TO ACCEPTED PLANS	☐ YES ☐ NO
	EQUIPMENT USED IS APPROVED	☐ YES ☐ NO
	IF NO, EXPLAIN DEVIATIONS	
INSTRUCTIONS	HAS PERSON IN CHARGE OF FIRE EQUIPMENT BEEN INSTRUCTED AS TO LOCATION OF CONTROL VALVES AND CARE AND MAINTENANCE OF THIS NEW EQUIPMENT? IF NO, EXPLAIN	☐ YES ☐ NO
	HAVE COPIES OF THE FOLLOWING BEEN LEFT ON THE PREMISES:	☐ YES ☐ NO
	1. SYSTEM COMPONENTS INSTRUCTIONS	☐ YES ☐ NO
	2. CARE AND MAINTENANCE INSTRUCTIONS	☐ YES ☐ NO
	3. NFPA 13A	☐ YES ☐ NO
LOCATION OF SYSTEM	SUPPLIES BUILDINGS	

	MAKE	MODEL	YEAR OF MANUFACTURE	ORIFICE SIZE	QUANTITY	TEMPERATURE RATING
SPRINKLERS						

PIPE AND FITTINGS Type of Pipe _____ Type of Fittings _____

	ALARM DEVICE			MAXIMUM TIME TO OPERATE THROUGH TEST CONNECTION	
ALARM VALVE OR FLOW INDICATOR	TYPE	MAKE	MODEL	MIN.	SEC.

	DRY VALVE			Q.O.D.		
	MAKE	MODEL	SERIAL NO.	MAKE	MODEL	SERIAL NO.

	TIME TO TRIP THRU TEST CONNECTION*		WATER PRESSURE	AIR PRESSURE	TRIP POINT AIR PRESSURE	TIME WATER REACHED TEST OUTLET*		ALARM OPERATED PROPERLY	
DRY PIPE OPERATING TEST	MIN.	SEC.	PSI	PSI	PSI	MIN.	SEC.	YES	NO
Without Q.O.D.									
With Q.O.D.									
IF NO, EXPLAIN									

* MEASURED FROM TIME INSPECTOR'S TEST CONNECTION IS OPENED.
85A (10-88)   PRINTED IN U.S.A.   (OVER)

Figure 2-1.2.1   Contractor's Material and Test Certificate for Aboveground Piping.

remove foreign materials that may have entered the underground piping during the course of the installation. For all systems, the flushing operation shall be continued until water is clear.

Experience has shown that stones, gravel, blocks of wood, bottles, work tools, work clothes, and other objects have been found in piping when flushing was performed. Also, objects in

DELUGE & PREACTION VALVES	OPERATION						
		☐ PNEUMATIC	☐ ELECTRIC		☐ HYDRAULIC		
	PIPING SUPERVISED	☐ YES	☐ NO	DETECTING MEDIA SUPERVISED		☐ YES	☐ NO
	DOES VALVE OPERATE FROM THE MANUAL TRIP AND/OR REMOTE CONTROL STATIONS					☐ YES	☐ NO
	IS THERE AN ACCESSIBLE FACILITY IN EACH CIRCUIT FOR TESTING			IF NO, EXPLAIN			
			☐ YES ☐ NO				
	MAKE	MODEL	DOES EACH CIRCUIT OPERATE SUPERVISION LOSS ALARM		DOES EACH CIRCUIT OPERATE VALVE RELEASE	MAXIMUM TIME TO OPERATE RELEASE	
			YES	NO	YES NO	MIN.	SEC.

TEST DESCRIPTION	HYDROSTATIC: Hydrostatic tests shall be made at not less than 50 psi (3.4 bars) above design pressure for two hours. Differential dry-pipe valve clappers shall be left open during test to prevent damage. All aboveground piping leakage shall be stopped. Systems with fire department connections shall be hydrostatically tested in accordance with NFPA 13, paragraph 1-11.2.

PNEUMATIC: Establish 40 psi (2.7 bars) air pressure and measure drop which shall not exceed 1-1/2 psi (0.1 bars) in 24 hours. Test pressure tanks at normal water level and air pressure and measure air pressure drop which shall not exceed 1-1/2 psi (0.1 bars) in 24 hours. |

TESTS	ALL PIPING HYDROSTATICALLY TESTED AT _____ PSI FOR _____ HRS.		IF NO, STATE REASON		
	DRY PIPING PNEUMATICALLY TESTED	☐ YES ☐ NO			
	EQUIPMENT OPERATES PROPERLY	☐ YES ☐ NO			
	DO YOU CERTIFY AS THE SPRINKLER CONTRACTOR THAT ADDITIVES AND CORROSIVE CHEMICALS, SODIUM SILICATE OR DERIVATIVES OF SODIUM SILICATE, BRINE, OR OTHER CORROSIVE CHEMICALS WERE NOT USED FOR TESTING SYSTEMS OR STOPPING LEAKS? ☐ YES ☐ NO				
	DRAIN TEST	READING OF GAGE LOCATED NEAR WATER SUPPLY TEST CONNECTION: _____ PSI		RESIDUAL PRESSURE WITH VALVE IN TEST CONNECTION OPEN WIDE _____ PSI	
	UNDERGROUND MAINS AND LEAD IN CONNECTIONS TO SYSTEM RISERS FLUSHED BEFORE CONNECTION MADE TO SPRINKLER PIPING.				
	VERIFIED BY COPY OF THE U FORM NO. 85B	☐ YES ☐ NO	OTHER	EXPLAIN	
	FLUSHED BY INSTALLER OF UNDER-GROUND SPRINKLER PIPING	☐ YES ☐ NO			

BLANK TESTING GASKETS	NUMBER USED	LOCATIONS	NUMBER REMOVED

WELDING	WELDED PIPING ☐ YES ☐ NO		
	IF YES...		
	DO YOU CERTIFY AS THE SPRINKLER CONTRACTOR THAT WELDING PROCEDURES COMPLY WITH THE REQUIREMENTS OF AT LEAST AWS D10.9, LEVEL AR-3	☐ YES	☐ NO
	DO YOU CERTIFY THAT THE WELDING WAS PERFORMED BY WELDERS QUALIFIED IN COMPLIANCE WITH THE REQUIREMENTS OF AT LEAST AWS D10.9, LEVEL AR-3	☐ YES	☐ NO
	DO YOU CERTIFY THAT WELDING WAS CARRIED OUT IN COMPLIANCE WITH A DOCUMENTED QUALITY CONTROL PROCEDURE TO INSURE THAT ALL DISCS ARE RETRIEVED, THAT OPENINGS IN PIPING ARE SMOOTH, THAT SLAG AND OTHER WELDING RESIDUE ARE REMOVED, AND THAT THE INTERNAL DIAMETERS OF PIPING ARE NOT PENETRATED	☐ YES	☐ NO

CUTOUTS (DISCS)	DO YOU CERTIFY THAT YOU HAVE A CONTROL FEATURE TO ENSURE THAT ALL CUTOUTS (DISCS) ARE RETRIEVED?	☐ YES	☐ NO

HYDRAULIC DATA NAMEPLATE	NAME PLATE PROVIDED	IF NO, EXPLAIN
	☐ YES ☐ NO	

REMARKS	DATE LEFT IN SERVICE WITH ALL CONTROL VALVES OPEN:

SIGNATURES	NAME OF SPRINKLER CONTRACTOR		
	TESTS WITNESSED BY		
	FOR PROPERTY OWNER (SIGNED)	TITLE	DATE
	FOR SPRINKLER CONTRACTOR (SIGNED)	TITLE	DATE

ADDITIONAL EXPLANATION AND NOTES

85A BACK

Figure 2-1.2.1   (Continued) Contractor's Material and Test Certificate for Aboveground Piping.

underground piping quite remote from the sprinkler installation, that otherwise would remain stationary, will sometimes be transported into sprinkler system piping when sprinkler systems operate. Sprinkler systems can draw greater flows than normal domestic systems. Fire department pumpers, when taking suction from hydrants for pumping into sprinkler systems, may compound the problem by increasing the velocity of water flow in underground piping.

Because of the inherent nature of sprinkler system design in which pipe sizes diminish from the point of connection of underground piping, objects that move from underground piping and enter sprinkler system risers may become lodged at a point in the system where they may totally obstruct the passage of water.

2-1.3.1.2 Underground mains and lead-in connections shall be flushed at the hydraulically calculated water demand rate of the system.

Since only the design flow rate is anticipated, the calculated water demand rate shall be used for flushing. It should be recognized, however, that overall field experience with NFPA 13 systems indicates that flushing is normally accomplished at the maximum flow rate available from the water supply.

2-1.3.1.3 To avoid property damage, provision shall be made for the disposal of water issuing from test outlets.

2-1.3.2* All systems shall be tested for leakage at 50 psi (3.4 bars) above maximum system design pressure.

All new systems, as well as additions to existing systems, shall be hydrostatically tested to a pressure which is at least 50 psi (3.4 bars) above the maximum system design pressure. The 50-psi factor has been in use for years within NFPA 13. This standard is set to ensure that pipe joints are made to withstand that pressure without coming apart or leaking. It is primarily a workmanship test and not a materials performance test. However, damaged materials (cracked fittings, leaky sprinklers, etc.) can be discovered during the hydrostatic test.

The measurement of the hydrostatic test pressure is taken at the lowest elevation within the system or portion of the system being tested. It is not considered necessary to test at the high point of the system due to the fact that application of pressure typically occurs at the lower elevation.

The measure of success for interior sprinkler piping under a hydrostatic test is that there is no visible leakage. Often a very small bead of water may form on a fitting during the test. Unless the bead continues to grow and drip, this is not considered visible leakage.

Exception: When a fire department connection is provided, hydrostatic pressure tests shall be provided in accordance with NFPA 13, Standard for the Installation of Sprinkler Systems.

In this case, the Exception is more restrictive than the base paragraph. When a fire department connection is installed, the system pressure can be increased using a fire department pumper and, therefore, it should be verified that the system has the integrity to withstand such pressures. Thus, a hydrostatic test should be conducted in accordance with NFPA 13. Normally, the required pressure for the hydrostatic test will be 200 psi (13.8 bars) when a fire department connection is provided.

In accordance with 2-4.2, a fire department connection is required on all systems with 20 or more sprinklers.

A-2-1.3.2 Testing of a system can be accomplished by filling the system with water and checking visually for leakage at each joint or coupling.

Fire department connections are not required for all systems covered by this standard, but may be installed at the discretion of the owner. In these cases, hydrostatic tests in accordance with NFPA 13, *Standard for the Installation of Sprinkler Systems*, are required.

Dry systems should also be tested by placing the system under air pressure. Any leak that results in a drop in system pressure greater than 2 psi (0.14 bar) in 24 hours should be corrected. Check for leaks using soapy water brushed on each joint or coupling. Leaks will be shown by the presence of bubbles. This test should be made prior to concealing of piping.

2-2 Design and Installation.

2-2.1 Devices and Materials.

2-2.1.1* Only new sprinklers shall be employed in the installation of sprinkler systems.

The sprinkler is one component of the system that must be depended upon to operate efficiently and effectively when a fire occurs. Also, it should not leak or rupture, or operate for any reason other than to extinguish a fire. For these reasons, and because of the relatively low cost of sprinklers themselves, new sprinklers are required.

A-2-2.1.1 At least three spare sprinklers of each type, temperature rating, and orifice size used in the system should be kept on the premises. When fused sprinklers are replaced by the owner, fire department, or others, care should be taken to assure that the replacement sprinkler has the same operating characteristics.

On those occasions when it is necessary to replace a sprinkler because it has operated (fused), or has been damaged, one should carefully verify that the orifice size, temperature rating, deflector configuration, and sprinkler type are exactly the same as those of the sprinkler being replaced. Many sprinklers have unique operating characteristics and some have special limitations on their area of coverage.

When multiple types of sprinklers are installed, at least three of each type should be maintained as the stock of spare sprinklers. It should also be noted that if standard or quick-response sprinklers are installed, the minimum number of spare sprinklers should be six (in accordance with NFPA 13) since more than three sprinklers may operate.

2-2.1.2 Only listed or approved devices and materials as indicated in this standard shall be used in sprinkler systems.

2-2.1.3 Sprinkler systems shall be designed for a maximum working pressure of 175 psi (12.1 bars).

Exception: Higher design pressures may be used when all system components are rated for pressures higher than 175 psi (12.1 bars).

Sprinklers are currently rated for a 175-psi (12.1-bar) working pressure. The Exception merely removes the 175-psi restriction in the event that higher pressure rated sprinklers are made available. However, in this instance one would not expect to find systems requiring a design pressure greater than 175 psi.

2-3 Water Supply.

2-3.1 General Provisions. Every automatic sprinkler system shall have at least one automatic water supply. When stored water is used as the sole source of supply, the minimum quantity shall equal the water demand rate times 30 minutes. (*See 2-5.1.3.*)

Since any sprinkler system is only as good as its water supply, the supply must be automatic and reliable. The adequacy of a public water supply is determined in accordance with 2-5.1.3, 2-5.2, and 2-5.3. When stored water is used, the amount should provide at least 30 minutes duration at the required flow rate. The 30-minute criteria is consistent with NFPA 13 criteria for Light Hazard Occupancies.

2-3.2* Water Supply Sources. The following water supply sources are acceptable:

(a) A connection to a reliable water works system with or without a booster pump, as required.

(b) An elevated tank.

(c) A pressure tank installed in accordance with NFPA 13, *Standard for the Installation of Sprinkler Systems*, and NFPA 22, *Standard for Water Tanks for Private Fire Protection*.

(d) A stored water source with an automatically operated pump, installed in accordance with NFPA 20, *Standard for the Installation of Centrifugal Fire Pumps*.

A-2-3.2 Connection for fire protection to city mains is often subject to local regulation concerning metering and backflow prevention require-ments. Preferred and acceptable water supply arrangements are shown in Figures A-2-3.2(a), (b), and (c). When a meter must be used between the city water main and the sprinkler system supply, an acceptable arrangement is shown in Figure A-2-3.2(c). Under these circumstances, the flow charac-teristics of the meter must be included in the hydraulic calculation of the system. When a tank is used for both domestic and fire protection purposes, a low water alarm acuated when the water level falls below 110 percent of the minimum quantity specified in 2-3.1 should be provided.

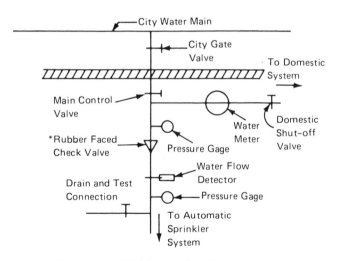

Figure A-2-3.2(a)   Preferable Arrangement.

Figures A-2-3.2(a), (b), and (c) show preferred and acceptable arrangements of connections to city mains and include control valves, meters, domestic take-offs, waterflow detectors, pres-sure gages, and check valves appropriately located. When back-flow preventers are required by local water company require-ments, they will normally be located in the position where the rubber-faced valve is shown on the diagrams. The piping ar-rangement, including valves and fittings, must be taken into

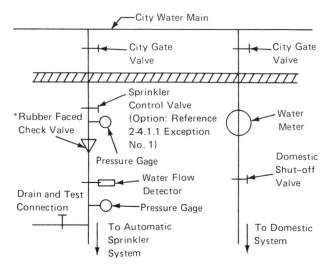

Figure A-2-3.2(b)   Acceptable Arrangement with Valve Supervision. (*See 2-4.1.1 Exception.***)**

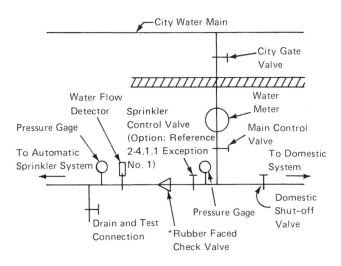

*Rubber Faced Check Valves Optional.

Figure A-2-3.2(c)   Acceptable Arrangement with Valve Supervision. (*See 2-4.1.1 Exception.***)**

account when performing the calculations for system adequacy.

From a fire protection standpoint, meters are not desirable because of their high friction loss characteristic, which must be included in the hydraulic calculations of the system. Some friction loss values for water meters can be found in Table

4-4.3(d) of NFPA 13D. Unfortunately, many water authorities will require metering of the water connection.

Occasionally, a system may be supplied from an elevated tank, and when this is done, the tank must have an elevation sufficient to provide adequate pressure to supply the system. This is determined in accordance with Section 2-4 of NFPA 13. When a pressure tank is used for water supply, the amount of water and air pressure in the tank are determined by following 2-6.3 and A-2-6.3 of NFPA 13.

A stored water source with an automatically operated pump must have a pump with a capacity sufficient to supply the system's demand as determined by 2-5.1.3 and 2-5.2 and must have adequate pressure determined in accordance with 2-5.3. It is conceivable that a connection to the water works system may require a booster pump to provide the pressures required in accordance with 2-5.3. It should be noted that NFPA 13R is more restrictive than NFPA 13D with respect to a stored water source with a pump in that the pump must comply with NFPA 20, *Standard for the Installation of Centrifugal Fire Pumps*. The Committee felt that the safeguard and reliability features of NFPA 20 were necessary for installation under the scope of this standard.

2-3.3 Multipurpose Piping System.

Since domestic demand might occur at the same time as operation of sprinklers, such demand must be taken into account in the system design. Tables A-2-3.3.1(a) and (b) provide guidance as to appropriate levels of domestic design demand. The tables are based on values found in plumbing codes published by the model code organizations. If the building being protected is of new construction, the domestic design demand should be readily available from the design documents and will not require additional effort on behalf of the installer.

As indicated in Figure A-2-3.2(a), the preferred arrangement is to combine piping for domestic and sprinkler systems. The Committee feels that such an arrangement will increase the reliability of the system. In a residential occupancy, if one does not have water for normal usage, it is likely that something will be done about it, and promptly. However, the Committee also recognizes that restrictions may be placed on such combined systems by local plumbing or health regulations because frequently there are trapped piping sections at the extremities of a sprinkler system that are not permitted by such codes. Further, some plumbing regulations permit piping methods of lesser quality than prescribed by this standard.

2-3.3.1* A common supply main to the building, serving both sprinklers and domestic uses, shall be acceptable when the domestic design demand is added to the sprinkler system demand.

Exception: Domestic design demand need not be added if provision is made to prevent flow on the domestic water system upon operation of sprinklers.

A-2-3.3.1 The following tables can be used to determine a domestic design demand. Using Table A-2-3.3.1(a), determine the total number of water supply fixture units downstream of any point in the piping serving both sprinkler and domestic needs. Using Table A-2-3.3.1(b), determine the appropriate total flow allowance, and add this flow to the sprinkler demand at the total pressure required for the sprinkler system at that point.

Table A-2-3.3.1(a)   Fixture Load Values

Private facilities (within individual dwelling units)	Unit
Bathroom group with flush tank (including lavatory, water closet, and bathtub with shower)	6
Bathroom group with flush valve	8
Bathtub	2
Dishwasher	1
Kitchen sink	2
Laundry trays	3
Lavatory	1
Shower stall	2
Washing machine	2
Water closet with flush valve	6
Water closet with flush tank	3

Public Facilities	
Bathtub	4
Drinking fountain	0
Kitchen sink	4
Lavatory	2
Service sink	3
Shower head	4
Urinal with 1 in. flush valve	10
Urinal with ¾ in. flush valve	5
Urinal with flush tank	3
Washing machine (8 lb)	3
Washing machine (16 lb)	4
Water closet with flush valve	10
Water closet with flush tank	5

Table A-2-3.3.1(b)   Total Estimated Domestic Demand

Total Fixture Load Units [from Table A-2-3.3.1(a)]	Total Demand in Gallons Per Minute	
	For Systems with Predominantly Flush Tanks	For Systems with Predominantly Flush Valves
1	3 gpm	
2	5	
5	10	15 gpm
10	15	25
20	20	35
35	25	45
50	30	50
70	35	60
100	45	70
150	55	80
200	65	90
250	75	100
350	100	125
500	125	150
750	175	175
1000	200	200
1500	275	275
2000	325	325
3500	500	500

2-3.3.2 Sprinkler systems with nonfire protection connections shall comply with Section 5-6 of NFPA 13, *Standard for the Installation of Sprinkler Systems*.

It has been considered feasible for many years to make use of the sprinkler piping, which normally stands idle, for a more active purpose. One such use is a circulating closed loop used for a heat pump system, in which water is circulated through heating and air conditioning equipment utilizing sprinkler pipe as the primary conductors. Section 5-6 of NFPA 13 provides the basic criteria so that the auxiliary functions can be added in such a manner and with sufficient controls to ensure that there will be no reduction in the effectiveness of the sprinkler system.

2-4 System Components.

2-4.1 Valve and Drains.

2-4.1.1 When a common supply main is used to supply both domestic and sprinkler systems, a single listed control valve shall be provided to shut off both the domestic and sprinkler systems, and a separate shutoff valve shall be provided for the domestic system only. [*See Figure A-2-3.2(a).*]

Exception: The sprinkler system piping may have a separate control valve when supervised by one of the following methods:

(a) Central station, proprietary, or remote station alarm service,

(b) Local alarm service that will cause the sounding of an audible signal at a constantly attended point, or

(c) Locking the valves open.

A control valve permits the system to be shut off in order to replace a sprinkler, but it must be in the open position at all other times. With only one control valve, there is little possibility of inadvertently shutting off the sprinkler system because, with the domestic water shut off, something would be done rather quickly to get the system back in operation. A separate valve for the sprinkler system is acceptable when the valve is supervised.

2-4.1.2   Each sprinkler system shall have a 1-in. (25.4-mm) or larger drain and test connection with valve on the system side of the control valve.

Test or drain pipes are used for both stated purposes. They are used to drain systems when repairs are necessary. They are also used to indicate that water supply is available at the system riser and on the system side of all check valves, the control valve, and underground piping. The preventive maintenance program should include proper documentation and test records so as to permit detection of possible deterioration of water supplies or valves that may have been closed.

2-4.1.3   Additional ½-in. (13-mm) drains shall be installed for each trapped portion of a dry system that is subject to freezing temperatures.

Because dry-pipe systems are subject to freezing, trapped areas must be provided with drain valves to remove water that has entered the system either because of a dry-pipe valve tripping or condensation of moisture from the pressurized air in the system. Additional guidance on auxiliary drains for dry-pipe systems can be found in 3-6.3.3 of NFPA 13. Condensate nipples are not required because it is assumed that the capacity of trapped sections will generally be small.

Auxiliary drains are not required for wet-pipe systems since trapped sections of piping in a wet system rarely need to be drained. However, if the building is used for seasonal occupancy, if the water system is shut off during periods of non-occupancy, and if the authority having jurisdiction has permitted a wet-pipe system to be installed in accordance with this standard, arrangements should be made to drain all portions of trapped piping. Additional guidance on auxiliary drains can be found in 3-6.2 of NFPA 13.

2-4.2   At least one 1½ in. (38 mm) or 2½ in. (64 mm) fire department connection shall be provided when the sprinkler system has 20 sprinklers or more.

Fire department connections are an important supplement to normal water supplies since they allow the fire department to bypass a closed control valve in most instances. Even when gravity tanks, pressure tanks, or other stored water sources are the sole source of supply, fire department connections should be included. The connection allows the fire department to pump from tankers or another water source into the system, thereby either increasing the supply, or adding to the tank's supply prior to its depletion, or both.

Additional guidance on the arrangement of the fire department connection can be found in Section 2-7 of NFPA 13. It should be noted that on wet-pipe systems with a single riser, as is common for systems installed in accordance with 13R, the connection should be made to the riser on the system side of indicating, check, and alarm valves (if provided). This reduces the likelihood that a valve can be closed and negate the benefit of a fire department connection. For this reason, there should be no shutoff valve in the fire department connection.

Approved check valves are to be provided in the fire department connection as near as practicable to the point where it joins the system. The check valve should be installed to permit flow into the system and restrict water flow from the system into the fire department connection. An approved automatic drip should be provided in the fire department connection to permit draining of water between the hose connection and the check valve. The drain should be located at the lowest point of the fire department connection piping to allow complete drainage.

All hose coupling threads on the fire department connection should match the threads used by the local fire department. The hose connections should be equipped with plugs or caps.

2-4.3 Pressure Gages.   Pressure gages shall be provided to indicate pressures on the supply and system sides of main check valves and dry-pipe valves, and to indicate pressure on water supply pressure tanks.

Pressure gages permit one to easily observe the presence of the proper water pressure and, in dry systems, the proper air pressure. Such a visual inspection is required by Section 2-7.

2-4.4 Piping Support.   Piping hanging and bracing methods shall comply with NFPA 13, *Standard for the Installation of Sprinkler Systems*.

Section 3-10 of NFPA 13 provides minimum requirements for pipe hangers. Paragraph 3-5.3.5 of NFPA 13 contains provisions for sway bracing which may be applicable. It should also be noted that nonmetallic pipe may require that sprinklers be secured to the ceiling. The pipe manufacturer's listing information should be consulted.

The Committee feels that pipe sizes, pipe lengths, short-term pressure surges, and water hammer anticipated in systems installed in accordance with this standard necessitate the use of pipe hanging and bracing requirements required in NFPA 13 rather than those permitted by NFPA 13D.

2-4.5 Sprinklers.

2-4.5.1   Listed residential sprinklers shall be used inside dwelling units. The basis of such a listing shall consist of tests to establish the ability of the sprinklers to control residential fires under standardized fire test conditions. The standardized room fires shall be based on a residential array of furnishings and finishes.

Both Underwriters Laboratories Inc. and Factory Mutual Research Corporation have standards for the testing and evaluation of residential sprinklers.[2,3] The standards require that the sprinklers pass room fire tests and have fast response characteristics. Standard sprinklers of the type used in NFPA 13 sprinkler systems do not meet these criteria and should not be used except as specifically permitted by this standard. (*See commentary to Section 1-2 for a further discussion of differences in the test criteria.*)

In addition to performance, the designs of residential sprinklers are typically more aesthetically pleasing. Examples of these newer designs are shown in Figures 2.1 and 2.2.

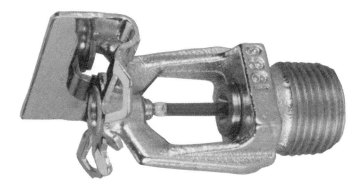

Figure 2.1. Viking Model M-1 horizontal sidewall residential sprinkler.

Figure 2.2. Viking Model M-1 pendent mounted residential sprinkler with glass bulb and orifice cover removed. (Sprinkler is shown in an inverted position.)

Figure 2.3. Viking Model M vertical sidewall quick-response sprinkler.

Figure 2.4 Viking Model M pendent mounted quick-response sprinkler.

Exception No. 1:   Residential sprinklers shall not be used in dry systems unless specifically listed for that purpose.

This Exception is based on the concern for the expected delay in water delivery inherent with dry-pipe systems and because there are no residential sprinklers listed for this application.

Exception No. 2:   Other types of listed sprinklers may be installed in accordance with their listing in dwelling units meeting the definition of a compartment (as defined in 2-5.1.2.2) provided no more than four sprinklers are located in the dwelling unit and at least one smoke detector is provided in each sleeping room.

This Exception permits the use of quick-response sprinklers and standard sprinklers in applications where the desired performance can be achieved without the need for direct wall wetting. While these sprinklers may not respond as quickly as a residential sprinkler, the presence of a smoke detector in each sleeping room should provide adequate protection for occupants who are capable of and respond to the alarm signal. Figures 2.3 and 2.4 show two types of quick-response sprinklers.

The Exception also permits systems to be installed in instances where room layout or ceiling configuration may restrict the use of residential sprinklers.[8] For example, hotel rooms are often protected with extended coverage sidewall sprinklers, which are not listed as residential sprinklers. The Committee felt that such an arrangement provides an acceptable level of protection, provided each room used for sleeping is also protected with a smoke detector.

2-4.5.2 Ordinary temperature rated sprinklers [135 to 170°F (57 to 77°C)] shall be installed where maximum ambient ceiling temperatures do not exceed 100°F (38°C).

The maximum temperature (ambient) in the residential environment should not exceed a value of 35 Fahrenheit degrees (2 Celsius degrees) below the temperature rating of the residential sprinkler which is used. The temperatures of residential sprinklers currently range from 145 to 165°F (63 to 74°C).

2-4.5.3 Intermediate temperature rated residential sprinklers [175 to 225°F (79 to 107°C)] shall be installed where maximum ambient ceiling temperatures are between 101 and 150°F (38 and 66°C).

Areas containing hot water heaters, building or room heating components, and the like require some special treatment. In addition, sprinklers that are installed in the vicinity of a fireplace may also be exposed to above average temperatures. The requirements of this paragraph, as well as the manufacturer's instructions, should be adhered to for placement of sprinklers in these areas.

2-4.5.4 The following practices shall be observed when installing residential sprinklers, unless maximum expected ambient temperatures are otherwise determined.

(a) Sprinklers under glass or plastic skylights exposed to direct rays of the sun shall be of intermediate temperature classification.

(b) Sprinklers in an unventilated concealed space under an uninsulated roof, or in an unventilated attic, shall be of intermediate temperature classification.

Temperatures of the air surrounding a sprinkler may exceed 100°F (38°C) when a skylight permits direct sun exposure to that area. Unventilated attics tend to develop rather high temperatures in the summer months; thus a similar precaution to prevent false activation should also be undertaken. It is always important to select the appropriate temperature rating for a given environment. This helps to insure that sprinklers will not operate prematurely and that there will not be an excessive number of sprinklers operating during the fire.

2-4.5.5 When residential sprinklers are installed within a compartment, as defined in 2-5.1.2.2, all sprinklers shall be from the same manufacturer and have the same heat-response element, including temperature rating.

The sensitivity of differing residential sprinklers varies, and that difference changes depending upon the velocity of airflow. To avoid the possibility of reverse order operation, all residential sprinklers in a compartment must be from the same manufacturer and have the same heat-responsive element.

Exception: Different temperature ratings are permitted when required by 2-4.5.4.

Sprinklers may be of a different temperature classification as permitted in 2-4.5.4.

2-4.5.6   Standard sprinklers shall be used in areas outside the dwelling unit.

Standard sprinklers are sprinklers which are listed, but not in accordance with the criteria for residential sprinklers.

Exception No. 1:   Residential sprinklers may be used in adjoining corridors or lobbies with flat, smooth ceilings and a height not exceeding 10 ft (3.0 m).

Residential sprinklers are appropriate for use where the fire threat is similar to that expected in a dwelling unit. Corridors and lobbies with ceiling heights of 10 ft (3.0 m) or less can be similar to hallways and living rooms within dwelling units. The criteria for a flat, smooth ceiling is consistent with the listing limitations of residential sprinklers.[8]

Exception No. 2:   Quick-response sprinklers may be used in accordance with 2-5.2, Exception No. 1.

Paragraph 2-5.2 does contain some limits on the number of quick-response sprinklers within a compartment. The principal concern with indiscriminate use of quick-response sprinklers is the possibility that too many sprinklers will operate, overtaxing the water supply and ultimately resulting in system failure. This issue is being further explored in ongoing research.

2-4.5.7   Operated or damaged sprinklers shall be replaced with sprinklers having the same performance characteristics as original equipment.

Underwriters Laboratories Inc. has conducted a program in which field samples of automatic sprinklers have been tested to determine their operating characteristics. Experience gained from this program has shown that, when a sprinkler has been damaged or painted after leaving the factory, it is unlikely to operate properly at the time of a fire, and thus should be replaced. Obviously, a sprinkler that has operated must also be replaced.

2-4.5.8 When nonmetallic ceiling plates (escutcheons) are used, they shall be listed. Escutcheon plates used to create a recessed or flush-type sprinkler shall be part of a listed sprinkler assembly.

> Although 2-5.1.7 presently provides guidance as to the minimum distance of sprinkler operating elements and deflectors below ceilings, similar guidance previously provided in NFPA 13 (*see 4-3.1*) was not fully observed in the field. New sprinklers with small frames are being used with recessing cups that can adversely affect sprinkler performance. Therefore, such cups shall be listed with the sprinkler they serve.

2-4.5.9 Painting and Ornamental Finishes.

2-4.5.9.1 Sprinkler frames may be factory painted or enameled as ornamental finish in accordance with 2-4.5.9.2; otherwise, sprinklers shall not be painted and any sprinklers that have been painted, except those with factory applied coatings, shall be replaced with new listed sprinklers.

2-4.5.9.2* Ornamental finishes shall not be applied to sprinklers by anyone other than the sprinkler manufacturer, and only sprinklers listed with such finishes shall be used.

> It is extremely important that only manufacturers who have included procedures for applying paint or ornamental finishes in the sprinkler listing be permitted to apply finishes. Applications made by others could seriously impair the sprinkler's operation or could render the sprinkler inoperable.

A-2-4.5.9.2 Decorative painting of a residential sprinkler is not to be confused with the temperature identification colors as referenced in 3-11.6 of NFPA 13, *Standard for the Installation of Sprinkler Systems*.

2-4.6 Alarms. Local waterflow alarms shall be provided on all sprinkler systems and shall be connected to the building fire alarm system, when provided.

> This standard requires a waterflow alarm to sound on the premises for all systems. The standard does not, however, require supplemental alarm systems.
> The purpose of the alarm is to indicate water flow and need not be considered a building evacuation alarm. As indicated in the Preface, it is assumed that smoke detectors will be provided, and such smoke detectors can give adequate warning for purposes of evacuation. Other codes, such as NFPA 101, will also govern whether evacuation alarms are necessary.

2-5 System Design.

Systems installed in accordance with this standard must be hydraulically designed. The calculation procedure must be as outlined in Chapter 7 of NFPA 13. The design of systems installed in accordance with this standard may consist of a two-part process. The design criteria for the sprinkler system protecting areas within a dwelling unit is prescribed in 2-5.1. In accordance with 2-5.2, the design criteria for sprinklers protecting areas outside the dwelling unit should be in accordance with NFPA 13. In many instances, the systems can be designed with a maximum of four sprinklers operating.

2-5.1 Design Criteria—Inside Dwelling Unit.

This section provides the design criteria for sprinklers protecting areas within the dwelling unit. Public corridors, lounges, and other areas outside the dwelling unit should be designed in accordance with 2-5.2.

2-5.1.1 Design Discharge. The system shall provide a discharge of not less than 18 gpm (68 L/min) to any single operating sprinkler and not less than 13 gpm (49 L/min) per sprinkler to the number of design sprinklers, but not less than the listing of the sprinkler(s).

This section of the standard prescribes a minimum discharge of 18 gpm (68 L/min) for a single operating sprinkler and 13 gpm (49 L/min) per sprinkler for multiple sprinklers within a compartment. The definition of a compartment is provided in 2-5.1.2.2. The water flow criteria translates to a minimum density of 0.125 gpm per sq ft [(5.1 L/min)/m²] for one sprinkler and 0.09 gpm per sq ft [(3.7 L/min)/m²] for multiple sprinklers, assuming the maximum spacing of 144 sq ft (13.4 m²) per sprinkler. The test data indicate that the primary criterion should be flow rate, rather than density, for a residential sprinkler system.[4,5,6,7]

Exception: Design discharge for sprinklers installed in accordance with Exception No. 2 of 2-4.5.1 shall be in accordance with sprinkler listing criteria.

Since Exception No. 2 to 2-4.5.1 permits sprinklers other than residential sprinklers to be provided within certain dwelling units, such sprinklers shall only be used in accordance with their listing. Depending on the type of sprinkler used, this may be a minimum operating pressure or a minimum flow rate.

2-5.1.2* Number of Design Sprinklers.

A-2-5.1.2 It is intended that the design area is to include up to four adjacent sprinklers producing the greatest water demand within the compartment.

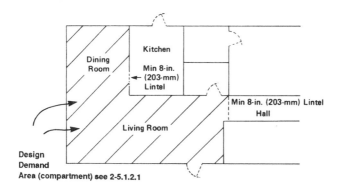

Figure A-2-5.1.2(a)   Sprinkler Design Areas for Typical Residential Occupancy.

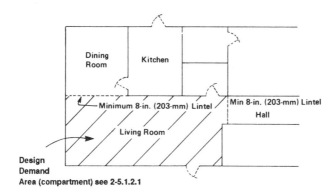

Figure A-2-5.1.2(b)   Sprinkler Design Areas for Typical Residential Occupancy.

2-5.1.2.1 The number of design sprinklers shall include all sprinklers within a compartment to a maximum of four sprinklers.

The Committee recognizes that NFPA 13D allows for a design to be based on two sprinklers operating. While this may be adequate for one- and two-family dwellings, the four-sprinkler design allows for a sufficient factor of safety taking into consideration additional variables, such as a broad range of evacuation capabilities.

The design is based on a maximum of four sprinklers. When all compartments are protected with fewer than four sprinklers, the maximum number of sprinklers within a single compartment shall define the design area. Paragraph 2-5.1.2.2 defines a compartment.

2-5.1.2.2   The definition of compartment for use in 2-5.1.2.1 to determine the number of design sprinklers is a space that is completely enclosed by walls and a ceiling. The compartment enclosure may have openings to an adjoining space if the openings have a minimum lintel depth of 8 in. (203 mm) from the ceiling.

A room compartment is determined by enclosing walls or the enclosing geometries of a soffet, lintel, or beam that forms a definite pattern in a smooth ceiling configuration. [*See Figures A-2-5.1.2(a) and (b).*] In open plan configurations, compartment boundaries may be determined by beams, soffets, or lintels over doorways, as well as by walls or partitions. The 8-in. (203-mm) minimum depth is to ensure that heat will be "banked," enhancing operation of sprinklers over the fire.

2-5.1.3 Water Demand.   The water demand for the system shall be determined by multiplying the design discharge of 2-5.1.1 by the number of design sprinklers of 2-5.1.2.

This calculation will produce a sprinkler system demand of 18 gpm (68 L/min) for a single sprinkler and between 26 (98 L/min) and 52 gpm (196 L/min) for multiple sprinklers. These values are used in performing the calculations to determine pipe sizing or adequacy of public water supply pressure. Different values may need to be used if the listing of the sprinkler provides different minimum flow rates.

2-5.1.4 Sprinkler Coverage.

2-5.1.4.1   Residential sprinklers shall be spaced so that the maximum area protected by a single sprinkler does not exceed 144 sq ft (13.4 m²).

This section indicates that the design area for a single residential sprinkler should not exceed 144 sq ft (13.4 m²). When tests and listing indicate greater spacing, this is permitted by 2-5.1.6.

2-5.1.4.2   The maximum distance between sprinklers shall not exceed 12 ft (3.7 m) and the maximum distance to a wall or partition shall not exceed 6 ft (1.8 m).

2-5.1.4.3 The minimum distance between sprinklers within a compartment shall be 8 ft (2.4 m).

2-5.1.5 The minimum operating pressure of any sprinkler shall be in accordance with the listing information of the sprinkler and shall provide the minimum flow rates specified in 2-5.1.1.

2-5.1.6 Application rates, design areas, areas of coverage, and minimum design pressures other than those specified in 2-5.1.1, 2-5.1.2, 2-5.1.4, and 2-5.1.5 may be used with special sprinklers that have been listed for such specific residential installation conditions.

These limitations on spacing were determined in the Los Angeles and North Carolina fire tests.[4,5,6,7] Since these tests were conducted with prototype fast response residential sprinklers, the Committee recognizes that other fast response residential sprinklers may be developed that will operate properly with different spacing limitations. In fact, some currently listed residential sprinklers have a coverage area of 18 ft by 18 ft (5.5 m by 5.5 m). Such sprinklers may be used provided that spacing which differs from those prescribed in this standard is justified based on testing conducted by a testing and inspection agency/laboratory.

It should be noted that when coverage areas are extended, the minimum flow rates which are consistent with the listing of the sprinkler must be used.

2-5.1.7 Position of Residential Sprinklers.

2-5.1.7.1 Pendent and upright sprinklers shall be positioned so that the deflectors are within 1 to 4 in. (25.4 to 102 mm) from the ceiling.

Exception: Special residential sprinklers shall be installed in accordance with the listing limitations.

This limitation is based on the results of the Los Angeles and North Carolina fire tests.[4,5,6,7]

2-5.1.7.2 Sidewall sprinklers shall be positioned so that the deflectors are within 4 to 6 in. (102 to 152 mm) from the ceiling.

Exception: Special residential sprinklers shall be installed in accordance with the listing limitations.

Dead air spaces in corners can affect a sprinkler's operation time. The 4-in. (102-mm) limitation ensures that the sprinkler will operate properly.

2-5.1.7.3* Sprinklers shall be positioned so that the response time and discharge are not unduly affected by obstructions such as ceiling slope, beams, or light fixtures.

A-2-5.1.7.3 Fire testing has indicated the need to wet walls in the area protected by residential sprinklers at a level closer to the ceiling than that accomplished by standard sprinkler distribution. Where beams, light fixtures, sloped ceilings, and other obstructions occur, additional residential sprinklers may be necessary to achieve proper response and distribution, and a greater water supply may be necessary.

Table A-2-5.1.7.3 and Figure A-2-5.1.7.3 provide guidance for location of sprinklers near ceiling obstructions.

Table A-2-5.1.7.3   Maximum Distance from Sprinkler Deflector to Bottom of Ceiling Obstruction

Distance from Sprinkler to Side of Ceiling Obstruction	Maximum Distance from Sprinkler Deflector to Bottom of Ceiling Obstruction
Less than 6 in.	Not permitted
6 in. to less than 1 ft	0 in.
1 ft to less than 2 ft	1 in.
2 ft to less than 2 ft 6 in.	2 in.
2 ft 6 in. to less than 3 ft	3 in.
3 ft to less than 3 ft 6 in.	4 in.
3 ft 6 in. to less than 4 ft	6 in.
4 ft to less than 4 ft 6 in.	7 in.
4 ft 6 in. to less than 5 ft	9 in.
5 ft to less than 5 ft 6 in.	11 in.
5 ft 6 in. to less than 6 ft	14 in.

For SI units: 1 in. = 25.4 mm; 1 ft = 0.3048 m.

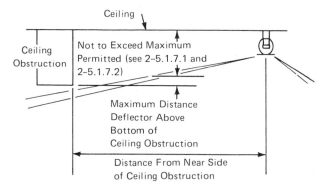

Figure A-2-5.1.7.3   Position of Deflector, Upright or Pendent, When Located Above Bottom of Ceiling Obstruction.

Location of sprinklers in violation of Table A-2-5.1.7.3 and Figure A-2-5.1.7.3 will cause obstruction of the discharge and may prevent the sprinkler from controlling a fire. Sprinklers designed to have different discharge characteristics may be positioned closer to a beam than is permitted by Table A-2-5.1.7.3. However, test data must be supplied to support such different positioning.

2-5.2 Design Criteria—Outside Dwelling Unit. The design discharge, number of design sprinklers, water demand of the system, sprinkler coverage, and position of sprinklers for areas to be sprinklered outside the dwelling unit shall comply with specifications in NFPA 13, *Standard for the Installation of Sprinkler Systems.*

This section provides the design criteria for sprinklers protecting areas outside the dwelling unit, such as lobbies, corridors, halls, and foyers which are not contained within the dwelling unit; basements and storage areas; inside stairwells; and equipment, furnace, trash, and linen rooms. The section refers the user of the standard to the applicable NFPA 13 criteria. However, it should be noted that 2-5.3 does require the system to be hydraulically calculated.

In many cases it will be possible to maintain a design criteria based on four sprinklers in accordance with the two following Exceptions.

Exception No. 1:   When compartmented into areas of 500 sq ft (46 m²) or less by 30-minute fire-rated construction, and the area is protected by standard or quick-response sprinklers not exceeding 130 sq ft (12 m²) per sprinkler, the system demand may be limited to the number of sprinklers in the compartment area, but not less than a total of four sprinklers. Openings from the compartments need not be protected provided such openings are provided with a lintel at least 8 in. (203 mm) in depth and the total area of such openings does not exceed 50 sq ft (4.6 m²) for each compartment. Discharge density shall be appropriate for the hazard classification as determined by NFPA 13.

This Exception is intended to provide an alternative to the design area criteria of NFPA 13. This is an adaptation of the largest room provisions of 2-2.3.1 of NFPA 13. The 130 sq ft (12 m²) area of coverage is based in part on the assumption that areas which will be protected in accordance with the Exception will be Ordinary Hazard areas.

Exception No. 2: Lobbies, in other than hotels and motels, foyers, corridors, and halls outside the dwelling unit, with flat, smooth ceilings and not exceeding 10 ft (3.0 m) in height, may be protected with residential sprinklers, with a maximum system demand of four sprinklers.

The purpose of this Exception is to clarify that residential sprinklers may be used when the limitations of ceiling height [10 ft (3.0 m)] and ceiling configuration (smooth), and the expected hazards, present a fire potential within the control range for residential sprinklers. Lobbies in hotels and motels are exempted since the fire hazard may no longer be residential in nature. Such lobbies often have adjacent or communicating spaces used for gift shops, restaurants, and small displays.

2-5.3 Pipe Sizing. Piping shall be sized in accordance with hydraulic calculation procedures to comply with NFPA 13, *Standard for the Installation of Sprinkler Systems.*

Hydraulic calculations must be prepared in accordance with Chapter 7 of NFPA 13, and all appropriate provisions of NFPA 13 must apply. It should be noted that the different criteria for inside and outside dwelling units (*see 2-5.1 and 2-5.2*) as well as the dual discharge criteria of 2-5.1.1 require at least two and often three separate sets of calculations.

2-6 Location of Sprinklers. Sprinklers shall be installed in all areas.

Exception No. 1: Sprinklers may be omitted from bathrooms not exceeding 55 sq ft (5.1 m²) with noncombustible plumbing fixtures.

Exception No. 2: Sprinklers may be omitted from small clothes closets where the least dimension does not exceed 3 ft (0.9 m) and the area does not exceed 24 sq ft (2.2 m²) and the walls and ceiling are surfaced with noncombustible or limited combustible materials as defined by NFPA 220, Standard on Types of Building Construction.

Exception No. 3: Sprinklers may be omitted from open attached: porches, balconies, corridors, and stairs.

Exception No. 4: Sprinklers may be omitted from attics, penthouse equipment rooms, crawl spaces, floor/ceiling spaces, elevator shafts, and other concealed spaces that are not used or intended for living purposes or storage.

The basis for omission of sprinklers in Exceptions No. 1 through 4 has been previously described under 1-2 and is based on NFPA statistics provided in Table A-1-2. These areas are the ones shown to have a low incidence of life loss from fires in dwellings.

With respect to Exception No. 1, the combustible load in bathrooms is normally extremely low, especially with noncombustible plumbing fixtures. While plastic plumbing fixtures must meet certain ignition test criteria (including that of ANSI Z124), the Committee is concerned with the fixture as a contributing fuel to the fire and not just as the first material ignited. The Committee is also concerned that involvement of such fixtures could permit the fire to enter concealed spaces that are not sprinklered.

With respect to Exceptions No. 3 and 4, mandatory sprinklering of these areas would necessitate the use of dry-pipe systems in areas where freezing weather is encountered.

2-7* Maintenance. The owner is responsible for the condition of a sprinkler system and shall keep the system in normal operating condition.

A-2-7 The responsibility for properly maintaining a sprinkler system is the obligation of the owner or manager, who should understand the sprinkler system operation. A minimum monthly maintenance program should include the following:

(a) Visual inspection of all sprinklers to ensure against obstruction of spray.

(b) Inspection of all valves to assure that they are open.

(c) Testing of all waterflow devices.

(d) Testing of the alarm system, if installed.

NOTE: When it appears likely that the test will result in a response of the fire department, notification to the fire department should be made prior to the test.

(e) Operation of pumps, where employed, should be operated. See NFPA 20, *Standard for the Installation of Centrifugal Fire Pumps*.

(f) Checking of the pressure of air used with dry systems.

(g) Checking of water level in tanks.

(h) Care should be taken to see that sprinklers are not painted either at the time of installation or during subsequent redecoration. When painting sprinkler piping or painting in areas next to sprinklers, the sprinklers may be protected by covering with a bag, which should be removed immediately after painting is finished.

For further information see NFPA 13A, *Recommended Practice for the Inspection, Testing and Maintenance of Sprinkler Systems.*

Table A-2-7
Inspection, Testing, and Maintenance Requirements

Component	Activity	Frequency	Reference
Control Valve	Inspection	Monthly	
	Maintenance	Annually	NFPA 13A, 2-7.1.4
Main Drain Valve	Flow Test	Annually	NFPA 13A, 2-6.1
Inspectors' Test Valve	Flow Test	Annually	
Waterflow Alarm	Flow Test	Annually	NFPA 13A, 4-5.3, 4-7.1
Sprinklers	Test	50 Yrs.	NFPA 13A, 3-3.3
Sprinklers, Res/QR	Test	20 Yrs.	NFPA 13A, 3-3.4
Pump	Flow Test	Annually	NFPA 13A, 2-4.2.5
Antifreeze Solutions	Test	Annually	NFPA 13A, 4-7.3

Proper maintenance of sprinkler systems is important for effective sprinkler system performance at the time a fire occurs. This is especially important due to the passive nature of the sprinkler system. It is mandated, therefore, that the owner be responsible for a proper preventive maintenance program involving inspection, testing, and maintenance. A system not maintained in normal operating condition is not in compliance with this standard.

NFPA 13A contains extensive material on the care and maintenance of automatic sprinklers and sprinkler system components. The minimum monthly maintenance program outlined in A-2-7 represents what the Committee considers to be a minimum acceptable program.

3

Referenced Publications

3-1 The following documents or portions thereof are referenced within this standard and shall be considered part of the requirements of this document. The edition indicated for each reference is the current edition as of the date of the NFPA issuance of this document.

3-1.1 NFPA Publications. National Fire Protection Association, Batterymarch Park, Quincy, MA 02269.

NFPA 13-1989, *Standard for the Installation of Sprinkler Systems*

NFPA 20-1987, *Standard for the Installation of Centrifugal Fire Pumps*

NFPA 22-1987, *Standard for Water Tanks for Private Fire Protection*

NFPA 74-1989, *Standard for the Installation, Maintenance, and Use of Household Fire Warning Equipment*

NFPA *101*-1988, *Life Safety Code*

NFPA 220-1985, *Standard on Types of Building Construction.*

3-1.2 Other Publications.

3-1.2.1 ANSI Publications. American National Standards Institute, Inc., 1430 Broadway, New York, NY 10018.

ANSI B16.1-1975, *Cast Iron Pipe Flanges and Flanged Fittings, Class 25, 125, 250 and 800*

ANSI B16.3-1985, *Malleable Iron Threaded Fittings, Class 150 and 300*

ANSI B16.4-1985, *Cast Iron Threaded Fittings, Classes 125 and 250*

ANSI B16.5-1981, *Pipe Flanges and Flanged Fittings*

ANSI B16.9-1986, *Factory-Made Wrought Steel Buttwelding Fittings*

ANSI B16.11-1980, *Forged Steel Fittings, Socket-Welding and Threaded*

ANSI B16.18-1984, *Cast Copper Alloy Solder Joint Pressure Fittings*

ANSI B16.22-1980, *Wrought Copper and Copper Alloy Solder Joint Pressure Fittings*

ANSI B16.25-1986, *Buttwelding Ends*

ANSI B36.10M-1985, *Welded and Seamless Wrought Steel Pipe.*

3-1.2.2 ASTM Publications. American Society for Testing and Materials, 1916 Race Street, Philadelphia, PA 19103.

ASTM A53-1987, *Standard Specification for Pipe, Steel, Black and Hot-Dipped, Zinc-Coated Welded and Seamless Steel Pipe*

ASTM A135-1986, *Standard Specification for Electric-Resistance-Welded Steel Pipe*

ASTM A234-1987, *Standard Specification for Piping Fittings of Wrought-Carbon Steel and Alloy Steel for Moderate and Elevated Temperatures*

ASTM A795-1985, *Specification for Black and Hot-Dipped Zinc-Coated (Galvanized) Welded and Seamless Steel Pipe for Fire Protection Use*

ASTM B32-1987, *Standard Specification for Solder Metal, 95-5 (Tin-Antimony-Grade 95TA)*

ASTM B88-1986, *Standard Specification for Seamless Copper Water Tube*

ASTM B251-1987, *Standard Specification for General Requirements for Wrought Seamless Copper and Copper-Alloy Tube*

ASTM E380-1986, *Standard for Metric Practice.*

3-1.2.3 AWS Publication. American Welding Society, 2501 N.W. 7th Street, Miami, FL 33125.

AWS A5.8-1981, *Specification for Brazing Filler Metal.*

Appendix A

The material contained in Appendix A of this standard is included within the text of this handbook, and therefore is not repeated here.

Appendix B
Referenced Publications

B-1 The following documents or portions thereof are referenced within this standard for informational purposes only and thus are not considered part of the requirements of this document. The edition indicated for each reference is the current edition as of the date of the NFPA issuance of this document.

B-1.1 NFPA Publications. National Fire Protection Association, Batterymarch Park, Quincy, MA 02269.

NFPA 13-1989, *Standard for the Installation of Sprinkler Systems*

NFPA 13A-1987, *Recommended Practice for the Inspection, Testing and Maintenance of Sprinkler Systems*

NFPA 20-1987, *Standard for the Installation of Centrifugal Fire Pumps.*

REFERENCES CITED IN COMMENTARY

[1]Nelson, H.E., *An Engineering Analysis of the Early Stages of Fire Development—The Fire at the Dupont Plaza Hotel and Casino*, December 31, 1986, National Bureau of Standards, Gaithersburg, MD, 1987.

[2]Factory Mutual Research Corporation, *Approval Standard — Residential Automatic Sprinklers, Class No. 2030*, Norwood, MA, September, 1983.

[3]Underwriters Laboratories Inc., *Standard for Residential Sprinklers for Fire Protection Service*, UL 1626, Northbrook, IL, April 1986.

[4]Cote, A.E., and Moore, D., *Field Test and Evaluation of Residential Sprinkler Systems, Los Angeles Test Series*. National Fire Protection Association, Boston, MA, April 1980.

[5]Moore, D., *Data Summary of the North Carolina Test Series of USFA Grant 79027 Field Test and Evaluation of Residential Sprinkler Systems*. National Fire Protection Association, Boston, MA, September 1980.

[6]Kung, H.-C., Spaulding, R.D., Hill, E.E. Jr., and Symonds, A.P., *Technical Report, Field Evaluation of Residential Prototype Sprinkler, Los Angeles Fire Test Program*, Factory Mutual Research Corporation, Norwood, MA, February 1982.

[7]Cote, A.E., *Final Report on Field Test and Evaluation of Residential Sprinkler Systems*, National Fire Protection Association, Quincy, MA, July 1982.

[8]Bill, R.G. Jr., Kung, H.-C., Brown, W.R., and Hill E.E. *Effects of Cathedral and Beamed Ceiling Construction on Residential Sprinkler Performance*, Factory Mutual Research Corporation, Norwood, MA, February 1988.

The following publications are available from the National Fire Protection Association, Batterymarch Park, Quincy, MA 02269.

NFPA 13-1989, *Standard for the Installation of Sprinkler Systems*
NFPA 13A-1987, *Recommended Practice for the Inspection, Testing and Maintenance of Sprinkler Systems*
NFPA 13D-1989, *Standard for the Installation of Sprinkler Systems in One- and Two-Family Dwellings and Mobile Homes*
NFPA 20-1987, *Standard for the Installation of Centrifugal Fire Pumps*

NFPA 72E-1987, *Standard on Automatic Fire Detectors*
NFPA 74-1989, *Standard for the Installation, Maintenance, and Use of Household Fire Warning Equipment*
NFPA *101*-1988, *Life Safety Code*
NFPA 172-1986, *Standard Fire Protection Symbols for Architectural and Engineering Drawings.*

The following publications are available from the American National Standards Institute, Inc., 1450 Broadway, New York, NY 10018.

ANSI Z124.1-1980, *Plastic Bathtub Units*
ANSI Z124.2-1980, *Plastic Shower Receptors and Shower Stalls.*

Supplement
Quick Response Sprinklers: A Technical Analysis

This supplement is not a part of the code texts of this Handbook, but is included for information purposes only. It is printed in black for ease of reading.

Preface

The supplement that follows was first included in the third edition of this handbook. Although the supplement has remained unchanged for 1989, the concepts, conclusions, and findings it contains are still valid and useful for the rapidly changing technology of "fast response sprinklers." Even as this book goes to press, the Sprinkler Committee is considering changes to the standard which will give the user specific guidance on the use and limitations of "quick-response sprinklers."

Much confusion abounds in the area of time constants (τ factors) and Response Time Index (RTI) and their relation to the control and extinguishment of fires. Such parameters concerning sprinkler sensitivity first became important during the development of the residential sprinkler program in the mid-1970s. Eventually, this concept was carried forward to other types of sprinklers such as large-drop, quick-response, and most recently the early suppression fast response (ESFR) sprinkler.

As shown in this supplement, the operating time of any sprinkler is dependent upon a variety of parameters related to fuel type, geometry, location, ceiling heights, and sprinkler materials among others. Collectively, these items come together to allow us to predict operating times of a sprinkler. Originally intending to determine the obvious benefits of improving life safety via fast response sprinklers, research has now been targeted which shows how property damage may be reduced with such a sprinkler.

Past, present, and future test programs have been and are now underway by the National Fire Protection Research Foundation. Through the efforts of Rick Mulhaupt, Vice President and Executive Director of the Research Foundation, testing programs on the family of fast response sprinklers continue to go forward.

Information compiled in this supplement is through the efforts of Russ Fleming of the National Fire Sprinkler Association. The NFSA has also been involved in the private testing of fast response sprinkler characteristics. The reader is directed to page 175 of this handbook for a further discussion of fast response sprinklers.

Robert E. Solomon, P.E.
Editor

IN APPRECIATION

The National Fire Protection Research Foundation wishes to thank Russell P. Fleming, P.E., for the insight and energy represented by this *Technical Analysis*, the gyroscope of the National Quick Response Sprinkler Research Project.

The Research Foundation also thanks the Factory Mutual Research Corporation, the National Bureau of Standards (Center for Fire Research), the National Fire Protection Association, Underwriters Laboratories, Worcester Polytechnic Institute and the Cobb County Fire Department for their generous assistance in assembling the database.

We are especially grateful to the diverse group of sponsors whose investment makes the National Quick Response Sprinkler Research Project and this document possible. Their foresight and faith will make its goal, "**Improved firesafety at reduced cost**," a reality. Through January, 1986, these included:

PRINCIPAL SPONSORS

American Fire Sprinkler Association, Inc.
The Boeing Company
Chemical Specialties Manufacturers Association
Eastman Kodak Company
Factory Mutual Research Corporation
Industrial Risk Insurers
National Fire Sprinkler Association, Inc.

SPONSORS

Allstate Insurance Company
American Paper Institute, Inc.
Automatic Sprinkler Corp. of America
Ball Corporation
Building Owners and Managers Association, International
Central Sprinkler Corporation
Dow Chemical Corporation
Dow Corning Corporation
General Motors Corporation
Grinnell Fire Protection Systems Company, Inc.
International Business Machines Corporation
IRM Insurance
Kemper Group
Marsh and McLennan Protection Consultants
Monsanto Company
The Reliable Automatic Sprinkler Company, Inc.
Schering-Plough
The Travelers Companies
Underwriters Laboratories, Inc.
The Viking Corporation
West Point Pepperell

ACKNOWLEDGMENTS

This research was sponsored by the National Fire Protection Research Foundation as part of its National Quick Response Sprinkler Research Project. The author appreciates the support and guidance given by the Foundation and its administrator, Frederick K. Mulhaupt, and to the following members of the Technical Advisory Committee for their helpful advice and assistance in the development of this report:

Earl J. Schiffhauer, Chairman
Eastman Kodak Company

Jack S. Barritt
Industrial Risk Insurers

Edward K. Budnick
National Bureau of Standards

James R. Beyreis
Underwriters Laboratories

Ron J. Coleman
Fullerton Fire Department

Paul M. Fitzgerald
Factory Mutual Research Corp.

David L. Fredrickson
S.C. Johnson & Son, Inc.

David Hilton
Cobb County Fire Department

Rolf H. Jensen
Rolf Jensen and Associates

John Nickles
American Fire Sprinkler Association

Gerald J. Rosicky
General Motors Corporation

Chester W. Schirmer, P.E.
Schirmer Engineering Corporation

Walter F. Schuchard
Electro Signal Lab

Harry Shaw
Harry Shaw and Associates

Tom Smith
United States Fire Administration

William Thomas
Kemper Group

John A. Viniello
National Fire Sprinkler Association

Reginald J. Wright
Underwriters Laboratories of Canada

The advice and support of Arthur E. Cote of the National Fire Protection Association, Miles R. Suchomel of Underwriters Laboratories, and Cheng Yao of Factory Mutual Research is also appreciated.

TABLE OF CONTENTS

INTRODUCTION

The National Fire Protection Research Foundation has undertaken a Quick Response Sprinkler (QRS) Research Project, intended to develop usable data which will provide for a timely yet orderly introduction of the quick response sprinkler technology into standards and codes. It is believed that this will ultimately result in improved life and property safety.

The project has been divided into several parts. Group 1 tests addressing high challenge fires are being conducted by the Factory Mutual Research Corporation in conjunction with the ESFR (Early Suppression Fast Response) sprinkler test program. Group 2 tests will address the broad range of occupancies between residential and industrial.

The National Fire Protection Research Foundation and its Technical Advisory Committee recognized a need for a literature survey and analysis of quick response sprinkler technology to assist in planning the Group 2 test program. This report represents the result of that effort.

There are several problems inherent in an analysis of quick response sprinkler technology. Probably the most fundamental is that there is no clear definition of "quick response" or of a "quick response sprinkler." The thermal sensitivity of the automatic sprinkler has been a subject of concern and debate throughout the 110-year history of the sprinkler in the United States. Since sensitivity is relative, even the most fundamental improvements in sprinkler sensitivity, such as preventing contact between the solder activating mechanism and the water in the pipes, were historically considered as having created a special class of "sensitive" sprinklers.[1] The definition problem is further compounded by the often-overlooked difference between the relative sensitivity of the sprinkler itself and the level of response it will demonstrate under specific conditions of spacing, position, and fire growth.

A second difficulty in an analysis of the impact of quick response sprinkler technology is the complex nature of fire growth and development, and of the interaction of sprinklers with fire. While much progress has been made, the level of understanding of how sprinkler systems control fires has not yet developed to a point that would permit a theoretical design approach. As a result, the basis of sprinkler system design criteria has been empirical at best.

This report is divided into several parts in an attempt to address the above difficulties:

I. A chronology of developments in quick response sprinkler technology. This provides some insight into the changing use of the term "quick response" by the product approval laboratories.

---

[1]Dana, Gorham, *Automatic Sprinkler Protection,* Second Edition, John Wiley and Sons, 1919.

601

II. A model of sprinkler response and suppression. The model is in the form of a flow chart or algorithm which, if the state of knowledge permitted accurate mathematical modeling of its components, would permit one to calculate whether successful system performance would be achieved under a given set of conditions.

III. A framework in which the elements of the model are assigned information representing state-of-the-art knowledge about the mechanism of that particular element, with particular emphasis being given to the impact of sprinkler sensitivity.

IV. Summaries of key reports providing information on laboratory or field testing of sprinkler sensitivity or of the effectiveness of quick response sprinklers.

V. Analysis and conclusions.

VI. A bibliography of documents reviewed in the course of this study. Numbers appearing in parentheses throughout this report are references to this bibliography.

The format of this report is based on the assumption that a literature survey and analysis for quick response sprinklers cannot consider response without also considering suppression. Response without suppression is detection.

It had been intended to include a summary of fire experience with quick response sprinklers, but it became evident that significant data is not available. A number of communities are employing the quick response technology through the use of NFPA 13D for dwellings and mobile homes, some others are experimenting with the use of NFPA 13D beyond its intended scope, and still other communities and corporations are encouraging the use of quick response sprinklers in NFPA 13 systems, often with local variations. The definition problem stated above would make a statistical analysis impossible even if data were available.

It should also be noted that a report of this type can never be considered complete. Additions to the framework which can bring us closer to a mathematical model of sprinkler response and suppression are expected as fire research continues.

I. CHRONOLOGY OF QUICK RESPONSE

1884 — C.J.H. Woodbury of the Factory Mutual Fire Insurance Companies conducts the first extensive testing of automatic sprinklers, including a test on "sensitiveness."[1] The response times of fifteen different types of sprinklers are recorded following a gradual temperature build-up to 112°F and then sudden immersion in steam. Activation times were found to range from 15 seconds to 85 seconds.

1931 — Cotton Marine Underwriters conduct tests which show that the newly-introduced quartz bulb sprinklers are considerably more sensitive than existing solder-type sprinklers (57).

1935 — The Grinnell Company introduces the "Duraspeed" sprinkler. With a larger ratio of surface area to mass in the actuating linkage, it is reportedly much faster than previous sprinklers (111).

May, 1966 — At the first meeting of the NFPA 13 Subcommittee on Fire Research, Chairman R. Russell suggests that faster sprinkler operation combined with better water distribution might solve the fire hazard of 20-ft high rolled paper storage (94).

September and November, 1971 — At the two initial meetings of the Quick Response Sprinkler Subcommittee of the NFPA Sprinkler Committee, it is agreed that:

"1. A quick response sprinkler design is technically feasible
  2. A quick response sprinkler would be of value to the Light-Hazard Life Safety situation
  3. Extensive fire testing would be required to establish optimum design criteria
  4. Approval laboratory standards will require modification or supplementation."[2]

The subcommittee can not decide if there would be any increase of performance in the high-hazard sprinkler applications.

1972 — Star Sprinkler introduces the "Quick-E," advertised as the "world's fastest operating sprinkler." Using heat collector fins to enhance response, the sprinkler exhibits an average response time of 52 seconds in the UL air oven test for the 135°F temperature rating, and 1 minute, 27 seconds for the

---

[1]Dana, Gorham, *Automatic Sprinkler Protection*, 2nd edition, John Wiley & Sons, New York, 1919, p. 67-69.
[2]Report of subcommittee meeting by Chairman Wayne E. Ault.

165°F rating, compared to average sprinkler response of more than 2 minutes.[3] The sprinkler is never given any special quick response designation. Plunge tests conducted in the 1980's indicate an RTI of about 72 m$^{1/2}$sec$^{1/2}$ (130 ft$^{1/2}$sec$^{1/2}$)for this sprinkler (94).

February, 1973 — Rolf Jensen & Associates propose the development of a QR (quick response) sprinkler for use in hotel rooms and condominiums in conjunction with the Water Tower Place project (88). The response criterion is an operating time of 1 minute or less for an ordinary temperature rated sprinkler in the UL air oven test. The specifications also include a requirement that the sprinkler, intended as a sidewall to cover an area up to 16 ft wide and 30 ft long, be able to respond to a fire condition at least as fast as a standard 165°F sprinkler located at the center of the room ceiling. Sprinklers are to produce a "reasonably uniform" distribution pattern, wet all walls to a point no less than 5 ft above the floor, and demonstrate effective distribution in a series of at least 4 fire tests employing small wood cribs located at various corners and side points of the test room.

December, 1973 — Underwriters Laboratories initiates an in-house project to develop a standard for Life Hazard Reduction (LHR) sprinklers (21). An original draft of an LHR sprinkler standard, including references to limiting values of CO, oxygen depletion, and smoke density, is distributed to UL's Industry Advisory Council.

March-July, 1974 — UL conducts its LHR test program, about 200 tests in a room expandable from 16 ft × 16 ft to 16 ft × 30 ft. A variety of fuel sources are used, including 12, 16, 20 and 44 lb wood cribs, alchohol pan fires, heptane pan fires, shredded paper in wire baskets, and foam buns representing stuffed furniture. Measurements are made of CO levels, oxygen content, smoke density, and temperature at various points in the room. Early into the testing, it is noted that sprinklers listed at 135°F were found to operate at air temperatures of 170°F "due to the retarding effect of the mass of metal adjacent to the link." Final data analysis reveals that smoldering fires in upholstered foamed plastic furniture can produce high levels of CO (>3000 ppm) prior to sprinkler activation. Because definitive values of survivable levels of toxic gases are not available, the concept of the LHR standard is dropped by UL.

July, 1974 — National Bureau of Standards issues report on tests of sprinkler systems actuated by smoke detectors for the protection of patient rooms (18).

September, 1974 — UL sends a proposed standard for "Quick Response — Extended Coverage Sprinklers, UL 199A" to its Industry Advisory Council.

[3]Manufacturer's literature.

The document does not contain references to limiting values of CO, oxygen depletion, or smoke density as had been envisioned for the LHR standard.

September, 1974 — A Grinnell model F931 sprinkler employing an "electronic squib" is given a listing by UL as a "quick response" (QR) sprinkler, operating in the air oven test in less than one minute. UL chooses a one minute limit as being twice as fast as a typical sprinkler operation in this test, making the QR sprinkler "at least 50 percent quicker than the presently listed link and lever sprinklers in the ordinary temperature rating."[4] The quick response attachment uses a mercury thermostat to complete a circuit and activate a squib, directing a heat-generating molten lead charge onto the fusible link of a standard sprinkler. The F931 gives a mean response time of 35.9 seconds in the air oven test. A later (1977) revision permits a separate listed detection device to activate the squib. Tests conducted much later find the standard sprinklers within the F931 to have an RTI of 116 $m^{1/2}sec^{1/2}$ (210 $ft^{1/2}sec^{1/2}$) (94).

March, 1975 — A first meeting is held by UL of its Ad Hoc Standard Development Committee on Quick Response-Extended Coverage Sprinklers. Various suggestions made include a proposal for separate standards for property, storage, and life safety protection, and also a proposal that separate standards be produced for quick response (QR) sprinklers, extended coverage (EC) sprinklers, and quick response-extended coverage (QREC) sprinklers. It is also suggested that further research be conducted to establish a representative fire source (other than the proposed 12 lb wood crib) to measure performance of QREC sprinklers.

May, 1975 — First edition approved of NFPA 13D — Sprinkler Systems for One- and Two-Family Dwellings and Mobile Homes. The standard calls for the use of standard sprinklers at coverage areas up to 256 sq ft and a water supply based on one sprinkler operating at a design density of 0.1 gpm/ft^2.

December, 1975 — The National Automatic Sprinkler and Fire Control Association, on behalf of its member manufacturers, conducts tests to determine the range of sprinkler sensitivities on the market using both the UL air oven and an early plunge test apparatus. It is found that the range of response times exhibited varies by at least a factor of two in the UL air oven method, and by a factor of three in the plunge test conditions. (26)

March, 1976 — Factory Mutual submits a proposal to the National Science Foundation for a 15-month research program to develop and evaluate design concepts for a low-cost residential sprinkler system.

---

[4]*Fire Protection Equipment Directory*, Underwriters Laboratories, January 1980.

March, 1976 — Manufacturers of listed EC sidewall sprinklers are advised by UL that questions exist as to the effect of ceilings on distribution. New testing is proposed in which limits of deflector distance to ceiling will be determined in addition to area limits, in a 16 ft wide × 24 ft long room with 8 ft ceiling height. Response time ratings are proposed as being the temperature rating of a standard sprinkler, spaced at 16 ft along the diagonal from the fire, operating immediately after the operation of the EC sidewall.

April, 1976 — As a result of manufacturer input, UL modifies its suggested test program for EC sidewall sprinklers. Rather than use a control sprinkler located 16 ft along a diagonal from the fire (which had been intended to simulate a standard sidewall spacing), the 10.5 ft diagonal associated with a standard pendent spaced at 15 ft × 15 ft is proposed. Although glass bulb control sprinklers are rejected, it is proposed that "fusible-alloy type sprinklers, having slower than average response times, in the 135°F, 160°F, and 212°F temperature ratings" be used as the control sprinklers. Instead of a minimum distribution of 0.03 gpm/ft², a proposal is made to conduct crib fire tests at the three locations of lightest distribution.

June, 1976 — As a result of additional input, UL modifies the proposed testing of EC sidewall sprinklers to permit group response testing. Moveable walls are proposed, starting with an area 16 ft wide × 24 ft long, and decreasing in 4 ft increments. Distribution testing, which must remain individual, is modified to include a requirement that at least 40 percent of each wall be wetted to a point 5 ft up on the wall.

August, 1976 — UL conducts group response testing of EC sidewall sprinklers, and in reviewing results proposes that "performance groupings" be established in conjunction with one, two, and three minute response times for each room size.

October, 1976 — UL proposes additional group testing of EC sidewall sprinklers to determine effects of enlarged rooms with multiple sprinklers per room. Manufacturers are advised that without such testing EC sprinkler listings will be restricted to one sprinkler per room.

December, 1976 — Factory Mutual issues report on development of plunge test to measure sprinkler sensitivity (36).

January, 1977 — UL reports on response tests it conducted at the request of manufacturers on 90 samples of the slowest 155°F bulb type sprinklers and 90 samples of the slowest 160°F solder type sprinklers, so as to use the highest average for the comparison requirement with EC sidewall sprinklers. Conducted in a 15 ft × 15 ft room with the deflector of the pendent sprinklers located 10 in. below the ceiling and 10 ft 7 in. from the fire, the glass bulb sprinklers exhibit an average operating time of 2:43, and the solder sprinklers an average time of 3:06, with respective standard deviations of 16 seconds and 12 seconds.

November, 1977 — The National Fire Prevention and Control Administration holds the first Low-Cost Residential Sprinkler Conference to obtain public input on the various research projects it has been funding.

March, 1978 — UL holds the first and only meeting of its Technical Advisory Group (TAG) for Special Purpose Sprinklers for Fire Protection Service. The TAG reviews proposed UL 199A, which includes requirements for quick response sprinklers, extended coverage sprinklers, and quick response-extended coverage sprinklers (126). A QR sprinkler is defined as a "sprinkler designed to open automatically by operation of a heat responsive releasing mechanism in less time than the standard sprinkler when both types are installed at the standard spacing. The QR sprinkler shall operate in the Operation — Air Oven Test with a computed statistical tolerance limit equal to or less than 57 seconds." An extended coverage sprinkler is defined as a sprinkler designed for use at greater than standard spacing, but with a response time equal to or less than a standard sprinkler used on standard spacing. A QREC sprinkler, in addition to being a QR sprinkler, is defined as a sprinkler given greater coverage than a standard or QR sprinkler. In discussing the proposed UL 199A, the TAG expresses concern that the EC and QREC sprinklers will only be listed for use in a fully enclosed room, and not in multiple installations. The TAG considers these restrictions to all but eliminate potential applications.

August, 1978 — UL conducts additional testing of EC sidewall sprinklers in a room having a 10 percent perimeter opening (110). With a sample of 90 control sprinklers giving a mean operating time of 8.49 minutes, an upper limit of 12 minutes is established for response of EC sidewall sprinklers. It is determined from the data that no correlation can be made between the response times of sidewall sprinklers vs. control sprinklers in the totally enclosed and partially open perimeter conditions. On this basis, a separate listing is proposed for EC sidewall sprinklers to be used in rooms with up to 10 percent perimeter opening.

August, 1978 — Residential sprinkler effort receives setback when fire fully involves draperies and spreads across ceiling prior to sprinkler operation in ventilated living room test at Factory Mutual. It is later determined that a more sensitive sprinkler is needed so as to operate prior to the rapid acceleration stage of the ventilated fire (43).

April, 1979 — UL conducts additional group testing on EC sidewall sprinklers to determine the effect on response time of multiple sprinkler installations within a large room, and for ceiling heights exceeding 8 ft (110). It is found that listings given to sprinklers based on standard dimensions are adequate for multiple applications and for ceiling heights up to 16 feet.

July, 1979 — Additional review of the April, 1979 test data causes UL to restrict allowable ceiling heights when EC sidewall sprinklers are used in multiple applications. The decision is later made to restrict ceiling height for multiple applications to 9 ft unless additional testing is provided.

August, 1979 — Residential sprinkler test program gets underway in Los Angeles (16,44).

October, 1980 — Underwriters Laboratories issues tentative requirements for residential sprinkler product listings (125).

November, 1980 — 1980 edition of NFPA 13D approved, requiring the use of listed quick response residential sprinklers for systems installed in one- and two-family dwellings and mobile homes, with design based on a coverage area of 144 sq ft per sprinkler and a water supply capable of providing 18 gpm to one sprinkler, 13 gpm to each of two sprinklers, based on the Los Angeles tests.

December, 1980 — UL informs ANSI it is withdrawing proposed standard 199A because "other standards are now being effectively used for the investigation of most of the products originally encompassed by the proposed first edition of UL 199A. These products involve three general types of special purpose sprinklers: residential, extended coverage (EC), and quick response (QR)." UL goes on to explain that a separate standard is being developed for residential sprinklers, EC sprinklers are included within UL 199, and a QR addition to UL 199 is under study by UL.

April, 1981 — At a special Technical Conference on Residential Sprinkler Systems, Factory Mutual Research introduces the concept of Response Time Index (RTI) to the sprinkler industry to improve the terminology of sprinkler sensitivity. Unlike the previously used "tau factor" or "time constant," RTI is independent of both gas temperature and velocity, and can be used as a basic property of a sprinkler. RTI values of FM-approved standard sprinklers are reported to vary by as much as 4:1 (87).

June, 1981 — Grinnell receives first listing for a residential sprinkler from UL, with availability scheduled for fall of 1981.

September, 1982 — A series of 11 full-scale hotel room fire tests are conducted in Fort Lauderdale to evaluate the effectiveness of quick response sprinklers used in a retrofit system installation employing polybutylene piping (17).

November, 1982 — Central Sprinkler Corporation receives UL listing for its

Omega series residential sprinklers, for use at spacings of up to 16 ft × 16 ft.

March, 1983 — Factory Mutual issues report to sponsors of 6 large-scale fire tests investigating the effects of sensitivity and temperature rating on performance of large drop sprinklers in high rack storage of plastics. Improved response is found to substantially reduce both fire damage and the number of sprinklers operating (27).

July, 1983 — UL reactivates the designation of "quick response," providing the QR listing at the request of Grinnell for its line of model FR-1 sprinklers, standard upright and pendent sprinklers newly equipped with the residential sprinkler link. A maximum response time of 13 seconds (plus 2 seconds tolerance = 15 seconds) is set as the limit for QR sprinklers in the UL plunge test, approximating the performance of the earlier Grinnell model F931 sprinkler with the quick response attachment (80).

September, 1983 — UL revises the listing on Grinnell's Model FR-1 QR sprinklers to include an extended coverage sidewall version. The listing is based in part on comparison tests made between the extended coverage sidewall using the residential sprinkler link (model FR-1) and the previously listed EC sidewall using the "quick response attachment" (model F931). The response of the FR-1 model is reported to have "compared favorably" to the response of the model F931, with neither sprinkler exceeding the 15 second tolerance limit in the UL plunge test.

September, 1983 — Factory Mutual Research publishes residential sprinkler approval standard (123).

October, 1983 — Operation San Francisco takes place, consisting of 16 full-scale tests conducted to explore the comparative effects of using standard vs. quick response sprinklers, as well as the effects of alternate HVAC system arrangements (7).

November, 1983 — Factory Mutual sponsors a special conference to introduce the concept of the Early Suppression Fast Response (ESFR) sprinkler to the sprinkler industry. Five U.S. and one European manufacturer agree to work with FM in the development of prototype hardware.

May, 1984 — The National Fire Protection Research Foundation announces plans to sponsor a two-part research program to examine and develop the concept of quick response sprinkler technology. The first part will incorporate the ESFR work begun by Factory Mutual. The second part will address quick response sprinkler use in multiple residence, light, and ordinary hazard occupancies.

II. MODEL OF SPRINKLER RESPONSE AND SUPPRESSION

The model was constructed to provide a way of organizing information, and also to evaluate the state-of-the-art of sprinkler protection, to make it possible to identify which elements are well understood and which are not.

The basic form of the model is of a cyclical algorithm. It is "time-driven" in that successive passes are made through the model as the fire progresses. The model recognizes three ways in which sprinklers can extinguish or control a fire:

1. Extinguishment or control by flame cooling
2. Early suppression by direct cooling of fuel surface
3. Control by prewetting to prevent fire spread

To some extent these three mechanisms are all theoretical, and it can be assumed that most successful sprinkler performances involve some combination of the three. The existence of the second and third mechanisms, however, forms the hypothesis of the Group 1 tests (Early Suppression Fast Response), and the existence of the first was investigated during the early part of the residential sprinkler program (40).

Within the model, individual elements are depicted as functions. Variables are identified in either capital letters or small letters. Capital letters are used for those variables which can be considered "decision" variables, i.e. those which a system designer might be expected to identify during the design of a sprinkler system for a particular occupancy or hazard. Small letters are used for "state" variables, those which develop during the course of a fire and over which the system designer exerts no direct control. In some cases the choice to identify a variable as decision or state is an arbitrary one, since design methods differ in their use of decision variables, and because not all variables which could be addressed are addressed.

Two variables, **res** and **pen**, are defined similarly to the variables RDD and ADD as used in the Group 1 test program. It should be noted that the first two are state variables, while RDD and ADD are considered decision variables, and defined basically as follows:

ADD — actual delivered density, i.e., the water application rate which would be delivered to a particular unit area (or average unit area) through the fire plume.

RDD — required delivered density, i.e., the water application rate required at the surface of the fuel to achieve early suppression of a fire in a specific storage commodity.

611

These terms have been associated with specific other variables and safety factors in the Group 1 program and are not used in the general model so as to avoid confusion. The terms **res** and **pen** do not relate to a specific set of other predetermined variables. The **res** factor is intended to represent the delivered density needed for early suppression at any stage of a fire, determined by the needs of the fuel and its geometry, with a consideration of the rate of heat release. The **pen** is the penetration ratio, i.e., the ratio of delivered density to **local density (ld)**, where local density is the water application rate which would be delivered to the fuel surface in the absence of a fire plume.

In addition to the above and the terms in the following list of variables used in the model, one additional type of density should be defined. This is **design density (DD)**:

DD — design density, i.e. the average water application rate assumed during system hydraulic design, determined by dividing the expected flow from a sprinkler by its assigned area of coverage.

LIST OF VARIABLES USED IN MODEL

a — area of fire

acool — atmospheric cooling factor from water spray

act — activation of sprinkler(s)

AMB — ambient room temperature

CEIL — ceiling configuration and compartmentation factor

cont — control of fire spread by prewetting fuel surfaces adjacent to the fire

CRIT — pre-established life safety or property loss criteria

DELAY — water transit time delay if dry system

dm — median drop diameter of sprinkler spray

fcool — flame cooling caused by water evaporation in the flame structure, diluting reactants with water vapor, replacing oxygen with steam, and otherwise reducing the heat feeding the combustible fuel

flow — flow from sprinkler(s)

FUEL — type of combustible material(s), including characteristic flame spread and smoke development

gcs — gas conditions at sprinkler (temperature and velocity)

GEOM — fuel arrangement factor, including burnable surface area per unit floor area

H — ceiling height above burning fuel

IGN — ignition point and source factor

K — orifice discharge coefficient of sprinkler

ld — local density, i.e., the water application rate which would be delivered to a particular unit area under non-fire conditions

LOAD — quantity (extent) of fuel available

MAIN — system integrity factor

pc — products of combustion, including ceiling and eye level temperatures, toxic gases and smoke obscuration

pen — penetration ratio, i.e., percentage of local density penetrating fire plume

PRES — initial operating pressure at sprinkler

pres — operating pressure following sprinkler activation(s)

q — rate of heat release of burning fuel

R — radial distance to sprinkler from fire axis

RAT — sprinkler temperature rating

res — rate of water application at fuel surface needed for early suppression

RTI — response time index, i.e., sprinkler thermal sensitivity

SPR — sprinkler deflector and distribution characteristics

supp — early suppression of fire by direct cooling of burning fuel surface with water

time — time from ignition

TYPE — type of system (wet vs. dry)

VENT — ventilation factor

Y — vertical distance from ceiling to sprinkler sensing element

MODEL OF SPRINKLER RESPONSE AND SUPPRESSION
(Expanded Framework)

Time = t_o + dt

A. Area of fire

    a = f (time, IGN, FUEL, LOAD, GEOM, q, cont)

B. Rate of heat release

    q = f (a, FUEL, GEOM, VENT, fcool, supp)

C. Products of combustion

    pc = f (a, q, FUEL, flow, SPR)

D. Gas conditions at sprinkler

    gcs = f (q, H, R, Y, VENT, CEIL, acool)

E. Is sprinkler activated?

    act = f (gcs, RTI, RAT, AMB, acool)

F. Is water available?

    flow = f (TYPE, pres, K, DELAY, MAIN)

MODEL OF SPRINKLER RESPONSE AND SUPPRESSION

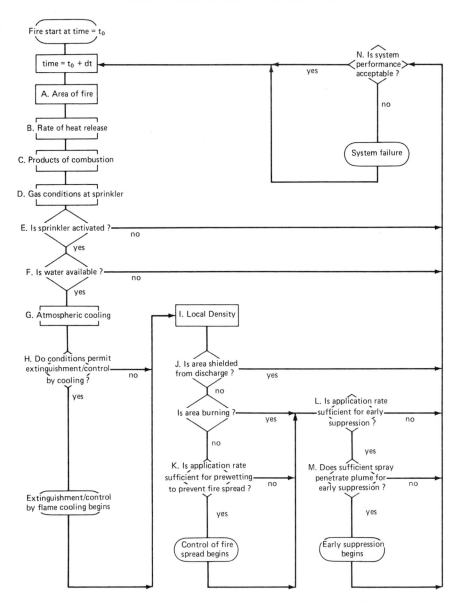

G. Atmospheric cooling

   acool = f (act, flow, dm, (H-Y), SPR, VENT)

H. Do conditions permit extinguishment/control by cooling?

   fcool = f (q, acool, GEOM)

  Flame cooling begins (fcool)

I.   Local density

$Id = f \text{ (act, flow, SPR, R, H, VENT)}$

J.   Is area shielded from discharge?

$= f \text{ (GEOM, Id, CEIL)}$

Is area burning?

$= f \text{ (a)}$

K.   Is application rate sufficient for prewetting to prevent spread?

$= f \text{ (FUEL, GEOM, Id, q)}$

Control of fire spread begins (cont)

L.   Is application rate sufficient for early suppression?

$Id > res?$ where $res = f \text{ (FUEL, GEOM, q)}$

M.   Does sufficient spray penetrate plume for early suppression?

$(Id \times pen) > res?$ where $pen = f \text{ (q, dm, PRES)}$

Early suppression begins (supp)

N.   Is system performance acceptable?

$= f \text{ (CRIT, a, pc, act)}$

III. INFORMATION FRAMEWORK

A. Area of Fire (a)

Area of fire is the plan area of the fire.

$$a = f \text{ (time, IGN, FUEL, LOAD, GEOM, q, cont)}$$

where IGN = ignition point and source factors
FUEL = the type of combustible material(s), including their properties such as flame spread rate
LOAD = amount (extent) of fuel available
GEOM = fuel arrangement factors, including surface area available to burn
q = rate of heat release of burning fuel
cont = control of flame spread by water (if available) to limit fire area

Several room fire growth computational models have been developed, the best known and most flexible being the Harvard Code. It permits consideration of up to five objects per room with the possibility of one object igniting another by either radiation or flame spread. Simulations can be provided for 1) a gas burner, which does not increase in area and whose burning rate is set by the gas flow rate, 2) a pool fire, which does not increase in area and whose burning rate is set by the heat flux reaching the fuel surface, and 3) a growing fire whose area is a function of time and whose burning rate per unit area is set by the heat flux reaching the fuel surface.[1]

Fire **LOAD** characteristics of some typical occupancies have been surveyed by the NBS, originally in the 1930s, but more recently for residences (1976) and office buildings (1973). Reported in terms of an equivalent 8000 Btu/lb fuel (wood), the live fire load of residences (not including structure or surface finish materials) was found to average 10 lb/sq ft.[2]

Due to differences in construction and finish materials, total weighted mean fire loads varied with type of residence:

single family attached = 13 lb/ft²
single family detached = 13 lb/ft²
mobile home = 18 lb/ft²

[1]Rockett, John A., "Modeling of NBS Mattress Tests With the Harvard Mark V Fire Simulation," NBSIR 81-2440, National Bureau of Standards, January 1982.
[2]Issen, Lionel A., "Single Family Residential Fire and Live Loads Survey," NBSIR 80-2155, National Bureau of Standards, December 1980.

Office buildings were found[3] to have a mean fire load, including interior finish, of 7.8 lb/ft² with a standard deviation of 6.8 lb/ft², although individual areas within office buildings averaged higher loadings, such as file rooms (approx. 17 lb/ft²) and library rooms (approx. 25-30 lb/ft²). A definite tendency was found within offices for fire loads to be concentrated around the perimeter of the room.

Large-scale tests of rack storage (18) showed that irregularly sized transverse and longitudinal flue spaces throughout a test array did not significantly affect fire severity. Addition of an extra tier of commodity significantly increased fire severity.

Author's Comments:

1. There are no known fire simulation models which contain a suppression algorithm to take into account the **cont** factor of the water from sprinklers. It is conceivable that existing models could be adapted to this purpose, i.e., to switch from a growing fire to a pool fire at the time of sprinkler operation.

2. The common assumption that the residential fire is a less severe challenge to sprinklers than most light hazard occupancies may not be true.

B. Rate of Heat Release (q)

Rate of heat release is the product of the fuel weight loss and the heating value of a unit mass of fuel.

$$q = f (a, FUEL, GEOM, VENT, fcool, supp)$$

where   VENT = ventilation factors
   fcool = flame cooling factor from evaporating water droplets (if available) to reduce fire intensity
   supp = early suppression action by direct cooling of fuel surface with water (if available) to reduce fire intensity

Available fire simulation models can be considered to model heat release rates only up to the point of water application. See above discussion.

For rack storage of Standard Class I and Standard Plastic commodities, the burning rate has been observed to be proportional to storage height (45,170).

C. Products of Combustion (pc)

Products of combustion include temperatures, toxic gases, and smoke obscuration produced by the fire, both before and after sprinkler operation.

---

[3]Culver, Charles G., "Characteristics of Fire Loads in Office Buildings," *Fire Technology*, Vol. 14, No. 1, February 1978.

$$pc = f (a, q, FUEL, flow, SPR)$$

where   flow =   flow from sprinkler(s) = f (K, PRES)
     and SPR =   sprinkler deflector and distribution factor

Takahashi is reported to have investigated smoke and steam generation from wood crib fires suppressed by water spray (113). He found that smoke density increases to a maximum value and afterward decreases with increased application rate of water.

In 9 smoldering-start limited ventilation bedroom tests (80 cfm of outside air), transition to flaming typically occurred after 60 minutes (FM-1978). Only in the 15 minutes prior to flaming did CO and smoke production become vigorous, with CO concentration at 5-ft levels never exceeding 360 ppm prior to flaming (41). In general, as soon as fire spread stopped, CO generation ceased. In 3 tests CO concentrations were checked. In the others, CO concentrations continued to increase, but at much smaller rates than those of typical room fires.

In wood crib (34 pieces of 22-inch long 2 $\times$ 4 framing lumber) suppression tests in a small room with door open and windows closed, using various nozzles (Battelle-1978), maximum CO levels ranged from 500 to 1400 ppm when sprinkler operation commenced after 5-minute freeburn. Maximum CO readings always occurred during periods of water application (32). In a test where water was turned off during suppression, CO level rose much higher. Two upholstered-furniture fires were extinguished with lower levels of CO than the wood crib fires.

In 14 flaming-start combustible finish living room tests, 2 flaming-start fires in noncombustible finish living room fires, a flaming-start kitchen fire, a flaming-start closet fire, 6 smoldering-start living room fires, and 3 smoldering-start bedroom fires (Los Angeles-1980), the life-safety protection provided by residential sprinklers was found to be dependent on the fire control capability (15,16,44,74). This held true for an additional 11 flaming-start combustible finish living room fires, 3 flaming-start combustible finish bedroom fires, and a flaming-start kitchen fire conducted in a mobile home (North Carolina-1980) (16,74). In the flaming-start tests involving noncombustible finish (including the kitchen and closet fires) fire control was established and CO and temperature levels were maintained within established limits. In the smoldering-start fires, adequate CO and temperature levels were maintained in fires where there was a transition to flaming. In one smoldering-start bedroom test with both windows and door closed, transition to flaming never occurred. Heat build-up operated the sprinkler after more than 4 hours, but CO limits were exceeded within 2 hours. In the other smoldering-start bedroom tests (with door and windows open), transition to flaming occurred after 2 to 3 hours, but adequate CO levels were maintained. In the flaming-start combustible finish living room fires, fire control (and CO and temperature conditions as a result) depended on the number of sprinklers opening and the degree of decay of sprinkler discharge rate.

In 2 residential sprinkler tests in mobile homes (Springdale, Arkansas-1981), combustion gas monitoring indicated that levels of CO, HCl, and HCN produced were considerably below those generally provided as life safety criteria, and far below those recorded in a third unsprinklered fire permitted to grow to total involvement of the mobile home (5). Levels of oxygen depletion, benzene, and particulate matter were also low. Test atmospheres were also monitored for acetaldehyde and acrolein, but none was detected.

In 8 flaming-start and fast flaming-start fires in hotel scenarios (Ft. Lauderdale-1982), QR sprinklers maintained survivable levels of CO, HCN, HCl, smoke obscuration and eye-level temperatures (17,6).

In 8 flaming-start tests in simulated exhibit areas with 9 ft ceiling heights (San Francisco-1983), both standard and QR sprinklers maintained conditions for "human viability," while viability was not maintained in a corresponding test without sprinklers. QR sprinklers had a measurable effect on the fire size, ceiling temperatures, and amount of smoke produced as compared to standard sprinklers (7). Improving the response time of the sprinklers in the exhibit area tests proved more effective in reducing smoke obscuration than altering the intake/exhaust modes of the HVAC system with 6 air changes per hour. In a slow-developing flaming-start hotel room bed-linen fire, room door soffit temperatures were considered excessive (174°F) prior to QR sprinkler activation, and smoke obscuration levels in the room doorway were close to critical, although CO was not a problem.

Author's Comments:

1. The **flow** and **SPR** factors are included in the **pc** function to account for "mixing effects" caused by sprinkler operation. Toxic concentration of a particular product of combustion found in smoke has been expressed as $Y_{tox} = M / d . A . C_{tox}$, where M is the mass burned, d is the depth of smoke layer, A is the area of smoke layer [(d . A) being the volume of smoke generated], and C_{tox} being the toxic threshold in mass per unit volume. It can be seen that limiting the mass burned through suppression limits toxic concentration, and that spray distribution can have an effect through "mixing," enlarging the effective volume of smoke.

2. In flaming fires or smoldering fires where transition to flaming takes place, several test programs have indicated both standard and QR sprinklers are capable of maintaining acceptable levels of toxic gases if sprinklers are capable of fire control. Smoldering fires which do not make a transition to flaming can result in unacceptable levels of toxic gases after long periods of time.

D. Gas Conditions at Sprinkler (gcs)

Gas conditions at sprinkler are the ceiling flow conditions, both temperature and velocity, at each individual sprinkler under consideration.

$$gcs = f(q, H, R, Y, VENT, CEIL, acool)$$

where
- H = ceiling height above burning fuel
- R = radial distance to sprinkler from fire axis
- Y = vertical distance from ceiling to sprinkler sensing element
- $CEIL$ = ceiling configuration and compartmentation factors
- $acool$ = atmospheric cooling factor resulting from water spray (if available)

Alpert, Heskestad, Delichatsios and others (1,8,33,34,69) have established a method of predicting gas velocity, temperature, and dimensions of the near-ceiling flow induced by constant-intensity and growing flaming fires under large, open, relatively smooth ceilings. It has been found that fixed-temperature detectors should be located a distance below the ceiling of no more than 6 percent of ceiling height to be sure of remaining within the ceiling jet. For optimum response, spacings of ¼ of ceiling height are recommended on the basis that closer spacings produce no improvement in response time.

Evans (22,24) developed a method to apply the above method of predicting gas temperatures and velocities to small fires in compartments, adjusting for the effects of a warm upper layer build-up. The compartmentation effects are reported capable of theoretically speeding up response by 40 percent or more.

Following the work of Alpert (1) and Heskestad and Smith (36), Beyler conducted a sensitivity analysis with the model of sprinkler activation (9). Graphical representations were made of the effects on sprinkler activation time of radial distance of the sprinkler from the fire, slow vs. fast fire development, sprinkler temperature rating, RTI, ceiling height, sprinkler to ceiling clearance, and rate of fire growth. The impact of some variables is affected by the impact of others. For example, the effects of most variables, such as temperature rating, sprinkler sensitivity, and radial distance from the fire, become less significant in a fast-developing fire. The graphs also show that ceiling height and clearance between sprinklers and ceiling play a major role in determining time of actuation.

During the residential sprinkler program (43), it was observed that because a nonventilated fire grows slowly, heat is transferred over time, and a sprinkler will then operate faster when it encounters air temperatures rising above its rated temperature. This has the net effect of requiring a ventilated fire to grow to a larger size to operate a sprinkler of the same sensitivity.

Relatively few studies have been done to investigate the effects of ceiling configuration factors affecting gas flow conditions and heat build-up. Fire alarm thermostat tests conducted by the National Board of Fire Underwriters in 1956 showed that spacing limitations under open-joisted construction must be substantially reduced below that permitted for smooth ceilings in order to provide the same degree of thermal sensitivity. Under similar fire conditions, operation times were found to be 2 to 2-½ times as long under the joisted ceilings when heat travel perpendicular to the joists was required (25). Thompson (111) described tests which indicate the origin of NFPA 13 allowances for sprinkler distance below ceilings. Campbell et al (13) tested the effects of placement within and below beam channels on sprinkler response. Work performed for the Fire Detection Institute in the development of Appendix C of NFPA 72E (133) included an investigation of the effects of beams in large ceilings (33). The effect was found to be proportional to the ratio of the depth of the beam to the ceiling height and also to the spacing of the beams in relation to their depth. In general, temperatures decrease very rapidly from the fire source in a direction across the beams, but decrease very slowly in a direction along the length of the beams. The effect of the beams is to channel the flow of heated gases. The study also yielded information relative to optimal location of detectors within channels vs. under beams.

Studies of the heat build-up leading to activation of sidewall sprinklers in compartments conducted at the National Bureau of Standards, Battelle, Factory Mutual and in Cobb County (31,32,38,58) give somewhat conflicting information as to optimal positioning of sidewall sprinklers below the ceiling. NBS testing indicated that substantially slower response could be expected for sidewall sprinklers as opposed to pendent sprinklers, especially those sidewall sprinklers located on the wall opposite the fire and for slow fire growth rates. The Battelle thermal mapping tests indicated that for corner or sidewall fire locations, sprinkler location anywhere within 12 inches of the ceiling was acceptable, and anywhere within 6 inches of the ceiling preferable. For a central fire location, the corner locations were less preferable for locating sidewall sprinklers. Factory Mutual testing using a corner fire scenario has indicated sidewall sprinklers located 12 inches below the ceiling are likely to operate prior to sprinklers closer to the ceiling. Tests conducted in Cobb County led authorities to restrict maximum distance of sidewall sprinklers below the ceiling to 4.5 inches.

E. Sprinkler Activation (act)

Sprinkler activation occurs when there is operation of any individual sprinkler in response to the fire. On a cumulative basis, **act** determines the total number of sprinklers operated during the fire, representing the area of operation, which can be compared to a design area.

$$act = f \text{ (gcs, RTI, RAT, AMB, acool)}$$

where   RTI = sprinkler sensitivity
RAT = sprinkler temperature rating
AMB = ambient temperature

In the early 1970s, attempts to improve sprinkler sensitivity included consideration of sprinkler operation by smoke detectors or alternate electronic means, including frangible discs with miniature detonators (18,49). During the late 1970s and early 1980s, as part of the U.S. Fire Administration's effort to develop a low-cost residential sprinkler system, additional research included the attempted development of sprinkler activating mechanisms made highly sensitive through the use of electronic sensors, the use of nitinol, and other means (3,4,20). Increased sprinkler sensitivity was ultimately achieved using conventional sensing mechanisms.

Heskestad and Smith (36,37) opened the door for calculation methods of sprinkler response times through development of the plunge test. In the plunge test, a sprinkler is immersed in a constant temperature hot air stream moving at a known velocity, and the time for the sprinkler to heat from ambient temperature to its operating temperature is recorded. This gives a relative measure of its thermal inertia, which can then be used as a property of the sprinkler, along with its operating temperature, in determining when the sprinkler will operate under given conditions of gas temperature and velocity. For any given velocity, a sprinkler has a characteristic "tau factor" or time constant (units of seconds) which represents its sensitivity. The model of sprinkler response is as follows:

$$d\,(\Delta T_L)\,/\,dt = \tau^{-1}\,(\Delta T_g - \Delta T_L)$$

where ΔT_L is the excess temperature of the sprinkler link above ambient, ΔT_g is the excess temperature of the gas above ambient, and τ is the time constant. The time constant is related to properties of the sprinkler by the equation:

$$\tau = mc\,/\,h_c A$$

where m is the mass of the sensing element or link, c is the specific heat of the element, A is the surface area through which heat is transferred, and h_c is the coefficient of convective heat transfer. Reproducibility of measurements in the plunge test is estimated to be within 3 percent.

Heskestad and Smith also verified some of the assumptions necessary to support a convective heating model of sprinkler operation:

a) that the effects of radiation are small — shown to be in the order of $\frac{1}{8}$ to $\frac{1}{5}$ of the convective contribution.

b) that conduction between the sprinkler body and sensing mechanism can be ignored.

c) that the heat of activation (fusion for a solder link) is small relative to the heat needed to bring the sensing mechanism to the activation temperature.

d) that the sensing element can be considered "thermally thin."

NBS researchers developed a two-parameter model to be able to take into account the error introduced by assumption (c) above (23). Its use introduces additional complexity to the plunge test procedure, however, and the error was estimated to be only on the order of 1 to 3 percent in realistic fire scenarios. This report also examined the validity of assumptions (a) and (b) and found that radiation and conduction effects could increase measured time constants by up to 18 and 23 percent respectively.

Since the time constant of a sprinkler is dependent on the velocity at which it is determined, Factory Mutual researchers later introduced the concept of the response time index (**RTI**) (37,87). Defined as the product of the time constant and the square root of its corresponding gas velocity, **RTI** is independent of both temperature and velocity and can be thought of as a basic property of a sprinkler activating mechanism. It is expressed in either metric units (m$^{1/2}$sec$^{1/2}$) or English units (ft$^{1/2}$sec$^{1/2}$). In English units, an **RTI** is 1.81 times its metric value. **RTI** values of standard response sprinklers were reported to be in the range of 100 to 400 (metric units).

British researchers have developed a corresponding plunge test method to determine the time constant of sprinklers, but their method incorporates a rising air temperature rather than a constant air temperature (112).

British and Australian research has indicated bulb-type sprinklers are more sensitive to radiant heat than solder-type sprinklers (28,91). Blackening the sprinkler sensing mechanisms substantially improved response to radiant fires, by up to 10 times for solder-type sprinklers.

Comparisons of actual test results to mathematical predictions of sprinkler response using the convection model have been reported successful in both residential fires (7,31) and high challenge fires (45), although the theoretical method has been reported to underpredict measured link temperatures for rack storage fires with ceiling clearance less than 2.9 m in which radiation appeared to be significant.

The estimated number of sprinklers which will activate during the course of a fire is commonly used as a design factor, either directly as in NFPA 13D (132) or indirectly through the use of a design area as in NFPA standards 13, 231, and 231C (131,129, 130). British installation standards (127), which serve as the basis for sprinkler design in most parts of the world where U.S. installation standards are not followed, are similar in this characteristic. Based on maximum allowable sprinkler spacing and the specified design area, the following number of operating sprinklers are used for design purposes for the occupancy hazard classifications of the British rules and NFPA 13:

Occ. Class.	FOC 29th Ed.(2)	NFPA 13
Extra Light Hazard	4	
Light Hazard		7-18
Ordinary Hazard Group I	6	
Ordinary Hazard Group 1		12-39
Ordinary Hazard Group II	12	
Ordinary Hazard Group 2		12-39
Ordinary Hazard Group III	18	
Ordinary Hazard Group 3		12-39
Ordinary Hazard Group IIIS	30	
Extra Hazard Group 1		25-53
Extra High Hazard (process)	29	
Extra Hazard Group 2		25-60

Regression analysis of combined U.S. and British statistics (2) indicates that the proportion q(N) of fires in which N or more sprinklers will operate is:

$$q(N) = N^{-0.68 - 0.12 \log N}$$

Loss analysis by Factory Mutual in the late 1950's gave evidence that there was an inverse relationship between the number of sprinklers opening in a fire and the sprinkler temperature rating (**RAT**). In 1963, Newman showed that this relationship existed despite variables of sprinkler spacing and pressure, ceiling height, extent of combustibles, and brands of sprinkler (50).

In 1964, Suchomel confirmed that the number of sprinklers opening in fire tests of high-piled (11 to 14 ft) stock decreased as the temperature rating of the sprinklers increased (62,108). The number of sprinklers opening also decreased with an increase in system operating pressure.

In 1983, Kung et al (42) found that the likelihood of excessive sprinkler activations in a large room using residential sprinklers decreased as the link sensitivity increased. Empirical correlations were established between the maximum temperature increase of simulated remote sprinklers and the following variables: link sensitivity, rated link temperature, and distance between the remote link and the operating sprinkler. As the sprinkler discharge rate was increased (from 18 gpm to 24.5 gpm), significant reduction occurred in the maximum temperature of the remote simulated links due to superior cooling.

Author's Comments:

For purposes of the model, it is of interest to try to classify the various fire tests and determine if the sprinklers were effective because of early suppression, control by prewetting, or flame cooling. The 1963 Factory Mutual tests (50) did experiment with increased pressure, and it is of interest why early suppression was not achieved at some point. Although some of the tests were conducted using an unextinguishable gasoline spray, and some were conducted using an extended array of pallets creating a severe shielding condition, others were conducted using a single 8 ft stack of 4 ft × 4 ft wood pallets. With a spacing of 10 ft × 10 ft the sprinklers were operated at pressures ranging from 15 psi to 25 psi. Assuming a K-factor of 5.6, this corresponds to a range of design densities from 0.22 gpm/ft² to 0.28 gpm/ft². British test results (52) would indicate that early suppression of wood fires might be achieved at the highest density. It must be assumed, however, that insufficient water penetrated the fire plume in these tests [(**pen** × **ld**) > **res** or ADD < RDD] and early suppression was not achieved. Even with 165°F sprinklers, the first sprinklers did not operate until 45 to 80 seconds after flames reached the top of the pallet stack. Also of note is the fact that the sprinklers used at pressures above 15 psi were two models which averaged 70 and 75 seconds, as opposed to the most sensitive, which averaged 48 seconds. In any event, early fire suppression was not observed, and fire control within the pallet stack was presumably accomplished by prewetting to control spread.

In the corresponding UL tests of 1964 (62), the ability of increased pressure (and corresponding delivered density) to reduce the number of sprinklers opening was observed, and declared an alternate to using high temperature rated sprinklers. Of 17 total tests conducted, the one with the lowest crib weight loss and lowest average ceiling temperatures involved the use of 160°F sprinklers and a flowing pressure of 93 psi, producing an average local density of 0.463 gpm/ft². The first sprinkler activated at 2:00. A total of 4 sprinklers opened, two of them over the crib. This particular test might have been of the early suppression type, but the others would fall into the control category.

In the 1983 residential sprinkler tests, fire control may likely have been by means of the flame cooling effect, inasmuch as it was at least partially a shielded fire condition. Increasing sensitivity did not increase but rather reduced the likelihood of multiple sprinklers opening.

F. Water Availability

Water availability occurs when there is discharge of water from the sprinkler following its operation.

$$\text{flow} = f \text{ (TYPE, pres, K, DELAY, MAIN)}$$

where   TYPE = type of system

pres = operating pressure, where (pres = f (PRES, act))

K = orifice discharge coefficient of sprinkler

DELAY = water transit delay if dry system

MAIN = system integrity factor

For a wet system this element provides no obstacle to extinguishment, assuming system integrity and an operating pressure, but merely determines the flow per activated sprinkler. For a dry system, the length of the delay in getting water to the sprinkler(s) will determine continued fire growth and additional sprinkler openings **act** (by not initiating the **cont, fcool**, or **supp** functions to limit the increase in **a** and **q**).

Several studies (2) have indicated that more sprinklers operate in dry pipe systems. Based on a combination of U.S. and British data, regression analysis produces the following equations for the proportion q(N) of fires in which N or more sprinklers operate:

For wet systems $q(N) = N^{-0.78 - 0.16 \log N}$

For dry systems $q(N) = N^{-0.27 - 0.22 \log N}$

This indicates that a higher proportion of fires in buildings protected by dry systems will attain any given size than in buildings with wet systems. On the average, twice as many sprinklers operate in dry systems as in wet systems.

Factory Mutual modeling and subsequent full-scale tests (10,35) comparing the water demands of wet and dry systems have demonstrated that, as long as the water delay time of a dry-pipe system is less than or equal to the freeburn time needed for the operation of the number of sprinklers equal to the number which would eventually operate on a wet system, the number of dry-system operating sprinklers will not exceed the number of wet-system operating sprinklers. The full-scale testing involved the protection of a plastics commodity stored 5m high and protected with a non-decaying design density of 0.6 gpm/ft².

Heskestad and Kung (35) have developed a theoretical formulation of dry pipe valve actuation time and water flow transit time for tree systems.

Author's Comment:

The Factory Mutual studies (10,35) of dry system delays may indicate that where water densities delivered by sprinklers are not capable of early suppression but are capable of prewetting to control fire spread, the time delay of a dry system is unimportant so long as water arrives prior to the opening of the number of sprinklers which would eventually operate if the system were wet.

G. Atmospheric Cooling (acool)

Atmospheric cooling results from the ability of the sprinkler discharge to absorb heat.

$$acool = f \text{ (act, flow, dm, (H-Y), SPR, VENT)}$$

where dm = median drop diameter = f (SPR, pres)

Using a one-dimensional analysis, Liu found it possible to predict the net reduction in corridor exit gas temperature adjacent to an unsuppressed compartment fire (153). The analysis showed that a spray with a smaller average droplet diameter was more effective in cooling. Correlation between full-scale tests and a reduced scale model was also successful.

Standard ½-inch sprinklers were found to be highly inefficient in cooling hot air streams in 1981 NBS stairwell tests (139).

H. Flame Cooling (fcool)

Flame cooling occurs when conditions permit the sprinkler(s) to control or extinguish the fire by means of water evaporation in the flame structure, diluting reactants with water vapor, replacing oxygen with steam to smother the fire, and otherwise reducing the heat feeding the combustible fuel. It can be thought of as a level of atmospheric cooling which begins to markedly reduce the rate of heat release.

$$fcool = f (q, acool, GEOM)$$

Studies of fire extinguishment by water spray have found cooling to be an efficient means of extinguishment, provided the degree of ventilation in the space surrounding the fire can be controlled (137, 159).

1975 work in the residential sprinkler program by Kung (40) investigated control and extinguishment of fires by the cooling mechanism. Hexane pool fires were used, preventing direct cooling by water contact with the burning fuel (since water is denser than heptane, unevaporated droplets passed through the burning surface). It was found that the ratio of the heat-absorption rate by sprinkler water versus the heat-release rate of the fire, normalized with the water discharge rate, varied as the -0.68 power of the relative median drop diameter. (This was established in tests with the room window open but door closed.) For a constant discharge rate, smaller orifice sprinklers (producing smaller droplets) were more effective in controlling temperatures. Smaller orifices combined with higher discharge rates were able to produce extinguishment. In tests conducted with both window and door open, however, no extinguishment took place, and the hexane continued to burn until it ran out.

1984 work by Kung et al (151) led to an empirical correlation between heat absorption, rate of spray discharge, and the convective heat flux through room openings based on:

1. heat release rate of fire
2. heat loss to ceilings, walls, and floor
3. room opening geometry
4. sprinkler discharge rate
5. spray median drop size

Author's Comments:

Despite the fact that the empirical correlation developed by Kung includes heat loss to walls and ceilings, no factor is included in the model. It can be considered conservative to ignore such heat losses where heat absorption by sprinkler spray is desired. The factors for sprinkler discharge rate and median drop size are included through the **acool** factor.

I. Local Density (ld)

Local density is the rate at which water would be applied to a specific portion of coverage area under non-fire conditions.

$$ld = f (act, flow, SPR, R, H, VENT)$$

where SPR = sprinkler deflector and distribution characteristics

Sprinkler activation (**act**) is significant to local density because it determines whether the rate application is by a single sprinkler or the overlapping patterns of multiple sprinklers.

Design density (**DD**) is not included as a variable within the model although this quantity is widely controlled through NFPA standards or recommendations of the authority having jurisdiction. Design density is calculated from the sprinkler **flow** (or pressure and K-factor) in combination with the intended coverage area of the sprinkler (based roughly on R). The standards or product listings sometimes contain specific minimum values of **pres** or **flow** which override consideration of design density.

Minimum requirements of local density are set by product standards using pan collection for specific non-fire test conditions. Actual distribution patterns are primarily affected by choice of sprinkler in terms of deflector design (**SPR**) and orifice size (**K**). Other major factors include flow rate (**flow**) determined by pressure (**pres**), horizontal distance R from the sprinkler to a particular pan, the height (**H**) of the sprinkler over the horizontal collection area, and ventilation factors (**VENT**) which could affect spray movement.

Beyler (136) found additional factors influencing spray pattern distribution, including the pipe diameter feeding the sprinkler, the clearance between deflector and ceiling, the orientation of frame arms, the vertical alignment of upright and pendent sprinklers (and presumably horizontal alignment of sidewalls), and the obstruction caused by sprinkler guards.

Large-scale rack storage tests (19) showed that there was considerable difference in the discharge characteristics and resulting effectiveness of protection provided by two ½-inch orifice sprinklers of different brands, even though both provided the same discharge density. Increasing the operating pressure of one sprinkler improved its performance back into line with that of the other.

For residential sprinklers, UL 1626 (125) requires horizontal collection in 1 ft² pans, extending two feet beyond the area of proposed coverage. A minimum individual application rate of 0.02 gpm/ft² is required both at the 1-sprinkler flow rate and the 2-sprinkler flow rate, using an 8-ft ceiling height (7 ft 9 in. between deflector and top of pans). In the vertical direction, walls are to be wet to within 28 inches of the ceiling with a 1-sprinkler flow and to within 36 inches with a 2-sprinkler flow. Each wall which surrounds the coverage area is to be wetted with at least 5 percent of the sprinkler flow at both rates. (These requirements are intended for both pendent and sidewall sprinklers of any intended coverage area.)

For residential sprinklers, FM 2030 (123) requires horizontal collection in 1 ft² pans over a 6 ft × 6 ft area with the sprinkler located over one corner. Using the 1-sprinkler flow and an 8-ft ceiling height, a minimum of 0.03 gpm/ft2 is required in each pan. In the vertical direction, six rows of eight stacked pans, ½ ft high × 1 ft wide, are used with a vertical space between pans of ½ ft, and then shifted upward ½ ft. Using the 1-sprinkler flow rate, each pan located 0.75 to 5.75 ft above the floor must collect at least 0.010 gpm/ft², and each pan located 5.75 to 6.25 ft above the floor must collect a mimimum of 0.002 gpm/ft². Sprinklers deficient in the distribution requirements may make up for this deficiency if their sensitivity exceeds minimum requirements. (These requirements are intended for pendent sprinklers with a 12 ft × 12 ft coverage area.)

For standard (and QR) upright and pendent sprinklers, UL 199 (128) requires a 10-pan (turntable) test and a 16-pan test. In the turntable test, ten 1 ft² pans are rotated 4 ft below the sprinkler deflector. At a flow rate of 15 gpm for standard and small-orifice sprinklers (21 gpm for large-orifice), distribution beyond an 8 ft radius may not exceed 0.01 gpm/ft². Minor ceiling wetting is considered acceptable. In the 16-pan test, 1 ft² pans are centered in a 4 ft × 4 ft square 7 ft 6 in. below four sprinklers spaced at 10 ft × 10 ft (this simulates the location of the top of the crib in the crib fire test). For each orifice diameter a flow rate is specified (15 gpm for ½-inch orifice) for the sprinklers, along with a minimum average density (0.150 gpm/ft² for ½-inch) to be collected in the pans. No individual pan may have less than 75 percent of the required average.

For sidewall sprinklers, UL 199 requires two sprinklers to be located 10 feet apart, with the deflectors 4 in. below a ceiling and 6 ft 8 in. above a set of 100 1 ft² pans. The pans form a square with one side being a line 1.5 ft in front of the sprinklers. For ½-inch and smaller orifice sprinklers, a flow rate of 15 gpm must produce a minimum average collection of 0.05 gpm/ft², with no individual pan less than 0.03 gpm/ft². For large-orifice sprinklers, higher densities are required using a higher flow. Each sidewall sprinkler must also wet the wall behind the sprinkler to a height within 4 ft of the deflector (apex directly below sprinkler), and must discharge at least 3.5 percent of its flow against this back wall.

For extended coverage sprinklers, UL 199 requires full distribution measurements in 1 ft² pans for the enclosed area recommended for coverage by the manufacturer, at a flow rate recommended by the manufacturer (minimum flow rate must correspond to a design density of 0.1

gpm/ft²). The three pans with the least water collection are designated as locations for the 1 ft³ wood cribs to be used in the fire tests.

For old-style sprinklers, UL 199 requires a single upright or pendent to be installed 10 ft above the center of a 10 ft × 10 ft arrangement of 1 ft² pans. At least 90 percent of a 15 gpm flow is to be contained within this area. While there are no specific requirements, individual collections are recorded to determine the uniformity of discharge.

For standard upright and pendent sprinklers, FM 2016/2017 (121) requires a turntable test similar to UL, but with flow rates of both 15 gpm and 30 gpm for ½-inch orifice sprinklers, and other flow rates for small and large orifice sprinklers. The results are recorded only for purposes of comparison during subsequent product re-examination. During the test, a tag is placed 7 ft from the sprinkler at the elevation of the sprinkler to check the potential for impingement which could lead to "cold-soldering." Horizontal 16-pan distribution testing is also conducted for arrays of 4 and 6 sprinklers. The 4-sprinkler array is nearly identical to the UL 16-pan test, except that different average waterflows per sprinkler are used, with different collection requirements. For the ½-inch orifice sprinkler, flows of 12.8, 16.6, and 24.0 gpm are used, with corresponding minimum average collections of 0.128, 0.166, and 0.240 gpm/ft². In the 6-sprinkler array, the 16-pan grid is located on the centerline between the two center sprinklers of a 2 × 3 sprinkler arrangement, all sprinklers spaced at 10 feet. For the ½-inch orifice sprinkler, a flow of 16.6 gpm is required to produce a minimum average collection of 0.166 gpm/ft².

For sidewall sprinklers, FM 2012 (121) includes a 2-sprinkler test nearly identical to UL, including pass/fail criteria, except that the tops of pans are located only 6 ft below the deflectors.

O'Dogherty and Nash proposed a concept for determining the basis of minimum sprinkler distribution uniformity (51). Knowing the minimum density needed to prevent the spread of fire for a given fuel, a sprinkler array (4 sprinklers overlapping) would be required to demonstrate that the total area of collection pans with a density less than the absolute minimum needed to prevent spread was less than that needed to support a fire of a size capable of activating sprinklers in the 2nd ring (the 16 sprinklers beyond the original 4). This would vary with ceiling height, however.

Author's Comments:

1. Local density as distribution. The distribution testing by laboratories is actually the determination of local density for specific values of **act, flow, R, H** and **VENT**. It is therefore a control on the **SPR** factor, checking on angular variations in spray discharge, frame arm shadows, and other variations caused by nozzle, deflector and frame design differences.

2. Uniformity. An attempt at minimum levels of uniformity is being made with the residential sprinklers, although the minimum local densities required for the conditions of the distribution testing differ between the laboratories, 0.02 gpm/ft² for UL and 0.03 gpm/ft² for FM. The same is true for the extended coverage sprinklers at UL, although the minimum local density is determined as a consequence of later performance in a fire test, and there is no evidence that the presence of the fire does not alter the spray pattern such that the true areas of lowest density are then being tested. The laboratory requirements for sidewall sprinklers appear to be very consistent, and specifically establish minimum levels of uniform discharge. In the laboratory standards for standard upright and pendent sprinklers, however, the requirements actually discourage uniformity of discharge. Since a minimum average collection in the 16 pans of the 16-pan tests is required to be the average of the flow over the protected area, it is a necessary consequence that other areas within the coverage area show less than the average. The laboratories do not appear to be concerned with how much less, since there are no specific requirements for minimum collection in these areas. Futhermore, it should be noted that for all standard sprinklers except the extended coverage sprinklers, distribution testing is conducted on multiple-sprinkler flows, permitting the spray from one sprinkler to "fill in the gaps" of another's spray pattern.

3. Factors.   The additional factors studied by Beyler are not listed as components of the local density function because they are considered minor when compared to the complications which would be introduced by their inclusion.

4. Vertical collection.   With the residential sprinklers the wall-wetting ability of a sprinkler at the perimeter of its coverage area has been identified as a concern. Previously, this was considered a concern only for the back wall protection ability of sidewall sprinklers. The philosophy of multiple sprinkler overlap in large areas, which has strongly influenced the distribution test requirements for standard sprinklers, does not recognize advantages of extensive wall-wetting, and in fact considers it a potential problem in terms of cold-soldering.

J. Discharge Shielding

Discharge shielding occurs when obstructions intervene between the sprinkler discharge and the burning fuel surface to prevent early suppression, or between the discharge and fuel surfaces adjacent to the burning fuel so as to prevent control of flame spread.

$$= f \, (GEOM, \, SPR, \, R, \, H, \, FLOW, \, CEIL)$$

Distribution testing of old-style sprinklers in 1926 concluded that there was excessive variation among sprinklers in the quantity and direction of water discharged at any given pressure. The coverage over planes parallel to the ceiling and 12 in. and 24 in. below the deflectors was not felt to differ sufficiently to warrant a requirement that stock not be closer than 24 in. to the sprinklers, however, and a compromise of 18 in. was suggested (145).

Tests to determine acceptable levels of obstruction caused by clearance less than 18 in. between sprinklers and the tops of patient room privacy curtains were based on the need to obtain a minimum local density of 0.033 gpm/ft² averaged over the area of a bed using a design density of 0.17 gpm/ft². The average local density over the area of the bed without the privacy curtain obstruction was 0.135 gpm/ft² (55).

Author's Comments:

1. Very few studies have been done to quantify the degree to which obstructions interfere with sprinkler spray, presumably because the differences are considerable among various brands of sprinklers, compounded by the non-uniformity and irregularity of spray patterns and the changes which take place due to changes in **flow** and other factors discussed for element I — Local Density.

2. The patient room privacy curtain test results were incorporated into NFPA 13 beginning with the 1983 edition to address privacy curtains, freestanding partitions, and room dividers.

K. Control of Fire Spread (cont)

Control of fire spread occurs when the local density is sufficient to suppress the spread of fire along the fuel surface.

$$cont = f (FUEL, GEOM, ld, q)$$

For mid-size wood cribs, a density of 0.075 igpm/ft² (0.094 gpm/ft²) applied directly to the top of the burning surface was found to be not capable of extinguishing the fire, but capable of limiting flame spread. In some cases an application rate of 0.05 igpm/ft² (0.0625 gpm/ft²) was found sufficient. Higher application rates, up to 0.2 igpm/ft² (0.25 gpm/ft²), markedly reduced flame spread, at which point increased application rates produced diminishing returns. For a given application rate, fire spread after beginning water application was found to be independent of the size of the fire when water was first applied. Even when the lowest rates of water application used (0.05 igpm/ft²) were not enough to stop flame spread, they did have the effect of slowing flame spread. For faster-growing wood crib fires (25.8 Btu/sec²) at least 0.1 igpm/ft² (0.125 gpm/ft²) was needed to limit fire spread (52).

In limited-ventilation smoldering-start fires on a couch with foam mattress and urethane foam bolsters, a local density of 0.033 gpm/ft² appeared to be a critical application rate for halting flame spread to the underside of the mattress (41).

L. Application Rate for Early Suppression

The application rate for early suppression is achieved when the local density available from the sprinkler, if delivered to the surface of the fire, would result in rapid knockdown of the fire, presumably through direct cooling of the fuel surface. Local density is compared to **res**, the delivered density which would be required to provide early suppression at any given stage of the fire. Combined with criteria concerning extent of damage to the fuel and the potential to operate more than four sprinklers, this **res** could be considered an RDD.

$$Id > res? \text{ where } res = f \text{ (FUEL, GEOM, q)}$$

1967 British tests indicated (52) that, for mid-size wood cribs (capable of producing a heat release of 8.7 Btu/sec^2), water applied through impinging jet nozzles at rates less than 0.23 igpm/ft^2 (0.275 gpm/ft^2) had little effect on the actual burning zone. Larger application rates tended to result in rapid extinction. Extinction time depended on the rate of water application but was independent of the size of the fire when water was first applied. Extinction time was also dependent on the fuel geometry affecting water access, the quantity of burning fuel per unit area, and the expected duration of the fire in the burning zone. Total quantity of water per unit area needed to extinguish the fire was found to be independent of the rate of water application.

In 1979 Takahashi (113) was reported to have developed a relationship for extinction time of wood cribs caused by dripping water on their upper surface:

$$t_e = (No/\alpha) \, P^{m/n}$$

where t_e = time to extinction
No = number of layers of crib
α = function of the weight loss of crib before water applied
P = application rate in liters/min
m/n = a constant

Also in Japan, Kida is reported as having investigated the water discharge rate for extinction of wood cribs from 0.5 to 30 kg and found:

$$Q = 0.9 \, (G \, V)^{0.6}$$

where Q = the amount of water required for extinguishment
G = weight loss in kg prior to water application
V = burning rate of cribs prior to water application in kg/min

In 1982 FM researchers suggested (84) that the critical water density for simple fuel geometries is simply related to a material's heat of gasification, which is the heat required to vaporize a unit mass of a combustible material initially at ambient temperature. They noted that this needed to be substantiated and measured for a variety of materials and examined for more complex geometries such as pallet stacks and boxes.

Heskestad (146) has summarized the findings of Bryan and Smith, 1945, Kida, 1973, Kung and Hill, 1975 (150), and Tamanini, 1976 (163), with respect to critical water application rate on wood cribs of various species, sizes, and preburns. The results indicate a critical water rate between 1.5 to 3 gm/m²/sec (0.0022 to 0.0044 gpm/ft²), mostly between 2 and 2.5 gm/m²/sec. He notes that the rate increases, to a degree, with preburn.

Magee and Reitz (154) have studied fire suppression of several plastics and found that, for situations with no external radiation, critical water application rates for extinction varied from 1.2 gm/m²/sec for horizontal polymethyl methacrylate to 4.4 gm/m²/sec (0.0176 to 0.0645 gpm/ft²) for horizontal polyethylene. With external radiation, considerably higher critical rates resulted.

In 1984 Factory Mutual defined the RDD as the water density required at the top of a burning array such that the fire will not redevelop, and such that a second ring of sprinklers (beyond 4) will not operate, and with an allowable combustion consumption limit of 10 percent (46,98,116-120). The critical water rate needed to control rack storage of a standard plastic commodity (polystyrene tubs open end down in eight $\frac{5}{32}$ inch cardboard containers forming 42 in. × 42 in. × 40 in. pile on wood pallet) in this manner with an **RTI** of 50 (English units) and sprinkler spacing assumed 10 ft × 10 ft was found to be roughly proportional to the height of storage (number of tiers high): (46)

> 2 × 2 × 3 array - 0.3 gpm/ft² (15 ft clearance)
> 2 × 2 × 4 array - 0.4 gpm/ft² (10 ft clearance)
> 2 × 2 × 5 array - 0.55 gpm/ft² (5 ft clearance)

RDD values proposed for these arrangments were 0.35, 0.45, and 0.65 gpm/ft² respectively.

Author's Comments:

1. **Critical water densities.** A comparison of the above reports might lead one to the conclusion that the rates of water application needed for extinguishment vary widely. It is important to note, however, that in some of the research (146, 154) the water application rates were determined under laboratory conditions and are expressed per unit of burning surface area, while in others (46,52) the water application is less controlled and the rate is expressed per unit of floor area. Thus, it appears that the rates of water application to the burning surface of combustibles needed for extinguishment under laboratory conditions are on the order of 0.002 to 0.007 gpm/ft². In large-scale testing, however, the rates needed to effect early suppression are on the order of 0.25 gpm/ft² and above.

2. **Role of preburn.** There appears to be a question as to whether the extent of burning prior to water application makes any difference when it

comes to the amount of water per unit area needed for extinguishment of the burning area. While it is certainly a factor if the burning proceeds to the point where the geometry is complicated (more difficult for water access), the British tests indicated a set amount of water was needed to extinguish the burning within a unit of surface area, and that the rate of water application determined how fast extinguishment occurred. The relationships developed by the Japanese include a preburn factor (weight loss of crib prior to water application) but the water application factors are not in terms of unit area. Factory Mutual data on RDD test results indicates the required rate for extinguishment is not sensitive to the rate of heat release at the time of sprinkler operation for rates of heat release up to 120,000 Btu/min, although there is an effect at higher rates of heat release, i.e., for values of q ranging from 120,000 Btu/min to 250,000 Btu/min (46).

3. Total amount of water.   The test reports all tend to agree that the total amount of water needed for extinguishment tends to be constant, regardless of rate of application. This is within limits, however, since there is a minimum rate below which the water appears to have no impact whatsoever on the fire.

M. Spray Penetration for Early Suppression

Spray penetration for early suppression is achieved when the early suppression capability is not lost due to the fact that smaller droplets can be carried away by the upward momentum of the fire plume. A penetration ratio factor accounts for the loss of a percentage of the local density based on the upward fire draft associated with the heat release rate, the mean droplet size and the momentum due to operating pressure, and then determines if the remaining application rate (which can be thought of as equivalent to the actual delivered density or ADD) is still adequate for early suppression.

$$(ld \times pen) > res? \quad \text{where} \quad pen = f(q, DM, PRES)$$

For sprinklers of similar geometry, the mean drop size generated is proportional to the $\frac{2}{3}$ power of the orifice diameter and the minus $\frac{1}{3}$ power of pressure (142).

Factory Mutual researchers have found that either gravitational penetration or momentum penetration may be used to propel water drops through the fire plume (86,116,119). Terminal velocity of gravitational penetration varies with temperatures and drop diameter. Burning rates from 50,000 to 150,000 Btu/min produce maximum gas velocities from 24 to 30 ft/sec, proving impenetrable for drops with diameters less than about 1.5 mm. Penetration ratio of a given fire plume is defined as the ratio of actual delivered density (ADD) through the plume to local density (**ld**) which would be delivered in the absence of the plume. A decrease in penetration ratio with increased pressure is caused by smaller drop diameters, but is followed by an increase in penetration ratio as the drops enter the momentum regime. In the gravitational regime, ADD is proportional to the quantity:

$$\frac{Id}{DD} \quad \frac{K^{1.37} \, PRES^{0.13}}{q^{0.22} \, S}$$

where **DD** is design density and **S** is the sprinkler spacing (a function of **R**). The proportionality factor differs between sprinkler designs. The gravity regime terminates and momentum regime begins at:

$$\frac{Id}{DD} \frac{1}{q^{2/5}} \quad \frac{K \bullet PRES_c}{S^2} = 0.05 \text{ (English units)}$$

where PRESc represents a critical transition pressure. To operate in the momentum regime, a pressure of 1.5 times the critical pressure is suggested. The critical pressure can be reduced by 1) using quick response sprinklers to reduce the rate of heat release at the time of sprinkler operation (**RTI** and **RAT**), 2) narrowing the sprinkler discharge pattern (**SPR**), 3) increasing the orifice size (**K**), 4) reducing the spacing (**R**), or 5) reducing the fire-to-sprinkler clearance (**H**).

Author's Comment:

Although the quantity (Id × pen) can be considered equivalent to ADD, a specific definition of ADD has not yet been finalized. This is because final determination has not been made as to the need for uniformity of delivered density over the top of the fuel surface. Various averaging methods are being considered.

N. System Performance

Sprinkler performance involves a pass/fail consideration of whether the sprinkler system has successfully controlled the fire to that point in the fire. It does so by comparing any pre-established life safety or property loss criteria against instantaneous or integrated values of products of combustion (including temperatures at various levels), size of fire, or even the number of sprinklers which have operated (although this can be considered irrelevant except in terms of potential water damage).

$$= f \, (CRIT, \, a, \, pc, \, act)$$

where CRIT = pre-established life safety or property loss criteria

In the impact assessment of sprinklers completed early in the residential sprinkler program by Johns Hopkins University, it was estimated that a properly installed and maintained sprinkler system which alerted the occupants prior to the development of critical levels of smoke density, toxic gases, or temperatures, and was capable of either extinguishing the fire or controlling it until the arrival of the fire department, installed throughout all buildings with water and telephone service, could reduce deaths and injuries due to fire by up to 90 percent, and property loss by 86 percent (29).

In residential sprinkler tests conducted in 1980, the Ad Hoc Insurance Committee on Residential Sprinklers (39) did not attempt to create arbitrary levels of acceptable performance in terms of limiting property damage, but used comparison testing to evaluate the impact of the systems. Three sets of comparative testing resulted in loss ratios ranging from 1.5:1 to 10.5:1 for unsprinklered vs. sprinklered scenarios.

Budnick (12) has noted that, while the state-of-the-art does not permit a precise assessment of the direct hazard to humans of exposure to fire conditions, a number of studies provide estimates of levels at which adverse effects occur. For his study of the effectiveness of sprinklers on life safety in residential occupancies he used the following approximations:

Temperature $> 100°C$ (212°F)
Carbon Monoxide $> 8,000$ ppm or 50% COHb
Oxygen $< 12\%$ by volume
Smoke Density $> .25 - .50$ OD/m

Using NFIRS data on residential fire deaths in combination with results of sprinkler tests, it was estimated that, used in combination with smoke detectors, properly operating standard response sprinklers have the capability of reducing residential fire deaths by 56 percent, 73 percent for residential sprinklers.

IV. TEST REPORT SUMMARIES

Several test reports were selected for individual summaries. Selection was based on a number of factors. The tests may have been:

1. Especially significant to the development or testing of quick response sprinkler technology.

2. Especially helpful to the understanding and support of the model presented in Section II.

3. Of general importance but not widely disseminated. Several of the summaries are not based on published reports but on unpublished memoranda or eyewitness accounts. While not qualifying for reference in a strict interpretation of a literature survey, it was felt that the information may be useful.

It was also the intent to choose tests representative of occupancies to be addressed in the Group 2 tests, as opposed to high challenge fire test programs.

An index of reports follows. Within this section of the analysis, the reports are arranged in order of their reference in the Bibliography. These reference numbers are included in the index within parentheses following each title.

The comments which follow the statement of summary/conclusions for each report are those of the author of this analysis.

INDEX OF TEST REPORT SUMMARIES

Report: Evaluation of Combustion Atmospheres (5)

Author: Beitel, Jesse J., Southwest Research Institute

Sponsor: The Foundation for Fire Safety, Rosslyn, Virginia

Date: September, 1981

Basic Description: Combustion atmospheres were monitored during two residential sprinkler tests in mobile homes in Springdale, Arkansas on August 1, 1981.

Scenarios: The two sprinkler tests consisted of a living room scenario and a bedroom scenario. In the living room, a newspaper-filled wastebasket was ignited near the arm of an upholstered chair, situated in a corner with an end table and sofa. In the bedroom fire, the newspaper-filled wastebasket was placed near a bed. A third fire test was conducted in an unsprinklered "kitchen scenario," with a sofa placed in the center of the kitchen. A urethane foam cushion from the sofa was used along with "lumber" and "paper" to provide the fuel source, and the fire was permitted to fully involve the mobile home.

Data Collected: Collection boxes were used to sample the atmospheres for carbon monoxide (CO), oxygen (O_2), acid gases (HCl and HCN), acetaldehydes, acrolein, benzene and particulates. In the two sprinklered tests, the sample boxes were opened by remote control. In the living room test, two of the boxes (one 6 in. below the living room ceiling and one at 5 ft 6 in. in an intermediate corridor) were turned on 30 seconds into the test and turned off at sprinkler activation, 7 minutes into the test. The other two boxes (same locations) were turned on at sprinkler activation and remained on for 2 minutes. In the bedroom test, 3 boxes were turned on 30 seconds into the test (additional box located 6 in. below bedroom ceiling) and turned off at sprinkler activation 45 seconds into the test. Three corresponding boxes were started at that point and remained on for 2 minutes. In the unsprinklered test, samples were taken by placing collection boxes into the mobile home through windows using long poles.

Sprinkler System Information: The report does not provide information on the sprinkler system. A corresponding article (80) on the tests, however, indicates a design meeting the requirements of the 1980 edition of NFPA 13D, with the exception that polybutylene pipe was used for the installation.

Report Summary/Conclusions: The report draws no conclusions but merely summarizes data in tables.

The report does state that high HCl and HCN values in the unsprinklered test were "probably due to the vinyl covering and the foam padding of the sofa and chair in the mobile home."

Comments: In the two sprinklered tests, the highest carbon monoxide concentration recorded was 255 ppm in the living room during the 2 minutes following sprinkler activation in the living room test. In 7 of the other 9 samples taken during the sprinklered tests, no carbon monoxide was reported detected, and in the other 2 the concentration was less than 100 ppm. By comparison, one 15-second sample taken in the living room 10 minutes into the unsprinklered test showed 13,358 ppm of CO.

Levels of HCl in the sprinklered tests ranged as high as 79 ppm in the bedroom prior to sprinkler operation in the bedroom test. In the unsprinklered test HCl levels reached 141 ppm in the living room after 10 minutes and 345 ppm in the rear bedroom after 25 minutes. Levels of HCN in the sprinklered tests ranged as high as 5 ppm in the living room during the 2 minutes following sprinkler operation in the living room test. In the unsprinklered test HCN levels reached 624 ppm in the living room after 10 minutes and 164 ppm in the rear bedroom after 25 minutes.

Levels of acetaldehyde and acrolein were reported as "none detected" in all tests.

In the living room sprinkler test, the highest concentration of particulate matter was reported to be 0.173 g/m³ in the living room during the 2 minutes following sprinkler activation. In the bedroom test, the highest concentration of 0.233 g/m³ was recorded in the bedroom following sprinkler activation. In the unsprinklered test, the highest concentration of 7.445 g/m³ was recorded in the living room 10 minutes into the test.

Report: "Operation San Francisco Smoke/Sprinkler Test Technical Report" (7)

Author: Benjamin/Clarke Associates

Date: April, 1984 report on tests conducted October, 1983

Basic Description: A total of 16 tests were conducted to explore the comparative effects of using standard and quick-response sprinklers as opposed to no sprinklers, as well as the effects of three alternate HVAC systems in simulated exhibit areas.

Scenarios: Hotel exhibit area with 9 ft ceiling height (4 QR, 3 standard, 1 unspr.), Hotel powder room (2 QR varying locations), Hotel room (1 QR, 1 unspr.), Corridor (1 QR, 1 unspr.), Apartment (1 QR, 1 unspr.), and Jail Cell (1 standard, 1 unspr.).

Data Collected: Varied with scenario. In general: Ceiling and 5-foot temperatures, temperature profiles at doorways, optical density at room centers and doorways, gas concentrations (CO, HCL and HCN) at central 5-foot levels and doorways.

Sprinkler System Information: RTI values given as 22(English) for QR sprinklers, 400 for standard sprinklers. Design densities estimated at between .21 gpm/ft to 1.92 gpm/ft, based on hydraulic calculations of system. (Note: RTI value for QR sprinklers can be presumed metric).

Report Summary/Conclusions: The report cautions that laboratory conditions were not possible, and that all conclusions are general qualitative observations. Nevertheless, it determined that 1.) Improving the response time of sprinklers in the exhibit area tests was more effective in smoke control than alterations of the HVAC modes investigated, 2.) There was a marked improvement in life safety conditions in sprinklered fires as compared to unsprinklered fires, 3.) In the exhibit area tests the quick response sprinklers had a measureable effect on the fire size and amount of smoke generated as compared to standard sprinklers, 4.) Quick response sprinklers do not guarantee "viability" in a slow-developing hotel room fire, although corridors are maintained tenable.

Comments: The report is valuable in terms of its qualitative analysis of the smoke control abilities of quick response sprinklers, but is not definitive as to the type of sprinkler used in each test, its sensitivity, orientation, or discharge density. Local densities are reported as having been estimated using hydraulic calculations, but information is not available as to water supply characteristics, sprinkler K-factors, or piping friction loss factors employed.

Report: "Home Sprinkler and Fire Alarm Tests" (14)

Author: Donald R. Ciardelli

Sponsor: Bloomington, Minnesota Fire Department

Date: June 21, 1975

Basic Description: Tests were conducted in a single family dwelling to investigate the effectiveness of smoke detectors, sprinklers, and sprinklers with quick response attachments.

Scenarios: Seven tests were conducted: 1a) basement with "U.L." wood crib, 1b and 2) basement with a second wood crib stacked on top of another, 3) miscellaneous trash and diesel fuel underneath basement stairs to simulate arson fire, 4) wastebasket (with fuel oil added) ignition of mattress in bedroom, 5) Kitchen range grease fire, 6) flaming ignition of living room couch.

No outside ventilation was used in the tests, but doors within the building were left open.

Data Collected: Test chronology and times of operation of detectors and sprinklers, amount of water needed to extinguish fire, temperature profiles from 28 thermocouples within the house, smoke obscuration readings on each of 3 floors, gas sampling devices in the basement and in the first floor kitchen to record levels of CO, O_2, CO_2, and burnable hydrocarbons.

Sprinkler System Information: The water flow to the sprinklers was metered and throttled to maintain 10 to 15 psi at a flowing sprinkler. Sprinklers used in the tests were standard 165°F pendent sprinklers and Grinnell model Q-34 extended coverage horizontal sidewall sprinklers. Some models of each type were equipped with Grinnell quick response attachments.

Report Summary/Conclusions: In the case of quick burning fires the (standard response) sprinklers proved "unexcelled." In "the slow smouldering fire they lose some of their effectiveness and smoke detection is needed."

In several tests, the effects of open stairways on heat and smoke movement were evident in either delaying or speeding up the operation of sprinklers. The report contains the statement that this "verifies and strengthens previous knowledge that locations of detectors and sprinklers must be engineered."

Comments: These tests are of interest in that they included some comparison of standard sprinklers with sprinklers equipped with the "quick response actuators." A summary of tests results is as follows:

Test 1A — Fire never operated 165°F pendent sprinkler due to escape of heat up open stairway, but an adjacent "telltale" 165°F sprinkler with 135°F QR attachment operated 9 minutes into the test. At this point ceiling temperature was 140°F. Hazardous levels of smoke obscuration, carbon monoxide, etc. did not develop.

Test 1B — Although "telltale" sprinklers with the QR attachment operated at the 4:07 and 6:51, the nearest 165°F standard sprinkler operated at 8:07 to control the fire and prevent the development of hazardous conditions. Sprinkler operation "increased smoke but pushed it to the floor. Smoke did not seem to travel upstairs after sprinkler operated."

Test 2 — repeat of Test 2 but with water supplied to QR sprinklers. Operation at 4:50. An observer within the basement (room of origin) stood throughout the test and reported no difficulty breathing.

Test 3 — A QR sprinkler located at the top of the basement stairs operated at 1:23, pushing the heat back into the basement and holding the fire under the stairs, where it was shielded from sprinkler discharge. The only other sprinkler in the test, a QR sprinkler within the basement and nearer to the fire, operated at 2:45.

Test 4 — A "telltale" 165°F sidewall sprinkler operated at 2:55, and the adjacent 165°F extended coverage sidewall supplied with water operated at 3:30, at which time smoke had banked 4 ft down from the ceiling. Due to the fast growth of the fire and small size of the room, test observers indicated the QR sprinklers would have had a significant impact in preventing the 400°F ceiling temperature and 55 percent/ft light obscuration. CO levels in the adjacent kitchen reached about 750 ppm.

Test 5 — Although the "telltale" sprinkler with QR attachment operated at 2:15, the adjacent 165°F pendent did not operate to control the fire until 4:07, by which time the main kitchen cabinets were burning. The fire was extinguished for the most part within 10 seconds.

Test 6 — An extended coverage sprinkler with QR attachment operated at 1:10 in this fast burning fire. The fire was nearly completely extinguished within 1 minute, and hazardous conditions did not develop.

This is the only known field report of testing comparing the response of standard sprinklers to that of standard sprinklers equipped with the quick response attachment. The performance of this attachment in a plunge test was later used by Underwriters Laboratories to provide the definition of "quick response."

Report: "Field Test and Evaluation of Residential Sprinkler Systems" (16)[1]

Author: Arthur E. Cote, National Fire Protection Association

Sponsor: U.S. Fire Administration

Date: July 1982 report (revised December 1982) on tests conducted from 1979 to 1981.

Basic Description: Report on the Los Angeles residential sprinkler test series and Charlotte, North Carolina mobile home test series, both of which were intended to validate the use of the "quick response" residential sprinkler in controlling residential home fires. In addition, the tests were intended to answer questions regarding the influence of certain variables on residential sprinkler system performance.

---

[1]This report includes the test results also reported in "Field Evaluation of Residential Prototype Sprinkler - Los Angeles Fire Test Program," by Factory Mutual Research (44)

Scenarios: Sixty full-scale tests were conducted in the Los Angeles single family home test series, and sixteen full-scale tests were conducted in the Charlotte mobile home test series.

Los Angeles Phase I (Tests 1-36) — included 20 tests corresponding to the L.A. smoke detector tests plus 16 combustible shielded living room corner tests.

Los Angeles Phase II (Tests 37-60) — investigated the effects of smaller room of origin, lintel, fire centered between sprinklers, and effect of standard sprinklers.

North Carolina Mobile Home Series (Tests NC 1-16) — tested the design found successful in the L.A. series in a mobile home.

Data Collected: Both the dwelling and mobile home were instrumented to obtain temperatures 36 in. and 60 in. above the floor and 3 in. below the ceiling, CO levels at 36 in. and 60 in. above the floor, O_2 at 60 in. above the floor, and optical density 60 in. above the floor. Gas velocity and temperature probes were located adjacent to simulated sprinkler links at strategic points.

Sprinkler System Information: The sprinkler used in all tests was the residential prototype developed by the Grinnell company. Testing of 18 samples used in the Los Angeles test series showed an average RTI value of 26.75 $m^{1/2}sec^{1/2}$, although the RTI values of the samples ranged from approximately 21 to 30. For the mobile home tests, the mean value of measured RTI was 28 $m^{1/2}sec^{1/2}$. The temperature rating was 140°F, and the sprinklers had a K-factor of 2.9. Because the sprinklers were prototypes, variation in water distribution patterns were reported to be considerable.

In the Los Angeles tests, sprinklers were tested at spacings of approximately 100, 150 and 144 sq ft per sprinkler. In the mobile home series spacings of 144 and 150 sq ft were investigated.

Report Summary/Conclusions: The results of the various test series were reported as follows:

Los Angeles Phase I. Three flaming-started kitchen fires — 2 were controlled by a single sprinkler spaced at 135 sq ft flowing 12 gpm. In the other the fire self-extinguished. Limits for tenability were not exceeded.

Three smoldering-start bedroom fires — 2 were extinguished by the activation of two sprinklers, each protecting 64 sq ft, with a 1-sprinkler flow of 12 gpm, and a 2-sprinkler flow of 9 gpm each. With windows and door open, tenability was maintained. In the third test transition to flaming never occurred within the 4 hours it took to activate the first sprinkler, and the CO concentration exceeded the critical integrated value after 112 minutes. This series demonstrated the need to have smoke detectors in conjunction with residential sprinklers.

Two flaming-start closet fires — 1 with a sprinkler in the closet flowing 12 gpm was extinguished while maintaining tenability. Without a sprinkler in the closet, 6 sprinklers activated and manual extinguishment was required. Critical limits were exceeded. This showed that sprinklers outside the closet are ineffective in controlling a closet fire.

Six flaming-start noncombustible living room fires — 5 were controlled by a single sprinkler protecting 147.3 sq ft flowing 12 gpm, and 1 was controlled by 3 sprinklers, each flowing 9 gpm. Critical limits were not exceeded.

Six smoldering-start noncombustible living room fires — all 6 were controlled by the activation of a single sprinkler flowing 12 gpm. Smoldering periods ranged from 23 minutes to more than 5 hours, but critical limits for tenability were not exceeded. A large amount of water could be delivered to the burning surface since the sofa was located less than 3 ft from the sprinkler axis.

Sixteen flaming-start combustible living room fires — 2 tests conducted at sprinkler spacing of 9 ft × 14 ft 11 in. were unsuccessful in controlling the fire. (Note: Additional tests were conducted at various spacings and flows to find a successful combination. Ultimately, this combination was determined to be a spacing of 144 sq ft (12 ft × 12 ft) with a 1-sprinkler flow of 18 gpm and a 2-sprinkler flow of 13 gpm each.) In tests in which the sprinklers successfully controlled the fire, critical limits for tenability were not exceeded. In some cases the maximum eye-level temperatures and CO concentration at the upstairs landing exceeded the values in the room of origin due to the open stairway. Closed bedroom doors were effective in maintaining low temperature and CO levels in those rooms.

Los Angeles Phase II. Four flaming-start living room tests with fire centered between 4 sprinklers — in 3 of the tests, with a total flow to the sprinklers of 18 gpm to 23 gpm, the fire was controlled by 2, 3, and 4 sprinklers. In these tests, critical tenability limits were not exceeded. Having more sprinklers closer to the fire appeared to cause more sprinklers to operate, but allowed water to be delivered sooner, controlling the fire. In the only test where the fire was shielded, 5 sprinklers opened and the test had to be terminated.

Four flaming-start combustible living room fires with the size of the room of origin reduced — in all 4 tests sprinklers spaced at 85.5 sq ft controlled the fire with a 1st-sprinkler flow of between 12 and 19 gpm, and with 1 to 3 sprinklers activating. Critical limits were not exceeded.

Five flaming-start combustible living room fires with 8-inch lintels — in 2 tests with 85.5 sq ft spacing the fire was controlled by 1 sprinkler flowing 18.7 gpm. In 2 tests with 144 sq ft spacing the fire was controlled by 1 sprinkler flowing between 15.6 and 18.7 gpm. In these four tests it had been expected that the lintel would prevent more than 2 sprinklers from operating. In 1 test with spacing of 9 ft 1 in. × 14 ft 11 in. (135 sq ft) and total flow of 18.1 gpm, 5 sprinklers operated and the fire was not controlled. Comparison of these test results with earlier tests did not permit conclusions to be drawn about the ability of 8-in. lintels to prevent additional sprinklers from operating.

Four flaming-start combustible living room tests using sidewall sprinklers — in 1 test with a sprinkler spacing of 15 ft × 18 ft and flow of 29 gpm to an extended coverage sidewall sprinkler with standard response 165°F link, a fire 18 ft opposite the sprinkler was not controlled. In a second test identical to the first but with the fire 4 ft from the sprinkler along the same wall, the fire was controlled. With a spacing of 10 ft × 15 ft and a flow of 26 gpm, a fire located opposite the sprinkler was not controlled when the sprinkler was equipped with a 135°F standard link, but was controlled when the sprinkler was equipped with a 140°F quick response link. In the 2 tests where the fire was controlled, temperatures briefly exceeded the 150°F limit (by 8° to 13°F) at the upstairs landing. Critical limits were exceeded in the other 2 tests. Water supply to the sidewall sprinkler was not reduced when other sprinklers operated in these tests.

Seven flaming-start combustible living room tests using standard response sprinklers (135°F, RTI = 136 m$^{1/2}$sec$^{1/2}$) — 4 tests using a sprinkler spacing of 12 ft × 12 ft were unsuccessful. In 2 tests, spacing at 85.5 sq ft with total system flow of 22 gpm "achieved marginal success," but an identical third test using a concealed model of the sprinkler did not.

North Carolina Mobile Home Series. One flaming-start kitchen fire — 1 sprinkler spaced at 149 sq ft and flowing 18 gpm was not able to control the fire.

Three flaming-start combustible bedroom fires — in all 3 tests 1 sprinkler protecting a 117 sq ft room and flowing between 15 gpm and 18 gpm controlled the fire. Critical limits were not exceeded.

Eleven flaming-start combustible living room fires — at a spacing of 11 ft 8 in. × 13 ft (an area of 151.7 sq ft) and a total system flow of 22 gpm, 1 sprinkler controlled the fire in 2 tests, the fire was not controlled in 1 test, and 1 other test was ruled invalid. As in the Los Angeles Phase I tests, a number of additional tests investigated variations of spacing and flows to determine the optimum combination. Again it was found that where the sprinklers controlled the fire, critical limits for tenability were not exceeded.

The following conclusions were drawn as a result of the above tests:

In the flaming-start tests, the critical limits for tenability were not exceeded for those tests in which the sprinkler system controlled the fire.

In the smoldering-start fire tests, the critical limits established for tenability were not exceeded for those tests in which the transition to flaming occurred and the sprinkler system controlled the fire.

In one smoldering-start fire the transition to flaming never occurred and the critical limits for tenability were exceeded.

In all other tests in which the sprinkler system was not able to control the fire the critical limits established for tenability were exceeded.

The report additonally discusses a kitchen fire in the North Carolina mobile home series which was ruled invalid when water was inadvertently held back for a period of 17 seconds following the opening of the first sprinkler due to a closed valve. A critical eye-level temperature of 151°F was reached in the kitchen 110 seconds after the sprinkler activated, and the sprinkler was considered incapable of controlling the fire.

Comments: These tests served a dual purpose in the residential sprinkler program. Originally intended to confirm the effectiveness of the residential sprinkler prototype and proposed NFPA 13D design criteria, the tests also took on a basic research role when it was found that the proposed design criteria (.08 gpm/ft² for the first sprinkler operating, .06 gpm/ft² for each of two sprinklers operating, with each sprinkler protecting an area of up to 150 ft²) were not effective in controlling the "worst case scenario" of the shielded corner fire in a living room with combustible walls and ceiling.

A significant finding of this test series was that, where the sprinkler system controlled the fire, critical limits for tenability were not exceeded. These tests were conducted in areas with ceiling heights of from 7.5 to 8 ft.

This test series was also important in establishing the design criteria for residential sprinkler systems approved as NFPA 13D-1980 at the Fall Meeting of the National Fire Protection Association on November 20, 1980.

Report: Field Test of a Retrofit Sprinkler System (17)

Author: Arthur E. Cote, National Fire Protection Research Foundation

Sponsors: FEMA, Grinnell, Shell, Simplex, FM, Marriott

Date: February 1983 report on tests conducted September 1982

Basic Description: A series of 11 full-scale hotel room fire tests were conducted in a vacant 5-story fire resistive hotel in Fort Lauderdale, Florida, to evaluate the effectiveness of quick response sidewall sprinklers used in conjunction with a polybutylene piping system in a retrofit system installation.

Scenarios: Fast-flaming (flammable liquid) fire in guest room (2 QR), flaming-start fire (wastepaper basket) in guest room (2 QR), fast-flaming (luggage) fire in guest room (1 QR, 1 unspr.), fast-flaming (flammable liquid) fire in corridor (1 QR), flaming-start (maid cart) fire in corridor (1 QR), flaming-start (maid cart) fire in storage room (1 QR), smoldering-start fire in guest room (2 QR). Most tests involved no ventilation, although the room doors were left open in two flaming-start guest room fires.

Data Collected: CO at 3 ft and 5 ft levels, HCN at 5 ft level, HCl at 5 ft level, smoke obscuration at 5 ft level, temperatures at 5 ft level, 3 in. from ceiling, at sprinklers, and at surface of polybutylene piping.

Sprinkler System Information: RTI value for QR sprinklers given as 47 (English units), with 140°F temperature rating. Single sidewall sprinklers in guest rooms covered areas of 242 square feet. With one sprinkler flowing, the flow of 37 to 40 gpm produced design densities averaging 0.15 to 0.165 gpm/ft². Pendent sprinklers spaced 12 to 14 feet on center in corridor protected 54 square feet per sprinkler. Flow from single sprinkler of 35 to 38 gpm produced design densities averaging 0.65 to 0.70 gpm/ft².

Report Summary/Conclusions: In the 8 fast-flaming and flaming-start fires, the QR sprinklers controlled the fire and critical limits for survivability were not exceeded. In the 2 smoldering-start fires, insufficient heat was generated to activate sprinklers. In the fast-flaming-start fire without sprinklers, the critical limits for survivability were exceeded.

Comments: These tests were useful in demonstrating the effectiveness of QR sprinklers in maintaining survivable conditions in flaming and fast-flaming fires. The scenarios were limited to a hotel occupancy, however, and the information on water application rates on fire hazards under different ventilation conditions was limited. Information on number of sprinklers operating is also limited, since only in the corridor tests were there multiple sprinklers in the same area.

The smoldering-start fires provided almost no information. In one, the test was terminated after 2 hours without even a smoke detector actuation. In the other, maximum CO concentrations reached 160 ppm at 77 minutes into the test, at which time there was an equipment malfunction. The first smoke detector operated at 36 minutes.

Unpublished Memoranda: Life Hazard Reduction (LHR) Tests (21)

Author: Dosedlo, Lee

Sponsor: Underwriters Laboratories

Date: March-October, 1974

Basic Description: Unpublished test work dealing with about 200 tests conducted during the attempted development of a standard for Life Hazard Reduction (LHR) sprinklers.

Scenarios: In the initial 75 tests, various sizes of wood cribs (44, 20, 16, and 12 lb) in different locations within a 16 ft × 16 ft × 8 ft room were ignited using various means, including excelsior and alchohol. Tests were repeated 3 times to establish repeatability. Freeburn tests were also conducted for comparison purposes. Tests involving alchohol pan fires, shredded newspaper in wire baskets, and foam buns (to represent stuffed furniture) were also conducted. In later tests, room size was varied from 12 ft × 12 ft to 16 ft × 28 ft. Additional fuels were investigated, such as heptane pans, and additional ignition sources were employed, including wicks, burning boards, and matches.

Data Collected: In the early tests, continuous recordings of temperature at 3 separate levels and 6 locations in the room were made, along with recordings of smoke density and oxygen content. Height of flames was observed and recorded, as was sprinkler operating time, time to fire extinguishment, and crib weight loss. The responses of smoke detectors and rate of rise detectors in the test room were also recorded. Later in the test program, measurements of carbon monoxide levels were initiated. Temperatures continued to be recorded at 16 points in the room, including the 5-½ ft level at the "quarter points" of the room.

Sprinkler System Information: Five different styles of listed sidewall and pendent sprinklers were tested for distribution and response time. The sprinkler flow rates were varied between 25, 20 and 15 gpm. For a corner location of a 16 lb wood crib, response times of sprinklers were also varied, using open sprinklers and manual control of the water supply.

Summary/Conclusions: No report on this work was published. Difficulties were reportedly encountered in setting threshold values for tolerable limits of toxic gases due to the state of the art at that time. As such, it was not practical to attempt to use test results to justify the use of the term "Life Hazard Reduction." The final internal correspondence notes that a "second generation LHR sprinkler may come along at a later date to incorporate smoke detection after definite limits on tenable values of toxic gases are established." This correspondence also included the statement that "...it was determined that in order to detect a small smoldering fire in upholstered foamed plastic furniture which produced excessive amounts of toxic gases (i.e., in excess of 3,000 ppm) prior to detection and activation of any sprinkler tested to date, several sensitive smoke detectors would be necessary."

Comments: It appears that the need for improved sprinkler sensitivity to control fires in small compartments prior to the development of unacceptable levels of products of combustion was perceived from the results of this test program, but that the drastic change in product technology that would be required was not considered feasible at the time. Early into the testing, it was noted that sprinklers listed at 135°F were found to operate at air temperatures of 170°F "due to the retarding effect of the mass of metal adjacent to the link." While data from the later portion of the test program is not available, some of the early results of crib fire tests in the 16 ft × 16 ft room show that the effects of varying sprinkler response time were recognized:

PARTIAL TEST RESULTS FROM 16LB WOOD CRIB LOCATED AT BACK LEFT CORNER
(ignition by excelsior, flow rate 25 gpm)

Time Water On	Time Water Off	Time to 150 °F at 5½ ft	Max Temp at 5½ ft	Time to 17% O$_2$	Min O$_2$	Crib Wt Loss %
0:30	10:00	—	80 (0:30)	—	20%	7.4
1:00	10:00	—	95 (1:00)	—	20%	10.3
1:30	10:00	—	110 (1:30)	—	18.5%	23.0
1:50	10:00	—	135 (1:50)	—	18%	22.4
2:00	10:00	—	140 (2:00)	8:25	17%	27.0
2:15 (auto)	6:00	—	150 (2:15)	*	17.5%	18.6*
2:30	10:00	2:10	185 (2:30)	7:15	16.5%	29.8
(freeburn to 10:00)		2:20	260 (5:00)	—	17%	43.1

* Note that this test, in which the sprinkler was permitted to activate automatically, was conducted for only 6 minutes as opposed to 10 minutes for the others. Had the test continued, it is likely that results would have fallen into line with the others.

Report: Large-Scale Tests to Study Sprinkler Sensitivity (27)

Authors: David G. Goodfellow and Joan M. A. Troup

Sponsors: Factory Mutual Research Corp., Allendale Mutual Insurance Co., Arkwright-Boston Manufacturers Mutual Insurance Co., Digital Equipment Co., Eastman Kodak Co., The Viking Corporation

Date: March 1983 report

Basic Description: Six large-scale tests were conducted in rack storage of a standard plastic commodity to study the effect of sprinkler sensitivity and temperature rating on sprinkler performance.

Scenarios: For tests 1 through 4, two double-row rack storage segments were separated by an 8 ft aisle. The 19 ft high array had four tiers of palletized commodity, crystalline polystyrene tubs in compartmented corrugated paper cartons. Clearance to the ceiling was 10 ft. Ordinary wooden pallets supported 8 cartons each, and each carton was a 21-inch cube. Each of the main racks consisted of five 8 ft long bays with a loaded length of about 32 ft. Target racks consisted of a 4-bay rack and a 2-bay rack centered on the longer rack. Longitudinal flue 6 in. wide separated the racks, transverse flues 6 in. wide separated the loaded pallets, and spaces between tiers were 12 in. Test 5 was identical except that the double-row racks were separated by a 6 ft aisle. Test 6 also used a 6 ft aisle, but with a fifth tier of storage forming a 24 ft high array. The main racks were shortened to 3 bays, and clearance to the ceiling was 5 ft. All doors and windows in the test center were closed. Ignition utilized two 3 in. diameter 6 in. long cellucotton rolls each soaked in 8 oz of gasoline, located at the base of the main array in the center transverse flue space, centered among 4 sprinklers.

Data Collected: The main array racks were placed on a weighing platform. Measurements were also made of water flow through the sprinkler system, water pressure at the system crossmain and a selected branch line, operating time of each sprinkler, and environmental conditions. Three calorimeters were installed in each test, located on the aisle face of the target rack opposite the ignition transverse flue space of the main array at the second, third, and fourth tier levels so as to measure heat transmission across the aisle space in terms of a combination of radiative and convective heat.

Sprinkler System Information: Sprinklers were spaced at 10 ft × 10 ft with deflectors 7 in. below the ceiling, and operated at a constant water pressure of 50 psi for all except Test 5, in which the sprinklers were operated at 40 psi. Sprinklers were Viking 0.64-inch orifice upright, resulting in a flow at each sprinkler of approximately 80 gpm for most tests, 72 gpm for Test 5. Sprinkler temperature rating and RTI were varied independently. Values which follow are approximate.

Test No.	Temp. Rating (°F)	RTI $(ft^{1/2}sec^{1/2})$
1	286	300
2	165	50
3	286	50
4	165	300
5	165	50
6	165	50

Report Summary/Conclusions: Although the report does not include a specific statement of conclusions, the abstract notes that "for the fire challenges and levels of protection studied, the use of sprinklers rated at 163°F, 48 $ft^{1/2}sec^{1/2}$ in place of sprinklers rated at 286°F, 309 $ft^{1/2}sec^{1/2}$ reduced control time from 16 min to 2 min; fuel consumption from 24 percent to 3 percent by weight; and water demand from 1400 gpm to 250 gpm."

Comments: Perhaps the most dramatic result of the comparative tests was found in the number of sprinklers operated. The high temperature rating/low sensitivity (high RTI) combination of Test 1 operated 18 sprinklers. Reducing only the RTI in Test 3 or only the temperature rating in Test 4 resulted in operation of 14 and 12 sprinklers respectively. Reducing both the temperature rating and RTI in Test 2 resulted in operation of only 3 sprinklers. This same combination of temperature rating and RTI resulted in only 4 and 3 sprinklers opening as aisle width, application rate and fuel loading were changed somewhat in Tests 5 and 6.

This test series, using the "large drop" sprinkler with its plume penetration abilities in combination with quick response, provided the primary impetus for the Factory Mutual program to develop an Early Suppression Fast Response sprinkler.

Interview: Cobb County Residential Sprinkler Tests (38)

Investigators: D. Hilton, J. Grier et al, with support from Georgia Institute of Technology

Sponsor: Cobb County, Georgia Fire Department

Date: 1982 - 1984

Basic Description: Over 200 tests of residential sprinklers conducted to demonstrate effectiveness of residential sprinklers and to investigate variables of spacing, location and obstructions.

Scenarios: Most scenarios involved wastebasket-ignition living room fires or grease pan kitchen fires.

Data Collected: Data collected on temperatures, sprinkler operating times and other factors has not been released.

Sprinkler System Information: Sprinkler systems tested are in basic conformance with the 1980 edition of NFPA 13D, with the exception that polybutylene piping was used even prior to its listing. Also, some of the tests involved the use of the original Grinnell model F954 pendent residential sprinkler at a spacing of 16 ft × 16 ft, beyond its listed spacing of 12 ft × 12 ft.

Summary/Conclusions: There has not been any report issued of these tests and therefore are no official conclusions. Some of the unofficial conclusions are found in local modifications of NFPA 13D, such as the use of the above-mentioned Grinnell residential sprinkler at spacings of up to 16 ft × 16 ft, beyond its UL listing. The Cobb County officials felt that the performance of the sprinkler was adequate without the high wall-wetting achieved by the 12 ft × 12 ft spacing. They are not concerned with the potential perimeter fire due to the common use of noncombustible finish and fire-retardant draperies. Other unofficial conclusions are reflected in proposed amendments to the "sprinkler amendment" originally enacted in February of 1983 to encourage the use of sprinklers in multi-residential occupancies. Under the proposed changes, for rooms or areas equipped with more than 2 sprinklers, the design area must include all of the sprinklers up to a maximum of 4 sprinklers. Sidewall sprinklers are limited to a maximum distance of 4-½ inches from the ceiling, although the minimum remains at the NFPA 13D limit of 4 inches. Where ceiling fans are used, rooms larger than 9 ft × 9 ft require at least 2 sprinklers.

Comments: Until such time as a report of testing is written (if ever) the extensive Cobb County testing will be of little more than anecdotal value. There are some exceptions, where the qualitative information is interesting, such as the ceiling fan tests. It was found that under summer conditions (windows open), the fan action did not delay the sprinkler response as expected by the Cobb County officials but seemed to hasten the operation of the sprinklers by blowing the heat toward the sprinklers (fan located midway between 2 sprinklers). Under simulated winter conditions (windows closed) the fan again did not have the expected effect. The Cobb County officials felt that the fans would hasten operation, but instead operation was delayed. Rather than push the heat in the direction of the sprinklers, the fan reportedly maintained an even mixing of the air temperature in the upper part of the room. When the sprinklers finally operated, they did so at almost the exact same time. The limit of 9 ft × 9 ft coverage for a sprinkler in a room containing a ceiling fan is based on concern for disruption of the sprinkler spray pattern by the moving fan blades. The limit of 4-½ inches proposed for maximum distance of sidewall sprinklers below the ceiling is also useful. Since it reflects testing in which sprinklers located at a greater depth were very slow to respond to a remote fire, it points out a concern which can be given further consideration by the testing laboratories.

Report: Report on 1980 Property Loss Comparison Fires (39)

Author: Jackson, Ralph J.

Sponsor: Ad Hoc Insurance Committee on Residential Sprinklers

Date: November, 1980

Basic Description: Fire tests were conducted to investigate the impact of residential sprinklers on property losses.

Scenarios: Three scenarios were used: a cigarette ignition of a living room couch, a wastebasket ignition of a bed and nightstand, and an electric coffee maker ignition of a kitchen. Each scenario was conducted both sprinklered and unsprinklered for a total of 6 tests.

Data Collected: Qualitative assessments of the fire development over time were accompanied by total amounts of water used for extinguishment and estimated claim payments which would result from the fire incident.

Sprinkler System Information: Sprinkler systems were installed essentially in compliance with the 1980 edition of NFPA 13D.

Report Summary/Conclusions: It was the observation of the ad hoc committee that these (1980 13D) sprinkler systems definitely have the ability to reduce claim payment expenses. The committee noted further that the amount of discount recommended in 1977 by the Insurance Services Office (5 percent) was less than what seemed to be indicated as a result of these tests.

Claim estimates were assembled by juries of three experienced insurance claim representatives. The cost of additional living expenses was included at $150 per day where the house could not be used immediately after a brief clean-up.

The ratio of total damage of the non-sprinklered to sprinklered fire scenarios were as follows:

Living Room	4:1	($34,110 to $8,617)
Bedroom	1.5:1	($8,650 to $6,077)
Kitchen	10.5:1	($35,245 to $3,364)

Comments: On the average, the sprinklered scenarios sustained less than one-fourth the damage of the unsprinklered scenarios. The wide range, however, shows that the particular conditions of the fire determine the property loss role played by sprinklers. The sprinklered fires had a much more uniform cost of damage, presumably due to the size of the fire at sprinkler actuation in combination with a standard duration of water flow prior to fire department arrival and system shut-down. The unsprinklered fires showed a wider variation due to the potential for rapid fire growth.

Report: Residential Sprinkler Protection Study (40)

Author: H. C. Kung, Factory Mutual Research

Sponsor: National Bureau of Standards

Date: November 1975

Basic Description: Research into the ability of sprinklers to control residential fires through the cooling mechanism. Some 36 hexane pool fires were used to evaluate effects of water discharge rate and droplet size, following which a realistic bedroom fire test was used to evaluate the optimum sprinkler orifice/discharge rate suggested by the hexane pool tests.

Scenarios: A 10 ft × 12 ft × 8 ft high test room was constructed, with a window and a door leading to a simulated hallway. Hexane pool fires with diameters ranging from 12 to 36 inches were used, with location of fire, degree of ventilation and sprinkler operation time varied along with water discharge rate and sprinkler orifice diameter. In the bedroom test, a flaming ignition of a bedding fire was used.

Data Collected: Heat release rate of fire was calculated from continuously-recorded weight loss of fuel. Various energies lost to walls, floors, ceiling, and through openings were computed, using 69 thermocouple measurements. Six measurements of gas velocity through the window, a single point measurement of CO, CO_2, and O_2 in the window, and one radiation measurement in the window were also included.

Sprinkler System Information: Five geometrically similar sprinklers were fabricated, with orifice diameters ranging from 0.164 in. to 0.437 in. (½-inch nominal). Various operating pressures were used to investigate cooling ability.

Report Summary/Conclusions: The conclusions resulting from the hexane pool tests were that 1) the ratio of the heat-absorption rate by sprinkler water versus the heat-release rate of the fire, normalized with the water discharge rate, varies as the minus 0.68 power of the relative median drop diameter, a relationship established in the series of tests with window open but door closed and the fuel pan located at the corner of the room; 2) gas temperatures inside the room and outflow gas temperature were lower with the use of smaller orifice sprinklers for a constant discharge rate; 3) an experimental sprinkler with a 0.218-inch diameter orifice operating at 39.0 psi and discharging 7.1 gpm was capable of satisfying the adopted test criteria for temperature control under all tested ventilation conditions, and was recommended as optimum for a residential sprinkler system, while awaiting further tests with realistic solid fuels.

The results of the bedroom test were concluded to demonstrate that the 0.218-inch orifice sprinkler, operating at the suggested flow and pressure, was capable of controlling a realistic bedroom fire, protecting the structure, preventing fire spread to other rooms, and providing adequate life safety protection for the flaming fires investigated. It was noted that a peak in the gas temperature at eye level to 150°C could have been prevented by use of a faster responding sprinkler.

The report recommended further investigation of the importance of 1) prewetting of combustibles to prevent further fire spread and 2) direct contact of water droplets with the burning surface to prevent the further generation of combustible vapor.

Comments: Since the object of the tests was to effect extinguishment by cooling, the use of hexane as a fuel source was considered appropriate. Being immiscible with water, and having a lower density and boiling point, it cannot be extinguished through contact. Large water droplets passing through the fire plume and into the pans of hexane were totally ineffective, since they could not even be converted to steam.

Three tests were conducted with all conditions equal (36-in diameter fuel pan, 11.2 gpm discharge, 0.274-inch orifice) except the sprinkler operation delay times, which were 20, 40 and 60 seconds. Five seconds prior to sprinkler operation, the heat release rates for the three were calculated as 42,000 Btu/min, 80,000 Btu/min, and 82,500 Btu/min. The average heat release rate of the fire for the minute beginning 50 seconds after sprinkler operation was 42,000 Btu/min, 40,500 Btu/min, and 44,500 Btu/min. The total convective heat flux out the window 80 seconds after sprinkler operation was 17 percent lower for the test with the 20-second delay than for the other two tests. As a result of the above, the report noted that the heat release rate and convective heat flux through the room opening during a quasi-steady state after sprinkler operation appeared insensitive to the sprinkler operation delay for these pool fires.

In eight tests conducted with both the window and door open and the fuel pan located in the corner of the room, none of the fuel pans was extinguished by the sprinkler, but continued to burn until the hexane was totally consumed. In fires with the window open but the door closed, or with both closed, extinguishment could be obtained with decreasing discharge rates if smaller orifice diameters and higher discharge pressures were used.

With the window open in the realistic bedroom fire test, a discharge of 7.1 gpm over 120 square feet (0.06 gpm/ft²) was able to control the fire with standard sprinkler response and a fire heat release rate which averaged 60,100 cal/sec (14,400 Btu/min) during the 20-second period just prior to sprinkler operation. For the one-minute period starting 58 seconds after sprinkler operation the average heat release rate was calculated as 5,500 Btu/min.

Report: Development of Low-Cost Residential Sprinkler Protection: A Technical Report (41)

Authors: H. C. Kung, David M. Haines, and Robert E. Greene, Jr. (FMRC)

Sponsor: National Fire Prevention and Control Administration

Date: April 1978

Basic Description: A program was undertaken to develop a low-cost residential sprinkler system requiring small discharge rates yet providing protection for life and property.

Scenarios: A series of 9 living room fire tests were conducted within a 741 sq ft apartment area constructed for the tests, varying sprinkler orifice size and water discharge pressure. A 58-ft corridor served the apartment. Living room furnishings consisted of a couch and a room divider. Smoldering-start fires were set on the couch using cigarettes or heating coils. All windows and doors were closed except doors to the bedroom and bathroom. Air conditioners in the living room and bedroom were run in the heating mode, and exhaust fans in the bathroom and kitchen were on. Walls and ceiling were noncombustible.

Data Collected: Sixty-two thermocouples provided temperature information, CO, CO_2, and O_2 were sampled at eye level, optical density was measured at eye level and 3 in. below the ceiling, and continuous weight loss of fuel was monitored.

Sprinkler System Information: Seven geometrically similar experimental sprinklers with orifice sizes ranging from 0.110 in. to 0.438 in. were activated by a fast response commercial link by means of a solenoid valve. The link had a temperature rating of 165°F and a time constant of about 120 seconds at a 5 ft/sec gas velocity. Three discharge pressures were investigated: 8, 28.5, and 88.5 psig. Three different orifice sizes were used at each pressure, resulting in three discharge rates. Design densities ranged from 0.023 gpm/ft² to 0.1 gpm/ft² for the 150 sq ft protection area per sprinkler. Local densities over the area of the couch ranged from 0.007 gpm/ft² to 0.081 gpm/ft².

Report Summary/Conclusions: Sprinklers were shown capable of life safety and property protection in a common fire scenario leading to deaths in residential fires. Water discharge rates considerably below those of NFPA 13D (0.1 gpm/ft²) were shown to be adequate. Water density distribution was identified as an important factor in controlling CO generation. One low-pressure and one high-pressure combination of orifice diameter and operating pressure were recommended for consideration in low-cost residential systems: a 0.274-in. sprinkler operating at 8 psig (6 gpm flow) and a 0.137-in. sprinkler operating at 88.5 psig (4.7 gpm flow).

Comments: In the tests, transition to flaming typically occurred after a smoldering interval of 60 minutes. The generation of smoke and CO is reported as becoming vigorous only in the final 15 minutes before flaming, with CO concentrations prior to flaming never exceeding 360 ppm at eye level.

All 9 tests were considered successful, yet there is visual evidence of less damage to the couch in the case of the .438-inch orifice sprinkler operating at 8 psi. The fire was reported to be "extinguished rapidly" and was one of only two tests in which the fire did not burn through and spread underneath the mattress. This led to a conclusion that a local density of more than 0.033 gpm/ft² was critical in certain regions of the couch to prevent fire spread underneath the mattress.

The links used to operate the sprinklers had an RTI of about 268 (English) or 148 (metric), which today would not be considered as quick response. Both a low-pressure and high-pressure recommendation were made because the low-cost residential program was still contemplating self-contained systems at this time. The ventilation conditions were estimated to have produced about three air changes per hour, but were not felt to have materially affected CO concentrations. The fire scenario was considered typical for a residential environment, and the absence of other fire loads aside from the couch was not considered a factor since the fire did not spread.

It should be noted that the 0.218-in. orifice sprinkler recommended in the November 1975 "Residential Sprinkler - Protection Study" was not recommended in these tests. A 0.219-in. orifice was tested at a pressure of 28.5 psi, somewhat less than the 39 psi recommended in the earlier report. It performed as well as the two recommeded sprinklers in terms of the two evaluation criteria. The 1975 13D protection level also performed satisfactorily (0.438-in. orifice at 8 psi discharging 15 gpm):

	Cumulative CO in 20 min.	Max. eye level temperature (°F)
.273 in. at 8 psi	27,080	147
.137 in. at 88.5 psi	29,650	159
.219 in. at 28.5 psi	28,920	147
.438 in. at 8 psi	8,980	150

It was presumably on the basis of potential low cost system installation that the .273 in. and .137 in. sprinklers were recommended, since the flows of 6 gpm and 4.7 gpm were thought more economical than the .219 in. sprinkler flow of 7 gpm or the .438 in. sprinkler flow of 15 gpm.

The Proceedings of the Third Conference on Low-Cost Residential Sprinklers (97) included an interim report on this project, as well as other reports identified separately within the literature search, and a draft performance standard for residential sprinklers and their installation, prepared by Factory Mutual. In the discussion of the draft performance standard, it was indicated that FM had chosen to concentrate on the smoldering-start fire on the basis that the smoldering-start fire was the most difficult for sprinklers to handle, since many people felt life-threatening conditions would develop prior to sprinkler operation. Kung indicated a few tests had been run with flaming starts, had activated the sprinkler early, and caused no problems. He reported that rooms with windows and doors closed had produced more problems because of the build-up of toxic gases.

In the discussion of the fire test scenario proposed for inclusion in the standard, FM representatives commented that it was not necessary to follow the approval procedure for industrial sprinklers, which are designed to put the fire out. The emphasis was instead on saving lives in residential occupancies, and it was decided to use a residential scenario "...rather than, say, to burn a crib and put the fire out."

It was admitted that a smoldering fire which remained smoldering would not activate a sprinkler, but that a smoldering fire reaching a certain size was likely to burst into flame, resulting in sprinkler operation. The sprinkler, controlling the burning rate, would then control the carbon monoxide production.

It was acknowledged that the draft performance standard had no criteria for fire spread. Conceivably, the fire could spread so long as life safety criteria were not violated.

In response to concerns that a ventilated fire would be more difficult to handle, several participants indicated their confidence that a fast-developing ventilated fire could be suppressed by the system. It was noted that the water densities specified were sufficient to prevent flashover.

Report: Sprinkler Performance in Living Room Fire Tests: Effect of Link Sensitivity and Room Size (42)

Authors: H. C. Kung, Robert D. Spaulding, and Edward E. Hill (FMRC)

Sponsor: United States Fire Administration

Date: July 1983 report

Basic Description: Simulated links of three different sensitivities were installed at various locations in rooms with a fast-developing living room fire scenario to investigate the effects of sensitivity, water discharge rate and room size on the likelihood of excessive sprinkler openings.

Scenarios: Three series of tests were conducted. In the first two series the fast-developing living room scenario consisted of a fire ignited in a newspaper-filled plastic wastebasket located beneath an end table in a corner between a vinyl-upholstered reclining chair and a sofa. The area was carpeted, and drapes hung at the wall behind the chair. In this scenario, the fire spread to the back and side of the chair, where it was shielded from direct sprinkler water impingement. The third test series was identical except for the ignition source. Two cotton wicks were soaked with a total of 0.8 oz of alchohol and placed under the bottom edge of the chair side. Also, the number of double-layer drapery panels was reduced from three to two. The first test series was conducted in a large (24 ft × 36 ft × 8 ft) test room with a door opening, a simulated staircase opening, and four 16 in. × 36 in. window openings. The second and third series were conducted in a smaller (12 ft × 24 ft × 8 ft) room with a 4.4 ft wide doorway opening and two of the window openings.

Data Collected: Eye-level (60 in.) readings were made of gas temperatures and carbon monoxide concentrations. Gas temperatures were also measured at elevations 3 in. below ceiling and 36 in. above the floor. A thermocouple was installed in the vicinity of each of 12 sets of simulated links, set 3 in. below the ceiling at various locations relative to the test sprinkler. Water distribution tests at a flow of 18 gpm were conducted within the same structure used for the fire tests, using thirty-six 1 ft² pans arranged in a square on the floor in the room corner, and also elevated to 22-½ inches to simulate the tops of the furniture. Vertical water distribution was also measured using three racks of collection pans.

Sprinkler System Information: Only one sprinkler was installed in the test room, located 6 ft from each of the corner walls. The sprinkler used was the prototype sprinkler used in the Los Angeles residential sprinkler test series. In all tests except one, the discharge rate was maintained at 18 gpm. The sprinkler was actuated by a simulated link with an RTI value varied between 22, 58, and 100 ft$^{1/2}$sec$^{1/2}$.

Report Summary/Conclusions: For the tests in the large room, when the simulated link with an RTI of 100 ft$^{1/2}$sec$^{1/2}$ was used to actuate the sprinkler at a rated temperature about 60°F above its initial temperature, both the drapes and chair were burning vigorously and flames had started to sweep across the ceiling at the time of sprinkler actuation. All simulated links at 8, 10, 12, and 14 ft north and west of the sprinkler (except one at 14 ft west) exceeded their rated temperature differences. When the simulated links of smaller RTI values were used to actuate the sprinkler at about the same temperature, the fire involvement at the time of actuation was less. Empirical correlations were established between the simulated link RTI value, the rated temperature difference, and the distance between the sprinkler and the simulated link for the particular scenario investigated. It was determined that the RTI value of the simulated link should be less than 23 ft$^{1/2}$sec$^{1/2}$ to prevent operation of a second link located 8 feet away with a temperature rating 70°F above its initial temperature.

For the second test series, in the smaller room, the correlation showed that the RTI value needed to be less than 10 ft$^{1/2}$sec$^{1/2}$ to prevent operation of the second link.

In the third test series, increasing the flow rate of the sprinkler to 24.5 gpm (discharge pressure of 76 psi) reduced the temperature rise of the simulated links. Using an RTI of 58 ft$^{1/2}$sec$^{1/2}$ it was found that all simulated links 10 feet or more from the sprinkler were maintained below their rated temperatures, due probably to the superior cooling ability of the sprinkler spray at higher flow rate and discharge pressure.

The measurements of carbon monoxide and gas temperature at eye level showed that the room of fire origin was survivable for all tests conducted. However, the sprinkler flow was held at 18 gpm despite the fact that multiple sprinkler actuations would have occurred in the tests in which the simulated links with RTI values of 58 and 100 ft$^{1/2}$sec$^{1/2}$ were used for sprinkler actuation at a temperature 60°F above ambient.

It was shown for the prototype sprinkler nozzle tested *in this kind* of fire that the likelihood of excessive sprinkler actuations decreased as the link sensitivity increased.

Comments: While these tests indicated that increasing sprinkler sensitivity tends to reduce the likelihood of multiple sprinkler operations, the report authors put forward a caution by suggesting that these results might be limited to a particular sprinkler and type of fire. It should also be noted that the level of sensitivities determined to be needed to prevent the operation of a second sprinkler were in the range of 10 to 23 $ft^{1/2}sec^{1/2}$, whereas the RTI of the prototype sprinkler used in the Los Angeles residential sprinkler tests was 48.4 (in English units).

Due to the shielded nature of this fire scenario, it can be presumed that early suppression was not fully possible even if water application rates were capable. The cooling effect of the sprinkler spray thus becomes a major means of fire control and the mechanism preventing multiple sprinkler operations.

Report: Sprinkler Performance in Residential Fire Tests (43)

Authors: H. C. Kung, R. D. Spaulding, and E. E. Hill, Jr., FMRC

Sponsor: FEMA/U.S. Fire Administration

Date: December 1980

Basic Description: Effect on sprinkler performance of link sensitivity, temperature rating and water distribution using flaming-start residential fire scenarios.

Scenarios: Three series of tests were conducted: 1) nonventilated fires (bedroom, small bedroom, and living room), 2) 3 ventilated living room fires with noncombustible walls and ceiling, and 3) 7 ventilated living room fires with combustible walls and ceiling. The most sensitive commercially available standard sprinklers were used in all nonventilated fires and in three ventilated fires. Sprinkler prototypes with increased sensitivity were tested in the ventilated tests.

In the bedroom fires, a newspaper was used to produce a flaming-start fire at the side of the bed. In the living room tests, a newspaper was ignited in a plastic wastebasket placed underneath an end table situated in a corner between a sofa and a vinyl-upholstered reclining chair. The rooms were furnished with carpeting, drapes, and other furniture.

Data Collected: CO, CO_2, and O_2 at 63 in. above floor in room of origin, gas temperatures at five elevations in one location, ceiling surface temperatures above ignition point, and gas velocities and temperatures near the sprinklers. Also, for the ventilated living room fires, additional measurements of gas temperatures and velocities at window and entry door.

Average water application densities were determined for sprinklers tested in the horizontal plane (w_c) for the 5 ft × 5 ft floor area in the corner of a 15 ft × 15 ft coverage area without walls, and in the vertical plane (w_v) for a 1 ft wide × 2.5 ft high area on the wall above the recliner (3.5 ft to 6 ft above the floor) in the living room scenario.

Sprinkler System Information: The sprinklers used in the first test series, the three nonventilated tests, were $\frac{5}{16}$-inch orfice sprinklers discharging 0.04 gpm/ft². The RTI of the sprinklers is calculated to be 201 (English) or 111 (metric), considered to be the most sensitive commercially available sprinkler links (time constant of 73.9 seconds measured at 7.4 ft/sec.).

In the second test series, the same sprinklers were used. The third test of the second series, however, employed a simulated 140°F sprinkler link of higher sensitivity, with an RTI of 46 (English) or 25.4 (metric) (time constant of 16.9 seconds at 7.4 ft/sec.).

In the third test series, four different types of sprinkler were investigated: 1) experimental 0.275-in. orifice, 2) experimental 0.329-in. orifice, 3) Grinnell prototype 1, and 4) Grinnell prototype 2. The commercial links were used in one of the 0.275-in. sprinkler tests, but the other tests employed the high sensitivity link used in the second test series. The manufacturer's prototypes used a 136°F link with an RTI of 41 (English) or 22.7 (metric) (time constant of 15.1 seconds at 7.4 ft/sec.). The Grinnell prototype 2 was a design which discharged a fairly uniform pattern in the angular distribution, even behind the frame arms.

Report Summary/Conclusions: The non-ventilated tests were considered successful, except for the fact that the eye level temperature in the small bedroom test exceeded 200°F prior to sprinkler operation due to the quick descension of heated gases caused by the small room size. It was noted that only a more sensitive sprinkler could have prevented this. In these fires, a local density of 0.02 gpm/ft² was adequate to suppress the fire, which was not shielded.

In the shielded living room fire scenario considerably more water was needed. In the nonventilated living room test, a w_v of 0.011 gpm/ft² was sufficient to prevent fire spread to the wall, but the w_c of 0.015 gpm/ft² was not enough to suppress the fire or cool gases sufficiently, since temperatures were judged capable of operating sprinklers in the hallway and kitchen. The lower heat release of the unventilated fire, however, prevented ceiling temperatures exceeding limits for structural damage.

In the ventilated living room tests, activation with a commercial sprinkler occurred when the fire was already in a fast accelerating stage, with the entire corner in flames. It was found that, for a sprinkler link with a certain sensitivity to achieve the same temperature rise in both a ventilated fire and a nonventilated fire, the ventilated fire has to grow to a higher intensity. A more sensitive link was considered necessary.

Using a more sensitive link, less water was required to control the fire. The 140°F sensitive sprinklers with a w_c of 0.015 gpm/ft² were capable of maintaining life safety conditions within the room of fire origin, but did not suppress the fire and cool gases enough to prevent the operation of sprinklers in the hallway and kitchen.

With combustible walls and ceiling in the ventilated living room, it was found to be even more imperative to use a sensitive sprinkler. Using the manufacturer's prototype , a critical value of 0.035 gpm/ft² was determined for w_c to suppress the corner fire and prevent combustible walls and ceiling from igniting, and a value of 0.012 gpm ft² for w_v to prevent fire jump from the back of the chair to the wall.

The report includes the following summary statement: "In conclusion, significantly more sensitive links than the commercially available link, representative of the most sensitive on the market, are essential to provide adequate life-safety and property protection in residential fires."

Comments: On August 30, 1978, the residential sprinkler program had received a shock when the fourth in series of five demonstration tests had produced a failure. The first three tests, involving two bedroom fires and one living room fire, had been largely successful, using a $\frac{5}{16}$ in. orifice sprinkler flowing 6 gpm at 8 psi over 150 square feet. The fourth test was identical to the earlier living room test except that the room was ventilated for the first time. This failure led to a new series of tests to evaluate prototype residential sprinklers in ventilated flaming-start fires, the subject of this report. This report considers itself as covering the second phase of the FMRC residential sprinkler program. It reports the first phase as having covered the smoldering-start living room couch scenarios, and noted that the "required discharge rate to control such a fire is quite small."

This is the report of the tests which turned the residential sprinkler effort to quick response sprinkler technology.

Report: A Study of the Performance of Automatic Sprinkler Systems (52)

Authors: M. J. O'Dogherty, P. Nash, and R. A. Young

Sponsor: Joint Fire Research Organization

Date: 1967

Basic Description: Investigations of sprinkler response and water requirements needed for control or extinguishment of wood crib fires.

Scenarios: Wood cribs with heights up to 4 ft and three different rates of fire development were used in an investigation of sprinkler response and fire extinction capabilities. The three sizes of cribs produced fire growth rates approximated in the analysis by the straight-line portion of their growth curves, corresponding to the portions between 45 and 75 percent of their maximum rates of convective heat output: 3.3 Btu/sec², 9.2 Btu/sec², and 27.2 Btu/sec². All fire tests were conducted in a laboratory 130 feet × 50 feet, with a 40 foot ceiling height. The ceiling was divided into bays by joists 4 ft 6 in. on center which were 1 foot wide and 19 inches deep. The test cribs were raised to simulate lower ceiling heights.

Impinging jet nozzles were used to deliver water in flame spread and extinction tests.

Data Collected: Ceiling air temperatures, water application rates, sprinkler bulb surface temperatures, crib weight loss, rates of fire spread along surface of cribs, and time to extinction. Convective heat output was calculated using a calorific value of 8000 Btu/lb for wood and assuming 75 percent of heat output appeared as convective heat.

Sprinkler System Information: In the response investigations, glass bulb sprinklers were mounted at various depths within the center of bays formed by beams 1 ft wide × 19 in. deep spaced 15 ft on center. In the suppression tests, water was not applied through sprinklers but through nozzles nearer the crib surface.

Report Summary/Conclusions: Temperature rise of the air was found to be almost directly proportional to the rate of convective heat output of the fire, regardless of the rate of fire growth. The rise in temperature of the glass bulb sprinklers, however, was dependent on the rate of fire growth due to thermal lag. The rates of air temperature change at ceiling level produced by the cribs were found to approximate fairly well the ceiling air temperature increase rates produced in previous fire tests of practical fire risks.

More rapid fire growth resulted in a larger size fire at sprinkler operation. Rate of convective heat output was found to be proportional to the plan area of the fire.

The size of the fire at the sprinkler operating time increased in proportion to the square of the ceiling height where the fire was located directly beneath a sprinkler, and linearly with height where the fire was at least 5 feet horizontally from the sprinkler. The size of fire at sprinkler operation also increased continuously with increased horizontal distance from the fire, and with the depth of sprinkler location below the ceiling, except that depth was not a factor where the sprinkler was located directly over the fire. The size of the fire at sprinkler operation increased when the sprinklers were located below the joists. The lateral positioning of sprinklers within a bay was found to be unimportant.

In the extinction tests, application of water was delayed until a predetermined rate of heat release was reached, as measured by ceiling air temperature.

For the mid-size crib, water application rate of 0.075 igpm/ft² (0.09 gpm/ft²) was usually required to limit flame spread within the crib length, but in some cases a rate of 0.05 igpm/ft² (0.06 gpm/ft²) was sufficient. Fire spread was markedly reduced with higher application rates. For a given water application rate, fire spread after beginning water application was independent of the size of the fire when water was first applied. Water application rates of less than 0.23 igpm/ft² (0.276 gpm/ft²) appeared to have little effect on a well-established burning zone, but inhibited further spread. Application rates above 0.23 igpm/ft² tended to result in rapid extinction. Extinction time depended on the rate of water application but was independent of the size of the fire at initial water application. Total quantity of water was not found to depend on the rate of water application. Even when the lowest rates of water application used (0.06 gpm/ft²) were not sufficient to control the spread of fire within the crib length, the rate of spread was reduced below that of a freely burning crib.

For the crib with the faster fire growth, control was found to be more difficult, with at least 0.1 igpm/ft² (0.12 gpm/ft²) required to hold fire spread within the length of the crib. Fire spread was larger at low application rates, but unchanged at high application rates. For the crib with the slowest fire growth rate, the rates of fire spread were the least for a given application rate.

Total extinction time was similar for the two slower-developing cribs, comprised of 1 in. × 1 in. sticks. The fastest-developing fire crib, consisting of ½ in. × ½ in. sticks, required less total water for extinction.

Over the range of fire development rates used in the experiments, it was deduced that the size of a fire at sprinkler operation could vary by a factor of 2 to 1 for any given ceiling height.

Comments: The observation that the time for extinction and the total fire spread during water application did not depend on the size of the fire when water was first applied was explained by noting that the spread of fire within the crib was linear, whatever its original size, for the range of fires used. In the time taken for the wood to become sufficiently wetted to prevent further fire spread (which is dependent on crib geometry), the distance travelled by the fire was constant.

The linear fire spread in both length and width would result in a fire growing as the square of time. The rates of convective heat output calculated from crib weight loss reflect this type of fire growth within the range of available fuel for continued growth.

The rate of convective heat output at the time of sprinkler operation in the response tests ranged from a low of about 40,000 Btu/min to a high of about 200,000 Btu/min, depending on the rate of fire growth and the ceiling height.

Report: "Full-Scale Fire Tests with Automatic Sprinklers in a Patient Room" — Phase I and II (54 and 55)

Authors: John G. O'Neill, Warren D. Hayes, and Richard H. Zile (Phase II only), National Bureau of Standards

Sponsor: U.S. Dept. of Health and Human Services

Date: June 1979 and July 1980 reports on tests conducted from 1977 through 1980.

Basic Description: Full-scale fire tests were conducted to examine the use of sprinklers in patient rooms of health care facilities.

Scenarios: A total of 21 tests were conducted. In the first series of 8 tests, a patient bed and bedding were ignited by means of an adjacent flaming wastebasket. Two types of mattresses (cotton and polyurethane) were investigated, as was the effect of combustible interior walls and the effect of a privacy curtain placed between the sprinkler and the bed.

In the second series of tests, the impact was examined of clothing wardrobe fires, horizontal sidewall sprinklers, quick response sprinklers, and ventilation provided by a fan coil unit, and further examination of the privacy curtain obstruction.

Data Collected: Temperatures were recorded throughout the patient room. Gas velocities were recorded entering and leaving the burn room and moving along the corridor. Sampling tubes for CO, O_2, and CO_2 were located in the doorway at 2 in. and 5 ft above the floor, and at a location simulating a patient in an adjacent bed. Light obscuration was measured in the doorway and in the corridor and lobby.

Sprinkler System Information: While most of the testing was conducted with standard response sprinklers, some tests investigated the performance of simulated fast response sprinklers. The operation of discs with time constants of 9, 14.4, and 21.5 seconds (measured at a velocity of 5 ft/sec) were used to activate adjacent sprinklers.

Report Summary/Conclusions: Primary conclusions were that the type and location of privacy curtain should be coordinated with the location of the sprinkler protecting the room, that the clothing wardrobe fire represents an extremely severe fire that can result in room flashover without any additional fuel contribution from other combustibles in the room. Automatic sprinklers were able to limit some of the hazards of the wardrobe fire, but unacceptable CO levels were spread throughout the test area in tests with pendent sprinklers. Horizontal sidewall sprinklers were more effective in controlling these fires.

The use of a fast response sprinkler (pendent or horizontal sidewall) significantly reduced smoke obscuration in the patient room doorway and adjacent corridor in a flaming mattress and bedding fire, but did not improve overall performance of either type of sprinkler in the combustible wardrobe fire test.

A Class D rated combustible ceiling contributed to the energy release and CO concentrations, but did not adversely affect the ability of sprinklers to control fire growth. Similarly, combustible wall finish did not adversely affect overall performance of sprinklers.

Comments: It was observed in these tests that quick response was not beneficial in a fire in which the burning fuel was shielded from sprinkler discharge. Lacking "telltale" sprinklers throughout the test area, it could not be determined if quick response would have a negative net effect on system performance in such a case.

V. ANALYSIS AND CONCLUSIONS

Analysis

The chronology reveals that the term "quick response" requires further definition. It is not yet clear if a sprinkler is considered to be quick response on the basis of performance in a plunge test, or if the term will relate to the expected capability of the sprinkler in its installed setting.

Underwriters Laboratories has to this point taken the first approach with regard to the listing of quick response sprinklers. Sprinklers which operate in less than 13 seconds in the plunge test (15 seconds with tolerance) are declared QR (80). The manner in which these sprinklers are listed for use then determines the sensitivity which they will exhibit in the field.

The residential sprinkler program has taken the second approach. Although both Factory Mutual and Underwriters Laboratories have effectively included a maximum RTI in their residential sprinkler product standards (123,125), it is the fire test which actually determines the needed sensitivity to obtain a residential sprinkler listing. Compared against the original prototype residential sprinkler at a 144 sq ft spacing, residential sprinklers seeking listings at larger spacings require either a lower RTI or a more effective suppression capability.

Unless quick response is clarified, there will continue to be confusion in the field. It is observed that the extended coverage sidewall sprinklers were originally labeled "quick response extended coverage" sprinklers because their extended spacing was justified on the basis that their response was quicker than the slowest listed sprinkler installed under the least favorable, yet permissible, conditions. While this superior response might have been evident in a plunge test, it was the nature of the extended coverage sprinkler program to equalize overall performance. Thus the "quick response extended coverage" sprinkler could be expected to respond under some fire conditions no faster than the slowest listed standard pendent sprinkler, installed at maximum spacing and maximum distance below the ceiling. Similarly, it is still possible for a manufacturer to demonstrate sensitivity in a plunge test which obtains QR designation for a sprinkler, and then "use up" this sensitivity advantage by expanding its coverage area, again reaching the lowest common denominator of fire response. This results in a "quick response sprinkler" which could be less sensitive to a developing fire than most standard response sprinklers installed at standard spacing.

It should be recognized that RTI is simply a property of a sprinkler like temperature rating, and together these properties determine when a sprinkler will operate under given air temperature and velocity conditions. Neither RTI nor temperature rating, nor even both together, are able to give an indication of when a sprinkler will respond to a fire, except in a relative way. In other words, all other things being equal, a sprinkler with a lower

671

RTI and/or lower temperature rating will respond faster to a fire. However, variables such as ceiling height, distance of the sprinkler to the fire and to the ceiling, and other factors affecting gas conditions at the sprinkler are all important in the determination of response time.

The above argument can also be made through the use of the model. For the early stages of the fire, prior to the activation of any sprinkler, the terms **acool, fcool, cont,** and **supp** are equal to zero due to the fact that there is no water flow. The sprinkler activation function therefore reduces to:

$$act = f (gcs, RTI, RAT, AMB)$$

where $gcs = f (q, H, R, Y, VENT, CEIL)$
$q = f (a, FUEL, GEOM, VENT)$
$a = f (time, IGN, FUEL, LOAD, GEOM, q)$

Therefore, expressed in terms of decision variables, sprinkler activation time can still be seen to be a function of many variables:

$$act = f (RTI, RAT, R, Y, H, CEIL, VENT, AMB, IGN, FUEL, LOAD, GEOM)$$

Figures 1 and 2 show that sprinkler response can be thought of in three dimensions, where response time is the distance from the origin, and the three axes represent RTI, temperature rating and, collectively, the other factors.

Figure 1 represents a given fire in a given enclosure. All variables are therefore established except for RTI, RAT, and two installation factors: R, the spacing factor, and Y, the sprinkler-to-ceiling clearance. In this manner the installation factors can be seen to be important in determining sprinkler response time along with the two sprinkler properties of sensitivity and temperature rating.

Figure 2 demonstrates the same "sphere of response" concept, but for a given fire only. Now it can be seen that additional factors such as ceiling height H, ceiling configuration CEIL, and ventilation characteristics VENT must also be considered.

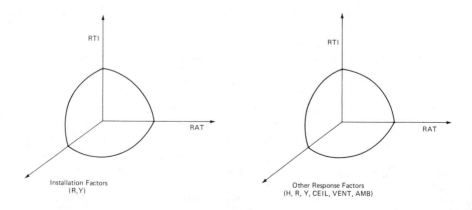

It is apparent that, if sprinkler response is to be a factor in system design, "quick response" should designate a design method, such as that of NFPA 13D, rather than individual sprinklers, so that the other factors are not overlooked.

It is also noted that the units of RTI (length$^{1/2}$time$^{1/2}$) do not lend themselves to easy recitation, and the continual mixing of English and metric units has caused confusion, even among researchers (7). Since sprinkler sensitivity is an international concern, it might be practical to treat RTI only as a metric term, expressed as m$^{1/2}$sec$^{1/2}$ or simply as RTI.

The Information Framework

The information framework, based on the model, attempts to organize what is known about sprinkler response and suppression. The framework summarizes research findings which express the level of understanding of the various elements. An analysis of the framework shows that the level of understanding of an element can be expressed as being at one of four levels:

1. Theoretical understanding. Can be calculated.
2. Limited calculation ability. Some empirical understanding.
3. Can be measured. No empirical correlation.
4. Cannot be calculated or measured.

A rough assignment of these levels to the elements of the model would show the following:

A. Area of fire — 2
B. Rate of heat release — 2
C. Products of combustion — 3
D. Gas conditions at sprinkler — 2
E. Sprinkler activation — 1
F. Water availability — 1
G. Atmospheric cooling — 2
H. Flame cooling application rate — 2
I. Local density — 2
J. Fuel shielding — 2
K. Control by prewetting application rate — 3
L. Early suppression application rate — 3
M. Plume penetration — 3
N. System performance — 3

It can be seen that, while much progress has been made in the area of predicting gas conditions at the sprinkler and the activation of the sprinkler itself, calculation methods have not yet been developed for other elements of the model. This indicates that design methods for quick response and other sprinklers must continue to be based on empirical data obtained from testing.

Analysis of the framework also permits some observations to be made regarding the net effect of quick response on the performance of sprinklers. It would appear that the effect of quick response varies with the manner in which the sprinklers extinguish or control the fire:

1. All other factors being equal, quick response sprinklers operate at an earlier stage of fire development than standard sprinklers, resulting in a smaller fire and lower levels of products of combustion at the time of sprinkler operation.

2. Where sprinkler spray and ventilation conditions permit extinguishment or control by flame cooling, it does not appear to be known if quick response has any effect. Heptane pool fires in which control or extinguishment was by cooling indicated the mechanism was insensitive to sprinkler operation delay time differences for the same size fire (40). Tests with different size fires using the same sprinkler sensitivity and spray discharge showed that smaller fires were easier to control, however, giving credibility to a quick response advantage. These fires produced heat release rates in the range of 10,000 to 42,000 Btu/min in the seconds prior to sprinkler operation.

3. Where water flow rates are not capable of early suppression but are capable of prewetting to control fire spread, quick response results in a smaller area of fire involvement at the time of first sprinkler operation. It also creates the potential for a large number of sprinklers to operate, depending on the atmospheric cooling ability of the sprinkler spray to counter the heat release of the unsuppressed burning area. It is not known, therefore, what the net effect of quick response is on the "control" concept, particularly in large areas involving multiple sprinklers. Testing by Factory Mutual (50) and Underwriters Laboratories (62,108) indicates additional sprinklers will open, and this phenomenon was also observed during the residential sprinkler program (43) when shielded fires were encountered by highly sensitive sprinklers with low discharge rates.

4. Where water densities are capable of early suppression, quick response can make early suppression possible by avoiding fire plume updrafts which might develop later in the fire and prevent sufficient droplet penetration. This was demonstrated by Factory Mutual in the "penetration factor" tests of 24 brands of ½-inch upright sprinklers measured against a 2.3 megawatt fire plume (130,000 Btu/min) (82). The penetration factors (early ADD measurements) are reported to have ranged from 23 to 65 among the 24 brands tested. The concept is being further investigated in the ADD testing of the Group 1 program.

5. Regardless of which suppression/control concept will be employed to handle the fire, quick response can prevent the spread of fire from the originally involved fuel to a more dangerous or less controllable fuel. This appears to be one of the major advantages of quick response in the ventilated living room fire test scenario developed during the residential sprinkler program (43,42). It was found necessary to operate the first sprinkler prior to full involvement of the curtains or spread to the combustible wall paneling in order for life safety conditions to be maintained in the room of fire origin.

6. In cases where quick response results in a smaller fire area, or in rapid extinguishment, the products of combustion are consequently reduced in volume, resulting in a more survivable environment. Even with quick response, however, sprinklers will not provide life safety conditions in the room of fire origin in all cases. The slowly smoldering fire has been shown capable of producing toxic conditions over long periods of time without generating sufficient heat to operate even quick response sprinklers.

Of the three extinguishment/control concepts, it appears that the most promising for the effective use of quick response sprinklers is early suppression. While flame cooling appears to result in the most efficient use of water, it also appears to require controlled environmental conditions, because it is the most sensitive to ventilation effects. The residential sprinkler program indicated (43) that it was not sound to rely upon flame cooling where environmental conditions are not controlled. Use of quick response sprinklers for control of fire spread by prewetting appears to be disadvantageous, as the potential for opening more sprinklers would require the use of a larger water supply, making system installation less economical.

The use of quick response sprinklers in an early suppression concept promises to produce more effective and efficient system performance, as demonstrated by the residential sprinkler program and ESFR program to date. The reduction in the expected number of operating sprinklers has the potential to reduce water supply demands, resulting in more economical system installation.

If quick response sprinklers are to be used in an early suppression concept for a broad range of occupancies, what are the decision variables which need to be addressed in a design method, and which would need to be investigated in the Group 2 test program? As a first step in this determination, it is of use to examine the decision variables used in currently accepted sprinkler design methods. In Table 1 on the pages following, each of the decision variables included in the framework is listed, along with an indication of whether it is used as a design consideration in the design method cited. Six design methods are examined:

NFPA 13 - 1985 (hydraulic design using area/density method)
NFPA 13D - 1984 (including current residential listings)
ESFR (design method contemplated by Group 1 test program)
NFPA 231 - 1984
NFPA 231C - 1980
Large Drop Sprinkler (Chapter 9 of NFPA 13 - 1985)

The state variables **q** and **act**, representing the rate of heat release and the number of sprinklers activating, are also included since some design approaches attempt to employ them as design factors.

COMPARISON OF DECISION VARIABLES

VARIABLE	NFPA 13D	NFPA 13	ESFR Program
FUEL type of fuel	Assumes "worst case" of polyurethane furniture, combustible walls and ceiling, and flammable draperies.	Occupancy hazard classifications contain vague guidelines re. combustibility and expected rates of heat release. Also consideration of presence of flammable and combustible liquids (EH).	Fuel, ignition point, load, and geometry lumped together as design factor in RDD testing of specific "worst case" commodity. Possibility exists to determine RDD for other specific fuel loadings and arrangements as a result of individual RDD testing. Geometry to be regulated in field. Height of storage (tiers) to be a design factor.
IGN ignition source and point	Not a design consideration. Flaming ignition of corner wastebasket used in "worst case" scenario.	Not a design consideration. Ignition assumed intimate with major fuel loading.	
LOAD quantity of fuel	Not a design consideration. Specific chair, sofa, and end table used in "worst case" scenario.	Occupancy hazard classification guidelines reference "quantity" of combustibles as a factor for consideration, as well as stock pile heights (OH).	See above.
GEOM geometry of fuel	Not a design consideration except for beam obstructions (Table A-4.2.2). Specific living room corner arrangement used in "worst case" scenario.	Not a design consideration except for obstructions per Tables 4-2.4.6 and 4-2.5.2 and Section 4-4.	See above.
VENT ventilation factors	Not a design consideration. Specific "worst case" living room scenario includes open window for ventilation.	Not a design consideration.	See above.
H height to ceiling	Not a design consideration. Ceiling height of 8' is assumed.	Not a design consideration.	Not a design consideration to this point. Intent is to provide ESFR sprinkler capable at variable clearances. Clearance lumped with fuel, load and other factors in RDD testing.
R radial distance to sprinkler	Not a design consideration. Limited by maximum sprinkler spacing, i.e., choice of SPR.	Sprinkler spacing used as a design factor within limits based on occupancy hazard classification.	Not a design consideration. To be regulated by 10' × 10' or alternate spacing.
Y sprinkler to ceiling clearance	Not a design consideration. Limited to a maximum of 4" except by special listing.	Not a design consideration. Limits depend on ceiling configuration and combustibility, and range up to 22".	Not a design consideration. To be regulated.

ADDRESSED BY DESIGN STANDARDS (Table 1a)

NFPA 231	NFPA 231C	Large Drop (Chap. 9)
Classification based on guidelines re combustibility of commodity, packaging, and pallets, and percentage of plastic.	Classification based on guidelines re combustibility of commodity, packaging, and pallets, and percentage of plastic.	Use restricted to specific fuels, loads, and geometries contained in Table A-9-3. These factors lumped together in full-scale testing.
Not a design consideration. Ignition assumed at bottom center of stored commodity.	Not a design consideration. Ignition assumed at bottom center of stored commodity.	Ignition assumed intimate with center base of stored commodity.
Height of storage a design factor per Figure 6-2.2.	Height of storage a design factor per Figure 6-8.2.	See above.
For Group A plastics only, open vs. closed array, stable vs. unstable piles, and free-flowing properties of commodity are design factors. Aisle widths regulated somewhat, as for idle pallet storage per 4-4.1.2.	Type of rack (double or single vs. multiple-row) and width of aisles are design factors. Flue spaces are regulated.	See above. Aisle width and flue spaces regulated.
Not a design consideration.	Not a design consideration. Design based on roof vents and draft curtains not being used per Section 3-3.	Not a design consideration.
Clearance to ceiling a design factor for Group A plastics only.	Clearance to ceiling a design factor per Figure 6-8.8. Presence of in-rack sprinklers a design factor.	Clearance lumped with fuel, load and other factors in determining scope of application.
Minor design factor. Limited by maximum sprinkler spacing, except that R may be increased for densities less than .25 gpm/ft².	Minor design factor. Limited by spacing of ceiling and in-rack sprinklers. R may be increased for densities less than .25 gpm/ft².	Not a design consideration. Limited by minimum and maximum spacings.
Not a design consideration. Follows NFPA 13.	Not a design consideration. Follows NFPA 13.	Not a design consideration. Minimum and maximum limits depend on ceiling configuration, and range up to 14".

COMPARISON OF DECISION VARIABLES

VARIABLE	NFPA 13D	NFPA 13	ESFR Program
CEIL ceiling configu- ration factors	Not a design considera- tion. Smooth, flat ceil- ing is assumed.	Ceiling configuration and combustibility used as a design factor in controlling R and Y.	Not a design considera- tion to this point.
RTI reponse time index	Not a design considera- tion except for special EC listings. Max. RTI limited to 55 (metric units) by FM, varies with RAT by UL.	Not a design factor ex- cept for special EC list- ings.	Not a design considera- tion. Max. RTI limited to 50 (English units) by FM.
RAT temperature rating	Not a design considera- tion. Ordinary rated sprinklers assumed.	Not a design factor ex- cept for EH occupancies per 2-2.1.2.12. Ordinary rated sprinklers required per 3-16.6.2 and excep- tions.	Not a design considera- tion. To be limited to ordinary rated sprinklers.
AMB ambient temperature	Not a design considera- tion.	Not a design considera- tion.	Not a design considera- tion.
TYPE wet/dry	Not a design considera- tion. Dry systems not permitted.	Dry system increases de- sign area by 30% per 2-2.1.2.9.	Not a design considera- tion to this point. Ques- tionable use of dry systems.
PRES operating pressure	Available pressure a de- sign factor in range of sprinkler choice, affect- ing R in turn.	Available pressure a de- sign factor, affecting R and number of design sprinklers act_{des}.	Available pressure a design factor affecting ADD of sprinkler.
DELAY water delay	Not a design considera- tion. Dry systems not permitted.	Large system volume re- quires q.o.d. per 5-2.3. Maximum volume lim- ited unless 1-minute max. delay proven.	Not a design considera- tion to this point. Ques- tionable use of dry sys- tems.
MAIN mainten- ance	Not a design considera- tion. Proper system maintenance assumed.	Not a design considera- tion. Proper system maintenance assumed.	Not a design considera- tion. Proper system main- tenance assumed.
K discharge coefficient or flow	Limited design considera- tion. Minimum flow rates for both 1-sprinkler and 2-sprinkler design areas established by choice of SPR.	Used as a design factor by consideration of PRES, K, and coverage area (affecting R) to sat- isfy design density re- quirements.	Not a design considera- tion. Determined by PRES to provide ADD at set values of K.
SPR distribution factors	Used as a design factor as a result of listing. Determines R and flow.	Not a design considera- tion, except for special listings. Regulated in listings.	Not a design considera- tion. Regulated in individ- ual listing through ADD testing.
RDD required delivered density	Not a design considera- tion.	Not a design considera- tion.	Used as a design factor. Established by large scale testing.

ADDRESSED BY DESIGN STANDARDS (cont. 1b)

NFPA 231	NFPA 231C	Large Drop (Chap. 9)
Not a design consideration, except that minor change to R permitted for 25' bays.	Not a design consideration, except that minor change to R permitted for 25' bays.	Minor design factor. Minimum pressure to be increased if insufficient firestopping in open wood joist construction. Ceiling configuration used in controlling Y.
Not a design consideration. Design based on standard response sprinklers per A-5.1.5.	Not a design consideration. Design based on standard response sprinklers.	Not a design consideration. Standard response assumed, or as demonstrated in large scale fire testing.
High temperature ratings permitted to reduce design area per Figure 6-1.2.	High temperature ratings permitted to reduce design areas.	Not a design consideration. Rating must match that used in large scale fire testing.
Not a design consideration.	Not a design consideration.	Not a design consideration.
Dry system increases design area 30% per 6-2.4.	Not a design consideration. Wet systems recommended by A-5-2.1.	Not a design consideration. Dry systems not permitted.
Available pressure a design factor, affecting R and number of design sprinklers act$_{des}$.	Available pressure a design factor, affecting R and number of design sprinklers act$_{des}$.	Available pressure a design factor, affecting number of design sprinklers act$_{des}$.
Follows NFPA 13.	Not a design consideration. Wet systems assumed.	Not a design consideration. Dry systems not permitted.
Not a design consideration. Proper system maintenance assumed.	Not a design consideration. Proper system maintenance assumed.	Not a design consideration. Proper system maintenance assumed.
Used as a design factor by consideration of PRES and coverage area (affecting R) to satisfy design density. K limited to that of 1/2 and 17/32 in. sprinklers.	Used as a design factor by consideration of PRES and coverage area (affecting R) to satisfy design density. K limited to that of 1/2 and 17/32 in. sprinklers.	Not a design consideration. Minimum flow rates established by levels of available PRES and regulated K of 11.0–11.5.
Not a design consideration. Regulated in listings.	Not a design consideration. Regulated in listings.	Not a design consideration. Regulated indirectly in sprinkler listing.
Not a design consideration.	Not a design consideration.	Not a design consideration.

COMPARISON OF DECISION VARIABLES

VARIABLE	NFPA 13D	NFPA 13	ESFR Program
ADD actual delivered density	Not a design considera-tion.	Not a design considera-tion.	Used as a design factor. Established in sprinkler listing.
q heat release rate	Not a design considera-tion.	Occupancy hazard clas-sifications contain vague references to expected rates of heat release.	Indirectly a design consid-eration through use of RDD.
act_{des} number of design sprinklers	act_1 and act_2 used as limited design factors, based on multiple sprinklers in compart-ment.	act_{des} used indirectly as design factor through concept of design area. act_{des} affects PRES.	May be used as a design factor. Possibility of 4-sprinkler design area (act_4).

ADDRESSED BY DESIGN STANDARDS (cont. 1c)

NFPA 231	NFPA 231C	Large Drop (Chap. 9)
Not a design consideration.	Not a design consideration.	Not a design consideration.
Not a design consideration.	Not a design consideration.	Not a design consideration.
act_{des} used indirectly as design factor through concept of design area. act_{des} affects PRES.	act_{des} used indirectly as design factor through concept of design area. act_{des} affects PRES.	act_{des} used as a design factor based on specific levels of PRES.

Based on the previous table, a second summary table can be made of the decision variables, indicating which are used as design factors (identified by X) and which are used as minor design factors (identified by x) under the various design methods:

SUMMARY OF DECISION VARIABLES USED AS DESIGN FACTORS

	13D	13	ESFR	231	231C	L. Drop
FUEL		X	?	X	X	X
IGN						
LOAD		X	X	X	X	
GEOM	x	x		x	X	
VENT						
H				x	X	
R		X		x	x	
Y						
CEIL		X		x	x	x
RTI						
RAT		x		X	X	
AMB						
TYPE		X		X		
PRES	X	X	X	X	X	X
DELAY		x		x		
MAIN						
K	x	X		X	X	
SPR	X	x				
RDD			X			
ADD			X			
q		x	x			
act	x	X		X	X	X

From this summary table, a number of observations can be made:

1. There are a variety of approaches being applied in sprinkler system design. These approaches involve differences in the decision variables addressed by a designer when dealing with a specific occupancy or hazard.

2. Virtually the only decision variable considered under every approach is the available operating pressure of the water supply. (Even this would not be true if pipe schedule systems were included in the tables.)

3. Other control variables which might seem to invite common use, such as FUEL[1] and LOAD are eliminated from some design approaches by one or both of the following means:

a. Using a "worst case scenario" to develop the design method and then applying it universally throughout the scope of the design approach. (Examples: NFPA 13D and possibly proposed ESFR)

b. Limiting the scope of the design approach to specific applications proven through large-scale testing. (Examples: Large drop and possibly NFPA 231C)

4. The large number of decision variables almost mandates some degree of simplification. Each variable not standardized or ignored introduces factorial complexities.

5. Some decision variables are not addressed because the impact is not sufficiently understood or measurable, such as VENT. Other decision variables, such as MAIN, simply cannot be accommodated in a design approach.

It is also interesting to look at how the decision variables are being handled in the two design approaches which currently involve the use of quick response sprinklers, NFPA 13D and the ESFR program. NFPA 13D relies heavily on the "worst case scenario" developed during the residential sprinkler program; the ESFR program relies upon large-scale testing to determine the RDD of a hazard, and individual sprinkler testing to determine ADD. In the following table, those variables which are handled via the worst case scenario (WCS), RDD, and ADD are identified, along with an indication of how the other variables are addressed:

---

[1]A ? appears for the FUEL variable for the ESFR program because it is not known whether design methods will include consideration of RDD test results for specific fuels or whether, as has been suggested, the design approach will be "aimed at protecting a broad class of high-hazard storage occupancies whose RDD requirement might be less than. . . rack storage of a plastic commodity in a 30-ft high building" (118).

HOW DECISION VARIABLES ARE ADDRESSED IN QR DESIGNS

	NFPA 13D	ESFR (Proposed)
FUEL	WCS	RDD
IGN	WCS	RDD
LOAD	WCS	RDD
GEOM	WCS	RDD
VENT	WCS	RDD
H	WCS/SPRLIST (8 ft)	RDD/ADD
R	SPRLIST	To be standardized
Y	SPRLIST	To be standardized
CEIL	WCS/SPRLIST (smooth)	To be standardized
RTI	SPRLIST	To be standardized
RAT	SPRLIST	To be standardized
AMB	Assumed normal	Assumed normal
TYPE	Standardized wet	Standardized wet?
PRES	SPRLIST	ADD
DELAY	Standardized wet	Standardized wet?
MAIN	Assumed functional	Assumed functional
K	SPRLIST	ADD
SPR	SPRLIST	ADD
q	(WCS)	(RDD)
act	Standardized at 2	Standardized at 4?

WCS = Standardized in "worst case scenario"
RDD = Included in RDD determination
ADD = Included in ADD determination
SPRLIST = Included in sprinkler listing determination

It can be seen that the manners in which decision variables are handled in the residential sprinkler and proposed ESFR programs are similar to the extent that WCS generally appears opposite RDD and SPRLIST generally appears opposite ADD. In other words, the decision to use the combustible living room corner arrangement as the design fire for the residential sprinkler program was equivalent to selecting a specific standardized commodity arrangement with a characteristic RDD. This leaves it to the testing laboratories, in their testing of individual sprinklers for residential listings, to conduct the equivalent of ADD tests. They determine if the sprinkler, at its proposed spacing and flow, has the capability to suppress the fire of equivalent size and strength. Rather than study the amount of water passing through a simulated fire plume as in ADD testing, however, the residential product standards call for the suppression of a fire in a simulated living room corner arrangement or, in the case of the FM standard, a realistic living room corner arrangement.

This design approach, then, may be recognized as state-of-the-art for design methods incorporating quick response sprinklers. Since both QR design methods limit the number of sprinklers which can operate, the design area is eliminated as a variable. Testing to determine how many sprinklers will open to control the fire, as was done in the development of NFPA 231C, is rejected.

How does one adapt this type of design approach to the wide range of occupancies and hazards represented within the scope of NFPA 13? Since the state-of-the-art does not permit a full calculation scheme, it remains necessary to use some type of occupancy classification system to create various levels of suppression capability. The existing hazard levels within NFPA 13 could be adapted to this purpose. However, the number of hazard categories should be reduced if possible, eliminating the need to draw fine distinctions in an area not well understood.

The following is suggested as a procedure for adapting the QR design approach to NFPA 13 occupancies:

1. Identify representative "worst case scenarios" for LH, OH1, OH2, OH3, EH1, and EH2 occupancies.

2. Conduct RDD testing of the above scenarios, along with the residential scenario used in the development of NFPA 13D. (The RDD apparatus may need to be adapted to investigate vertical distribution as well as horizontal distribution for compartmented occupancies.) In the RDD testing, water flow should be commenced at various intervals following the start of the fire (30 sec, 1 min, 2 min, etc.) which can later be correlated to expected sprinkler operating times under a variety of spacing, ceiling height, ceiling configuration, and compartmentation conditions.

3. Attempt to combine some of the hazard classifications on the basis of the RDD tests. (It may be found that light hazard and residential scenarios, for instance, have similar RDD demands regardless of the water delivery delay time.)

4. Analyze the levels of heat release generated during the RDD tests to determine when and if a significant ADD impact is likely. (Is the level of q high enough to create a significant difference between local density and ADD?)[2] Some ADD testing of standard sprinklers will be necessary for these determinations.

5. Base sprinkler designs on the ability of local density to meet RDD for situations where the level of heat release does not warrant consideration of ADD. Some minimum degree of spray pattern uniformity will need to be established by the testing laboratories. Where ADD needs to be considered, it can be determined as a property of a sprinkler as is being done in the ESFR program.

Appendix C to NFPA 72E provides the following approximate values of heat release rates for purposes of gaining a perspective:

---

[2] It is not known if the ADD concept is valid for other than the high-challenge fire, or in other words, if droplet penetration of the fire plume poses a problem for the heat release rates associated with the broad range of occupancies contemplated in the Group 2 test program.

Material	Btu/min
Medium wastebasket with milk cartons	6,000
Large barrel with milk cartons	8,400
Upholstered chair with polyurethane foam	21,000
Latex foam mattress (heat at room door)	72,000
Furnished living room (heat at open door)	240,000 - 480,000

In the Group 1 RDD tests, the first sprinkler is expected to operate when the heat release rate is no more than 120,000 Btu/min, although RDD tests have reportedly been conducted in which the convective heat release rates were as high as 280,000 Btu/min at the time of water application.

In 1967 British tests (52), 4-foot high wood cribs resulted in convective heat release rates ranging from 40,000 to 200,000 Btu/min at the time of first sprinkler operation. The 200,000 Btu/min rate resulted from a combination of slow (glass bulb) sprinkler response, high (40 ft) ceiling height, and a fast-growing fire.

The ADD concept would appear to be of greatest significance in occupancies with high, large open ceilings and fast-developing fires.

6. Develop a design method which addresses the variety of conditions found in NFPA 13 occupancies, based on the rates of heat release observed in the RDD testing. This might be done using charts or nomographs as in Appendix C of NFPA 72E (133), and would need to take into consideration the following as design factors:

- ceiling heights
- spacings
- clearance from sprinkler to ceiling
- ceiling configuration and ventilation factors[3]
- dry system delays
- size of room (upper layer effect)
- increased RDD resulting from response delays caused by the above factors (if observed)
- potential for life hazard conditions to develop based on the above factors and type of occupancy

7. Conduct full-scale confirmation testing.

Under the concept envisioned above, the spacing of a sprinkler would be based on its ability to deliver a uniform density over a given area, which could be tied into a minimum operating pressure. The spacing might be acceptable for one hazard classification but not for another because of differences in the droplet size make-up of the spray. In other words, what

---

[3]Since ventilation and some ceiling configuration factors are not well understood, these would probably need to be addressed in terms of estimated safety factors.

might be a density of 0.2 gpm/ft² for a light hazard challenge would be only a density of 0.15 gpm/ft² for an extra hazard challenge because of differences in the ADD.

As envisioned, the design process would consist of the following:

1. Classify the occupancy into the appropriate hazard classification.[4]

2. Propose a spacing and positioning of sprinklers for a given situation.

3. Determine a response coefficient for the first operating sprinkler(s) under the most adverse conditions, based on the factors identified above.

4. Determine a "second ring" heating coefficient (and a products of combustion coefficient if life safety is a concern) which represent the integrated values of heat release and toxic gases during the fire development and suppression sequence, based on the response coefficient and a proposed delivered density.

5. Use the "second ring" heating coefficient as in step 3 to ensure that additional sprinklers will not operate during the suppression sequence. Where life safety is a concern, the products of combustion coefficient can likewise be used to make sure critical levels are not exceeded during the fire.

6. Choose a sprinkler listed for the hazard with a spacing capable of delivering the desired density at the available system pressure, provided the proposed spacing and positioning is satisfactory to this point. If no such sprinkler is available, the process must begin again with a more conservative spacing and positioning.

Admittedly, the design procedure outlined above is considerably more complex than present NFPA 13 design methods. It is also considerably more complex than existing QR design methods. It should be recognized, however, that a great many simplifications have been introduced into present QR design methods, made possible by the narrow scopes of occupancy conditions they represent and the corresponding standardization of many otherwise significant variables.

---

[4]For those attempting to make occupancy hazard classifications, there would be detailed information available on the typical heat release rates associated with the hazard levels. It must be remembered, however, that it is not possible to characterize hazard levels by their heat release rate only. There is also presumably a "suppressibility factor" involved, related to the heat of gassification (84). Therefore, the guidance provided by examples of the hazard classifications would still be necessary unless special RDD testing were performed.

Conclusions

In summary, conclusions are as follows:

1. At present, there is no universal definition nor understanding of the term "quick response" as applied to automatic sprinklers, and the measure of sprinkler sensitivity by means of RTI is not well understood.

2. While considerable progress has been made in the development of quantitative methods for predicting gas conditions at the sprinkler and the activation point of the sprinkler itself, the state-of-the-art does not permit the use of a full mathematical model of sprinkler response and suppression.

3. The net effect of quick response varies with the manner in which sprinklers control or extinguish the fire:

- All other factors being equal, QR sprinklers operate at an earlier stage of fire development.
- Quick response enhances the potential for early suppression of fires by avoiding fire plume updrafts which might occur later in the fire.
- Quick response increases the potential for multiple sprinkler operations where water application rates are not sufficient for early suppression but control is accomplished through prewetting to prevent spread.
- It is not known if quick response has any effect where extinguishment takes place by flame cooling.
- Regardless of the suppression/control method, quick response can prevent the spread of fire to adjacent, potentially more dangerous, fuels.
- Where quick response results in a smaller fire area or in rapid extinguishment, the products of combustion are reduced in volume, resulting in a more survivable environment. QR sprinklers will not provide life safety in all cases due to the potential for a slowly smoldering fire.

4. Of the three methods of extinguishment or control, early suppression appears to be the most promising for effective use of quick response technology.

5. Existing sprinkler system design approaches vary significantly in terms of which decision variables are addressed. For simplification, decision variables are eliminated through the use of a "worst case scenario" or by narrowing the scope of a design approach to specific applications proven through large-scale testing.

6. Existing design methods using QR sprinklers (NFPA 13D and proposed ESFR) are similar in their elimination of many decision variables, made possible by reliance upon testing and through standardization of occupancy conditions.

7. A more complex design method than presently exists may be necessary for the proper incorporation of quick response technology into the broad range of occupancy conditions found within the scope of NFPA 13. If response is to play a major role in determining the effectiveness of sprinklers, it becomes necessary to address those factors which affect response.

Recommendations

The following recommendations are made based on the preceeding analysis and conclusions:

1. The designation of sprinklers as "quick response" should be discontinued in favor of the development of a quick response sprinkler design approach which addresses the expected response of installed sprinklers to a potential fire.

2. Use of the term RTI should be limited to metric units ($m^{1/2}sec^{1/2}$) to avoid confusion.

3. A quick response design method should be developed for NFPA 13 occupancies based on the early suppression mode of fire extinguishment.

VI. BIBLIOGRAPHY OF REVIEWED DOCUMENTS

Research Reports Relating to Response

1.   Alpert, R. L., "Calculation of Response Time of Ceiling-Mounted Fire Detectors," *Fire Technology*, Vol. 8, No. 3, August 1972.

2.   Baldwin, R. and North, M. A., "The Number of Sprinkler Heads Opening in Fires," Fire Research Station, Borehamwood, England, August 1971.

3.   Battelle Columbus Laboratories, "Development of an Experimental Prototype Low-Cost Electronic Sensor/Actuator for a Residential Automatic Sprinkler Head," Federal Emergency Management Agency, 1982.

4.   Battelle Columbus Laboratories, "Final Report on Development of a Nitinol-Actuated Fire Sprinkler," Federal Emergency Management Agency, September 1982.

5.   Beitel, Jesse J., "Evaluation of Combustion Atmospheres," Southwest Research Institute, September, 1981.

6.   Beitel, Jesse J., Priest, Deggary N., and Tauch, Donald R., "Combustion By-Products Analysis of Full-Scale Hotel Room Fire Tests," Southwest Research Institute, San Antonio, November 1982.

7.   Benjamin/Clarke Associates, "Operation San Francisco Smoke/Sprinkler Test Technical Report," International Association of Fire Chiefs, Washington, April 1984.

8.   Benjamin, I., Heskestad, G., Bright, R., and Hayes, T., "An Analysis of the Report on Environments of Fire Detectors," Fire Detection Institute, 1979.

9.   Beyler, Craig L., "The Interaction of Fire and Sprinklers," National Bureau of Standards/Center for Fire Research, NBS-GCR-78-121, September, 1977.

10.   Brown, J. Richard, Jr., "Fire Tests to Compare Sprinkler Demands of Wet-Pipe and Dry-Pipe Systems," Society of Fire Protection Engineers Technology Report 80-1, May, 1980.

11.   Bukowski, R. W., Waterman, T. E., and Christian, W. J., "Detector Sensitivity and Siting Requirements for Dwellings," IIT Research Institute, 1975.

12.   Budnick, Edward K., "Estimating Effectiveness of State-of-the-Art Detectors and Automatic Sprinklers on Life Safety in Residential

Occupancies," National Bureau of Standards/Center for Fire Research, NBSIR 84-2819, January, 1984.

13. Campbell, Layard E., "Study of Sprinkler Sensitivity," National Automatic Sprinkler and Fire Control Association, May, 1968.

14. Ciardelli, Donald R., "Home Sprinkler and Fire Alarm Tests," Bloomington Fire Department, Bloomington, MN, June, 1975.

15. Cote, Arthur E. and Moore, David A., "Field Test and Evaluation of Residential Sprinkler Systems, Los Angeles Test Series," National Fire Protection Association, April, 1980.

16. Cote, Arthur E., "Final Report on Field Test and Evaluation of Residential Sprinkler Systems," National Fire Protection Association, July, 1982 (Revised December, 1982).

17. Cote, Arthur E., "Final Report on Field Test of a Retrofit Sprinkler System," National Fire Protection Research Foundation, February, 1983.

18. Custer, Richard L. P., "Detector Actuated Automatic Sprinkler Systems - A Preliminary Evaluation," National Bureau of Standards, July, 1974.

19. Dean, Ronald K., "Investigation of Conditions Potentially Affecting Rack Storage Fire Severities," Factory Mutual Research Corp., FMRC J.I. OEOJ1.RR, October, 1980.

20. Doerschuk, David C. and Kleszczelski, Stan E., "Investigation of Improved Sensor/Actuator Concepts for Residential Fire Sprinkler Systems - Final Report," Battelle Columbus Laboratories, January 31, 1980.

21. Dosedlo, Lee J., Unpublished memoranda on the Life Hazard Reduction (LHR) Sprinkler Test Program, Underwriters Laboratories, March-October 1974.

22. Evans, David D., "Calculating Sprinkler Activation Time in Compartments," National Bureau of Standards/Center for Fire Research, 1984.

23. Evans, David D., and Madrzykowski, Daniel, "Characterizing the Thermal Response of Fusible-Link Sprinklers," National Bureau of Standards/Center for Fire Research, NBSIR 81-2329, August, 1981.

24. Evans, David D., "Thermal Actuation of Extinguishing Systems," National Bureau of Standards/Center for Fire Research, NBSIR 83-2807, March, 1984.

25. "Fire Alarm Thermostat Tests," National Board of Fire Underwriters, 1956.

26. Fleming, Russell P., "Air Oven Testing Project Report," Unpublished report for National Automatic Sprinkler and Fire Control Association, January, 1976.

27. Goodfellow, David G., and Troup, Joan M. A., "Large-Scale Fire Tests to Study Sprinkler Sensitivity," Factory Mutual Research Corporation, FMRC J.I. OH4R7.RR, March, 1983.

28. Grubits, S. J. and Moulen, A. W., "Operation of Sprinkler Heads By Radiant Heat," Experimental Building Station, New South Wales, Australia, March, 1983.

29. Halpin, B. M., Dinan, J., and Deters, O., "Assessment of the Potential Impact of Fire Protection Systems on Actual Fire Incidents," Johns Hopkins University/Applied Physics Laboratory, October, 1978, (Reprinted by Federal Emergency Management Agency /U.S. Fire Administration, May, 1980).

30. Harpe, S. W., Waterman, T. E. and Christian, W. J., "Detector Sensitivity and Siting Requirements for Dwellings - Phase 2," NBS-GCR 77-82, National Bureau of Standards, February, 1977.

31. Hayes, Warren D. and Zile, Richard H., "Full-Scale Study of the Effect of Pendent and Sidewall Location on the Activation Time of an Automatic Sprinkler," NBSIR 82-2521, National Bureau of Standards/Center for Fire Research, July, 1982.

32. Henderson, Norman C., Riegel, Peter S., Patton, Richard M., and Larcomb, D. Bruce, "Investigation of Low-Cost Residential Sprinkler Systems," Battelle Columbus Laboratories, January, 1978.

33. Heskestad, G. and Delichatsios, M. A., "Environments of Fire Detectors - Phase I: Effect of Fire Size, Ceiling, Height and Material," July, 1977, "Phase II: Effect of Ceiling Configuration," June, 1978, FMRC 22427 and 22584, Factory Mutual Research Corp., 1977.

34. Heskestad, G. and Delichatsios, M. A., "The Initial Convective Flow in Fire," *Seventeenth Symposium (International) on Combustion*, The Combustion Institute, Pittsburgh, 1979.

35. Heskestad, G. and Kung, H. C., "Relative Water Demands of Wet and Dry-Pipe Sprinkler Systems," FMRC No. 15918, Factory Mutual Research Corporation, November, 1973.

36.   Heskestad, Gunnar, and Smith, Herbert F., "Investigation of a New Sprinkler Sensitivity Approval Test: The Plunge Test," FMRC No. 22485, Factory Mutual Research Corporation, December, 1976.

37.   Heskestad, Gunnar, and Smith, Herbert F., "Plunge Test for Determination of Sprinkler Sensitivity," FMRC J.I. 3A1E2.RR, Factory Mutual Research Corporation, December, 1980.

38.   Hilton, D. and Grier, J., Interview re. Residential Sprinkler Tests, Cobb County Fire Department, Georgia, 1982-1984.

39.   Jackson, Ralph J., "Report on 1980 Property Loss Comparison Fires," Ad Hoc Insurance Committee on Residential Sprinklers, Federal Emergency Management Agency/U.S.Fire Administration, November, 1980.

40.   Kung, H. C., "Residential Sprinkler - Protection Study," Factory Mutual Research Corporation, November 1975.

41.   Kung, H. C., Haines, David M., and Greene, Robert E., Jr., "Development of Low-Cost Residential Sprinkler Protection: A Technical Report," Factory Mutual Research Corporation, April, 1978.

42.   Kung, H. C., Spaulding, Robert D., and Hill, Edward E., "Sprinkler Performance in Living-Room Fire Tests: Effect of Link Sensitivity and Room Size," Factory Mutual Research Corporation, July, 1983.

43.   Kung, H. C., Spaulding, Robert D., and Hill, Edward E., Jr., "Sprinkler Performance in Residential Fire Tests," FMRC No.22574, Factory Mutual Research Corporation, December, 1980.

44.   Kung, H. C., Spaulding, Robert D., Hill, Edward E., Jr., and Symonds, Alan P., "Field Evaluation of Residential Prototype Sprinkler - Los Angeles Fire Test Program," Factory Mutual Research Corporation, February, 1982.

45.   Kung, Hsiang-Cheng, Spaulding, Robert D., and You, Hong-Zeng, "Response of Sprinkler Links to Rack Storage Fires," FMRC J.I. OG2E7.RA (2), Factory Mutual Research, November, 1984.

46.   Lee, James L., "Early Suppression Fast Response (ESFR) Program -Phase I: Determination of Required Delivered Density (RDD) in Rack Storage Fires of Plastic Commodity," FMRC J.I. OJOJ5.RA, Factory Mutual Research Corporation, May, 1984.

47.   Maier, A. R., "Sprinkler Tests Answer Heat Rating Questions," *Fire Engineering*, Vol. 132, No. 4, April, 1979.

48. Moore, David A., "Data Summary of the North Carolina Test Series of USFA Grant 79027 Field Test and Evaluation of Residential Sprinkler Systems," National Fire Protection Association, Boston, Mass., September, 1980.

49. Nash, P., Bridge, N. W., and Young, R. A., "The Rapid Extinction of Fires in High-Racked Storages," Fire Research Station, Borehamwood, England, June, 1974.

50. Newman, R. M., "Effect of Temperature Rating on Number of Sprinklers Opened," Factory Mutual Engineering Division, FM No. 14537, May 24, 1963.

51. O'Dogherty, M. J. and Nash, P., "Considerations of the Fire Size to Operate, and the Water Distribution From Automatic Sprinklers," Fire Research Note 899, Fire Research Station, Borehamwood, November, 1971.

52. O'Dogherty, M. J., Nash, P., and Young, R. A., "A Study of the Performance of Automatic Sprinkler Systems," Fire Research Technical Paper No. 17, Joint Fire Research Organization, Her Majesty's Stationery Office, London, 1967.

53. O'Dogherty, M. J. and Young, R. A., "The Performance of Automatic Sprinkler Systems - Part I - The Effect of Ceiling Height, Rate of Fire Developmenmt, and Sprinkler Position on Response to a Growing Fire," F. R. Note No. 601, Joint Fire Research Organization.

54. O'Neill, John G. and Hayes, Warren D., Jr., "Full-Scale Fire Tests with Automatic Sprinklers in a Patient Room," NBSIR 79-1749, National Bureau of Standards/Center for Fire Research, June, 1979.

55. O'Neill, John G., Hayes, Warren D., Jr., and Zile, Richard H., "Full-Scale Fire Tests With Automatic Sprinklers in a Patient Room. Phase II," NBSIR 80-2097, National Bureau of Standards/Center for Fire Research, July, 1980.

56. Patton, Richard M., "Life Safety System Fire Test Program," Patton Fire Protection and Research, Inc., December 17, 1971.

57. Pratt, R. H., "Report to Cotton Marine Underwriters of Tests to Determine Relative Sensitiveness of Quartz Bulb and Solder Type Sprinkler Heads," Unpublished report of tests made at the Public Cotton Warehouse of the Board of Commissioners, Port of New Orleans on June 3-6, 1931, July, 1931.

58. Spaulding, Robert D., "Sidewall Sprinkler Position in Residential

Fires," Term paper submitted to Worcester Polytechnic Institute, December, 1984.

59.   Spaulding, Robert D. and Hill, Edward E., Jr., "Communities Sprinkler Fire Tests - Cobb County, Georgia," FMRC J.I. OG5NO.RA, Factory Mutual Research Corporation, March, 1982.

60.   Spaulding, Robert D. and Hill, Edward E., Jr., "Communities Sprinkler Fire Tests - Scottsdale, Arizona," FMRC J.I. OG5NO.RA, Factory Mutual Research Corporation, July, 1982.

61.   Spaulding, Robert D., Kung, Hsiang-Cheng, Hill, Edward E., Jr., and Symonds, Alan P., "Field Evaluation of Residential Prototype Sprinkler - Charlotte, North Carolina Mobile Home Test Series," (draft report), FMRC J.I. OEOR3.RA(2), Factory Mutual Research Corporation, May, 1981.

62.   Suchomel, Miles R., "Study of Factors Influencing the Use of High-Temperature Sprinklers," National Board of Fire Underwriters, New York, July 29, 1964.

63.   Troup, Joan M. A., "Exploratory, Intermediate-Scale Fire Tests With Early-Suppression Fast Response (ESFR) Prototype Sprinklers," FMRC J.I. OMOJ2.RR, Factory Mutual Research Corporation, April, 1985.

64.   Veselov, A. I., "Studies With Respect to Creation of High Speed Automatic Fire Extinguishers," *Fire Science and Technology*, September, 1978.

65.   Webb, William A., "Effectiveness of Automatic Sprinkler Systems in Exhibition Halls," *Fire Technology*, Vol. 4, No. 2, May, 1968.

66.   Young, J. R., "Spacing Automatic Sprinklers," *NFPA Quarterly*, October, 1960.

67.   Yurkonis, Peter R., "Study to Establish the Existing Automatic Fire Suppression Technology for Use in Residential Occupancies," Rolf Jensen and Associates, August 31, 1977.

Articles and Books Relating to Response

68.   "ADD and the Problem of Early Fire Suppression," *ESFR Update*, Factory Mutual Engineering Corp., Vol. 2, No. 1, 1985.

69.   Beyler, Craig L., "A Design Method for Flaming Fire Detection," *Fire Technology*, Vol. 20, No. 4, November, 1984.

70. Brown, Phillip A., "Residential Sprinklers Perform in Scottsdale," *Sprinkler Quarterly*, National Fire Sprinkler Association, No. 42, Summer 1982.

71. Chopra, Iqbal S., "Subject 1626: Residential Sprinklers," *Lab Data*, Underwriters Laboratories, 1984.

72. Claiborne, Gaylon R., "Automatic Sprinkler - It Is, It Isn't," *The Building Official and Code Administrator*, June, 1971.

73. Coleman, R. J., "Operation San Francisco: A Demonstration of State-of-the-Art Sprinkler Technology," *Fire Chief*, Vol. 27, No. 12, December, 1983.

74. Cote, Arthur E., "Field Test and Evaluation of Residential Sprinkler Systems: Parts I-III," *Fire Technology*, Vol. 19, No. 4, Vol. 20 Nos. 1 and 2, Nov. 1983, Feb. and May 1984.

75. Cote, Arthur E., "Highlights of a Field Test of a Retrofit Sprinkler System," *Fire Journal*, Vol. 77, No. 3, May 1983.

76. Cote, Arthur E., "Update on Residential Sprinkler Protection," *Fire Journal*, Vol. 77, No. 6, November, 1983.

77. "Effect of Thermal Sensitivity on the Performance of Automatic Sprinklers," Proposal to National Automatic Sprinkler and Fire Control Association, Inc., by Factory Mutual Research Corporation, FMRC No. 22485.1, December, 1975.

78. "Early Suppression - Fast Response: A Revolution in Sprinkler Technology," *ESFR Update*, Factory Mutual Engineering Corporation, Vol. 1, No. 1., 1984.

79. Feldstein, Lee M., "Operation San Francisco," (Parts I-III), *International Fire Chief*, Vol. 49, No. 11 and Vol. 50, Nos. 1 and 3, November 1983, January 1984, and March 1984.

80. Fleming, Russell P., "Understanding the New Sprinklers," *Sprinkler Quarterly*, National Fire Sprinkler Association, No. 47, Fall 1983.

81. Friedman, Raymond, "Automatic Sprinklers: Recent U.S. Progress," Paper presented at 6th Joint Panel Meeting UJNR Panel on Fire Research and Safety, May, 1982.

82. Friedman, Raymond, "In Search of a Better Sprinkler," *Sprinkler Quarterly*, National Fire Sprinkler Association, No. 48, Winter 1983-84.

83. "Sprinkler Technology: Research on Sensitivity," *Record*, Factory Mutual Engineering Corporation, Vol. 60, No. 3, Fall 1983.

84. Friedman, Raymond and de Ris, John, "Fire Modeling Research at Factory Mutual," Paper presented at Sixth U.S.-Japan Natural Resources Panel Meeting on Fire Research and Safety, March, 1982.

85. Gerard, John C., "Fort Lauderdale Retrofit Sprinkler Tests," *Sprinkler Quarterly*, National Fire Sprinkler Association, No. 45, Spring 1983.

86. Heskestad, Gunnar, "Improving 'ADD'," Paper (RC 83-TP-16) presented at the Sprinkler Technology Seminar, Factory Mutual Research Corporation, November 2, 1983.

87. Heskestad, Gunnar, "The Sprinkler Response Time Index (RTI)," Paper (RC 81-TP-3) presented at the Technical Conference on Residential Sprinkler Systems, Factory Mutual Research Corporation, April 28-29, 1981.

88. Jensen, Rolf, "Specifications and a Proposed Test Program of a QR Apartment Sprinkler for Water Tower Place," Rolf Jensen & Associates, February 22, 1973.

89. Levine, Robert S., "Proceedings: Ad Hoc Mathematical Fire Modeling Working Group," *Fire Technology*, Vol. 20, No. 2, May, 1984.

90. Moore, David A., "Field Test and Evaluation of Residential Sprinkler Systems," *Fire Journal*, Vol. 74, No. 6, November, 1980.

91. Nash, P. and Young, R. A., *Automatic Sprinklers for Fire Protection*, Victor Green Publications Limited, London, 1978.

92. "New Sprinklers - Russian Style," *Fire Technology*, Vol. 2, No. 2, May, 1966.

93. O'Neil, John G., "Fast Response Sprinklers in Patient Room Fires," *Fire Technology*, Vol. 17, No. 4, November, 1981.

94. Pepi, Jerome S., "Concept and Development of the Residential and Fast Response Sprinklers," *Sprinkler Quarterly*, National Fire Sprinkler Association, Spring 1985.

95. Pietrzak, L. M., "Refinement and Optimization of Residential Sprinkler Technology - A Concept Paper," Mission Research Corporation, Santa Barbara, California, February, 1978.

96. *Proceedings of the Fourth Conference on Low-Cost Residential Sprin-*

kler Systems, September 26-27, 1979, Federal Emergency Management Agency/U.S. Fire Administration, August, 1980.

97. *Proceedings of the Third Conference on Low-Cost Residential Sprinklers*, November 29-30, 1977, National Fire Prevention and Control Administration, June, 1978.

98. "RDD: A New Measure of Water Supply," *ESFR Update*, Factory Mutual Engineering Corp., Vol. 1, No. 3, 1984.

99. Reiss, M. H., "Operation San Francisco From the Smoke Detection and Alarm Point of View," *International Fire Chief*, Vol. 50, No. 5, May, 1984.

100. *Report to Congress on Fire Protection Systems: Detectors, Remote Alarm Systems, and Sprinklers*, Federal Emergency Management Agency/U.S. Fire Administration, June, 1981.

101. "The Residential Sprinkler System: A New Slant on Old Technology," *Record*, Factory Mutual Engineering Corporation, Vol. 58, No. 4, July-August, 1981.

102. "Response Time Index: The Key to Fast Response," *ESFR Update*, Factory Mutual Engineering Corp., Vol. 1, No. 2, 1984.

103. Riseden, J. Frank, "Cobb County, Georgia: Is Their 'Life Safety System' Leading The Way?" *Sprinkler Quarterly*, National Fire Sprinkler Association, No. 47, Fall 1983.

104. "The Role of Ignition Location in the Development of the ESFR Sprinkler," *ESFR Update*, Factory Mutual Engineering Corporation, Vol. 2, No. 1, 1985.

105. Rolf Jensen & Associates, "Study to Establish the Existing Automatic Fire Suppression Technology for Use in Residential Occupancies: A Summary Report," U.S. Department of Commerce/National Fire Prevention and Control Administration, December, 1978.

106. Shaw, Harry, "Fast Response Sprinklers - The Answer to the Toxicity Problem," *Sprinkler Quarterly*, National Fire Sprinkler Association, No. 47, Fall 1983.

107. Shell Chemical Company, "Springdale Mobile Home Test," *Sprinkler Quarterly*, Natl. Fire Sprinkler Association, No. 43, Fall 1982.

108. Suchomel, Miles R., "Factors Influencing the Use of High-Temperature Sprinklers," *Fire Technology*, Vol. 1, No. 1, February, 1965.

109. Suchomel, Miles R., "UL Listing Procedures for Residential Sprinklers," (article to be published in 1985).

110. Suchomel, Miles R., and Castino, G. T., "Sprinkler Performance Tests -The Extended Coverage Panacea," *Fire Technology*, Vol. 16, No. 2, May, 1980.

111. Thompson, Norman J., *Fire Behavior and Sprinklers*, National Fire Protection Association, 1964.

112. Thorn, P. F., "The Thermal Response of Sprinkler Heads," Building Research Establishment/Fire Research Station, Borehamwood, England, 1984.

113. Unoki, Juzo, "Sprinkler Technology and Design in Japan," Paper presented at U.S.-Japan Panel on Fire Research, National Bureau of Standards, October 15-17, 1980.

114. Watson, Barry, "Fast Response Sprinkler Heads," *Fire Surveyor*, Victor Green Publications Ltd., London, Vol. 13, No. 4, August, 1984.

115. Watson, Barry, "Sprinklers - Early Suppression and Fast Response?" *Fire Prevention*, Fire Protection Association, London, No. 174, November, 1984.

116. Yao, Cheng, "Applications of Sprinkler Technology - Early Suppression of High-Challenge Fires With Fast Response Sprinkler," Paper presented at the ASTM/SFPE Symposium on Application of Fire Science to Fire Engineering, June, 1984.

117. Yao, Cheng, "Early Fire Suppression With Fast-Response Sprinklers," *Sprinkler Quarterly*, National Fire Sprinkler Association, Patterson, NY, No. 48, Winter 1983-84.

118. Yao, Cheng, "ESFR Sprinkler System - A New Approach to High-Challenge Storage Protection," *Fire Journal*, Vol. 79, No. 2, March, 1985.

119. Yao, Cheng, "Extinguishing Role of Water Sprinklers," *Proceedings -Engineering Applications of Fire Technology Workshop*, April 16-18, 1980, Society of Fire Protection Engineers, Boston, 1980.

120. Yao, Cheng, and Marsh, William S., "Early Suppression - Fast Response: A Revolution in Sprinkler Technology," *Fire Journal*, Vol. 78, No.1, January, 1984.

Standards

121. *Approval Standard (Draft) - Automatic Sprinklers For Fire Protection Service*, Factory Mutual Research Corporation, December, 1983.

122. *Approval Standard (Draft) - Early Suppression Fast Response Automatic Sprinklers*, Factory Mutual Research Corporation, January, 1985.

123. *Approval Standard - Residential Automatic Sprinklers*, Factory Mutual Research Corporation, Class No. 2030, September, 1983.

124. *Factory Mutual Laboratories Approval Standard for Automatic Upright and Pendent Standard Sprinklers (Formerly Designated as Spray Sprinklers)*, Factory Mutual Engineering Division - Factory Mutual Laboratories, February 7, 1956.

125. *Outline of Proposed Investigation for Residential Automatic Sprinklers for Fire Protection Service*, Subject 1626, Underwriters Laboratories, October 27, 1980.

126. *Proposed First Edition of the Standard for Special Purpose Sprinklers for Fire Protection Service*, Subject 199A, Underwriters Laboratories, February, 1978.

127. *Rules for Automatic Sprinkler Installations*, 29th Edition, Fire Offices' Committee, London, December 1968.

128. *Standard for Automatic Sprinklers for Fire Protection Service*, UL 199, Seventh Edition, Underwriters Laboratories, July 20, 1982.

129. *Standard for General Storage*, NFPA 231, National Fire Protection Association, 1985 Edition.

130. *Standard for Rack Storage*, NFPA 231C, National Fire Protection Association, 1980 Edition.

131. *Standard for the Installation of Sprinkler Systems*, NFPA 13, National Fire Protection Association, 1985 Edition.

132. *Standard for the Installation of Sprinkler Systems in One- and Two-Family Dwellings and Mobile Homes*, NFPA 13D, National Fire Protection Association, 1984 Edition.

133. *Standard on Automatic Fire Detectors*, NFPA 72E, National Fire Protection Association, 1984 Edition.

Research Reports on Sprinkler Suppression

134. Alpert, R. L., and Mathews, M. K., "Calculation of Large-Scale Flow Fields Induced by Droplet Sprays," FMRC J.I. OEOJ4.BU, Factory Mutual Research Corporation, December, 1979.

135. Ball, J. A. and Pietrzak, L. M., "A Computer Simulation to Estimate Water Application Effectiveness in Suppression of Compartment Fires," Paper presented at Fall Technical Meeting 1976 of the Eastern / Combustion Institute, November, 1976.

136. Beyler, C. L., "Effect of Selected Variables on the Distribution of Water From Automatic Sprinklers," NBS-GCR-77-105, National Bureau of Standards/Center for Fire Research, June, 1977.

137. Braidech, Matthew M. and Neale, John A., "The Mechanism of Extinguishment of Fire by Finely Divided Water," NBFU Research Report No. 10, National Board of Fire Underwriters, 1955.

138. Bridge, N. W., and Young, R. A., "Experimental Appraisal of an American Sprinkler System for the Protection of Goods in High-Racked Storages," Fire Research Note No. 1003, Fire Research Station, Borehamwood, England, February, 1974.

139. Cooper, Leonard Y. and O'Neill, John G, "Fire Tests of Stairwell-Sprinkler Systems," NBSIR 81-2202, National Bureau of Standards, April, 1981.

140. Custer, Richard L. P. and Wahle, Klaus, "Distribution of Water Through a Vertical Plane From Automatic Sprinkler Heads," NBSIR 75-920, National Bureau of Standards/Center for Fire Research, December, 1975.

141. Dean, Ronald K., "Stored Plastics Test Program," FMRC Serial No. 02062, Factory Mutual Research Corporation, June, 1975.

142. Dundas, P. H., "Optimization of Sprinkler Fire Protection Progress Report No. 10 - The Scaling of Sprinkler Discharge: Prediction of Drop Size," FMRC Serial No. 18792, Factory Mutual Research Corporation, June, 1974.

143. Fryburg, George, "Review of Literature Pertinent to Fire-Extinguishing Agents and to Basic Mechanisms Involved in Their Action," Technical Note 2102, National Advisory Committee For Aeronautics, May, 1950.

144. Goodfellow, David G., "Apparatus and Procedure for Measuring the Actual Delivered Density (ADD) of Sprinklers in Simulated Rack Storage Fires - Volume II - Data From Measurement of the Actual Delivered Density

(ADD) of a Large-Drop Sprinkler," FMRC J.I. OKOJ4.RR, Factory Mutual Research Corporation, January, 1985.

145.   Hendricks, R. W., "The Distribution of Water by Automatic Sprinkler Systems," Underwriters Laboratories, 1927.

146.   Heskestad, Gunnar, "The Role of Water in Suppression of Fire: A Review," CIB/W14/80-35, Factory Mutual Research.

147.   Jensen, Rolf H., "Fire Tests in High-Piled Combustible Stock," Underwriters Laboratories, June, 1962.

148.   Kalelkar, A. S., "Understanding Sprinkler Performance: Modeling of Combustion and Extinction," *Fire Technology*, 1971.

149.   Kirchhoff, Peter, "European Standards for Automatic Sprinklers," Technology Report 75-7, Society of Fire Protection Engineers, 1975.

150.   Kung, H. C. and Hill, John P., "Extinction of Wood Crib and Pallet Fires," *Combustion and Flame*, Vol. 24, p. 305-317, 1975.

151.   Kung, H. C., You, H. Z., and Han, Z., "Effects of Water Discharge Rate and Drop Size on Spray Cooling in Room Fires," Paper presented to Eastern Section: The Combustion Institute, December 3-5, 1984, Factory Mutual Research Corporation RC84-TP-17.

152.   Labes, Willis G., "Evaluation of Fire Protection Spray Devices The State-Of-The-Art," NBS-GCR 76-72, National Bureau of Standards/Center for Fire Research, June 30, 1976.

153.   Liu, Stanley T., "Analytical and Experimental Study of Evaporative Cooling and Room Fire Suppression by Corridor Sprinkler System," NBSIR 77-1287, National Bureau of Standards/Center for Fire Research, November, 1977.

154.   Magee, Richard S. and Reitz, Rolf D., "Extinguishment of Radiation Augmented Plastic Fires by Water Sprays," *Fifteenth Symposium (International) on Combustion*, The Combustion Institute, 1975.

155.   Nash, P. and Young, R. A., "The Performance of the Sprinkler in the Extinction of Fire," *Fire Surveyor*, Vol.3, No.3, 1974.

156.   Newman, R. M., "Fire Protection Requirements for High-Piled Palletized Storage of Cased Distilled Spirits," FMRC Serial No. 17792, Factory Mutual Research Corporation, June, 1970.

157. "Observations on the Most Favorable Size of Drops for Extinguishing Fires With Atomized Water and on the Range of a Stream of Water Spray," Karlsruhe Polytechnical Institute, 1957.

158. O'Dogherty, M. J., Southgate, E. B., and Barnes, D., "Measurements of Water Entrainment in a Square Sprinkler Array," Fire Research Note 661, Fire Research Station, Borehamwood, England, April, 1967.

159. Rasbash, D. J. and Rogowski, Z. W., "The Extinction of Open Fires With Water Spray, Part II: The Effect of Drop Size on the Extinction of Different Liquid Fires," Fire Research Note No. 162, Fire Research Station, Borehamwood, England, February, 1955.

160. Rogers, S. P. and Young, R. A., "The Performance of an Extra Light Hazard Sprinkler Installation," Fire Research Note No. 1065, Fire Research Station, Borehamwood, England, April, 1977.

161. Stolp, Martin, "The Extinction of Small Wood Crib Fires by Water," Paper presented at 5th International Fire Protection Seminar, Karlsruhe, Germany, September, 1976.

162. Suchomel, Miles R., "Fire Tests of Cardboard Boxes Stored on Steel Racks," Underwriters Laboratories, January 22, 1968.

163. Tamanini, F., "The Extinguishment of Wood Crib Fires With Water," Paper presented at Airlie House Conference on Firesafety for Buildings, July, 1973.

164. Thomas, P. H. and Smart, P. H. T., "Fire Extinction Tests in Rooms," Fire Research Station, Borehamwood, England, 1954.

165. Thompson, Norman J., "New Developments in Upright Sprinklers," *NFPA Quarterly*, July, 1952.

166. Thompson, Norman J., "Proving Spray Sprinkler Efficiency," *NFPA Quarterly*, January, 1954.

167. Webb, William A., "Fire Tests of High-Piled Rolled Wadding," Underwriters Laboratories, December 7, 1965.

168. Yao, Cheng, "Development of Large-Drop Sprinklers," FMRC Serial No. 22476, Factory Mutual Research Corporation, June, 1976.

169. Yao, C. and Kalelkar, A. S., "Effect of Drop Size on Sprinkler Performance," *Fire Technology*, 1970.

170.   You, Hong-Zeng and Kung, Hsiang-Cheng, "Strong Buoyant Plumes of Growing Rack Storage Fires," FMRC J.I. OG2E7.RA(1), Factory Mutual Research Corporation, February, 1984.

171.   Young, J. R., "Spacing Automatic Sprinklers," *NFPA Quarterly*, October, 1960.

Index

705

Reference

Reference

Reference

Reference